Applied Numerical Methods

With MATLAB® for Engineers and Scientists

Second Edition

Steven C. Chapra
Berger Chair in Computing and Engineering
Tufts University

Boston   Burr Ridge, IL   Dubuque, IA   New York   San Francisco   St. Louis
Bangkok   Bogotá   Caracas   Kuala Lumpur   Lisbon   London   Madrid   Mexico City
Milan   Montreal   New Delhi   Santiago   Seoul   Singapore   Sydney   Taipei   Toronto

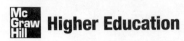

The McGraw·Hill Companies

Higher Education

APPLIED NUMERICAL METHODS WITH MATLAB FOR ENGINEERS AND SCIENTISTS
SECOND EDITION

Published by McGraw-Hill, a business unit of The McGraw-Hill Companies, Inc., 1221 Avenue of the Americas, New York, NY 10020.

This book is printed on acid-free paper.

1 2 3 4 5 6 7 8 9 0 DOC/DOC 0 9 8 7 6

ISBN 978-0-07-313290-7
MHID 0-07-313290-X

Senior Sponsoring Editor: *Bill Stenquist*
Developmental Editor: *Megan Hoar*
Executive Marketing Manager: *Michael Weitz*
Lead Project Manager: *Peggy J. Selle*
Lead Production Supervisor: *Sandy Ludovissy*
Media Project Manager: *Laurie Lenstra*
Associate Media Producer: *Christina Nelson*
Designer: *Rick D. Noel*
Cover Designer: *Studio Montage, St. Louis, Missouri*
(USE) Cover Image: *© Vol. 27/PhotoDisc/Getty Images*
Compositor: *Interactive Composition Corporation*
Typeface: *10/12 Times Roman*
Printer: *R. R. Donnelley Crawfordsville, IN*

MATLAB™ is a registered trademark of The MathWorks, Inc.

Library of Congress Cataloging-in-Publication Data

Chapra, Steven C.
  Applied numerical methods with MATLAB, for engineers and scientists / Steven C. Chapra. - 2nd ed.
     p.   cm.
  Includes bibliographical references and index.
  ISBN 978-0-07-313290-7 — ISBN 0-07-313290-X (acid-free paper)
  1. Numerical analysis—Data processing—Textbooks. 2. MATLAB—Textbooks. I. Title.

QA297.C4185   2008
518-dc22

2006026364
CIP

www.mhhe.com

To

My brothers,
John and Bob Chapra

ABOUT THE AUTHOR

Steve Chapra teaches in the Civil and Environmental Engineering Department at Tufts University, where he holds the Louis Berger Chair in Computing and Engineering. His other books include *Numerical Methods for Engineers* and *Surface Water-Quality Modeling.*

Steve received engineering degrees from Manhattan College and the University of Michigan. Before joining the faculty at Tufts, he worked for the Environmental Protection Agency and the National Oceanic and Atmospheric Administration, and taught at Texas A&M University and the University of Colorado. His general research interests focus on surface water-quality modeling and advanced computer applications in environmental engineering.

He has received a number of awards for his scholarly contributions, including the 1993 Rudolph Hering Medal (ASCE) and the 1987 Meriam/Wiley Distinguished Author Award (American Society for Engineering Education). He has also been recognized as the outstanding teacher among the engineering faculties at both Texas A&M University (1986 Tenneco Award) and the University of Colorado (1992 Hutchinson Award).

Steve was originally drawn to environmental engineering and science because of his love of the outdoors. He is an avid fly fisherman and hiker. An unapologetic nerd, his love affair with computing began when he was first introduced to Fortran programming as an undergraduate in 1966. Today, he feels truly blessed to be able to meld his love of mathematics, science, and computing with his passion for the natural environment. In addition, he gets the bonus of sharing it with others through his teaching and writing!

Beyond his professional interests, he enjoys art, music (especially classical music, jazz, and bluegrass), and reading history. Despite unfounded rumors to the contrary, he never has, and never will, voluntarily bungee jump or sky dive.

If you would like to contact Steve, or learn more about him, visit his home page at http://ase.tufts.edu/cee/faculty/chapra/bio.asp or e-mail him at steven.chapra@tufts.edu.

CONTENTS

About the Author iv

Preface xiii

Guided Tour xvii

PART ONE Modeling, Computers, and Error Analysis 1

1.1 Motivation 1
1.2 Part Organization 2

CHAPTER 1

**Mathematical Modeling, Numerical Methods,
and Problem Solving** 4

1.1   A Simple Mathematical Model   5
1.2   Conservation Laws in Engineering and Science   12
1.3   Numerical Methods Covered in This Book   13
Problems   17

CHAPTER 2

MATLAB Fundamentals 20

2.1   The MATLAB Environment   21
2.2   Assignment   22
2.3   Mathematical Operations   27
2.4   Use of Built-In Functions   30
2.5   Graphics   33
2.6   Other Resources   36
2.7   Case Study: Exploratory Data Analysis   37
Problems   39

CHAPTER 3

Programming with MATLAB 42

3.1   M-Files   43
3.2   Input-Output   47

3.3   Structured Programming   51
3.4   Nesting and Indentation   63
3.5   Passing Functions to M-Files   66
3.6   Case Study: Bungee Jumper Velocity   71
Problems   75

CHAPTER 4

Roundoff and Truncation Errors   79

4.1   Errors   80
4.2   Roundoff Errors   84
4.3   Truncation Errors   92
4.4   Total Numerical Error   103
4.5   Blunders, Model Errors, and Data Uncertainty   108
Problems   109

PART TWO   Roots and Optimization   111

2.1 Overview   111
2.2 Part Organization   112

CHAPTER 5

Roots: Bracketing Methods   114

5.1   Roots in Engineering and Science   115
5.2   Graphical Methods   116
5.3   Bracketing Methods and Initial Guesses   117
5.4   Bisection   122
5.5   False Position   128
5.6   Case Study: Greenhouse Gases and Rainwater   132
Problems   135

CHAPTER 6

Roots: Open Methods   139

6.1   Simple Fixed-Point Iteration   140
6.2   Newton-Raphson   144
6.3   Secant Methods   149
6.4   MATLAB Function: `fzero`   151
6.5   Polynomials   154
6.6   Case Study: Pipe Friction   157
Problems   162

CHAPTER 7

Optimization   166

7.1   Introduction and Background   167
7.2   One-Dimensional Optimization   170
7.3   Multidimensional Optimization   179
7.4   Case Study: Equilibrium and Minimum Potential Energy   181
Problems   183

PART THREE   Linear Systems   189

3.1 Overview   189
3.2 Part Organization   191

CHAPTER 8

Linear Algebraic Equations and Matrices   193

8.1   Matrix Algebra Overview   194
8.2   Solving Linear Algebraic Equations with MATLAB   203
8.3   Case Study: Currents and Voltages in Circuits   205
Problems   209

CHAPTER 9

Gauss Elimination   212

9.1   Solving Small Numbers of Equations   213
9.2   Naive Gauss Elimination   218
9.3   Pivoting   225
9.4   Tridiagonal Systems   227
9.5   Case Study: Model of a Heated Rod   229
Problems   233

CHAPTER 10

***LU* Factorization   236**

10.1 Overview of *LU* Factorization   237
10.2 Gauss Elimination as *LU* Factorization   238
10.3 Cholesky Factorization   244
10.4 MATLAB Left Division   246
Problems   247

CHAPTER 11

Matrix Inverse and Condition 249

11.1 The Matrix Inverse 249
11.2 Error Analysis and System Condition 253
11.3 Case Study: Indoor Air Pollution 258
Problems 261

CHAPTER 12

Iterative Methods 264

12.1 Linear Systems: Gauss-Seidel 264
12.2 Nonlinear Systems 270
12.3 Case Study: Chemical Reactions 277
Problems 279

PART FOUR   Curve Fitting 281

4.1 Overview 281
4.2 Part Organization 283

CHAPTER 13

Linear Regression 284

13.1 Statistics Review 286
13.2 Linear Least-Squares Regression 292
13.3 Linearization of Nonlinear Relationships 300
13.4 Computer Applications 304
13.5 Case Study: Enzyme Kinetics 307
Problems 312

CHAPTER 14

General Linear Least-Squares and Nonlinear Regression 316

14.1 Polynomial Regression 316
14.2 Multiple Linear Regression 320
14.3 General Linear Least Squares 322
14.4 QR Factorization and the Backslash Operator 325
14.5 Nonlinear Regression 326
14.6 Case Study: Fitting Sinusoids 328
Problems 332

CHAPTER 15

Polynomial Interpolation   335

15.1 Introduction to Interpolation   336
15.2 Newton Interpolating Polynomial   339
15.3 Lagrange Interpolating Polynomial   347
15.4 Inverse Interpolation   350
15.5 Extrapolation and Oscillations   351
Problems   355

CHAPTER 16

Splines and Piecewise Interpolation   359

16.1 Introduction to Splines   359
16.2 Linear Splines   361
16.3 Quadratic Splines   365
16.4 Cubic Splines   368
16.5 Piecewise Interpolation in MATLAB   374
16.6 Multidimensional Interpolation   379
16.7 Case Study: Heat Transfer   382
Problems   386

PART FIVE   Integration and Differentiation   389

5.1 Overview   389
5.2 Part Organization   390

CHAPTER 17

Numerical Integration Formulas   392

17.1 Introduction and Background   393
17.2 Newton-Cotes Formulas   396
17.3 The Trapezoidal Rule   398
17.4 Simpson's Rules   405
17.5 Higher-Order Newton-Cotes Formulas   411
17.6 Integration with Unequal Segments   412
17.7 Open Methods   416
17.8 Multiple Integrals   416
17.9 Case Study: Computing Work with Numerical Integration   419
Problems   422

CHAPTER 18

Numerical Integration of Functions   426

18.1 Introduction   426
18.2 Romberg Integration   427

18.3 Gauss Quadrature   432
18.4 Adaptive Quadrature   439
18.5 Case Study: Root-Mean-Square Current   440
Problems   444

CHAPTER 19

Numerical Differentiation   448

19.1 Introduction and Background   449
19.2 High-Accuracy Differentiation Formulas   452
19.3 Richardson Extrapolation   455
19.4 Derivatives of Unequally Spaced Data   457
19.5 Derivatives and Integrals for Data with Errors   458
19.6 Partial Derivatives   459
19.7 Numerical Differentiation with MATLAB   460
19.8 Case Study: Visualizing Fields   465
Problems   467

PART SIX   Ordinary Differential Equations   473

6.1 Overview   473
6.2 Part Organization   477

CHAPTER 20

Initial-Value Problems   479

20.1 Overview   481
20.2 Euler's Method   481
20.3 Improvements of Euler's Method   487
20.4 Runge-Kutta Methods   493
20.5 Systems of Equations   498
20.6 Case Study: Predatory-Prey Models and Chaos   504
Problems   509

CHAPTER 21

Adaptive Methods and Stiff Systems   514

21.1 Adaptive Runge-Kutta Methods   514
21.2 Multistep Methods   521
21.3 Stiffness   525
21.4 MATLAB Application: Bungee Jumper with Cord   531
21.5 Case Study: Pliny's Intermittent Fountain   532
Problems   537

CHAPTER 22

Boundary-Value Problems 540

22.1 Introduction and Background 541
22.2 The Shooting Method 545
22.3 Finite-Difference Methods 552
Problems 559

APPENDIX A: EIGENVALUES 565

APPENDIX B: MATLAB BUILT-IN FUNCTIONS 576

APPENDIX C: MATLAB M-FILE FUNCTIONS 578

BIBLIOGRAPHY 579

INDEX 580

PREFACE

This book is designed to support a one-semester course in numerical methods. It has been written for students who want to learn and apply numerical methods in order to solve problems in engineering and science. As such, the methods are motivated by problems rather than by mathematics. That said, sufficient theory is provided so that students come away with insight into the techniques and their shortcomings.

MATLAB® provides a great environment for such a course. Although other environments (e.g., Excel/VBA, Mathcad) or languages (e.g., Fortran 90, C++) could have been chosen, MATLAB presently offers a nice combination of handy programming features with powerful built-in numerical capabilities. On the one hand, its M-file programming environment allows students to implement moderately complicated algorithms in a structured and coherent fashion. On the other hand, its built-in, numerical capabilities empower students to solve more difficult problems without trying to "reinvent the wheel."

This second edition differs from the first edition in four key ways:

1. **Organization.** The first edition consisted of a series of 20 chapters. For the second edition, I have clustered these chapters into Parts as outlined in Fig. P.1. Aside from organizing the material in a more coherent fashion, this has allowed me to include an introduction/overview at the beginning of each part to orient students to the general topic area.

2. **New Chapters.** As shown in Fig. P.1, I have developed three new chapters. The primary rationale for adding these chapters is to give students a more complete coverage of numerical methods as well as MATLAB'S capabilities. The new chapters deal with the following topics:

   - **Optimization.** This chapter is placed just after the chapters dealing with roots of nonlinear equations. Although standard MATLAB (i.e., excluding Toolboxes) does not have comprehensive optimization capabilities, it has a few built-in functions that can be used to introduce the topic and solve some nice engineering and scientific problems. Although the focus of the chapter is on one-dimensional optimization, a brief introduction to multivariable optimization is included.

PART ONE Modeling, Computers, and Error Analysis	PART TWO Roots and Optimization	PART THREE Linear Systems	PART FOUR Curve Fitting	PART FIVE Integration and Differentiation	PART SIX Ordinary Differential Equations
CHAPTER 1 Mathematical Modeling, Numerical Methods, and Problem Solving	CHAPTER 5 Roots: Bracketing Methods	CHAPTER 8 Linear Algebraic Equations and Matrices	CHAPTER 13 Linear Regression	CHAPTER 17 Numerical Integration Formulas	CHAPTER 20 Initial-Value Problems
CHAPTER 2 MATLAB Fundamentals	CHAPTER 6 Roots: Open Methods	CHAPTER 9 Gauss Elimination	CHAPTER 14 General Linear Least-Squares and Nonlinear Regression	CHAPTER 18 Numerical Integration of Functions	CHAPTER 21 Adaptive Methods and Stiff Systems
CHAPTER 3 Programming with MATLAB	CHAPTER 7 Optimization	CHAPTER 10 LU Factorization	CHAPTER 15 Polynomial Interpolation	CHAPTER 19 Numerical Differentiation	CHAPTER 22 Boundary-Value Problems
CHAPTER 4 Roundoff and Truncation Errors		CHAPTER 11 Matrix Inverse and Condition	CHAPTER 16 Splines and Piecewise Interpolation		
		CHAPTER 12 Iterative Methods			

FIGURE P.1
The shaded areas represent new material. In addition, several of the original chapters have been supplemented with new topics, homework problems, and case studies.

- **Numerical Differentiation.** This chapter is placed after the last chapter on numerical integration. I added it for completeness, as well as to illustrate some of the inherent difficulties of numerical differentiation. It is also included because of the role of finite differences in the solution of boundary-value problems.
- **Boundary-Value Problems.** This chapter is placed at the end of the section on ordinary differential equations. Again, I think that this addition is important for completeness. In addition, it allows me to illustrate the finite-difference approach for solving ODEs. I think that this is important because, although PDEs are not included explicitly in this edition, finite-difference solutions for ODE boundary-value problems provide students with an idea of how PDEs are solved numerically. The subject area also provides nice fodder for more challenging and interesting homework problems.

3. **Case Studies.** These consist of engineering and science applications that are more complex and richer than the standard examples presented in the chapters. They are placed at the ends of selected chapters with the intention of (1) illustrating the nuances of the methods, and (2) showing more realistically how the methods along with MATLAB are used in engineering and science.

4. **New Homework Problems.** Most of the end-of-chapter problems have been modified, and a variety of new problems have been added. In particular, an effort has been made to include several new problems for each chapter that are more challenging and difficult than the problems in the first edition.

Aside from these additions, the second edition is very similar to the first. In particular, I have endeavored to maintain most of the features contributing to its pedagogical effectiveness including extensive use of worked examples and engineering and scientific applications. As with the previous edition, I have made a concerted effort to make this book as "student-friendly" as possible. Thus, I've tried to keep my explanations straightforward and practical.

Although my primary intent is to empower students by providing them with a sound introduction to numerical problem solving, I have the ancillary objective of making this introduction exciting and pleasurable. I believe that motivated students who enjoy engineering and science, problem solving, mathematics—and yes—programming, will ultimately make better professionals. If my book fosters enthusiasm and appreciation for these subjects, I will consider the effort a success.

Acknowledgements.   Several members of the McGraw-Hill team have contributed to this project. Special thanks are due to Amanda Green, Suzanne Jeans, Peggy Selle, Bill Stenquist, and Megan Hoar for their encouragement, support, and direction. Rick Noel developed a clean, clear, and aesthetically pleasing design. Naman Mahisauria of Interactive Composition Corporation also did an outstanding job in the book's final production phase. Last, but not least, Beatrice Sussman once again demonstrated why she is the best copy-editor in the business.

During the course of this project, the folks at The MathWorks, Inc., have truly demonstrated their overall excellence as well as their strong commitment to engineering and science education. In particular, Courtney Esposito and Naomi Fernandes of The MathWorks, Inc., Book Program have been especially helpful.

The generosity of the Berger family, and in particular Fred Berger, has provided me with the opportunity to work on creative projects such as this book dealing with computing and engineering. In addition, my colleagues in the Civil and Environmental Engineering Department at Tufts, notably Noelle Brooker, Ilse Allen, Jim Limbrunner, and Masoud Sanayei, have been very supportive and helpful.

Significant suggestions were also given by a number of colleagues. In particular, Dave Clough (University of Colorado–Boulder), Mike Gustafson (Duke University), Jim Guilkey (University of Utah), Laura Goadrich (University of Wisconsin, Madison), and Douglas Harder (University of Waterloo) provided valuable ideas and suggestions. In addition, a number of reviewers provided useful feedback and advice including Prabhakar Clement (Auburn University), John Cotton (Virginia Polytechnic Institute and State University), Deji Demuren (Old Dominion University), Ali Elkamel (University of Waterloo), Leon Gerber (St. John's University), Dalia M. Gil (Polytechnic University of P.R.–Orlando Campus), Naira Hovakimyan (Virginia Polytechnic Institute and State University), Egwu E. Kalu (FAMU-FSU College of Engineering), Ian H. Leslie (New Mexico State University), Xin Li (University of Central Florida), Leslie Loo (Nanyang Technological University–Singapore), Betty Mayfield (Hood College), Clare McCabe (Vanderbilt University), John Medige (University at Buffalo, The State University of New York), Robert R. Meyer (University of Wisconsin), Jeff Moehlis (University of California–Santa Barbara), Dan Nguyen (University of Alberta), J. Walt Oler (Texas Tech University), Luke Olson (University of Illinois at Urbana–Champaign), Jeffrey J. Potoff (Wayne State University), David Rappaport (Queen's University), Charles Schwartz (University of Maryland), Dipendra K. Sinha (San Francisco State University), Brian Vick (Virginia Polytechnic Institute and State University), and Ralph Wilkerson (University of Missouri–Rolla).

It should be stressed that although I received useful advice from the aforementioned individuals, I am responsible for any inaccuracies or mistakes you may find in this book. Please contact me via e-mail if you should detect any errors.

Finally, I want to thank my family, and in particular my wife, Cynthia, for the love, patience, and support they have provided through the time I've spent on this project.

Steven C. Chapra
Tufts University

Medford, Massachusetts
steven.chapra@tufts.edu

GUIDED TOUR

Chapter Objectives   *Chapter Objectives* begin each chapter. The objectives provide students with the function of each chapter as well as the specific topics covered in each chapter. The objectives enable students to set tangible goals before they begin each chapter.

You've Got a Problem   A section entitled *You've Got a Problem* can be found on the first page of most chapters. Here Chapra poses a real-life problem that requires the type of numerical solution technique that is the subject of the chapter. The intent is to introduce the student to the topic via a tangible example rather than through abstract mathematics. After an exposition of the numerical methods, the problem is then revisited in order to demonstrate how the learned material provides the means to solve the problem.

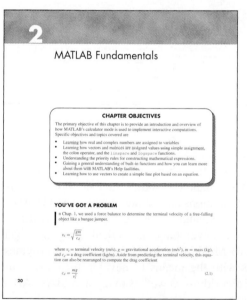

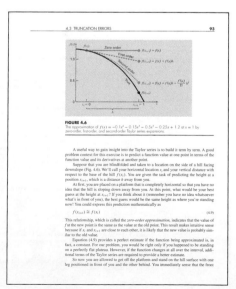

Theory Presented as It Informs Key Concepts
The text is intended for Numerical Methods users, not developers. Therefore, theory is not included for "theory's sake," for example no proofs. Theory is included as it informs key concepts such as the Taylor series, convergence, condition, etc. Hence, the student is shown how the theory connects with practical issues in problem solving.

Illustrations and Tables Illustrations and tables are clear and accurate in order to help students better visualize the important concepts presented in the text.

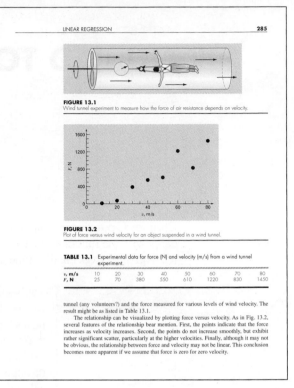

FIGURE 13.1
Wind tunnel experiment to measure how the force of air resistance depends on velocity.

FIGURE 13.2
Plot of force versus wind velocity for an object suspended in a wind tunnel.

TABLE 13.1 Experimental data for force (N) and velocity (m/s) from a wind tunnel experiment.

v, m/s	10	20	30	40	50	60	70	80
F, N	25	70	380	550	610	1220	830	1450

tunnel (any volunteers?) and the force measured for various levels of wind velocity. The result might be as listed in Table 13.1.

The relationship can be visualized by plotting force versus velocity. As in Fig. 13.2, several features of the relationship bear mention. First, the points indicate that the force increases as velocity increases. Second, the points do not increase smoothly, but exhibit rather significant scatter, particularly at the higher velocities. Finally, although it may not be obvious, the relationship between force and velocity may not be linear. This conclusion becomes more apparent if we assume that force is zero for zero velocity.

TABLE 2.2 Specifiers for colors, symbols, and line types.

Colors		Symbols		Line Types	
Blue	b	Point	.	Solid	-
Green	g	Circle	o	Dotted	:
Red	r	X-mark	x	Dashdot	-.
Cyan	c	Plus	+	Dashed	--
Magenta	m	Star	*		
Yellow	y	Square	s		
Black	k	Diamond	d		
		Triangle(down)	v		
		Triangle(up)	^		
		Triangle(left)	<		
		Triangle(right)	>		
		Pentagram	p		
		Hexagram	h		

The `plot` command displays a solid line by default. If you want to plot each point with a symbol, you can include a specifier enclosed in single quotes in the `plot` function. Table 2.2 lists the available specifiers. For example, if you want to use open circles enter

```
>> plot (t, v, 'o')
```

MATLAB allows you to display more than one data set on the same plot. For example, if you want to connect each data marker with a straight line you could type

```
>> plot (t, v, t, v, 'o')
```

It should be mentioned that, by default, previous plots are erased every time the `plot` command is implemented. The `hold on` command holds the current plot and all axis properties so that additional graphing commands can be added to the existing plot. The `hold off` command returns to the default mode. For example, if we had typed the following commands, the final plot would only display symbols:

```
>> plot (t, v)
>> plot (t, v, 'o')
```

In contrast, the following commands would result in both lines and symbols being displayed:

```
>> plot (t, v)
>> hold on
>> plot (t, v, 'o')
>> hold off
```

In addition to `hold`, another handy function is `subplot`, which allows you to split the graph window into subwindows or *panes*. It has the syntax

```
subplot (m, n, p)
```

This command breaks the graph window into an *m*-by-*n* matrix of small axes, and selects the *p*-th axes for the current plot.

Introductory MATLAB Material The text includes two introductory chapters on how to use MATLAB. Chapter 2 shows students how to perform computations and create graphs in MATLAB's standard command mode. Chapter 3 provides a primer on developing numerical programs via MATLAB M-file functions. Thus, the text provides students with the means to develop their own numerical algorithms as well as to tap into MATLAB's powerful built-in routines.

Algorithms Presented Using MATLAB M-files Instead of using pseudocode, this book presents algorithms as well-structured MATLAB M-files. Aside from being useful computer programs, these provide students with models for their own M-files that they will develop as homework exercises.

Thus, if we knew beforehand that an error of less than 0.5859 was acceptable, the formula tells us that eight iterations would yield the desired result.

Although we have emphasized the use of relative errors for obvious reasons, there will be cases where (usually through knowledge of the problem context) you will be able to specify an absolute error. For these cases, bisection along with Eq. (5.6) can provide a useful root location algorithm.

5.4.1 MATLAB M-file: `bisect`

An M-file to implement bisection is displayed in Fig. 5.7. It is passed the function (`func`) along with lower (`xl`) and upper (`xu`) guesses. In addition, an optional stopping criterion (`es`)

FIGURE 5.7
An M-file to implement the bisection method.

```
function [root,ea,iter]=bisect(func,xl,xu,es,maxit,varargin)
% bisect: root location zeroes
%   [root,ea,iter]=bisect(func,xl,xu,es,maxit,p1,p2,...):
%       uses bisection method to find the root of func
% input:
%   func = name of function
%   xl, xu = lower and upper guesses
%   es = desired relative error (default = 0.0001%)
%   maxit = maximum allowable iterations (default = 50)
%   p1,p2,... = additional parameters used by func
% output:
%   root = real root
%   ea = approximate relative error (%)
%   iter = number of iterations
if nargin<3,error('At least 3 input arguments required'),end
test = func(xl,varargin{:})*func(xu,varargin{:});
if test>0,error('no sign change'),end
if nargin<4|isempty(es), es=0.0001;end
if nargin<5|isempty(maxit), maxit=50;end
iter = 0; xr = xl;
while (1)
   xrold = xr;
   xr = (xl + xu)/2;
   iter = iter + 1;
   if xr ~= 0,ea = abs((xr - xrold)/xr) * 100;end
   test = func(xl,varargin{:})*func(xr,varargin{:});
   if test < 0
      xu = xr;
   elseif test > 0
      xl = xr;
   else
      ea = 0;
   end
   if ea <= es | iter >= maxit,break,end
end
root = xr;
```

16.7 CASE STUDY   HEAT TRANSFER

Background. Lakes in the temperate zone can become thermally stratified during the summer. As depicted in Fig. 16.11, warm, buoyant water near the surface overlies colder, denser bottom water. Such stratification effectively divides the lake vertically into two layers: the *epilimnion* and the *hypolimnion*, separated by a plane called the *thermocline*.

Thermal stratification has great significance for environmental engineers and scientists studying such systems. In particular, the thermocline greatly diminishes mixing between the two layers. As a result, decomposition of organic matter can lead to severe depletion of oxygen in the isolated bottom waters.

The location of the thermocline can be defined as the inflection point of the temperature-depth curve—that is, the point at which $d^2T/dz^2 = 0$. It is also the point at which the absolute value of the first derivative or gradient is a maximum.

The temperature gradient is important in its own right because it can be used in conjunction with Fourier's law to determine the heat flux across the thermocline.

$$J = -D\rho C \frac{dT}{dz} \qquad (16.33)$$

where J = heat flux [cal/(cm^2 · s)], α = an eddy diffusion coefficient (cm^2/s), ρ = density ($\cong$ 1 g/cm^3), and C = specific heat [$\cong$ 1 cal/(g · C)].

In this case study, natural cubic splines are employed to determine the thermocline depth and temperature gradient for Platte Lake, Michigan (Table 16.3). The latter is also used to determine the heat flux for the case where $\alpha = 0.01$ cm^2/s.

FIGURE 16.11
Temperature versus depth during summer for Platte Lake, Michigan.

TABLE 16.3 Temperature versus depth during summer for Platte Lake, Michigan.

z, m	0	2.3	4.9	9.1	13.7	18.3	22.9	27.2
T, °C	22.8	22.8	22.8	20.6	13.9	11.7	11.1	11.1

Worked Examples and Case Studies Extensive worked examples are laid out in detail so that students can clearly follow the steps in each numerical computation. The case studies consist of engineering and science applications which are more complex and richer than the worked examples. They are placed at the ends of selected chapters with the intention of (1) illustrating the nuances of the methods, and (2) showing more realistically how the methods along with MATLAB are applied for problem solving.

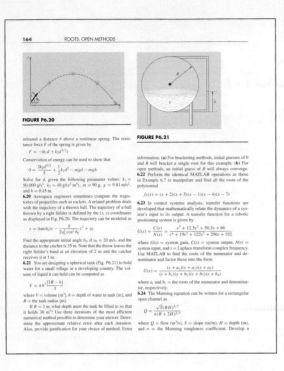

Problem Sets   The text includes a wide variety of problems. Many are drawn from engineering and scientific disciplines. Others are used to illustrate numerical techniques and theoretical concepts. Problems include those that can be solved with a pocket calculator as well as others that require computer solution with MATLAB.

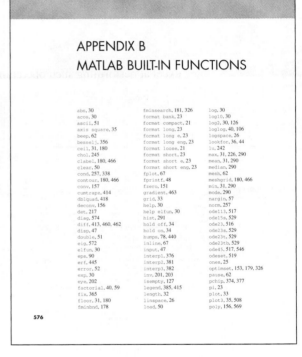

Useful Appendices and Indexes
Appendix A covers Eigenvalues, Appendix B contains MATLAB commands, and Appendix C contains M-file functions.

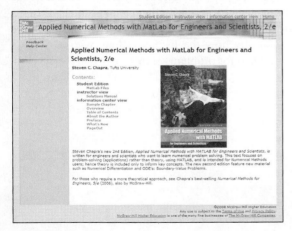

Supplements   A text website is available at http://www.mhhe.com/chapra. Resources include PowerPoint slides of text figures and chapter objectives, M-files and additional MATLAB resources.

PART ONE

Modeling, Computers, and Error Analysis

1.1 MOTIVATION

What are numerical methods and why should you study them?

Numerical methods are techniques by which mathematical problems are formulated so that they can be solved with arithmetic and logical operations. Because digital computers excel at performing such operations, numerical methods are sometimes referred to as *computer mathematics*.

In the pre–computer era, the time and drudgery of implementing such calculations seriously limited their practical use. However, with the advent of fast, inexpensive digital computers, the role of numerical methods in engineering and scientific problem solving has exploded. Because they figure so prominently in much of our work, I believe that numerical methods should be a part of every engineer's and scientist's basic education. Just as we all must have solid foundations in the other areas of mathematics and science, we should also have a fundamental understanding of numerical methods. In particular, we should have a solid appreciation of both their capabilities and their limitations.

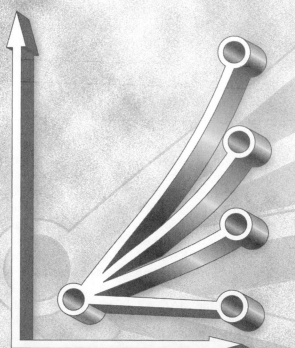

Beyond contributing to your overall education, there are several additional reasons why you should study numerical methods:

1. Numerical methods greatly expand the types of problems you can address. They are capable of handling large systems of equations, nonlinearities, and complicated geometries that are not uncommon in engineering and science and that are often impossible to solve analytically with standard calculus. As such, they greatly enhance your problem-solving skills.

2. Numerical methods allow you to use "canned" software with insight. During your career, you will

invariably have occasion to use commercially available prepackaged computer programs that involve numerical methods. The intelligent use of these programs is greatly enhanced by an understanding of the basic theory underlying the methods. In the absence of such understanding, you will be left to treat such packages as "black boxes" with little critical insight into their inner workings or the validity of the results they produce.

3. Many problems cannot be approached using canned programs. If you are conversant with numerical methods, and are adept at computer programming, you can design your own programs to solve problems without having to buy or commission expensive software.

4. Numerical methods are an efficient vehicle for learning to use computers. Because numerical methods are expressly designed for computer implementation, they are ideal for illustrating the computer's powers and limitations. When you successfully implement numerical methods on a computer, and then apply them to solve otherwise intractable problems, you will be provided with a dramatic demonstration of how computers can serve your professional development. At the same time, you will also learn to acknowledge and control the errors of approximation that are part and parcel of large-scale numerical calculations.

5. Numerical methods provide a vehicle for you to reinforce your understanding of mathematics. Because one function of numerical methods is to reduce higher mathematics to basic arithmetic operations, they get at the "nuts and bolts" of some otherwise obscure topics. Enhanced understanding and insight can result from this alternative perspective.

With these reasons as motivation, we can now set out to understand how numerical methods and digital computers work in tandem to generate reliable solutions to mathematical problems. The remainder of this book is devoted to this task.

1.2 PART ORGANIZATION

This book is divided into six parts. The latter five parts focus on the major areas of numerical methods. Although it might be tempting to jump right into this material, *Part One* consists of four chapters dealing with essential background material.

Chapter 1 provides a concrete example of how a numerical method can be employed to solve a real problem. To do this, we develop a *mathematical model* of a free-falling bungee jumper. The model, which is based on Newton's second law, results in an ordinary differential equation. After first using calculus to develop a closed-form solution, we then show how a comparable solution can be generated with a simple numerical method. We end the chapter with an overview of the major areas of numerical methods that we cover in Parts Two through Six.

Chapters 2 and 3 provide an introduction to the MATLAB® software environment. *Chapter 2* deals with the standard way of operating MATLAB by entering commands one at a time in the so-called *calculator mode*. This interactive mode provides a straightforward means to orient you to the environment and illustrate how it is used for common operations such as performing calculations and creating plots.

Chapter 3 shows how MATLAB's *programming mode* provides a vehicle for assembling individual commands into algorithms. Thus, our intent is to illustrate how MATLAB serves as a convenient programming environment to develop your own software.

Chapter 4 deals with the important topic of error analysis, which must be understood for the effective use of numerical methods. The first part of the chapter focuses on the *roundoff errors* that result because digital computers cannot represent some quantities exactly. The latter part addresses *truncation errors* that arise from using an approximation in place of an exact mathematical procedure.

Mathematical Modeling, Numerical Methods, and Problem Solving

CHAPTER OBJECTIVES

The primary objective of this chapter is to provide you with a concrete idea of what numerical methods are and how they relate to engineering and scientific problem solving. Specific objectives and topics covered are

- Learning how mathematical models can be formulated on the basis of scientific principles to simulate the behavior of a simple physical system.
- Understanding how numerical methods afford a means to generate solutions in a manner that can be implemented on a digital computer.
- Understanding the different types of conservation laws that lie beneath the models used in the various engineering disciplines and appreciating the difference between steady-state and dynamic solutions of these models.
- Learning about the different types of numerical methods we will cover in this book.

YOU'VE GOT A PROBLEM

Suppose that a bungee-jumping company hires you. You're given the task of predicting the velocity of a jumper (Fig. 1.1) as a function of time during the free-fall part of the jump. This information will be used as part of a larger analysis to determine the length and required strength of the bungee cord for jumpers of different mass.

You know from your studies of physics that the acceleration should be equal to the ratio of the force to the mass (Newton's second law). Based on this insight and your knowledge

Upward force due to air resistance

Downward force due to gravity

FIGURE 1.1
Forces acting on a free-falling bungee jumper.

of fluid mechanics, you develop the following mathematical model for the rate of change of velocity with respect to time,

$$\frac{dv}{dt} = g - \frac{c_d}{m}v^2$$

where v = vertical velocity (m/s), t = time (s), g = the acceleration due to gravity ($\cong 9.81 \text{ m/s}^2$), c_d = a second-order drag coefficient (kg/m), and m = the jumper's mass (kg).

Because this is a differential equation, you know that calculus might be used to obtain an analytical or exact solution for v as a function of t. However, in the following pages, we will illustrate an alternative solution approach. This will involve developing a computer-oriented numerical or approximate solution.

Aside from showing you how the computer can be used to solve this particular problem, our more general objective will be to illustrate (a) what numerical methods are and (b) how they figure in engineering and scientific problem solving. In so doing, we will also show how mathematical models figure prominently in the way engineers and scientists use numerical methods in their work.

1.1 A SIMPLE MATHEMATICAL MODEL

A *mathematical model* can be broadly defined as a formulation or equation that expresses the essential features of a physical system or process in mathematical terms. In a very general sense, it can be represented as a functional relationship of the form

$$\begin{matrix} \text{Dependent} \\ \text{variable} \end{matrix} = f \begin{pmatrix} \text{independent} \\ \text{variables} \end{pmatrix}, \text{parameters}, \begin{matrix} \text{forcing} \\ \text{functions} \end{matrix} \end{pmatrix} \qquad (1.1)$$

where the *dependent variable* is a characteristic that usually reflects the behavior or state of the system; the *independent variables* are usually dimensions, such as time and space, along which the system's behavior is being determined; the *parameters* are reflective of the system's properties or composition; and the *forcing functions* are external influences acting upon it.

The actual mathematical expression of Eq. (1.1) can range from a simple algebraic relationship to large complicated sets of differential equations. For example, on the basis of his observations, Newton formulated his second law of motion, which states that the time rate of change of momentum of a body is equal to the resultant force acting on it. The mathematical expression, or model, of the second law is the well-known equation

$$F = ma \qquad (1.2)$$

where F is the net force acting on the body (N, or kg m/s^2), m is the mass of the object (kg), and a is its acceleration (m/s^2).

The second law can be recast in the format of Eq. (1.1) by merely dividing both sides by m to give

$$a = \frac{F}{m} \tag{1.3}$$

where a is the dependent variable reflecting the system's behavior, F is the forcing function, and m is a parameter. Note that for this simple case there is no independent variable because we are not yet predicting how acceleration varies in time or space.

Equation (1.3) has a number of characteristics that are typical of mathematical models of the physical world.

- It describes a natural process or system in mathematical terms.
- It represents an idealization and simplification of reality. That is, the model ignores negligible details of the natural process and focuses on its essential manifestations. Thus, the second law does not include the effects of relativity that are of minimal importance when applied to objects and forces that interact on or about the earth's surface at velocities and on scales visible to humans.
- Finally, it yields reproducible results and, consequently, can be used for predictive purposes. For example, if the force on an object and its mass are known, Eq. (1.3) can be used to compute acceleration.

Because of its simple algebraic form, the solution of Eq. (1.2) was obtained easily. However, other mathematical models of physical phenomena may be much more complex, and either cannot be solved exactly or require more sophisticated mathematical techniques than simple algebra for their solution. To illustrate a more complex model of this kind, Newton's second law can be used to determine the terminal velocity of a free-falling body near the earth's surface. Our falling body will be a bungee jumper (Fig. 1.1). For this case, a model can be derived by expressing the acceleration as the time rate of change of the velocity (dv/dt) and substituting it into Eq. (1.3) to yield

$$\frac{dv}{dt} = \frac{F}{m} \tag{1.4}$$

where v is velocity (in meters per second). Thus, the rate of change of the velocity is equal to the net force acting on the body normalized to its mass. If the net force is positive, the object will accelerate. If it is negative, the object will decelerate. If the net force is zero, the object's velocity will remain at a constant level.

Next, we will express the net force in terms of measurable variables and parameters. For a body falling within the vicinity of the earth, the net force is composed of two opposing forces: the downward pull of gravity F_D and the upward force of air resistance F_U (Fig. 1.1):

$$F = F_D + F_U \tag{1.5}$$

If force in the downward direction is assigned a positive sign, the second law can be used to formulate the force due to gravity as

$$F_D = mg \tag{1.6}$$

where g is the acceleration due to gravity (9.81 m/s^2).

Air resistance can be formulated in a variety of ways. Knowledge from the science of fluid mechanics suggests that a good first approximation would be to assume that it is proportional to the square of the velocity,

$$F_U = -c_d v^2 \tag{1.7}$$

where c_d is a proportionality constant called the *drag coefficient* (kg/m). Thus, the greater the fall velocity, the greater the upward force due to air resistance. The parameter c_d accounts for properties of the falling object, such as shape or surface roughness, that affect air resistance. For the present case, c_d might be a function of the type of clothing or the orientation used by the jumper during free fall.

The net force is the difference between the downward and upward force. Therefore, Eqs. (1.4) through (1.7) can be combined to yield

$$\frac{dv}{dt} = g - \frac{c_d}{m} v^2 \tag{1.8}$$

Equation (1.8) is a model that relates the acceleration of a falling object to the forces acting on it. It is a *differential equation* because it is written in terms of the differential rate of change (dv/dt) of the variable that we are interested in predicting. However, in contrast to the solution of Newton's second law in Eq. (1.3), the exact solution of Eq. (1.8) for the velocity of the jumper cannot be obtained using simple algebraic manipulation. Rather, more advanced techniques such as those of calculus must be applied to obtain an exact or analytical solution. For example, if the jumper is initially at rest ($v = 0$ at $t = 0$), calculus can be used to solve Eq. (1.8) for

$$v(t) = \sqrt{\frac{gm}{c_d}} \tanh\left(\sqrt{\frac{gc_d}{m}} t\right) \tag{1.9}$$

where tanh is the hyperbolic tangent that can be either computed directly[1] or via the more elementary exponential function as in

$$\tanh x = \frac{e^x - e^{-x}}{e^x + e^{-x}} \tag{1.10}$$

Note that Eq. (1.9) is cast in the general form of Eq. (1.1) where $v(t)$ is the dependent variable, t is the independent variable, c_d and m are parameters, and g is the forcing function.

EXAMPLE 1.1 **Analytical Solution to the Bungee Jumper Problem**

Problem Statement.   A bungee jumper with a mass of 68.1 kg leaps from a stationary hot air balloon. Use Eq. (1.9) to compute velocity for the first 12 s of free fall. Also determine the terminal velocity that will be attained for an infinitely long cord (or alternatively, the jumpmaster is having a particularly bad day!). Use a drag coefficient of 0.25 kg/m.

---

[1] MATLAB® allows direct calculation of the hyperbolic tangent via the built-in function `tanh(x)`.

Solution. Inserting the parameters into Eq. (1.9) yields

$$v(t) = \sqrt{\frac{9.81(68.1)}{0.25}} \tanh\left(\sqrt{\frac{9.81(0.25)}{68.1}}\, t\right) = 51.6938 \tanh(0.18977t)$$

which can be used to compute

t, s	v, m/s
0	0
2	18.7292
4	33.1118
6	42.0762
8	46.9575
10	49.4214
12	50.6175
∞	51.6938

According to the model, the jumper accelerates rapidly (Fig. 1.2). A velocity of 49.4214 m/s (about 110 mi/h) is attained after 10 s. Note also that after a sufficiently long

FIGURE 1.2
The analytical solution for the bungee jumper problem as computed in Example 1.1. Velocity increases with time and asymptotically approaches a terminal velocity.

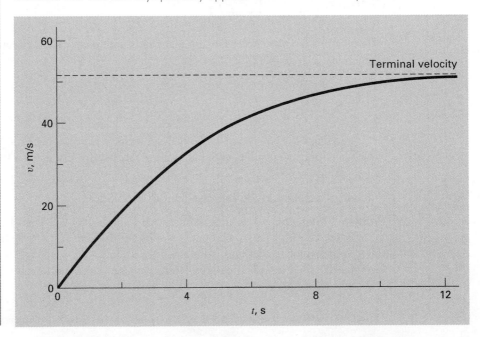

time, a constant velocity, called the *terminal velocity*, of 51.6983 m/s (115.6 mi/h) is reached. This velocity is constant because, eventually, the force of gravity will be in balance with the air resistance. Thus, the net force is zero and acceleration has ceased.

Equation (1.9) is called an *analytical* or *closed-form solution* because it exactly satisfies the original differential equation. Unfortunately, there are many mathematical models that cannot be solved exactly. In many of these cases, the only alternative is to develop a numerical solution that approximates the exact solution.

Numerical methods are those in which the mathematical problem is reformulated so it can be solved by arithmetic operations. This can be illustrated for Eq. (1.8) by realizing that the time rate of change of velocity can be approximated by (Fig. 1.3):

$$\frac{dv}{dt} \cong \frac{\Delta v}{\Delta t} = \frac{v(t_{i+1}) - v(t_i)}{t_{i+1} - t_i} \tag{1.11}$$

where Δv and Δt are differences in velocity and time computed over finite intervals, $v(t_i)$ is velocity at an initial time t_i, and $v(t_{i+1})$ is velocity at some later time t_{i+1}. Note that $dv/dt \cong \Delta v/\Delta t$ is approximate because Δt is finite. Remember from calculus that

$$\frac{dv}{dt} = \lim_{\Delta t \to 0} \frac{\Delta v}{\Delta t}$$

Equation (1.11) represents the reverse process.

FIGURE 1.3
The use of a finite difference to approximate the first derivative of v with respect to t.

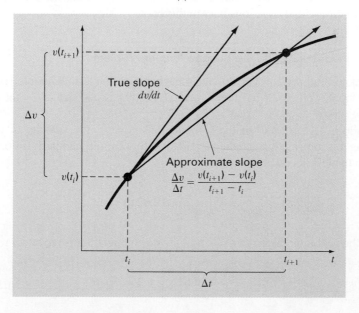

Equation (1.11) is called a *finite-difference approximation* of the derivative at time t_i. It can be substituted into Eq. (1.8) to give

$$\frac{v(t_{i+1}) - v(t_i)}{t_{i+1} - t_i} = g - \frac{c_d}{m} v(t_i)^2$$

This equation can then be rearranged to yield

$$v(t_{i+1}) = v(t_i) + \left[g - \frac{c_d}{m} v(t_i)^2 \right] (t_{i+1} - t_i) \tag{1.12}$$

Notice that the term in brackets is the right-hand side of the differential equation itself [Eq. (1.8)]. That is, it provides a means to compute the rate of change or slope of v. Thus, the equation can be rewritten as

$$v_{i+1} = v_i + \frac{dv_i}{dt} \Delta t \tag{1.13}$$

where the nomenclature v_i designates velocity at time t_i and $\Delta t = t_{i+1} - t_i$.

We can now see that the differential equation has been transformed into an equation that can be used to determine the velocity algebraically at t_{i+1} using the slope and previous values of v and t. If you are given an initial value for velocity at some time t_i, you can easily compute velocity at a later time t_{i+1}. This new value of velocity at t_{i+1} can in turn be employed to extend the computation to velocity at t_{i+2} and so on. Thus at any time along the way,

New value = old value + slope × step size

This approach is formally called *Euler's method*. We'll discuss it in more detail when we turn to differential equations later in this book.

EXAMPLE 1.2 | Numerical Solution to the Bungee Jumper Problem

Problem Statement. Perform the same computation as in Example 1.1 but use Eq. (1.13) to compute velocity with Euler's method. Employ a step size of 2 s for the calculation.

Solution. At the start of the computation $(t_0 = 0)$, the velocity of the jumper is zero. Using this information and the parameter values from Example 1.1, Eq. (1.13) can be used to compute velocity at $t_1 = 2$ s:

$$v = 0 + \left[9.81 - \frac{0.25}{68.1} (0)^2 \right] \times 2 = 19.62 \text{ m/s}$$

For the next interval (from $t = 2$ to 4 s), the computation is repeated, with the result

$$v = 19.62 + \left[9.81 - \frac{0.25}{68.1} (19.62)^2 \right] \times 2 = 36.4137 \text{ m/s}$$

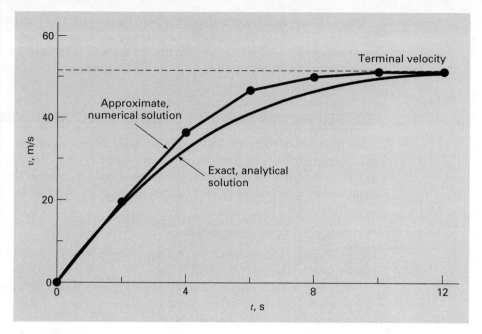

FIGURE 1.4
Comparison of the numerical and analytical solutions for the bungee jumper problem.

The calculation is continued in a similar fashion to obtain additional values:

t, s	v, m/s
0	0
2	19.6200
4	36.4137
6	46.2983
8	50.1802
10	51.3123
12	51.6008
∞	51.6938

The results are plotted in Fig. 1.4 along with the exact solution. We can see that the numerical method captures the essential features of the exact solution. However, because we have employed straight-line segments to approximate a continuously curving function, there is some discrepancy between the two results. One way to minimize such discrepancies is to use a smaller step size. For example, applying Eq. (1.13) at 1-s intervals results in a smaller error, as the straight-line segments track closer to the true solution. Using hand calculations, the effort associated with using smaller and smaller step sizes would make such numerical solutions impractical. However, with the aid of the computer, large numbers of calculations can be performed easily. Thus, you can accurately model the velocity of the jumper without having to solve the differential equation exactly.

As in Example 1.2, a computational price must be paid for a more accurate numerical result. Each halving of the step size to attain more accuracy leads to a doubling of the number of computations. Thus, we see that there is a trade-off between accuracy and computational effort. Such trade-offs figure prominently in numerical methods and constitute an important theme of this book.

1.2 CONSERVATION LAWS IN ENGINEERING AND SCIENCE

Aside from Newton's second law, there are other major organizing principles in science and engineering. Among the most important of these are the *conservation laws*. Although they form the basis for a variety of complicated and powerful mathematical models, the great conservation laws of science and engineering are conceptually easy to understand. They all boil down to

$$\text{Change} = \text{increases} - \text{decreases} \tag{1.14}$$

This is precisely the format that we employed when using Newton's law to develop a force balance for the bungee jumper [Eq. (1.8)].

Although simple, Eq. (1.14) embodies one of the most fundamental ways in which conservation laws are used in engineering and science—that is, to predict changes with respect to time. We will give it a special name—the *time-variable* (or *transient*) computation.

Aside from predicting changes, another way in which conservation laws are applied is for cases where change is nonexistent. If change is zero, Eq. (1.14) becomes

$$\text{Change} = 0 = \text{increases} - \text{decreases}$$

or

$$\text{Increases} = \text{decreases} \tag{1.15}$$

Thus, if no change occurs, the increases and decreases must be in balance. This case, which is also given a special name—the *steady-state* calculation—has many applications in engineering and science. For example, for steady-state incompressible fluid flow in pipes, the flow into a junction must be balanced by flow going out, as in

$$\text{Flow in} = \text{flow out}$$

For the junction in Fig. 1.5, the balance can be used to compute that the flow out of the fourth pipe must be 60.

For the bungee jumper, the steady-state condition would correspond to the case where the net force was zero or [Eq. (1.8) with $dv/dt = 0$]

$$mg = c_d v^2 \tag{1.16}$$

Thus, at steady state, the downward and upward forces are in balance and Eq. (1.16) can be solved for the terminal velocity

$$v = \sqrt{\frac{gm}{c_d}}$$

Although Eqs. (1.14) and (1.15) might appear trivially simple, they embody the two fundamental ways that conservation laws are employed in engineering and science. As such, they will form an important part of our efforts in subsequent chapters to illustrate the connection between numerical methods and engineering and science.

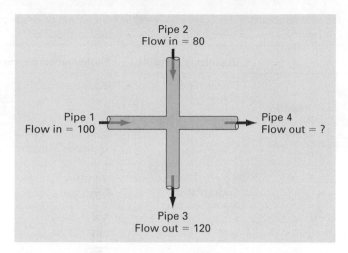

FIGURE 1.5
A flow balance for steady incompressible fluid flow at the junction of pipes.

Table 1.1 summarizes some models and associated conservation laws that figure prominently in engineering. Many chemical engineering problems involve mass balances for reactors. The mass balance is derived from the conservation of mass. It specifies that the change of mass of a chemical in the reactor depends on the amount of mass flowing in minus the mass flowing out.

Civil and mechanical engineers often focus on models developed from the conservation of momentum. For civil engineering, force balances are utilized to analyze structures such as the simple truss in Table 1.1. The same principles are employed for the mechanical engineering case studies to analyze the transient up-and-down motion or vibrations of an automobile.

Finally, electrical engineering studies employ both current and energy balances to model electric circuits. The current balance, which results from the conservation of charge, is similar in spirit to the flow balance depicted in Fig. 1.5. Just as flow must balance at the junction of pipes, electric current must balance at the junction of electric wires. The energy balance specifies that the changes of voltage around any loop of the circuit must add up to zero.

We should note that there are many other branches of engineering beyond chemical, civil, electrical, and mechanical. Many of these are related to the Big Four. For example, chemical engineering skills are used extensively in areas such as environmental, petroleum, and biomedical engineering. Similarly, aerospace engineering has much in common with mechanical engineering. We will endeavor to include examples from these areas in the coming pages.

1.3 NUMERICAL METHODS COVERED IN THIS BOOK

We chose Euler's method for this introductory chapter because it is typical of many other classes of numerical methods. In essence, most consist of recasting mathematical operations into the simple kind of algebraic and logical operations compatible with digital computers. Figure 1.6 summarizes the major areas covered in this text.

TABLE 1.1 Devices and types of balances that are commonly used in the four major areas of engineering. For each case, the conservation law on which the balance is based is specified.

Field	Device	Organizing Principle	Mathematical Expression
Chemical engineering	Reactors	Conservation of mass	Mass balance: Input → Output. Over a unit of time period $\Delta \text{mass} = \text{inputs} - \text{outputs}$
Civil engineering	Structure	Conservation of momentum	Force balance: $+F_V$, $-F_H$, $+F_H$, $-F_V$. At each node Σ horizontal forces $(F_H) = 0$ Σ vertical forces $(F_V) = 0$
Mechanical engineering	Machine	Conservation of momentum	Force balance: Upward force, $x = 0$, Downward force. $m \dfrac{d^2x}{dt^2} = \text{downward force} - \text{upward force}$
Electrical engineering	Circuit	Conservation of charge	Current balance: $+i_1 \longrightarrow$ • $\longrightarrow -i_3$, $+i_2$. For each node Σ current $(i) = 0$
		Conservation of energy	Voltage balance: i_1R_1, i_2R_2, i_3R_3, ξ. Around each loop Σ emf's $- \Sigma$ voltage drops for resistors $= 0$ $\Sigma \xi - \Sigma iR = 0$

(a) *Part 2*: **Roots and optimization**

Roots: Solve for x so that $f(x) = 0$

Optimization: Solve for x so that $f'(x) = 0$

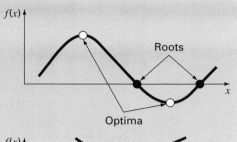

(b) *Part 3*: **Linear algebraic equations**

Given the a's and the b's, solve for the x's

$$a_{11}x_1 + a_{12}x_2 = b_1$$
$$a_{21}x_1 + a_{22}x_2 = b_2$$

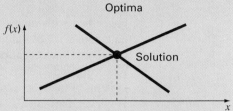

(c) *Part 4*: **Curve fitting**

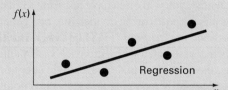
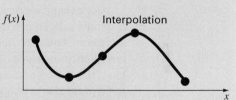

(d) *Part 5*: **Integration and differentiation**

Integration: Find the area under the curve

Differentiation: Find the slope of the curve

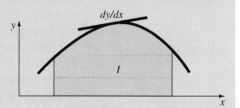

(e) *Part 6*: **Differential equations**

Given

$$\frac{dy}{dt} \approx \frac{\Delta y}{\Delta t} = f(t, y)$$

solve for y as a function of t

$$y_{i+1} = y_i + f(t_i, y_i)\Delta t$$

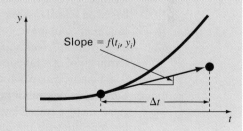

FIGURE 1.6
Summary of the numerical methods covered in this book.

Part Two deals with two related topics: root finding and optimization. As depicted in Fig. 1.6*a*, *root location* involves searching for the zeros of a function. In contrast, *optimization* involves determining a value or values of an independent variable that correspond to a "best" or optimal value of a function. Thus, as in Fig. 1.6*a*, optimization involves identifying maxima and minima. Although somewhat different approaches are used, root location and optimization both typically arise in design contexts.

Part Three is devoted to solving systems of simultaneous linear algebraic equations (Fig. 1.6*b*). Such systems are similar in spirit to roots of equations in the sense that they are concerned with values that satisfy equations. However, in contrast to satisfying a single equation, a set of values is sought that simultaneously satisfies a set of linear algebraic equations. Such equations arise in a variety of problem contexts and in all disciplines of engineering and science. In particular, they originate in the mathematical modeling of large systems of interconnected elements such as structures, electric circuits, and fluid networks. However, they are also encountered in other areas of numerical methods such as curve fitting and differential equations.

As an engineer or scientist, you will often have occasion to fit curves to data points. The techniques developed for this purpose can be divided into two general categories: regression and interpolation. As described in *Part Four* (Fig. 1.6*c*), *regression* is employed where there is a significant degree of error associated with the data. Experimental results are often of this kind. For these situations, the strategy is to derive a single curve that represents the general trend of the data without necessarily matching any individual points.

In contrast, *interpolation* is used where the objective is to determine intermediate values between relatively error-free data points. Such is usually the case for tabulated information. The strategy in such cases is to fit a curve directly through the data points and use the curve to predict the intermediate values.

As depicted in Fig. 1.6*d*, *Part Five* is devoted to integration and differentiation. A physical interpretation of *numerical integration* is the determination of the area under a curve. Integration has many applications in engineering and science, ranging from the determination of the centroids of oddly shaped objects to the calculation of total quantities based on sets of discrete measurements. In addition, numerical integration formulas play an important role in the solution of differential equations. Part Five also covers methods for *numerical differentiation*. As you know from your study of calculus, this involves the determination of a function's slope or its rate of change.

Finally, *Part Six* focuses on the solution of *ordinary differential equations* (Fig. 1.6*e*). Such equations are of great significance in all areas of engineering and science. This is because many physical laws are couched in terms of the rate of change of a quantity rather than the magnitude of the quantity itself. Examples range from population-forecasting models (rate of change of population) to the acceleration of a falling body (rate of change of velocity). Two types of problems are addressed: initial-value and boundary-value problems.

PROBLEMS

1.1 Use calculus to verify that Eq. (1.9) is a solution of Eq. (1.8).

1.2 The following information is available for a bank account:

Date	Deposits	Withdrawals	Balance
5/1			1512.33
	220.13	327.26	
6/1			
	216.80	378.61	
7/1			
	450.25	106.80	
8/1			
	127.31	350.61	
9/1			

Use the conservation of cash to compute the balance on 6/1, 7/1, 8/1, and 9/1. Show each step in the computation. Is this a steady-state or a transient computation?

1.3 Repeat Example 1.2. Compute the velocity to $t = 12$ s, with a step size of **(a)** 1 and **(b)** 0.5 s. Can you make any statement regarding the errors of the calculation based on the results?

1.4 Rather than the nonlinear relationship of Eq. (1.7), you might choose to model the upward force on the bungee jumper as a linear relationship:

$$F_U = -c'v$$

where c' = a first-order drag coefficient (kg/s).
(a) Using calculus, obtain the closed-form solution for the case where the jumper is initially at rest ($v = 0$ at $t = 0$).
(b) Repeat the numerical calculation in Example 1.2 with the same initial condition and parameter values. Use a value of 12.5 kg/s for c'.

1.5 For the free-falling bungee jumper with linear drag (Prob. 1.4), assume a first jumper is 70 kg and has a drag coefficient of 12 kg/s. If a second jumper has a drag coefficient of 15 kg/s and a mass of 75 kg, how long will it take her to reach the same velocity jumper 1 reached in 10 s?

1.6 For the free-falling bungee jumper with linear drag (Prob. 1.4), compute the velocity of a free-falling parachutist using Euler's method for the case where $m = 80$ kg and $c' = 10$ kg/s. Perform the calculation from $t = 0$ to 20 s with a step size of 1 s. Use an initial condition that the parachutist has an upward velocity of 20 m/s at $t = 0$. At $t = 10$ s, assume that the chute is instantaneously deployed so that the drag coefficient jumps to 50 kg/s.

1.7 The amount of a uniformly distributed radioactive contaminant contained in a closed reactor is measured by its concentration c (becquerel/liter or Bq/L). The contaminant decreases at a decay rate proportional to its concentration; that is

$$\text{Decay rate} = -kc$$

where k is a constant with units of day^{-1}. Therefore, according to Eq. (1.14), a mass balance for the reactor can be written as

$$\frac{dc}{dt} = -kc$$

$$\begin{pmatrix} \text{change} \\ \text{in mass} \end{pmatrix} = \begin{pmatrix} \text{decrease} \\ \text{by decay} \end{pmatrix}$$

(a) Use Euler's method to solve this equation from $t = 0$ to 1 d with $k = 0.2$ d^{-1}. Employ a step size of $\Delta t = 0.1$ d. The concentration at $t = 0$ is 10 Bq/L.
(b) Plot the solution on a semilog graph (i.e., ln c versus t) and determine the slope. Interpret your results.

1.8 A storage tank (Fig. P1.8) contains a liquid at depth y where $y = 0$ when the tank is half full. Liquid is withdrawn at a constant flow rate Q to meet demands. The contents are resupplied at a sinusoidal rate $3Q \sin^2(t)$. Equation (1.14) can be written for this system as

$$\frac{d(Ay)}{dt} = 3Q \sin^2(t) - Q$$

$$\begin{pmatrix} \text{change in} \\ \text{volume} \end{pmatrix} = (\text{inflow}) - (\text{outflow})$$

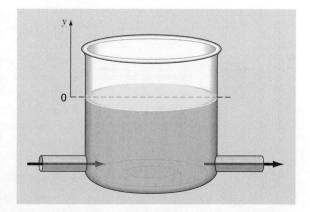

FIGURE P1.8

or, since the surface area A is constant

$$\frac{dy}{dt} = 3\frac{Q}{A}\sin^2(t) - \frac{Q}{A}$$

Use Euler's method to solve for the depth y from $t = 0$ to 10 d with a step size of 0.5 d. The parameter values are $A = 1200$ m^2 and $Q = 500$ m^3/d. Assume that the initial condition is $y = 0$.

1.9 For the same storage tank described in Prob. 1.8, suppose that the outflow is not constant but rather depends on the depth. For this case, the differential equation for depth can be written as

$$\frac{dy}{dt} = 3\frac{Q}{A}\sin^2(t) - \frac{\alpha(1 + y)^{1.5}}{A}$$

Use Euler's method to solve for the depth y from $t = 0$ to 10 d with a step size of 0.5 d. The parameter values are $A = 1200$ m^2, $Q = 500$ m^3/d, and $\alpha = 300$. Assume that the initial condition is $y = 0$.

1.10 The volume flow rate through a pipe is given by $Q = vA$, where v is the average velocity and A is the cross-sectional area. Use volume-continuity to solve for the required area in pipe 3 of Fig. P1.10.

1.11 A group of 30 students attend a class in a room which measures 10 m by 8 m by 3 m. Each student takes up about 0.075 m^3 and gives out about 80 W of heat (1 W = 1 J/s). Calculate the air temperature rise during the first 15 minutes of the class if the room is completely sealed and insulated. Assume the heat capacity C_v for air is 0.718 kJ/(kg K). Assume air is an ideal gas at 20 °C and 101.325 kPa. Note that the heat absorbed by the air Q is related to the mass of the air

m the heat capacity, and the change in temperature by the following relationship:

$$Q = m \int_{T_1}^{T_2} C_v dT = mC_v(T_2 - T_1)$$

The mass of air can be obtained from the ideal gas law:

$$PV = \frac{m}{\text{Mwt}}RT$$

where P is the gas pressure, V is the volume of the gas, Mwt is the molecular weight of the gas (for air, 28.97 kg/kmol), and R is the ideal gas constant [8.314 kPa m^3/(kmol K)].

1.12 Figure P1.12 depicts the various ways in which an average man gains and loses water in one day. One liter is ingested as food, and the body metabolically produces 0.3 liters. In breathing air, the exchange is 0.05 liters while inhaling, and 0.4 liters while exhaling over a one-day period. The body will also lose 0.2, 1.4, 0.2, and 0.35 liters through sweat, urine, feces, and through the skin, respectively. To maintain steady state, how much water must be drunk per day?

1.13 In our example of the free-falling parachutist, we assumed that the acceleration due to gravity was a constant value of 9.8 m/s^2. Although this is a decent approximation when we are examining falling objects near the surface of the earth, the gravitational force decreases as we move above sea level. A more general representation based on Newton's inverse square law of gravitational attraction can be written as

$$g(x) = g(0)\frac{R^2}{(R + x)^2}$$

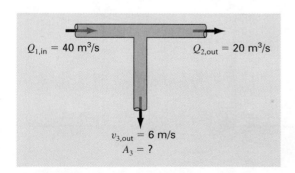

FIGURE P1.10

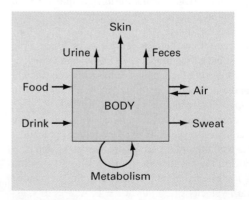

FIGURE P1.12

where $g(x)$ = gravitational acceleration at altitude x (in m) measured upward from the earth's surface (m/s^2), $g(0)$ = gravitational acceleration at the earth's surface ($\cong$ 9.8 m/s^2), and R = the earth's radius ($\cong$ 6.37 $\times$ 10^6 m).

(a) In a fashion similar to the derivation of Eq. (1.8), use a force balance to derive a differential equation for velocity as a function of time that utilizes this more complete representation of gravitation. However, for this derivation, assume that upward velocity is positive.

(b) For the case where drag is negligible, use the chain rule to express the differential equation as a function of altitude rather than time. Recall that the chain rule is

$$\frac{dv}{dt} = \frac{dv}{dx}\frac{dx}{dt}$$

(c) Use calculus to obtain the closed form solution where $v = v_0$ at $x = 0$.

(d) Use Euler's method to obtain a numerical solution from $x = 0$ to 100,000 m using a step of 10,000 m where the initial velocity is 1400 m/s upward. Compare your result with the analytical solution.

1.14 Suppose that a spherical droplet of liquid evaporates at a rate that is proportional to its surface area.

$$\frac{dV}{dt} = -kA$$

where V = volume (mm^3), t = time (hr), k = the evaporation rate (mm/hr), and A = surface area (mm^2). Use Euler's method to compute the volume of the droplet from $t = 0$ to 10 min using a step size of 0.25 min. Assume that $k = 0.1$ mm/hr and that the droplet initially has a radius of 3 mm. Assess the validity of your results by determining the radius

of your final computed volume and verifying that it is consistent with the evaporation rate.

1.15 Newton's law of cooling says that the temperature of a body changes at a rate proportional to the difference between its temperature and that of the surrounding medium (the ambient temperature),

$$\frac{dT}{dt} = -k(T - T_a)$$

where T = the temperature of the body (°C), t = time (min), k = the proportionality constant (per minute), and T_a = the ambient temperature (°C). Suppose that a cup of coffee originally has a temperature of 68 °C. Use Euler's method to compute the temperature from $t = 0$ to 10 min using a step size of 1 min if $T_a = 21$ °C and $k = 0.017$/min.

1.16 A fluid is pumped into the network shown in Fig. P1.16. If $Q_2 = 0.6$, $Q_3 = 0.4$, $Q_7 = 0.2$, and $Q_8 = 0.3$ m^3/s, determine the other flows.

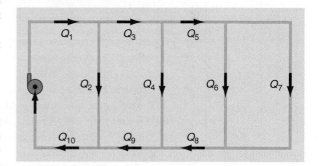

FIGURE P1.16

2

MATLAB Fundamentals

CHAPTER OBJECTIVES

The primary objective of this chapter is to provide an introduction and overview of how MATLAB's calculator mode is used to implement interactive computations. Specific objectives and topics covered are

- Learning how real and complex numbers are assigned to variables
- Learning how vectors and matrices are assigned values using simple assignment, the colon operator, and the `linspace` and `logspace` functions.
- Understanding the priority rules for constructing mathematical expressions.
- Gaining a general understanding of built-in functions and how you can learn more about them with MATLAB's Help facilities.
- Learning how to use vectors to create a simple line plot based on an equation.

YOU'VE GOT A PROBLEM

I n Chap. 1, we used a force balance to determine the terminal velocity of a free-falling object like a bungee jumper.

$$v_t = \sqrt{\frac{gm}{c_d}}$$

where v_t = terminal velocity (m/s), g = gravitational acceleration (m/s^2), m = mass (kg), and c_d = a drag coefficient (kg/m). Aside from predicting the terminal velocity, this equation can also be rearranged to compute the drag coefficient

$$c_d = \frac{mg}{v_t^2} \tag{2.1}$$

TABLE 2.1 Data for the mass and associated terminal velocities of a number of jumpers.

m, kg	83.6	60.2	72.1	91.1	92.9	65.3	80.9
v_t, m/s	53.4	48.5	50.9	55.7	54	47.7	51.1

Thus, if we measure the terminal velocity of a number of jumpers of known mass, this equation provides a means to estimate the drag coefficient. The data in Table 2.1 were collected for this purpose.

In this chapter, we will learn how MATLAB can be used to analyze such data. Beyond showing how MATLAB can be employed to compute quantities like drag coefficients, we will also illustrate how its graphical capabilities provide additional insight into such analyses.

2.1 THE MATLAB ENVIRONMENT

MATLAB is a computer program that provides the user with a convenient environment for performing many types of calculations. In particular, it provides a very nice tool to implement numerical methods.

The most common way to operate MATLAB is by entering commands one at a time in the command window. In this chapter, we use this interactive or *calculator mode* to introduce you to common operations such as performing calculations and creating plots. In Chap. 3, we show how such commands can be used to create MATLAB programs.

One further note. This chapter has been written as a hands-on exercise. That is, you should read it while sitting in front of your computer. The most efficient way to become proficient is to actually implement the commands on MATLAB as you proceed through the following material.

MATLAB uses three primary windows:

- Command window. Used to enter commands and data.
- Graphics window. Used to display plots and graphs.
- Edit window. Used to create and edit M-files.

In this chapter, we will make use of the command and graphics windows. In Chap. 3 we will use the edit window to create M-files.

After starting MATLAB, the command window will open with the command prompt being displayed

```
>>
```

The calculator mode of MATLAB operates in a sequential fashion as you type in commands line by line. For each command, you get a result. Thus, you can think of it as operating like a very fancy calculator. For example, if you type in

```
>> 55 - 16
```

MATLAB will display the result[1]

```
ans =
    39
```

---

[1] MATLAB skips a line between the label (ans =) and the number (39). Here, we omit such blank lines for conciseness. You can control whether blank lines are included with the `format compact` and `format loose` commands.

Notice that MATLAB has automatically assigned the answer to a variable, `ans`. Thus, you could now use `ans` in a subsequent calculation:

```
>> ans + 11
```

with the result

```
ans =
   50
```

MATLAB assigns the result to `ans` whenever you do not explicitly assign the calculation to a variable of your own choosing.

2.2 ASSIGNMENT

Assignment refers to assigning values to variable names. This results in the storage of the values in the memory location corresponding to the variable name.

2.2.1 Scalars

The assignment of values to scalar variables is similar to other computer languages. Try typing

```
>> a = 4
```

Note how the assignment echo prints to confirm what you have done:

```
a =
   4
```

Echo printing is a characteristic of MATLAB. It can be suppressed by terminating the command line with the semicolon (;) character. Try typing

```
>> A = 6;
```

You can type several commands on the same line by separating them with commas or semicolons. If you separate them with commas, they will be displayed, and if you use the semicolon, they will not. For example,

```
>> a = 4,A = 6;x = 1;

a =
   4
```

MATLAB treats names in a case-sensitive manner—that is, the name `a` is not the same as the name `A`. To illustrate this, enter

```
>> a
```

and then enter

```
>> A
```

See how their values are distinct. They are distinct names.

We can assign complex values to variables, since MATLAB handles complex arithmetic automatically. The unit imaginary number $\sqrt{-1}$ is preassigned to the variable i. Consequently, a complex value can be assigned simply as in

```
>> x = 2+i*4

x =
   2.0000 + 4.0000i
```

It should be noted that MATLAB allows the symbol j to be used to represent the unit imaginary number for input. However, it always uses an i for display. For example,

```
>> x = 2+j*4

x =
   2.0000 + 4.0000i
```

There are several predefined variables, for example, pi.

```
>> pi

ans =
   3.1416
```

Notice how MATLAB displays four decimal places. If you desire additional precision, enter the following:

```
>> format long
```

Now when pi is entered the result is displayed to 15 significant figures:

```
>> pi

ans =
   3.14159265358979
```

To return to the four decimal version, type

```
>> format short
```

The following is a summary of the format commands you will employ routinely in engineering and scientific calculations. They all have the syntax: format *type*.

type	Result	Example
short	Scaled fixed-point format with 5 digits	3.1416
long	Scaled fixed-point format with 15 digits for double and 7 digits for single	3.14159265358979
short e	Floating-point format with 5 digits	3.1416e+000
long e	Floating-point format with 15 digits for double and 7 digits for single	3.141592653589793e+000
short g	Best of fixed- or floating-point format with 5 digits	3.1416
long g	Best of fixed- or floating-point format with 15 digits for double and 7 digits for single	3.14159265358979
short eng	Engineering format with at least 5 digits and a power that is a multiple of 3	3.1416e+000
long eng	Engineering format with exactly 16 significant digits and a power that is a multiple of 3	3.14159265358979e+000
bank	Fixed dollars and cents	3.14

2.2.2 Arrays, Vectors and Matrices

An array is a collection of values that are represented by a single variable name. One-dimensional arrays are called *vectors* and two-dimensional arrays are called *matrices*. The scalars used in Section 2.2.1 are actually a matrix with one row and one column.

Brackets are used to enter arrays in the command mode. For example, a row vector can be assigned as follows:

```
>> a = [ 1 2 3 4 5 ]

a =
     1     2     3     4     5
```

Note that this assignment overrides the previous assignment of $a = 4$.

In practice, row vectors are rarely used to solve mathematical problems. When we speak of vectors, we usually refer to column vectors, which are more commonly used. A column vector can be entered in several ways. Try them.

```
>> b = [2;4;6;8;10]
```

or

```
>> b = [ 2;
4;
6;
8;
10 ]
```

or, by transposing a row vector with the ' operator,

```
>> b = [ 2 4 6 8 10 ]'
```

The result in all three cases will be

```
b =
     2
     4
     6
     8
    10
```

A matrix of values can be assigned as follows:

```
>> A = [1 2 3 ; 4 5 6 ; 7 8 9]

A =
     1     2     3
     4     5     6
     7     8     9
```

In addition, the Enter key (carriage return) can be used to separate the rows. For example, in the following case, the Enter key would be struck after the 3, the 6 and the] to assign the matrix:

```
>> A = [1 2 3
4 5 6
7 8 9]
```

At any point in a session, a list of all current variables can be obtained by entering the who command:

```
>> who

Your variables are:
A    a    ans   b    x
```

or, with more detail, enter the whos command:

```
>> whos

  Name        Size                Bytes   Class
  A           3x3                    72   double array
  a           1x5                    40   double array
  ans         1x1                     8   double array
  b           5x1                    40   double array
  x           1x1                    16   double array
(complex)
Grand total is 21 elements using 176 bytes
```

Note that subscript notation can be used to access an individual element of an array. For example, the fourth element of the column vector b can be displayed as

```
>> b(4)

ans =
     8
```

For an array, A(m,n) selects the element in mth row and the nth column. For example,

```
>> A(2,3)

ans =
     6
```

There are several built-in functions that can be used to create matrices. For example, the ones and zeros functions create vectors or matrices filled with ones and zeros, respectively. Both have two arguments, the first for the number of rows and the second for the number of columns. For example, to create a 2×3 matrix of zeros:

```
>> E = zeros(2,3)

E =
     0    0    0
     0    0    0
```

Similarly, the ones function can be used to create a row vector of ones:

```
>> u = ones(1,3)

u =
     1    1    1
```

2.2.3 The Colon Operator

The colon operator is a powerful tool for creating and manipulating arrays. If a colon is used to separate two numbers, MATLAB generates the numbers between them using an increment of one:

```
>> t = 1:5

t =
     1     2     3     4     5
```

If colons are used to separate three numbers, MATLAB generates the numbers between the first and third numbers using an increment equal to the second number:

```
>> t = 1:0.5:3

t =
    1.0000    1.5000    2.0000    2.5000    3.0000
```

Note that negative increments can also be used

```
>> t = 10:-1:5

t =
    10     9     8     7     6     5
```

Aside from creating series of numbers, the colon can also be used as a wildcard to select the individual rows and columns of a matrix. When a colon is used in place of a specific subscript, the colon represents the entire row or column. For example, the second row of the matrix A can be selected as in

```
>> A(2,:)

ans =
     4     5     6
```

We can also use the colon notation to selectively extract a series of elements from within an array. For example, based on the previous definition of the vector t:

```
>> t(2:4)

ans =
     9     8     7
```

Thus, the second through the fourth elements are returned.

2.2.4 The `linspace` and `logspace` Functions

The `linspace` and `logspace` functions provide other handy tools to generate vectors of spaced points. The `linspace` function generates a row vector of equally spaced points. It has the form

```
linspace(x1, x2, n)
```

which generates n points between $x1$ and $x2$. For example

```
>> linspace(0,1,6)

ans =
         0    0.2000    0.4000    0.6000    0.8000    1.0000
```

If the n is omitted, the function automatically generates 100 points.

The `logspace` function generates a row vector that is logarithmically equally spaced. It has the form

```
logspace(x1, x2, n)
```

which generates n logarithmically equally spaced points between decades 10^{x_1} and 10^{x_2}. For example,

```
>> logspace(-1,2,4)

ans =
    0.1000    1.0000    10.0000   100.0000
```

If n is omitted, it automatically generates 50 points.

2.3 MATHEMATICAL OPERATIONS

Operations with scalar quantities are handled in a straightforward manner, similar to other computer languages. The common operators, in order of priority, are

^	Exponentiation
−	Negation
* /	Multiplication and division
\	Left division[2]
+ −	Addition and subtraction

These operators will work in calculator fashion. Try

```
>> 2*pi

ans =
    6.2832
```

Also, scalar real variables can be included:

```
>> y = pi/4;
>> y ^ 2.45

ans =
    0.5533
```

Results of calculations can be assigned to a variable, as in the next-to-last example, or simply displayed, as in the last example.

As with other computer calculation, the priority order can be overridden with parentheses. For example, because exponentiation has higher priority then negation, the following result would be obtained:

```
>> y = -4 ^ 2

y =
    -16
```

Thus, 4 is first squared and then negated. Parentheses can be used to override the priorities as in

```
>> y = (-4) ^ 2

y =
    16
```

[2] Left division applies to matrix algebra. It will be discussed in detail later in this book.

Calculations can also involve complex quantities. Here are some examples that use the values of x $(2 + 4i)$ and y (16) defined previously:

```
>> 3 * x

ans =
   6.0000 +12.0000i

>> 1 / x

ans =
   0.1000 - 0.2000i

>> x ^ 2

ans =
 -12.0000 +16.0000i

>> x + y

ans =
  18.0000 + 4.0000i
```

The real power of MATLAB is illustrated in its ability to carry out vector-matrix calculations. Although we will describe such calculations in detail in Chap. 8, it is worth introducing some of those manipulations here.

The *inner product* of two vectors (dot product) can be calculated using the * operator,

```
>> a * b

ans =
   110
```

and likewise, the *outer product*

```
>> b * a

ans =
    2     4     6     8    10
    4     8    12    16    20
    6    12    18    24    30
    8    16    24    32    40
   10    20    30    40    50
```

To further illustrate vector-matrix multiplication, first redefine a and b:

```
>> a = [1 2 3];
```

and

```
>> b = [4 5 6]';
```

Now, try

```
>> a * A

ans =
     30    36    42
```

or

```
>> A * b
ans =
    32
    77
   122
```

Matrices cannot be multiplied if the inner dimensions are unequal. Here is what happens when the dimensions are not those required by the operations. Try

```
>> A * a
```

MATLAB automatically displays the error message:

```
??? Error using ==> mtimes
Inner matrix dimensions must agree.
```

Matrix-matrix multiplication is carried out in likewise fashion:

```
>> A * A
ans =
    30    36    42
    66    81    96
   102   126   150
```

Mixed operations with scalars are also possible:

```
>> A/pi
ans =
   0.3183    0.6366    0.9549
   1.2732    1.5915    1.9099
   2.2282    2.5465    2.8648
```

We must always remember that MATLAB will apply the simple arithmetic operators in vector-matrix fashion if possible. At times, you will want to carry out calculations item by item in a matrix or vector. MATLAB provides for that too. For example,

```
>> A ^ 2
ans =
    30    36    42
    66    81    96
   102   126   150
```

results in matrix multiplication of A with itself.

What if you want to square each element of A? That can be done with

```
>> A .^ 2
ans =
     1     4     9
    16    25    36
    49    64    81
```

The . preceding the ^ operator signifies that the operation is to be carried out element by element. The MATLAB manual calls these *array operations*. They are also often referred to as *element-by-element operations*.

MATLAB contains a helpful shortcut for performing calculations that you've already done. Press the up-arrow key. You should get back the last line you typed in.

```
>> A .^ 2
```

Pressing Enter will perform the calculation again. But you can also edit this line. For example, change it to the line below and then press Enter.

```
>> A .^ 3

ans =
       1       8      27
      64     125     216
     343     512     729
```

Using the up-arrow key, you can go back to any command that you entered. Press the up-arrow until you get back the line

```
b * a
```

Alternatively, you can type b and press the up-arrow once and it will automatically bring up the last command beginning with the letter b. The up-arrow shortcut is a quick way to fix errors without having to retype the entire line.

2.4 USE OF BUILT-IN FUNCTIONS

MATLAB and its Toolboxes have a rich collection of built-in functions. You can use online help to find out more about them. For example, if you want to learn about the log function, type in

```
>> help log

LOG    Natural logarithm.
   LOG(X) is the natural logarithm of the elements of X.
   Complex results are produced if X is not positive.

   See also LOG2, LOG10, EXP, LOGM.
```

For a list of all the elementary functions, type

```
>> help elfun
```

One of their important properties of MATLAB's built-in functions is that they will operate directly on vector and matrix quantities. For example, try

```
>> log(A)

ans =
          0    0.6931    1.0986
     1.3863    1.6094    1.7918
     1.9459    2.0794    2.1972
```

and you will see that the natural logarithm function is applied in array style, element by element, to the matrix A. Most functions, such as sqrt, abs, sin, acos, tanh, and exp, operate in array fashion. Certain functions, such as exponential and square root, have matrix definitions also. MATLAB will evaluate the matrix version when the letter m is appended to the function name. Try

```
>> sqrtm(A)

ans =
    0.4498 + 0.7623i   0.5526 + 0.2068i   0.6555 - 0.3487i
    1.0185 + 0.0842i   1.2515 + 0.0228i   1.4844 - 0.0385i
    1.5873 - 0.5940i   1.9503 - 0.1611i   2.3134 + 0.2717i
```

There are several functions for rounding. For example, suppose that we enter a vector:

```
>> E = [-1.6 -1.5 -1.4 1.4 1.5 1.6];
```

The round function rounds the elements of E to the nearest integers:

```
>> round(E)

ans =
    -2    -2    -1    1    2    2
```

The ceil (short for ceiling) function rounds to the nearest integers toward infinity:

```
>> ceil(E)

ans =
    -1    -1    -1    2    2    2
```

The floor function rounds down to the nearest integers toward minus infinity:

```
>> floor(E)

ans =
    -2    -2    -2    1    1    1
```

There are also functions that perform special actions on the elements of matrices and arrays. For example, the sum function returns the sum of the elements:

```
>> F = [3 5 4 6 1];
>> sum(F)

ans =
    19
```

In a similar way, it should be pretty obvious what's happening with the following commands:

```
>> min(F),max(F),mean(F),prod(F),sort(F)

ans =
    1

ans =
    6

ans =
    3.8000

ans =
    360

ans =
    1    3    4    5    6
```

A common use of functions is to evaluate a formula for a series of arguments. Recall that the velocity of a free-falling bungee jumper can be computed with [Eq. (1.9)]:

$$v = \sqrt{\frac{gm}{c_d}} \tanh\left(\sqrt{\frac{gc_d}{m}}t\right)$$

where v is velocity (m/s), g is the acceleration due to gravity (9.81 m/s^2), m is mass (kg), c_d is the drag coefficient (kg/m), and t is time (s).

Create a column vector t that contains values from 0 to 20 in steps of 2:

```
>> t = [0:2:20]'

t =
     0
     2
     4
     6
     8
    10
    12
    14
    16
    18
    20
```

Check the number of items in the t array with the length function:

```
>> length(t)

ans =
    11
```

Assign values to the parameters:

```
>> g = 9.81; m = 68.1; cd = 0.25;
```

MATLAB allows you to evaluate a formula such as $v = f(t)$, where the formula is computed for each value of the t array, and the result is assigned to a corresponding position in the v array. For our case,

```
>> v = sqrt(g*m/cd)*tanh(sqrt(g*cd/m)*t)

v =
          0
    18.7292
    33.1118
    42.0762
    46.9575
    49.4214
    50.6175
    51.1871
    51.4560
    51.5823
    51.6416
```

2.5   **GRAPHICS**

MATLAB allows graphs to be created quickly and conveniently. For example, to create a graph of the `t` and `v` arrays from the data above, enter

```
>> plot (t, v)
```

The graph appears in the graphics window and can be printed or transferred via the clip-board to other programs.

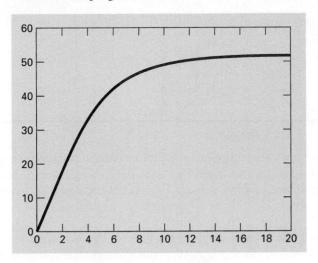

You can customize the graph a bit with commands such as the following:

```
>> title('Plot of v versus t')
>> xlabel('Values of t')
>> ylabel('Values of v')
>> grid
```

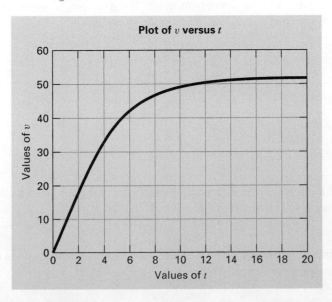

TABLE 2.2 Specifiers for colors, symbols, and line types.

Colors		Symbols		Line Types	
Blue	b	Point	.	Solid	–
Green	g	Circle	o	Dotted	:
Red	r	X-mark	x	Dashdot	-.
Cyan	c	Plus	+	Dashed	--
Magenta	m	Star	*		
Yellow	y	Square	s		
Black	k	Diamond	d		
		Triangle(down)	∨		
		Triangle(up)	∧		
		Triangle(left)	<		
		Triangle(right)	>		
		Pentagram	p		
		Hexagram	h		

The `plot` command displays a solid line by default. If you want to plot each point with a symbol, you can include a specifier enclosed in single quotes in the `plot` function. Table 2.2 lists the available specifiers. For example, if you want to use open circles enter

```
>> plot (t, v, 'o')
```

MATLAB allows you to display more than one data set on the same plot. For example, if you want to connect each data marker with a straight line you could type

```
>> plot (t, v, t, v, 'o')
```

It should be mentioned that, by default, previous plots are erased every time the `plot` command is implemented. The `hold on` command holds the current plot and all axis properties so that additional graphing commands can be added to the existing plot. The `hold off` command returns to the default mode. For example, if we had typed the following commands, the final plot would only display symbols:

```
>> plot (t, v)
>> plot (t, v, 'o')
```

In contrast, the following commands would result in both lines and symbols being displayed:

```
>> plot (t, v)
>> hold on
>> plot (t, v, 'o')
>> hold off
```

In addition to `hold`, another handy function is `subplot`, which allows you to split the graph window into subwindows or *panes*. It has the syntax

```
subplot (m, n, p)
```

This command breaks the graph window into an m-by-n matrix of small axes, and selects the p-th axes for the current plot.

We can demonstrate `subplot` by examining MATLAB's capability to generate three-dimensional plots. The simplest manifestation of this capability is the `plot3` command which has the syntax

```
plot3 (x, y, z)
```

where x, y, and z are three vectors of the same length. The result is a line in three-dimensional space through the points whose coordinates are the elements of x, y, and z.

Plotting a helix provides a nice example to illustrate its utility. First, let's graph a circle with the two-dimensional `plot` function using the parametric representation: $x = \sin(t)$ and $y = \cos(t)$. We employ the `subplot` command so we can subsequently add the three-dimensional plot.

```
>> t = 0:pi/50:10*pi;
>> subplot(1,2,1);plot(sin(t),cos(t))
>> axis square
>> title('(a)')
```

As in Fig. 2.1*a*, the result is a circle. Note that the circle would have been distorted if we had not used the `axis square` command.

FIGURE 2.1
A two-pane plot of (*a*) a two-dimensional circle and (*b*) a three-dimensional helix.

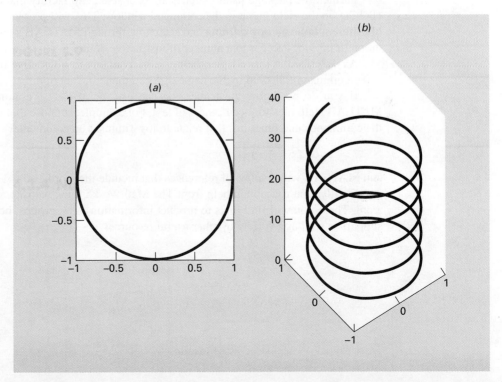

Now, let's add the helix to the graph's right pane. To do this, we again employ a parametric representation: $x = \sin(t)$, $y = \cos(t)$, and $z = t$

```
>> subplot(1,2,2);plot3(sin(t),cos(t),t);
>> title('(b)')
```

The result is shown in Fig. 2.1*b*. Can you visualize what's going on? As time evolves, the x and y coordinates sketch out the circumference of the circle in the x–y plane in the same fashion as the two-dimensional plot. However, simultaneously, the curve rises vertically as the z coordinate increases linearly with time. The net result is the characteristic spring or spiral staircase shape of the helix.

There are other features of graphics that are useful—for example, plotting objects instead of lines, families of curves plots, plotting on the complex plane, log-log or semilog plots, three-dimensional mesh plots, and contour plots. As described next, a variety of resources are available to learn about these as well as other MATLAB capabilities.

2.6 OTHER RESOURCES

The foregoing was designed to focus on those features of MATLAB that we will be using in the remainder of this book. As such, it is obviously not a comprehensive overview of all of MATLAB's capabilities. If you are interested in learning more, you should consult one of the excellent books devoted to MATLAB (e.g., Palm, 2005; Hanselman and Littlefield, 2005; and Moore, 2007).

Further, the package itself includes an extensive Help facility that can be accessed by clicking on the Help menu in the command window. This will provide you with a number of different options for exploring and searching through MATLAB's Help material. In addition, it provides access to a number of instructive demos.

As described in this chapter, help is also available in interactive mode by typing the `help` command followed by the name of a command or function.

If you do not know the name, you can use the `lookfor` command to search the MATLAB Help files for occurrences of text. For example, suppose that you want to find all the commands and functions that relate to logarithms, you could enter

```
>> lookfor logarithm
```

and MATLAB will display all references that include the word `logarithm`.

Finally, you can obtain help from The MathWorks, Inc., website at www.mathworks.com. There you will find links to product information, newsgroups, books, and technical support as well as a variety of other useful resources.

2.7 CASE STUDY   EXPLORATORY DATA ANALYSIS

Background. Your textbooks are filled with formulas developed in the past by renowned scientists and engineers. Although these are of great utility, engineers and scientists often must supplement these relationships by collecting and analyzing their own data. Sometimes this leads to a new formula. However, prior to arriving at a final predictive equation, we usually "play" with the data by performing calculations and developing plots. In most cases, our intent is to gain insight into the patterns and mechanisms hidden in the data.

In this case study, we will illustrate how MATLAB facilitates such exploratory data analysis. We will do this by estimating the drag coefficient of a free-falling human based on Eq. (2.1) and the data from Table 2.1. However, beyond merely computing the drag coefficient, we will use MATLAB's graphical capabilities to discern patterns in the data.

Solution. The data from Table 2.1 along with gravitational acceleration can be entered as

```
>> m=[83.6 60.2 72.1 91.1 92.9 65.3 80.9];
>> vt=[53.4 48.5 50.9 55.7 54 47.7 51.1];
>> g=9.81;
```

The drag coefficients can then be computed with Eq. (2.1). Because we are performing element-by-element operations on vectors, we must include periods prior to the operators:

```
>> cd=g*m./vt.^2

cd =
    0.2876   0.2511   0.2730   0.2881   0.3125   0.2815   0.3039
```

We can now use some of MATLAB's built-in functions to generate some statistics for the results:

```
>> cdavg=mean(cd),cdmin=min(cd),cdmax=max(cd)
cdavg =
    0.2854
cdmin =
    0.2511
cdmax =
    0.3125
```

Thus, the average value is 0.2854 with a range from 0.2511 to 0.3125 kg/m.

Now, let's start to play with this data by using Eq. (2.1) to make a prediction of the terminal velocity based on the average drag:

```
>> vpred=sqrt(g*m/cdavg)

vpred =
    53.6065   45.4897   49.7831   55.9595   56.5096   47.3774
52.7338
```

Notice that we do not have to use periods prior to the operators in this formula? Do you understand why?

We can plot these values versus the actual measured terminal velocities. We will also superimpose a line indicating exact predictions (the 1:1 line) to help assess the results.

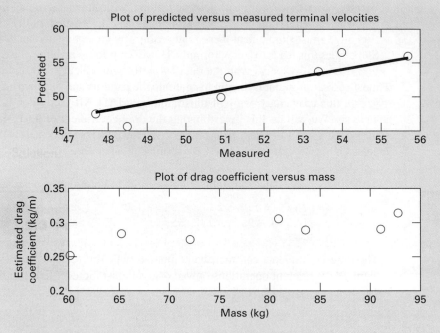

FIGURE 2.2
Two plots created with MATLAB.

Because we are going to eventually generate a second plot, we employ the subplot command:

```
>> subplot(2,1,1);plot(vt,vpred,'o',vt,vt)
>> xlabel('measured')
>> ylabel('predicted')
>> title('Plot of predicted versus measured terminal ...
velocities')
```

As in the top plot of Fig. 2.2, because the predictions generally follow the 1:1 line, you might initially conclude that the average drag coefficient yields decent results. However, notice how the model tends to underpredict the low velocities and overpredict the high. This suggests that rather than being constant, there might be a trend in the drag coefficients. This can be seen by plotting the estimated drag coefficients versus mass:

```
>> subplot(2,1,2);plot(m,cd,'o')
>> xlabel('mass (kg)')
>> ylabel('estimated drag coefficient (kg/m)')
>> title('Plot of drag coefficient versus mass')
```

The resulting plot, which is the bottom graph in Fig. 2.2, suggests that rather than being constant, the drag coefficient seems to be increasing as the mass of the jumper increases. Based on this result, you might conclude that your model needs to be improved. At the least, it might motivate you to conduct further experiments with a larger number of jumpers to confirm your preliminary finding.

In addition, the result might also stimulate you to go to the fluid mechanics literature and learn more about the science of drag. If you did this, you would discover that the parameter c_d is actually a lumped drag coefficient that along with the true drag includes other factors such as the jumper's frontal area and air density:

$$c_d = \frac{C_D \rho A}{2} \tag{2.2}$$

where C_D = a dimensionless drag coefficient, ρ = air density (kg/m^3), and A = frontal area (m^2), which is the area projected on a plane normal to the direction of the velocity.

Assuming that the densities were relatively constant during data collection (a pretty good assumption if the jumpers all took off from the same height on the same day), Eq. (2.2) suggests that heavier jumpers might have larger areas. This hypothesis could be substantiated by measuring the frontal areas of individuals of varying masses.

PROBLEMS

2.1 A simple electric circuit consisting of a resistor, a capacitor, and an inductor is depicted in Fig. P2.1. The charge on the capacitor $q(t)$ as a function of time can be computed as

$$q(t) = q_0 e^{-Rt/(2L)} \cos\left[\sqrt{\frac{1}{LC} - \left(\frac{R}{2L}\right)^2} \, t\right]$$

where t = time, q_0 = the initial charge, R = the resistance, L = inductance, and C = capacitance. Use MATLAB to generate a plot of this function from $t = 0$ to 0.7, given that $q_0 = 12$, $R = 50$, $L = 5$, and $C = 10^{-4}$.

2.2 The standard normal probability density function is a bell-shaped curve that can be represented as

$$f(z) = \frac{1}{\sqrt{2\pi}} e^{-z^2/2}$$

Use MATLAB to generate a plot of this function from $z = -4$ to 4. Label the ordinate as frequency and the abscissa as z.

2.3 Use the `linspace` function to create vectors identical to the following created with colon notation:
(a) `t = 5:6:30`
(b) `x = -3:4`
2.4 Use colon notation to create vectors identical to the following created with the `linspace` function:
(a) `v = linspace(-3,1,9)`
(b) `r = linspace(8,0,17)`
2.5 If a force F (N) is applied to compress a spring, its displacement x (m) can often be modeled by Hooke's law:

$$F = kx$$

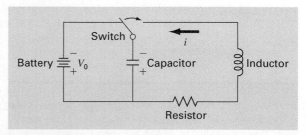

FIGURE P2.1

where k = the spring constant (N/m). The potential energy stored in the spring U (J) can then be computed as

$$U = \frac{1}{2}kx^2$$

Five springs are tested and the following data compiled:

F, N	11	12	15	9	12
x, m	0.013	0.020	0.009	0.010	0.012

Use MATLAB to store F and x as vectors and then compute vectors of the spring constants and the potential energies. Use the `max` function to determine the maximum potential energy.

2.6 The density of freshwater can be computed as a function of temperature with the following cubic equation:

$$\rho = 5.5289 \times 10^{-8}T_C^3 - 8.5016 \times 10^{-6}T_C^2$$
$$+ 6.5622 \times 10^{-5}T_C + 0.99987$$

where ρ = density (g/cm^3) and T_C = temperature (°C). Use MATLAB to generate a vector of temperatures ranging from 32 °F to 82.4 °F using increments of 3.6 °F. Convert this vector to degrees Celsius and then compute a vector of densities based on the cubic formula. Create a plot of ρ versus T_C. Recall that $T_C = 5/9(T_F - 32)$.

2.7 Manning's equation can be used to compute the velocity of water in a rectangular open channel:

$$U = \frac{\sqrt{S}}{n}\left(\frac{BH}{B + 2H}\right)^{2/3}$$

where U = velocity (m/s), S = channel slope, n = roughness coefficient, B = width (m), and H = depth (m). The following data is available for five channels:

n	S	B	H
0.035	0.0001	10	2
0.020	0.0002	8	1
0.015	0.0010	19	1.5
0.030	0.0008	24	3
0.022	0.0003	15	2.5

Store these values in a matrix where each row represents one of the channels and each column represents one of the parameters. Write a single-line MATLAB statement to compute a column vector containing the velocities based on the values in the parameter matrix.

2.8 It is general practice in engineering and science that equations be plotted as lines and discrete data as symbols. Here is some data for concentration (c) versus time (t) for the photodegradation of aqueous bromine:

t, min	10	20	30	40	50	60
c, ppm	3.4	2.6	1.6	1.3	1.0	0.5

This data can be described by the following function:

$$c = 4.84e^{-0.034t}$$

Use MATLAB to create a plot displaying both the data (using square symbols) and the function (using a dotted line). Plot the function for $t = 0$ to 75 min.

2.9 The `semilogy` function operates in an identical fashion to the `plot` function except that a logarithmic (base-10) scale is used for the y axis. Use this function to plot the data and function as described in Prob. 2.8. Explain the results.

2.10 Here is some wind tunnel data for force (F) versus velocity (v):

v, m/s	10	20	30	40	50	60	70	80
F, N	25	70	380	550	610	1220	830	1450

This data can be described by the following function:

$$F = 0.2741v^{1.9842}$$

Use MATLAB to create a plot displaying both the data (using diamond symbols) and the function (using a dotted line). Plot the function for $v = 0$ to 90 m/s.

2.11 The `loglog` function operates in an identical fashion to the `plot` function except that logarithmic scales are used for both the x and y axes. Use this function to plot the data and function as described in Prob. 2.10. Explain the results.

2.12 The Maclaurin series expansion for the sine is

$$\sin x = x - \frac{x^3}{3!} + \frac{x^5}{5!} - \frac{x^7}{7!} + \frac{x^9}{9!} - \cdots$$

Use MATLAB to create a plot of the sine (solid line) along with a plot of the series expansion (dashed line) up to and including the term $x^7/7!$. Use the built-in function `factorial` in computing the series expansion. Make the range of the abscissa from $x = 0$ to $3\pi/2$.

2.13 You contact the jumpers used to generate the data in Table 2.1 and measure their frontal areas. The resulting values, which are ordered in the same sequence as the corresponding values in Table 2.1, are

A, m²	0.454	0.401	0.453	0.485	0.532	0.474	0.486

(a) If the air density is $\rho = 1.225$ kg/m³, use MATLAB to compute values of the dimensionless drag coefficient C_D.

(b) Determine the average, minimum and maximum of the resulting values.

(c) Develop a side-by-side plot of A versus m (left side) and C_D versus m (right side). Include descriptive axis labels and titles on the plots.

2.14 The following parametric equations generate a helix that contracts exponentially as it evolves

$$x = e^{-0.1t} \sin t$$
$$y = e^{-0.1t} \cos t$$
$$z = t$$

Use `subplot` to generate a two-dimensional line plot of (x, y) in the top pane and a three-dimensional line plot of (x, y, z) in the bottom pane.

2.15 Exactly what will be displayed after the following MATLAB commands are typed?

(a)
```
>> x = 2;
>> x ^ 3;
>> y = 8 - x
```

(b)
```
>> q = 4:2:10;
>> r = [7 8 4; 3 6 -2];
>> sum(q) * r(2, 3)
```

2.16 The trajectory of an object can be modeled as

$$y = (\tan \theta_0)x - \frac{g}{2v_0^2 \cos^2 \theta_0}x^2 + y_0$$

where y = height (m), θ_0 = initial angle (radians), x = horizontal distance (m), g = gravitational acceleration (= 9.81 m/s²), v_0 = initial velocity (m/s), and y_0 = initial height. Use MATLAB to find the displacement for $y_0 = 0$ and $v_0 = 30$ m/s for initial angles ranging from 15 to 75° in increments of 15°. Employ a range of horizontal distances from $x = 0$ to 100 m in increments of 5 m. The results should be assembled in an array where the first dimension (rows) corresponds to the distances, and the second dimension (columns) corresponds to the different initial angles. Use this matrix to generate a single plot of the heights versus horizontal distances for each of the initial angles. Employ a legend to distinguish among the different cases, and scale the plot so that the minimum height is zero using the `axis` command.

2.17 The temperature dependence of chemical reactions can be computed with the *Arrhenius equation:*

$$k = Ae^{-E/(RT_a)}$$

where k = reaction rate (s⁻¹), A = the preexponential (or frequency) factor, E = activation energy (J/mol), R = gas constant [8.314 J/(mole · K)], and T_a = absolute temperature (K). A compound has $E = 1 \times 10^5$ J/mol and $A = 7 \times 10^{16}$. Use MATLAB to generate values of reaction rates for temperatures ranging from 273 to 333 K. Use `subplot` to generate a side-by-side graph of (a) k versus T_a and (b) $\log_{10} k$ versus $1/T_a$. Employ the `semilogy` function to create (b). Interpret your results.

3

Programming with MATLAB

CHAPTER OBJECTIVES

The primary objective of this chapter is to learn how to write M-file programs to implement numerical methods. Specific objectives and topics covered are

- Learning how to create well-documented M-files in the edit window and invoke them from the command window.
- Understanding how script and function files differ.
- Understanding how to incorporate help comments in functions.
- Knowing how to set up M-files so that they interactively prompt users for information and display results in the command window.
- Understanding the role of subfunctions and how they are accessed.
- Knowing how to create and retrieve data files.
- Learning how to write clear and well-documented M-files by employing structured programming constructs to implement logic and repetition.
- Recognizing the difference between `if...elseif` and `switch` constructs.
- Recognizing the difference between `for...end` and `while` structures.
- Understanding what is meant by vectorization and why it is beneficial.
- Understanding how anonymous functions can be employed to pass functions to function function M-files.

YOU'VE GOT A PROBLEM

In Chap. 1, we used a force balance to develop a mathematical model to predict the fall velocity of a bungee jumper. This model took the form of the following differential equation:

$$\frac{dv}{dt} = g - \frac{c}{m}v^2$$

We also learned that a numerical solution of this equation could be obtained with Euler's method:

$$v_{i+1} = v_i + \frac{dv_i}{dt}\Delta t$$

This equation can be implemented repeatedly to compute velocity as a function of time. However, to obtain good accuracy, many small steps must be taken. This would be extremely laborious and time consuming to implement by hand. However, with the aid of MATLAB, such calculations can be performed easily.

So our problem now is to figure out how to do this. This chapter will introduce you to how MATLAB M-files can be used to obtain such solutions.

3.1 M-FILES

The most common way to operate MATLAB is by entering commands one at a time in the command window. M-files provide an alternative way of performing operations that greatly expand MATLAB's problem-solving capabilities. An *M-file* contains a series of statements that can be run all at once. Note that the nomenclature "M-file" comes from the fact that such files are stored with a `.m` extension. M-files come in two flavors: script files and function files.

3.1.1 Script Files

A *script file* is merely a series of MATLAB commands that are saved on a file. They are useful for retaining a series of commands that you want to execute on more than one occasion. The script can be executed by typing the file name in the command window or by invoking the menu selections in the edit window: **Debug, Run.**

EXAMPLE 3.1   Script File

Problem Statement.   Develop a script file to compute the velocity of the free-falling bungee jumper.

Solution.   Open the editor with the menu selection: **File, New, M-file.** Type in the following statements to compute the velocity of the free-falling bungee jumper at a specific time [recall Eq. (1.9)]:

```
g = 9.81; m = 68.1; t = 12; cd = 0.25;
v = sqrt(g * m / cd) * tanh(sqrt(g * cd / m) * t)
```

Save the file as `scriptdemo.m`. Return to the command window and type

```
>>scriptdemo
```

The result will be displayed as

```
v =
   50.6175
```

Thus, the script executes just as if you had typed each of its lines in the command window.

As a final step, determine the value of g by typing

```
>> g

g =
    9.8100
```

So you can see that even though g was defined within the script, it retains its value back in the command workspace. As we will see in the following section, this is an important distinction between scripts and functions.

3.1.2 Function Files

Function files are M-files that start with the word function. In contrast to script files, they can accept input arguments and return outputs. Hence they are analogous to user-defined functions in programming languages such as Fortran, Visual Basic or *C*.

The syntax for the function file can be represented generally as

```
function outvar = funcname(arglist)
% helpcomments
statements
outvar = value;
```

where *outvar* = the name of the output variable, *funcname* = the function's name, *arglist* = the function's argument list (i.e., comma-delimited values that are passed into the function), *helpcomments* = text that provides the user with information regarding the function (these can be invoked by typing Help *funcname* in the command window), and *statements* = MATLAB statements that compute the *value* that is assigned to *outvar*.

Beyond its role in describing the function, the first line of the *helpcomments*, called the *H1 line,* is the line that is searched by the lookfor command (recall Sec. 2.6). Thus, you should include key descriptive words related to the file on this line.

The M-file should be saved as *funcname*.m. The function can then be run by typing *funcname* in the command window as illustrated in the following example. Note that even though MATLAB is case-sensitive, your computer's operating system may not be. Whereas MATLAB would treat function names like freefallvel and FreeFallVel as two different variables, your operating system might not.

EXAMPLE 3.2   Function File

Problem Statement.   As in Example 3.1, compute the velocity of the free-falling bungee jumper, but now we will use a function file for the task.

Solution.   Type the following statements in the file editor:

```
function v = freefallvel(t, m, cd)
% freefallvel: bungee velocity with second-order drag
% v=freefallvel(t,m,cd) computes the free-fall velocity
%                       of an object with second-order drag
input:
```

```
%   t = time (s)
%   m = mass (kg)
%   cd = second-order drag coefficient (kg/m)
% output:
%   v = downward velocity (m/s)

g = 9.81;      % acceleration of gravity
v = sqrt(g * m / cd)*tanh(sqrt(g * cd / m) * t);
```

Save the file as `freefallvel.m`. To invoke the function, return to the command window and type in

```
>> freefallvel(12,68.1,0.25)
```

The result will be displayed as

```
ans =
   50.6175
```

One advantage of a function M-file is that it can be invoked repeatedly for different argument values. Suppose that you wanted to compute the velocity of a 100-kg jumper after 8 s:

```
>> freefallvel(8,100,0.25)

ans =
   53.1878
```

To invoke the help comments type

```
>> help freefallvel
```

which results in the comments being displayed

```
freefallvel: bungee velocity with second-order drag
  v=freefallvel(t,m,cd) computes the free-fall velocity
                        of an object with second-order drag
input:
  t = time (s)
  m = mass (kg)
  cd = second-order drag coefficient (kg/m)
output:
  v = downward velocity (m/s)
```

If at a later date, you forgot the name of this function, but remembered that it involved bungee jumping, you could enter

```
>> lookfor bungee
```

and the following information would be displayed

```
freefall.m: % freefall: bungee velocity with second-order drag
```

Note that, at the end of the previous example, if we had typed

```
>> g
```

the following message would have been displayed

```
??? Undefined function or variable 'g'.
```

So even though g had a value of 9.81 within the M-file, it would not have a value in the command workspace. As noted previously at the end of Example 3.1, this is an important distinction between functions and scripts. The variables within a function are said to be *local* and are erased after the function is executed. In contrast, the variables in a script retain their existence after the script is executed.

Function M-files can return more than one result. In such cases, the variables containing the results are comma-delimited and enclosed in brackets. For example, the following function, stats.m, computes the mean and the standard deviation of a vector:

```
function [mean, stdev] = stats(x)
n = length(x);
mean = sum(x)/n;
stdev = sqrt(sum((x-mean).^2/(n-1)));
```

Here is an example of how it can be applied:

```
>> y = [8 5 10 12 6 7.5 4];
>> [m,s] = stats(y)

m =
    7.5000

s =
    2.8137
```

Because script M-files have limited utility, function M-files will be our primary programming tool for the remainder of this book. Hence, we will often refer to function M-files as simply M-files.

3.1.3 Subfunctions

Functions can call other functions. Although such functions can exist as separate M-files, they may also be contained in a single M-file. For example, the M-file in Example 3.2 (without comments) could have been split into two functions and saved as a single M-file[1]:

```
function v = freefallsubfunc(t, m, cd)
v = vel(t, m, cd);
end

function v = vel(t, m, cd)
g = 9.81;
v = sqrt(g * m / cd)*tanh(sqrt(g * cd / m) * t);
end
```

---

[1] Note that although end statements are optional in single-function M-files, we like to include them when subfunctions are involved to highlight the boundaries between the main function and the subfunctions.

This M-file would be saved as `freefallsubfunc.m`. In such cases, the first function is called the *main* or *primary function*. It is the only function that is accessible to the command window and other functions and scripts. All the other functions (in this case, `vel`) are referred to as *subfunctions*.

A subfunction is only accessible to the main function and other subfunctions within the M-file in which it resides. If we run `freefallsubfunc` from the command window, the result is identical to Example 3.2:

```
>> freefallsubfunc(12,68.1,0.25)

ans =
    50.6175
```

However, if we attempt to run the subfunction `vel`, an error message occurs:

```
>> vel(12,68.1,.25)
??? Undefined command/function 'vel'.
```

3.2   INPUT-OUTPUT

As in Section 3.1, information is passed into the function via the argument list and is output via the function's name. Two other functions provide ways to enter and display information directly using the command window.

The `input` **Function.**   This function allows you to prompt the user for values directly from the command window. Its syntax is

```
n = input('promptstring')
```

The function displays the *promptstring*, waits for keyboard input, and then returns the value from the keyboard. For example,

```
m = input('Mass (kg): ')
```

When this line is executed, the user is prompted with the message

```
Mass (kg):
```

If the user enters a value, it would then be assigned to the variable `m`.

The `input` function can also return user input as a string. To do this, an `'s'` is appended to the function's argument list. For example,

```
name = input('Enter your name: ','s')
```

The `disp` **Function.**   This function provides a handy way to display a value. Its syntax is

```
disp(value)
```

where *value* = the value you would like to display. It can be a numeric constant or variable, or a string message enclosed in hyphens. Its application is illustrated in the following example.

EXAMPLE 3.3   An Interactive M-File Function

Problem Statement.   As in Example 3.2, compute the velocity of the free-falling bungee jumper, but now use the `input` and `disp` functions for input/output.

Solution.   Type the following statements in the file editor:

```
function freefalli
% freefalli: interactive bungee velocity
%   freefalli interactive computation of the
%             free-fall velocity of an object
%             with second-order drag.
g = 9.81;     % acceleration of gravity
m = input('Mass (kg): ');
cd = input('Drag coefficient (kg/m): ');
t = input('Time (s): ');
disp(' ')
disp('Velocity (m/s):')
disp(sqrt(g * m / cd)*tanh(sqrt(g * cd / m) * t))
```

Save the file as `freefalli.m`. To invoke the function, return to the command window and type

```
>> freefalli

Mass (kg): 68.1
Drag coefficient (kg/m): 0.25
Time (s): 12

Velocity (m/s):
   50.6175
```

The `fprintf` Function.   This function provides additional control over the display of information. A simple representation of its syntax is

```
fprintf('format', x, ...)
```

where *format* is a string specifying how you want the value of the variable x to be displayed. The operation of this function is best illustrated by examples.

A simple example would be to display a value along with a message. For instance, suppose that the variable `velocity` has a value of 50.6175. To display the value using eight digits with four digits to the right of the decimal point along with a message, the statement along with the resulting output would be

```
>> fprintf('The velocity is %8.4f m/s\n', velocity)

The velocity is 50.6175 m/s
```

This example should make it clear how the format string works. MATLAB starts at the left end of the string and displays the labels until it detects one of the symbols: % or \. In our example, it first encounters a % and recognizes that the following text is a format code. As in Table 3.1, the *format codes* allow you to specify whether numeric values are

TABLE 3.1 Commonly used format and control codes employed with the `fprintf` function.

Format Code	Description
%d	Integer format
%e	Scientific format with lowercase e
%E	Scientific format with uppercase E
%f	Decimal format
%g	The more compact of %e or %f

Control Code	Description
\n	Start new line
\t	Tab

displayed in integer, decimal, or scientific format. After displaying the value of `velocity`, MATLAB continues displaying the character information (in our case the units: `m/s`) until it detects the symbol \. This tells MATLAB that the following text is a control code. As in Table 3.1, the *control codes* provide a means to perform actions such as skipping to the next line. If we had omitted the code \n in the previous example, the command prompt would appear at the end of the label `m/s` rather than on the next line as would typically be desired.

The `fprintf` function can also be used to display several values per line with different formats. For example,

```
>> fprintf('%5d %10.3f %8.5e\n',100,2*pi,pi);

    100      6.283 3.14159e+000
```

It can also be used to display vectors and matrices. Here is an M-file that enters two sets of values as vectors. These vectors are then combined into a matrix, which is then displayed as a table with headings:

```
function fprintfdemo
x = [1 2 3 4 5];
y = [20.4 12.6 17.8 88.7 120.4];
z = [x;y];
fprintf('      x         y\n');
fprintf('%5d %10.3f\n',z);
```

The result of running this M-file is

```
>> fprintfdemo

     x         y
     1     20.400
     2     12.600
     3     17.800
     4     88.700
     5    120.400
```

3.2.1 Creating and Accessing Files

MATLAB has the capability to both read and write data files. The simplest approach involves a special type of binary file, called a *MAT-file,* which is expressly designed for implementation within MATLAB. Such files are created and accessed with the `save` and `load` commands.

The `save` command can be used to generate a MAT-file holding either the entire workspace or a few selected variables. A simple representation of its syntax is

```
save filename var1 var2 ... varn
```

This command creates a MAT-file named `filename`.mat that holds the variables `var1` through `varn`. If the variables are omitted, all the workspace variables are saved. The `load` command can subsequently be used to retrieve the file:

```
load filename var1 var2 ... varn
```

which retrieves the variables `var1` through `varn` from `filename`.mat. As was the case with `save`, if the variables are omitted, all the variables are retrieved.

For example, suppose that you use Eq. (1.9) to generate velocities for a set of drag coefficients:

```
>> g=9.81;m=80;t=5;
>> cd=[.25 .267 .245 .28 .273]';
>> v=sqrt(g*m ./cd).*tanh(sqrt(g*cd/m)*t);
```

You can then create a file holding the values of the drag coefficients and the velocities with

```
>> save veldrag v cd
```

To illustrate how the values can be retrieved at a later time, remove all variables from the workspace with the `clear` command,

```
>> clear
```

At this point, if you tried to display the velocities you would get the result:

```
>> v
??? Undefined function or variable 'v'.
```

However, you can recover them by entering

```
>> load veldrag
```

Now, the velocities are available as can be verified by typing

```
>> who

Your variables are:
cd   v
```

Although MAT-files are quite useful when working exclusively within the MATLAB environment, a somewhat different approach is required when interfacing MATLAB with other programs. In such cases, a simple approach is to create text files written in the widely accessible ASCII format.

ASCII files can be generated in MATLAB by appending `-ascii` to the `save` command. In contrast to MAT-files where you might want to save the entire workspace, you would typically save a single rectangular matrix of values. For example,

```
>> A=[5 7 9 2;3 6 3 9];
>> save simpmatrix.txt -ascii
```

In this case, the `save` command stores the values in A in 8-digit ASCII form. If you want to store the numbers in double precision, just append `-ascii -double`. In either case, the file can be accessed by other programs such as spreadsheets or word processors. For example, if you open this file with a text editor, you will see

```
5.0000000e+000   7.0000000e+000   9.0000000e+000   2.0000000e+000
3.0000000e+000   6.0000000e+000   3.0000000e+000   9.0000000e+000
```

Alternatively, you can read the values back into MATLAB with the `load` command,

```
>> load simpmatrix.txt
```

Because `simpmatrix.txt` is not a MAT-file, MATLAB creates a double precision array named after the *filename*:

```
>> simpmatrix

simpmatrix =
     5     7     9     2
     3     6     3     9
```

Alternatively, you could use the `load` command as a function and assign its values to a variable as in

```
>> A = load(simpmatrix.txt)
```

The foregoing material covers but a small portion of MATLAB's file management capabilities. For example, a handy import wizard can be invoked with the menu selections: **File, Import Data.** As an exercise, you can demonstrate the import wizards convenience by using it to open `simpmatrix.txt`. In addition, you can always consult `help` to learn more about this and other features.

3.3 STRUCTURED PROGRAMMING

The simplest of all M-files perform instructions sequentially. That is, the program statements are executed line by line starting at the top of the function and moving down to the end. Because a strict sequence is highly limiting, all computer languages include statements allowing programs to take nonsequential paths. These can be classified as

- *Decisions* (or Selection). The branching of flow based on a decision.
- *Loops* (or Repetition). The looping of flow to allow statements to be repeated.

3.3.1 Decisions

The `if` Structure. This structure allows you to execute a set of statements if a logical condition is true. Its general syntax is

```
if condition
  statements
end
```

where *condition* is a logical expression that is either true or false. For example, here is a simple M-file to evaluate whether a grade is passing:

```
function grader(grade)
% grader(grade):
%    determines whether grade is passing
% input:
%    grade = numerical value of grade (0-100)
% output:
%    displayed message
if grade >= 60
  disp('passing grade')
end
```

The following illustrates the result

```
>> grader(95.6)

passing grade
```

For cases where only one statement is executed, it is often convenient to implement the `if` structure as a single line,

```
if grade > 60, disp('passing grade'), end
```

This structure is called a *single-line if*. For cases where more than one statement is implemented, the multiline if structure is usually preferable because it is easier to read.

Error Function.   A nice example of the utility of a single-line if is to employ it for rudimentary error trapping. This involves using the `error` function which has the syntax,

```
error(msg)
```

When this function is encountered, it displays the text message *msg* and causes the M-file to terminate and return to the command window.

An example of its use would be where we might want to terminate an M-file to avoid a division by zero. The following M-file illustrates how this could be done:

```
function f = errortest(x)
if x == 0, error('zero value encountered'), end
f = 1/x;
```

If a nonzero argument is used, the division would be implemented successfully as in

```
>> errortest(10)

ans =
    0.1000
```

However, for a zero argument, the function would terminate prior to the division and the error message would be displayed in red typeface:

```
>> errortest(0)

??? Error using ==> errortest
zero value encountered
```

TABLE 3.2 Summary of relational operators in MATLAB.

Example	Operator	Relationship
x == 0	==	Equal
unit ~= 'm'	~=	Not equal
a < 0	<	Less than
s > t	>	Greater than
3.9 <= a/3	<=	Less than or equal to
r >= 0	>=	Greater than or equal to

Logical Conditions.   The simplest form of the `condition` is a single relational expression that compares two values as in

> `value₁ relation value₂`

where the `values` can be constants, variables, or expressions and the `relation` is one of the relational operators listed in Table 3.2.

MATLAB also allows testing of more than one logical condition by employing logical operators. We will emphasize the following:

- ~ *(Not).* Used to perform logical negation on an expression.

  > `~expression`

  If the `expression` is true, the result is false. Conversely, if the `expression` is false, the result is true.
- & *(And).* Used to perform a logical conjunction on two expressions.

  > `expression₁ & expression₂`

  If both `expressions` evaluate to true, the result is true. If either or both `expressions` evaluates to false, the result is false.
- | *(Or).* Used to perform a logical disjunction on two expressions.

  > `expression₁ | expression₂`

  If either or both `expressions` evaluate to true, the result is true.

Table 3.3 summarizes all possible outcomes for each of these operators. Just as for arithmetic operations, there is a priority order for evaluating logical operations. These

TABLE 3.3 A truth table summarizing the possible outcomes for logical operators employed in MATLAB. The order of priority of the operators is shown at the top of the table.

x	y	Highest ———————————————————➤ Lowest		
		~x	x & y	x | y
T	T	F	T	T
T	F	F	F	T
F	T	T	F	T
F	F	T	F	F

are from highest to lowest: ~, & and |. In choosing between operators of equal priority, MATLAB evaluates them from left to right. Finally, as with arithmetic operators, parentheses can be used to override the priority order.

Let's investigate how the computer employs the priorities to evaluate a logical expression. If a = -1, b = 2, x = 1, and y = 'b', evaluate whether the following is true or false:

```
a * b > 0 & b == 2 & x > 7 | ~(y > 'd')
```

To make it easier to evaluate, substitute the values for the variables:

```
-1 * 2 > 0 & 2 == 2 & 1 > 7 | ~('b' > 'd')
```

The first thing that MATLAB does is to evaluate any mathematical expressions. In this example, there is only one: -1 * 2,

```
-2 > 0 & 2 == 2 & 1 > 7 | ~('b' > 'd')
```

Next, evaluate all the relational expressions

```
-2 > 0 & 2 == 2 & 1 > 7 | ~('b' > 'd')
   F    &   T    &   F  | ~          F
```

At this point, the logical operators are evaluated in priority order. Since the ~ has highest priority, the last expression (~F) is evaluated first to give

```
F & T & F | T
```

The & operator is evaluated next. Since there are two, the left-to-right rule is applied and the first expression (F & T) is evaluated:

```
F & F | T
```

The & again has highest priority

```
F | T
```

Finally, the | is evaluated as true. The entire process is depicted in Fig. 3.1.

The `if...else` **Structure.**   This structure allows you to execute a set of statements if a logical condition is true and to execute a second set if the condition is false. Its general syntax is

```
if condition
    statements₁
else
    statements₂
end
```

The `if...elseif` **Structure.**   It often happens that the false option of an `if...else` structure is another decision. This type of structure often occurs when we have more than two options for a particular problem setting. For such cases, a special form of decision structure, the `if...elseif` has been developed. It has the general syntax

```
if condition₁
    statements₁
elseif condition₂
    statements₂
```

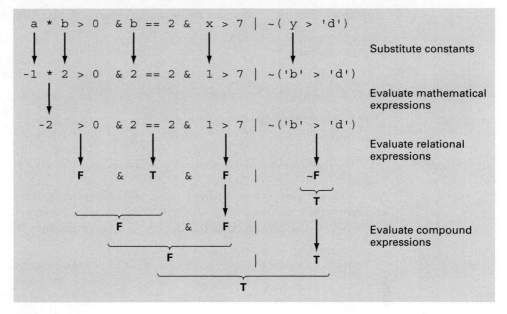

```
a * b > 0  & b == 2 &  x > 7 | ~( y > 'd')
```
Substitute constants

```
-1 * 2 > 0  & 2 == 2 &  1 > 7 | ~('b' > 'd')
```
Evaluate mathematical expressions

```
-2    > 0  & 2 == 2 &  1 > 7 | ~('b' > 'd')
```
Evaluate relational expressions

```
      F    &   T    &    F   |    ~F
                                   T
```
Evaluate compound expressions

```
           F        &    F   |    T
              F          |        T
                    T
```

FIGURE 3.1
A step-by-step evaluation of a complex decision.

```
    elseif condition₃
      statements₃
        .
        .
        .
    else
      statements_else
    end
```

EXAMPLE 3.4   `if` Structures

Problem Statement. For a scalar, the built-in MATLAB `sign` function returns the sign of its argument ($-1, 0, 1$). Here's a MATLAB session that illustrates how it works:

```
>> sign(25.6)
ans =
    1
>> sign(-0.776)
ans =
    -1
>> sign(0)
ans =
    0
```

Develop an M-file to perform the same function.

Solution. First, an `if` structure can be used to return `1` if the argument is positive:

```
function sgn = mysign(x)
% mysign(x) returns 1 if x is greater than zero.
if x > 0
  sgn = 1;
end
```

This function can be run as

```
>> mysign(25.6)

ans =
    1
```

Although the function handles positive numbers correctly, if it is run with a negative or zero argument, nothing is displayed. To partially remedy this shortcoming, an `if...else` structure can be used to display −1 if the condition is false:

```
function sgn = mysign(x)
% mysign(x) returns 1 if x is greater than zero.
%                 -1 if x is less than or equal to zero.
if x > 0
  sgn = 1;
else
  sgn = -1;
end
```

This function can be run as

```
>> mysign(-0.776)

ans =
   -1
```

Although the positive and negative cases are now handled properly, −1 is erroneously returned if a zero argument is used. An `if...elseif` structure can be used to incorporate this final case:

```
function sgn = mysign(x)
% mysign(x) returns 1 if x is greater than zero.
%                 -1 if x is less than zero.
%                 0 if x is equal to zero.
if x > 0
  sgn = 1;
elseif x < 0
  sgn = -1;
else
  sgn = 0;
end
```

The function now handles all possible cases. For example,

```
>> mysign(0)

ans =
    0
```

The `switch` **Structure.** The `switch` structure is similar in spirit to the `if...elseif` structure. However, rather than testing individual conditions, the branching is based on the value of a single test expression. Depending on its value, different blocks of code are implemented. In addition, an optional block is implemented if the expression takes on none of the prescribed values. It has the general syntax

```
switch testexpression
  case value₁
     statements₁
  case value₂
     statements₂
     .
     .
     .
  otherwise
     statementsₒₜₕₑᵣwᵢₛₑ
end
```

As an example, here is function that displays a message depending on the value of the string variable, `grade`.

```
grade = 'B';
switch grade
  case 'A'
    disp('Excellent')
  case 'B'
    disp('Good')
  case 'C'
    disp('Mediocre')
  case 'D'
    disp('Whoops')
  case 'F'
    disp('Would like fries with your order?')
  otherwise
    disp('Huh!')
end
```

When this code was executed, the message "Good" would be displayed.

Variable Argument List. MATLAB allows a variable number of arguments to be passed to a function. This feature can come in handy for incorporating default values into your functions. A *default value* is a number that is automatically assigned in the event that the user does not pass it to a function.

As an example, recall that earlier in this chapter, we developed a function `freefallvel`, which had three arguments:

```
v = freefallvel(t,m,cd)
```

Although a user would obviously need to specify the time and mass, they might not have a good idea of an appropriate drag coefficient. Therefore, it would be nice to have the program supply a value if they omitted it from the argument list.

MATLAB has a function called `nargin` that provides the number of input arguments supplied to a function by a user. It can be used in conjunction with decision structures like the `if` or `switch` constructs to incorporate default values as well as error messages into your functions. The following code illustrates how this can be done for `freefallvel`:

```
function v = freefallvelt(t, m, cd)
% freefallvel: bungee velocity with second-order drag
%    v=freefallvel(t,m,cd) computes the free-fall velocity
%                             of an object with second-order drag.
% input:
%    t = time (s)
%    m = mass (kg)
%    cd = drag coefficient (default = 0.27 kg/m)
% output:
%    v = downward velocity (m/s)
switch nargin
  case 0
    error('Must enter time and mass')
  case 1
    error('Must enter mass')
  case 2
    cd = 0.27;
end
g = 9.81;      % acceleration of gravity
v = sqrt(g * m / cd)*tanh(sqrt(g * cd / m) * t);
```

Notice how we have used a `switch` structure to either display error messages or set the default, depending on the number of arguments passed by the user. Here is a command window session showing the results:

```
>> freefallvel(12,68.1,0.25)

ans =
   50.6175

>> freefallvel(12,68.1)

ans =
   48.8747

>> freefallvel(12)

??? Error using ==> freefallvel
Must enter mass

>> freefallvel()

??? Error using ==> freefallvel
Must enter time and mass
```

Note that `nargin` behaves a little differently when it is invoked in the command window. In the command window, it must include a string argument specifying the function and it returns the number of arguments in the function. For example,

```
>> nargin('freefallvel')

ans =
   3
```

3.3.2 Loops

As the name implies, loops perform operations repetitively. There are two types of loops, depending on how the repetitions are terminated. A *for loop* ends after a specified number of repetitions. A *while loop* ends on the basis of a logical condition.

The `for...end` **Structure.** A `for` loop repeats statements a specific number of times. Its general syntax is

```
for index = start:step:finish
   statements
end
```

The `for` loop operates as follows. The `index` is a variable that is set at an initial value, `start`. The program then compares the `index` with a desired final value, `finish`. If the `index` is less than or equal to the `finish`, the program executes the `statements`. When it reaches the `end` line that marks the end of the loop, the `index` variable is increased by the `step` and the program loops back up to the `for` statement. The process continues until the `index` becomes greater than the `finish` value. At this point, the loop terminates as the program skips down to the line immediately following the `end` statement.

Note that if an increment of 1 is desired (as is often the case), the `step` can be dropped. For example,

```
for i = 1:5
   disp(i)
end
```

When this executes, MATLAB would display in succession, 1, 2, 3, 4, 5. In other words, the default `step` is 1.

The size of the `step` can be changed from the default of 1 to any other numeric value. It does not have to be an integer, nor does it have to be positive. For example, step sizes of 0.2, –1, or –5, are all acceptable.

If a negative `step` is used, the loop will "countdown" in reverse. For such cases, the loop's logic is reversed. Thus, the `finish` is less than the `start` and the loop terminates when the `index` is less than the `finish`. For example,

```
for j = 10:-1:1
   disp(j)
end
```

When this executes, MATLAB would display the classic "countdown" sequence: 10, 9, 8, 7, 6, 5, 4, 3, 2, 1.

EXAMPLE 3.5 Using a `for` Loop to Compute the Factorial

Problem Statement. Develop an M-file to compute the factorial.[2]

$0! = 1$
$1! = 1$
$2! = 1 \times 2 = 2$

---

[2] Note that MATLAB has a built-in function `factorial` that performs this computation.

$$3! = 1 \times 2 \times 3 = 6$$
$$4! = 1 \times 2 \times 3 \times 4 = 24$$
$$5! = 1 \times 2 \times 3 \times 4 \times 5 = 120$$
$$\vdots$$

Solution.   A simple function to implement this calculation can be developed as

```
function fout = factor(n)
% factor(n):
%   Computes the product of all the integers from 1 to n.
x = 1;
for i = 1:n
  x = x * i;
end
fout = x;
end
```

which can be run as

```
>> factor(5)

ans =
   120
```

This loop will execute 5 times (from 1 to 5). At the end of the process, x will hold a value of 5! (meaning 5 factorial or $1 \times 2 \times 3 \times 4 \times 5 = 120$).

Notice what happens if $n = 0$. For this case, the for loop would not execute, and we would get the desired result, $0! = 1$.

Vectorization.   The for loop is easy to implement and understand. However, for MATLAB, it is not necessarily the most efficient means to repeat statements a specific number of times. Because of MATLAB's ability to operate directly on arrays, *vectorization* provides a much more efficient option. For example, the following for structure:

```
i = 0;
for t = 0:0.02:50
  i = i + 1;
  y(i) = cos(t);
end
```

can be represented in vectorized form as

```
t = 0:0.02:50;
y = cos(t);
```

It should be noted that for more complex code, it may not be obvious how to vectorize the code. That said, wherever possible, vectorization is recommended.

Preallocation of Memory.   MATLAB automatically increases the size of arrays every time you add a new element. This can become time consuming when you perform actions such as adding new values one at a time within a loop. For example, here is some code that sets value of elements of y depending on whether or not values of t are greater than one:

```
t = 0:.01:5;
for i = 1:length(t)
  if t(i)>1
    y(i) = 1/t(i);
  else
    y(i) = 1;
  end
end
```

For this case, MATLAB must resize y every time a new value is determined. The following code preallocates the proper amount of memory by using a vectorized statement to assign ones to y prior to entering the loop.

```
t = 0:.01:5;
y = ones(size(t));
for i = 1:length(t)
  if t(i)>1
    y(i) = 1/t(i);
  end
end
```

Thus, the array is only sized once. In addition, preallocation helps reduce memory fragmentation, which also enhances efficiency.

The while Structure. A while loop repeats as long as a logical condition is true. Its general syntax is

```
while condition
  statements
end
```

The *statements* between the while and the end are repeated as long as the *condition* is true. A simple example is

```
x = 8
while x > 0
  x = x - 3;
  disp(x)
end
```

When this code is run, the result is

```
x =
      8
      5
      2
     -1
```

The while...break Structure. Although the while structure is extremely useful, the fact that it always exits at the beginning of the structure on a false result is somewhat constraining. For this reason, languages such as Fortran 90 and Visual Basic have special structures that allow loop termination on a true condition anywhere in the loop. Although such structures are currently not available in MATLAB, their functionality can be mimicked by a special version of the `while` loop. The syntax of this version, called a *while...break structure,* can be written as

```
while (1)
   statements
   if condition, break, end
   statements
end
```

where `break` terminates execution of the loop. Thus, a single line `if` is used to exit the loop if the condition tests true. Note that as shown, the `break` can be placed in the middle of the loop (i.e., with statements before and after it). Such a structure is called a *midtest loop.*

If the problem required it, we could place the `break` at the very beginning to create a *pretest loop.* An example is

```
while (1)
   If x < 0, break, end
   x = x - 5;
end
```

Notice how 5 is subtracted from `x` on each iteration. This represents a mechanism so that the loop eventually terminates. Every decision loop must have such a mechanism. Otherwise it would become a so-called *infinite loop* that would never stop.

Alternatively, we could also place the `if...break` statement at the very end and create a *posttest loop,*

```
while (1)
   x = x - 5;
   if x < 0, break, end
end
```

It should be clear that, in fact, all three structures are really the same. That is, depending on where we put the exit (beginning, middle, or end) dictates whether we have a pre-, mid- or posttest. It is this simplicity that led the computer scientists who developed Fortran 90 and Visual Basic to favor this structure over other forms of the decision loop such as the conventional `while` structure.

The `pause` Command.   There are often times when you might want a program to temporarily halt. The command `pause` causes a procedure to stop and wait until any key is hit. A nice example involves creating a sequence of plots that a user might want to leisurely peruse before moving on to the next. The following code employs a `for` loop to create a sequence of interesting plots that can be viewed in this manner:

```
for n = 3:10
   mesh(magic(n))
   pause
end
```

The `pause` can also be formulated as `pause(n)`, in which case the procedure will halt for `n` seconds. This feature can be demonstrated by implementing it in conjunction with several other useful MATLAB functions. The `beep` command causes the computer to emit a beep sound. Two other functions, `tic` and `toc`, work together to measure elapsed time.

The `tic` command saves the current time that `toc` later employs to display the elapsed time. The following code then confirms that `pause(n)` works as advertised complete with sound effects:

```
tic
beep
pause(5)
beep
toc
```

When this code is run, the computer will beep. Five seconds later it will beep again and display the following message:

```
Elapsed time is 5.006306 seconds.
```

By the way, if you ever have the urge to use the command `pause(inf)`, MATLAB will go into an infinite loop. In such cases, you can return to the command prompt by typing **Ctrl+C** or **Ctrl+Break.**

Although the foregoing examples might seem a tad frivolous, the commands can be quite useful. For instance, `tic` and `toc` can be employed to identify the parts of an algorithm that consume the most execution time. Further, the **Ctrl+C** or **Ctrl+Break** key combinations come in real handy in the event that you inadvertently create an infinite loop in one of your M-files.

3.4 NESTING AND INDENTATION

We need to understand that structures can be "nested" within each other. *Nesting* refers to placing structures within other structures. The following example illustrates the concept.

EXAMPLE 3.6   Nesting Structures

Problem Statement.   The roots of a quadratic equation

$$f(x) = ax^2 + bx + c$$

can be determined with the quadratic formula

$$x = \frac{-b \pm \sqrt{b^2 - 4ac}}{2a}$$

Develop a function to implement this formula given values of the coeffcients.

Solution.   *Top-down design* provides a nice approach for designing an algorithm to compute the roots. This involves developing the general structure without details and then refining the algorithm. To start, we first recognize that depending on whether the parameter *a* is zero, we will either have "special" cases (e.g., single roots or trivial values) or conventional cases using the quadratic formula. This "big-picture" version can be programmed as

```
function quadroots(a, b, c)
% quadroots: roots of quadratic equation
%   quadroots(a,b,c): real and complex roots
%                     of quadratic equation
% input:
%   a = second-order coefficient
```

```
%    b = first-order coefficient
%    c = zero-order coefficient
% output:
%    r1 = real part of first root
%    i1 = imaginary part of first root
%    r2 = real part of second root
%    i2 = imaginary part of second root
if a == 0
  %special cases
else
  %quadratic formula
end
```

Next, we develop refined code to handle the "special" cases:

```
%special cases
if b ~= 0
  %single root
  r1 = -c / b
else
  %trivial solution
  error('Trivial solution. Try again')
end
```

And we can develop refined code to handle the quadratic formula cases:

```
%quadratic formula
d = b ^ 2 - 4 * a * c;
if d >= 0
  %real roots
  r1 = (-b + sqrt(d)) / (2 * a)
  r2 = (-b - sqrt(d)) / (2 * a)
else
  %complex roots
  r1 = -b / (2 * a)
  i1 = sqrt(abs(d)) / (2 * a)
  r2 = r1
  i2 = -i1
end
```

We can then merely substitute these blocks back into the simple "big-picture" framework to give the final result:

```
function quadroots(a, b, c)
% quadroots: roots of quadratic equation
%    quadroots(a,b,c): real and complex roots
%                      of quadratic equation
% input:
%    a = second-order coefficient
%    b = first-order coefficient
%    c = zero-order coefficient
% output:
%    r1 = real part of first root
%    i1 = imaginary part of first root
```

```
%    r2 = real part of second root
%    i2 = imaginary part of second root
if a == 0
    %special cases
    if b ~= 0
        %single root
        r1 = -c / b
    else
        %trivial solution
        error('Trivial solution. Try again')
    end
else
    %quadratic formula
    d = b ^ 2 - 4 * a * c;      %discriminant
    if d >= 0
        %real roots
        r1 = (-b + sqrt(d)) / (2 * a)
        r2 = (-b - sqrt(d)) / (2 * a)
    else
        %complex roots
        r1 = -b / (2 * a)
        i1 = sqrt(abs(d)) / (2 * a)
        r2 = r1
        i2 = -i1
    end
end
```

As highlighted by the shading, notice how indentation helps to make the underlying logical structure clear. Also notice how "modular" the structures are. Here is a command window session illustrating how the function performs:

```
>> quadroots(1,1,1)

r1 =
    -0.5000
i1 =
     0.8660
r2 =
    -0.5000
i2 =
    -0.8660

>> quadroots(1,5,1)

r1 =
    -0.2087
r2 =
    -4.7913

>> quadroots(0,5,1)

r1 =
    -0.2000
```

```
>> quadroots(0,0,0)

??? Error using ==> quadroots
Trivial solution. Try again
```

3.5   PASSING FUNCTIONS TO M-FILES

Much of the remainder of the book involves developing functions to numerically evaluate other functions. Although a customized function could be developed for every new equation we analyzed, a better alternative is to design a generic function and pass the particular equation we wish to analyze as an argument. In the parlance of MATLAB, these functions are given a special name: *function functions*. Before describing how they work, we will first introduce anonymous functions, which provide a handy means to define simple user-defined functions without developing a full-blown M-file.

3.5.1 Anonymous Functions

Anonymous functions allow you to create a simple function without creating an M-file. They can be defined within the command window with the following syntax:

```
fhandle = @(arglist) expression
```

where *fhandle* = the function handle you can use to invoke the function, *arglist* = a comma separated list of input arguments to be passed to the function, and *expression* = any single valid MATLAB expression. For example,

```
>> f1=@(x,y) x^2 + y^2;
```

Once these functions are defined in the command window, they can be used just as other functions:

```
>> f1(3,4)

ans =
    25
```

Aside from the variables in its argument list, an anonymous function can include variables that exist in the workspace where it is created. For example, we could create an anonymous function $f(x) = 4x^2$ as

```
>> a = 4;
>> b = 2;
>> f2=@(x) a*x^b;
>> f2(3)

ans = 36
```

Note that if subsequently we enter new values for a and b, the anonymous function does not change:

```
>> a = 3;
>> f2(3)

ans = 36
```

Thus, the function handle holds a snapshot of the function at the time it was created. If we want the variables to take on new values, we must recreate the function. For example, having changed a to 3,

```
>> f2=@x a*x^b;
```

with the result

```
>> f2(3)

ans =
    27
```

It should be noted that prior to MATLAB 7, `inline` functions performed the same role as anonymous functions. For example, the anonymous function developed above, `f1`, could be written as

```
>> f1=inline('x^2 + y^2','x','y');
```

Although they are being phased out in favor of anonymous function, some readers might be using earlier versions, and so we thought it would be helpful to mention them. MATLAB help can be consulted to learn more about their use and limitations.

3.5.2 Function Functions

Function functions are functions that operate on other functions which are passed to it as input arguments. The function that is passed to the function function is referred to as the *passed function*. A simple example is the built-in function `fplot`, which plots the graphs of functions. A simple representation of its syntax is

```
fplot(fun, lims)
```

where `fun` is the function being plotted between the *x*-axis limits specified by `lims` = [*xmin xmax*]. For this case, `fun` is the passed function. This function is "smart" in that it automatically analyzes the function and decides how many values to use so that the plot will exhibit all the function's features.

Here is an example of how `fplot` can be used to plot the velocity of the free-falling bungee jumper. The function can be created with an anonymous function:

```
>> vel=@(t) ...
sqrt(9.81*68.1/0.25)*tanh(sqrt(9.81*0.25/68.1)*t);
```

We can then generate a plot from $t = 0$ to 12 as

```
>> fplot(vel,[0 12])
```

The result is displayed in Fig. 3.2.

Note that in the remainder of this book, we will have many occasions to use MATLAB's built-in function functions. As in the following example, we will also be developing our own.

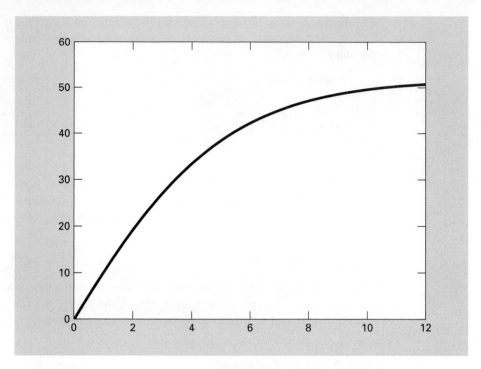

FIGURE 3.2
A plot of velocity versus time generated with the `fplot` function.

EXAMPLE 3.7   Building and Implementing a Function Function

Problem Statement.   Develop an M-file function function to determine the average value
of a function over a range. Illustrate its use for the bungee jumper velocity over the range
from $t = 0$ to 12 s:

$$v(t) = \sqrt{\frac{gm}{c_d}} \tanh\left(\sqrt{\frac{gc_d}{m}}\, t\right)$$

where $g = 9.81$, $m = 68.1$, and $c_d = 0.25$.

Solution.   The average value of the function can be computed with standard MATLAB
commands as

```
>> t=linspace(0,12);
>> v=sqrt(9.81*68.1/0.25)*tanh(sqrt(9.81*0.25/68.1)*t);
>> mean(v)

ans =
   36.0870
```

Inspection of a plot of the function (Fig. 3.2) shows that this result is a reasonable estimate
of the curve's average height.

We can write an M-file to perform the same computation:

```
function favg = funcavg(a,b,n)
% funcavg: average function height
%   favg=funcavg(a,b,n): computes average value
%                        of function over a range
% input:
%   a = lower bound of range
%   b = upper bound of range
%   n = number of intervals
% output:
%   favg = average value of function
x = linspace(a,b,n);
y = func(x);
favg = mean(y);
end

function f = func(t)
f=sqrt(9.81*68.1/0.25)*tanh(sqrt(9.81*0.25/68.1)*t);
end
```

The main function first uses `linspace` to generate equally spaced *x* values across the range. These values are then passed to a subfunction `func` in order to generate the corresponding *y* values. Finally, the average value is computed. The function can be run from the command window as

```
>> funcavg (0,12,60)

ans =
   36.0127
```

Now let's rewrite the M-file so that rather than being specific to `func`, it evaluates a nonspecific function name `f` that is passed in as an argument:

```
function favg = funcavg (f,a,b,n)
% funcavg: average function height
%   favg=funcavg(f,a,b,n): computes average value
%                          of function over a range
% input:
%   f = function to be evaluated
%   a = lower bound of range
%   b = upper bound of range
%   n = number of intervals
% output:
%   favg = average value of function
x = linspace(a,b,n);
y = f(x);
favg = mean(y);
```

Because we have removed the subfunction `func`, this version is truly generic. It can be run from the command window as

```
>> vel=@(t) ...
sqrt(9.81*68.1/0.25)*tanh(sqrt(9.81*0.25/68.1)*t);
>> funcavg(vel,0,12,60)
```

```
ans =
   36.0127
```

To demonstrate its generic nature, `funcavg` can easily be applied to another case by merely passing it a different function. For example, it could be used to determine the average value of the built-in `sin` function between 0 and 2π as

```
>> funcavg(@sin,0,2*pi,180)
```

```
ans =
 -6.3001e-017
```

Does this result make sense?

We can see that `funcavg` is now designed to evaluate any valid MATLAB expression. We will do this on numerous occasions throughout the remainder of this text in a number of contexts ranging from nonlinear equation solving to the solution of differential equations.

3.5.3 Passing Parameters

Recall from Chap. 1 that the terms in mathematical models can be divided into dependent and independent variables, parameters, and forcing functions. For the bungee jumper model, the velocity (v) is the dependent variable, time (t) is the independent variable, the mass (m) and drag coefficient (c_d) are parameters, and the gravitational constant (g) is the forcing function. It is commonplace to investigate the behavior of such models by performing a *sensitivity analysis*. This involves observing how the dependent variable changes as the parameters and forcing functions are varied.

In Example 3.7, we developed a function function, `funcavg`, and used it to determine the average value of the bungee jumper velocity for the case where the parameters were set at $m = 68.1$ and $c_d = 0.25$. Suppose that we wanted to analyze the same function, but with different parameters. Of course, we could retype the function with new values for each case, but it would be preferable to just change the parameters.

As we learned in Sec. 3.5.1, it is possible to incorporate parameters into anonymous functions. For example, rather than "wiring" the numeric values, we could have done the following:

```
>> m=68.1;cd=0.25;
>> vel=@(t) sqrt(9.81*m/cd)*tanh(sqrt(9.81*cd/m)*t);
>> funcavg(vel,0,12,60)
```

```
ans =
   36.0127
```

However, if we want the parameters to take on new values, we must recreate the anonymous function.

MATLAB offers a better alternative by adding the term `varargin` as the function function's last input argument. In addition, every time the passed function is invoked within the function function, the term `varargin{:}` should be added to the end of its argument list (note the curly brackets). Here is how both modifications can be implemented for `funcavg` (omitting comments for conciseness):

```
function favg = funcavg(f,a,b,n,varargin)
x = linspace(a,b,n);
y = f(x,varargin{:});
favg = mean(y);
```

When the passed function is defined, the actual parameters should be added at the end of the argument list. If we used an anonymous function, this can be done as in

```
>> vel=@(t,m,cd) sqrt(9.81*m/cd)*tanh(sqrt(9.81*cd/m)*t);
```

When all these changes have been made, analyzing different parameters becomes easy. To implement the case where $m = 68.1$ and $c_d = 0.25$, we could enter

```
>> funcavg(vel,0,12,60,68.1,0.25)

ans =
    36.0127
```

An alternative case, say $m = 100$ and $c_d = 0.28$, could be rapidly generated by merely changing the arguments:

```
>> funcavg(vel,0,12,60,100,0.28)

ans =
    38.9345
```

3.6 CASE STUDY   BUNGEE JUMPER VELOCITY

Background. In this section, we will use MATLAB to solve the free-falling bungee jumper problem we posed at the beginning of this chapter. This involves obtaining a solution of

$$\frac{dv}{dt} = g - \frac{c_d}{m}v^2$$

Recall that, given an initial condition for time and velocity, the problem involved iteratively solving the formula,

$$v_{i+1} = v_i + \frac{dv_i}{dt}\Delta t$$

Now also remember that to attain good accuracy, we would employ small steps. Therefore, we would probably want to apply the formula repeatedly to step out from our initial time to attain the value at the final time. Consequently, an algorithm to solve the problem would be based on a loop.

Solution. Suppose that we started the computation at $t = 0$ and wanted to predict velocity at $t = 12$ s using a time step of $\Delta t = 0.5$ s. We would therefore need to apply the iterative equation 24 times—that is,

$$n = \frac{12}{0.5} = 24$$

where $n =$ the number of iterations of the loop. Because this result is exact (i.e., the ratio is an integer), we can use a `for` loop as the basis for the algorithm. Here's an M-file to do this including a subfunction defining the differential equation:

```
function vend = velocity1(dt, ti, tf, vi)
% velocity1: Euler solution for bungee velocity
%   vend = velocity1(dt, ti, tf, vi)
%           Euler method solution of bungee
%           jumper velocity
% input:
%   dt = time step (s)
%   ti = initial time (s)
%   tf = final time (s)
%   vi = initial value of dependent variable (m/s)
% output:
%   vend = velocity at tf (m/s)
t = ti;
v = vi;
n = (tf - ti) / dt;
for i = 1:n
  dvdt = deriv(v);
  v = v + dvdt * dt;
  t = t + dt;
end
vend = v;
end

function dv = deriv(v)
dv = 9.81 - (0.25 / 68.1) * v^2;
end
```

This function can be invoked from the command window with the result:

```
>> velocity1(0.5,0,12,0)

ans =
   50.9259
```

Note that the true value obtained from the analytical solution is 50.6175 (Example 3.1). We can then try a much smaller value of dt to obtain a more accurate numerical result:

```
>> velocity1(0.001,0,12,0)

ans =
   50.6181
```

Although this function is certainly simple to program, it is not foolproof. In particular, it will not work if the computation interval is not evenly divisible by the time step. To cover such cases, a `while . . . break` loop can be substituted in place of the shaded area (note that we have omitted the comments for conciseness):

```
function vend = velocity2(dt, ti, tf, vi)
t = ti;
v = vi;
h = dt;
while(1)
  if t + dt > tf, h = tf - t; end
  dvdt = deriv(v);
  v = v + dvdt * h;
  t = t + h;
  if t >= tf, break, end
end
vend = v;
end

function dv = deriv(v)
dv = 9.81 - (0.25 / 68.1) * v^2;
end
```

As soon as we enter the `while` loop, we use a single line `if` structure to test whether adding `t + dt` will take us beyond the end of the interval. If not (which would usually be the case at first), we do nothing. If so, we would shorten up the interval—that is, we set the variable step `h` to the interval remaining: `tf - t`. By doing this, we guarantee that the last step falls exactly on `tf`. After we implement this final step, the loop will terminate because the condition `t >= tf` will test true.

Notice that before entering the loop, we assign the value of the time step `dt` to another variable `h`. We create this *dummy variable* so that our routine does not change the given value of `dt` if and when we shorten the time step. We do this in anticipation that we might need to use the original value of `dt` somewhere else in the event that this code were integrated within a larger program.

If we run this new version, the result will be the same as for the version based on the `for` structure:

```
>> velocity2(0.5,0,12,0)

ans =
   50.9259
```

Further, we can use a `dt` that is not evenly divisible into `tf - ti`:

```
>> velocity2(0.35,0,12,0)

ans =
   50.8348
```

We should note that the algorithm is still not foolproof. For example, the user could have mistakenly entered a step size greater than the calculation interval (e.g., `tf - ti = 5` and `dt = 20`). Thus, you might want to include error traps in your code to catch such errors and then allow the user to correct the mistake.

As a final note, we should recognize that the foregoing code is not generic. That is, we have designed it to solve the specific problem of the velocity of the bungee jumper. A more generic version can be developed as

```
function yend = odesimp(dydt, dt, ti, tf, yi)
t = ti; y = yi; h = dt;
while (1)
  if t + dt > tf, h = tf - t; end
  y = y + dydt(y) * h;
  t = t + h;
  if t >= tf, break, end
end
yend = y;
```

Notice how we have stripped out the parts of the algorithm that were specific to the bungee example (including the subfunction defining the differential equation) while keeping the essential features of the solution technique. We can then use this routine to solve the bungee jumper example, by specifying the differential equation with an anonymous function and passing its function handle to odesimp to generate the solution

```
>> dvdt=@(v) 9.81-(0.25/68.1)*v^2;
>> odesimp(dvdt,0.5,0,12,0)

ans =
  50.9259
```

We could then analyze a different function without having to go in and modify the M-file. For example, if $y = 10$ at $t = 0$, the differential equation $dy/dt = -0.1y$ has the analytical solution $y = 10e^{-0.1t}$. Therefore, the solution at $t = 5$ would be $y(5) = 10e^{-0.1(5)} = 6.0653$. We can use odesimp to obtain the same result numerically as in

```
>> odesimp(@(y) -0.1*y,0.005,0,5,10)

ans =
  6.0645
```

PROBLEMS

3.1 The cosine function can be evaluated by the following infinite series:

$$\cos x = 1 - \frac{x^2}{2!} + \frac{x^4}{4!} - \cdots$$

Create an M-file to implement this formula so that it computes and displays the values of cos x as each term in the series is added. In other words, compute and display in sequence the values for

$$\cos x = 1$$

$$\cos x = 1 - \frac{x^2}{2!}$$

$$\cos x = 1 - \frac{x^2}{2!} + \frac{x^4}{4!}$$

$$\vdots$$

up to the order term of your choosing. For each of the preceding, compute and display the percent relative error as

$$\%\text{error} = \frac{\text{true} - \text{series approximation}}{\text{true}} \times 100\%$$

As a test case, employ the program to compute cos(1.5) for up to and including eight terms—that is, up to the term $x^{14}/14!$.

3.2 An amount of money P is invested in an account where interest is compounded at the end of the period. The future worth F yielded at an interest rate i after n periods may be determined from the following formula:

$$F = P(1 + i)^n$$

Write an M-file that will calculate the future worth of an investment for each year from 1 through n. The input to the function should include the initial investment P, the interest rate i (as a decimal), and the number of years n for which the future worth is to be calculated. The output should consist of a table with headings and columns for n and F. Run the program for $P = \$100,000$, $i = 0.06$, and $n = 7$ years.

3.3 Economic formulas are available to compute annual payments for loans. Suppose that you borrow an amount of money P and agree to repay it in n annual payments at an interest rate of i. The formula to compute the annual payment A is

$$A = P\frac{i(1 + i)^n}{(1 + i)^n - 1}$$

Write an M-file to compute A. Test it with $P = \$55,000$ and an interest rate of 6.6% ($i = 0.066$). Compute results for $n = 1, 2, 3, 4,$ and 5 and display the results as a table with headings and columns for n and A.

3.4 The average daily temperature for an area can be approximated by the following function:

$$T = T_{\text{mean}} + (T_{\text{peak}} - T_{\text{mean}})\cos(\omega(t - t_{\text{peak}}))$$

where T_{mean} = the average annual temperature, T_{peak} = the peak temperature, ω = the frequency of the annual variation ($= 2\pi/365$), and t_{peak} = day of the peak temperature ($\cong 205$ d). Parameters for some U.S. towns are listed here:

City	T_{mean} (°C)	T_{peak} (°C)
Miami, FL	22.1	28.3
Yuma, AZ	23.1	33.6
Bismarck, ND	5.2	22.1
Seattle, WA	10.6	17.6
Boston, MA	10.7	22.9

Develop an M-file that computes the average temperature between two days of the year for a particular city. Test it for (a) January–February in Miami, FL ($t = 0$ to 59) and (b) July–August temperature in Boston, MA ($t = 180$ to 242).

3.5 Figure P3.5 shows a cylindrical tank with a conical base. If the liquid level is quite low, in the conical part, the volume is simply the conical volume of liquid. If the liquid level is midrange in the cylindrical part, the total volume of liquid includes the filled conical part and the partially filled cylindrical part.

Use decisional structures to write an M-file to compute the tank's volume as a function of given values of R and d. Design the function so that it returns the volume for all cases

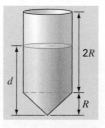

FIGURE P3.5

where the depth is less than $3R$. Return an error message ("Overtop") if you overtop the tank—that is, $d > 3R$. Test it with the following data:

R	1.5	1.5	1.5	1.5
d	1	2	4.5	4.6

Note that the tank's radius is R.

3.6 Two distances are required to specify the location of a point relative to an origin in two-dimensional space (Fig. P3.6):

- The horizontal and vertical distances (x, y) in Cartesian coordinates.
- The radius and angle (r, θ) in polar coordinates.

It is relatively straightforward to compute Cartesian coordinates (x, y) on the basis of polar coordinates (r, θ). The reverse process is not so simple. The radius can be computed by the following formula:

$$r = \sqrt{x^2 + y^2}$$

If the coordinates lie within the first and fourth coordinates (i.e., $x > 0$), then a simple formula can be used to compute θ:

$$\theta = \tan^{-1}\left(\frac{y}{x}\right)$$

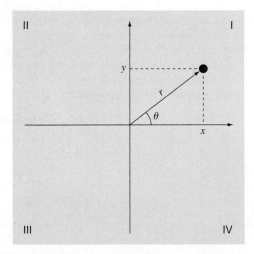

FIGURE P3.6

The difficulty arises for the other cases. The following table summarizes the possibilities:

x	y	θ
<0	>0	$\tan^{-1}(y/x) + \pi$
<0	<0	$\tan^{-1}(y/x) - \pi$
<0	=0	π
=0	>0	$\pi/2$
=0	<0	$-\pi/2$
=0	=0	0

Write a well-structured M-file to calculate r and θ as a function of x and y. Express the final results for θ in degrees. Test your program by evaluating the following cases:

x	y	r	θ
1	0		
1	1		
0	1		
−1	1		
−1	0		
−1	−1		
0	0		
0	−1		
1	−1		

3.7 Develop an M-file to determine polar coordinates as described in Prob. 3.6. However, rather than designing the function to evaluate a single case, pass vectors of x and y. Have the function display the results as a table with columns for x, y, r, and θ. Test the program for the cases outlined in Prob. 3.6.

3.8 Develop an M-file function that is passed a numeric grade from 0 to 100 and returns a letter grade according to the scheme:

Letter	Criteria
A	$90 \le$ numeric grade ≤ 100
B	$80 \le$ numeric grade < 90
C	$70 \le$ numeric grade < 80
D	$60 \le$ numeric grade < 70
F	numeric grade < 60

3.9 Manning's equation can be used to compute the velocity of water in a rectangular open channel:

$$U = \frac{\sqrt{S}}{n}\left(\frac{BH}{B + 2H}\right)^{2/3}$$

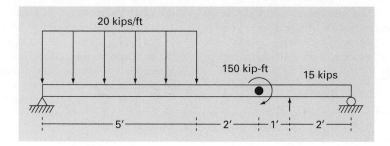

FIGURE P3.10

where U = velocity (m/s), S = channel slope, n = roughness coefficient, B = width (m), and H = depth (m). The following data is available for five channels:

n	S	B	H
0.035	0.0001	10	2
0.020	0.0002	8	1
0.015	0.0010	20	1.5
0.030	0.0007	24	3
0.022	0.0003	15	2.5

Write an M-file that computes the velocity for each of these channels. Enter these values into a matrix where each column represents a parameter and each row represents a channel. Have the M-file display the input data along with the computed velocity in tabular form where velocity is the fifth column. Include headings on the table to label the columns.

3.10 A simply supported beam is loaded as shown in Fig. P3.10. Using singularity functions, the displacement along the beam can be expressed by the equation:

$$u_y(x) = \frac{-5}{6}[\langle x - 0\rangle^4 - \langle x - 5\rangle^4] + \frac{15}{6}\langle x - 8\rangle^3$$
$$+ 75\langle x - 7\rangle^2 + \frac{57}{6}x^3 - 238.25x$$

By definition, the singularity function can be expressed as follows:

$$\langle x - a\rangle^n = \begin{cases} (x - a)^n & \text{when } x > a \\ 0 & \text{when } x \le a \end{cases}$$

Develop an M-file that creates a plot of displacement versus distance along the beam, x. Note that $x = 0$ at the left end of the beam.

3.11 The volume V of liquid in a hollow horizontal cylinder of radius r and length L is related to the depth of the liquid h by

$$V = \left[r^2 \cos^{-1}\left(\frac{r - h}{r}\right) - (r - h)\sqrt{2rh - h^2} \right] L$$

Develop an M-file to create a plot of volume versus depth. Test the program for $r = 2$ m and $L = 5$ m.

3.12 Develop a vectorized version of the following code:

```
tstart=0; tend=20; ni=5;
t(1)=tstart;
y(1)=10 + 5*cos(2*pi*t(1)/(tend-tstart));
for i=2:ni+1
  t(i)=t(i-1)+(tend-tstart)/ni;
  y(i)=10 + 5*cos(2*pi*t(i)/ ...
      (tend-tstart));
end
```

3.13 The "divide and average" method, an old-time method for approximating the square root of any positive number a, can be formulated as

$$x = \frac{x + a/x}{2}$$

Write a well-structured M-file function based on a `while...break` loop structure to implement this algorithm. Use proper indentation so that the structure is clear. At each step estimate the error in your approximation as

$$\varepsilon = \left| \frac{x_{new} - x_{old}}{x_{new}} \right|$$

Repeat the loop until ε is less than or equal to a specified value. Design your program so that it returns both the result and the error. Make sure that it can evaluate the square root of numbers that are equal to and less than zero. For the latter case, display the result as an imaginary number. For example, the square root of -4 would return $2i$. Test your program by evaluating $a = 0$, 2, 4, and -9 for $\varepsilon = 1 \times 10^{-4}$.

3.14 *Piecewise functions* are sometimes useful when the relationship between a dependent and an independent variable cannot be adequately represented by a single equation. For example, the velocity of a rocket might be described by

$$v(t) = \begin{cases} 11t^2 - 5t & 0 \le t \le 10 \\ 1100 - 5t & 10 \le t \le 20 \\ 50t + 2(t-20)^2 & 20 \le t \le 30 \\ 1520e^{-0.2(t-30)} & t > 30 \\ 0 & \text{otherwise} \end{cases}$$

Develop an M-file function to compute v as a function of t. Then, use this function to generate a plot of v versus t for $t = -5$ to 50.

3.15 Develop an M-file function called `rounder` to round a number x to a specified number of decimal digits, n. The first line of the function should be set up as

```
function xr = rounder(x, n)
```

Test the program by rounding each of the following to 2 decimal digits: $x = 467.9587, -467.9587, 0.125, 0.135, -0.125,$ and -0.135.

3.16 Develop an M-file function to determine the elapsed days in a year. The first line of the function should be set up as

```
function nd = days(mo, da, leap)
```

where `mo` = the month (1–12), `da` = the day (1–31), and `leap` = (0 for non–leap year and 1 for leap year). Test it for January 1, 1999, February 29, 2000, March 1, 2001, June 21, 2002, and December 31, 2004. Hint: A nice way to do this combines the `for` and the `switch` structures.

3.17 Develop an M-file function to determine the elapsed days in a year. The first line of the function should be set up as

```
function nd = days(mo, da, year)
```

where `mo` = the month (1–12), `da` = the day (1–31), and `year` = the year. Test it for January 1, 1999, February 29, 2000, March 1, 2001, June 21, 2002, and December 31, 2004.

3.18 Develop a function function M-file that returns the difference between the passed function's maximum and minimum value given a range of the independent variable. In addition, have the function generate a plot of the function for the range. Test it for the following cases:

(a) $f(t) = 10e^{-0.25t}\sin(t-4)$ from $t = 0$ to 6π.
(b) $f(x) = e^{5x}\sin(1/x)$ from $x = 0.01$ to 0.2.
(c) The built-in `humps` function from $x = 0$ to 3.

3.19 Modify the function function `odesimp` developed at the end of Sec. 3.6 so that it can be passed the arguments of the passed function. Test it for the following case:

```
>> dvdt=@(v,m,cd) 9.81-(cd/m)*v^2;
>> odesimp(dvdt,0.5,0,12,0,68.1,0.25)
```

4

Roundoff and Truncation Errors

CHAPTER OBJECTIVES

The primary objective of this chapter is to acquaint you with the major sources of errors involved in numerical methods. Specific objectives and topics covered are

- Understanding the distinction between accuracy and precision.
- Learning how to quantify error.
- Learning how error estimates can be used to decide when to terminate an iterative calculation.
- Understanding how roundoff errors occur because digital computers have a limited ability to represent numbers.
- Understanding why floating-point numbers have limits on their range and precision.
- Recognizing that truncation errors occur when exact mathematical formulations are represented by approximations.
- Knowing how to use the Taylor series to estimate truncation errors.
- Understanding how to write forward, backward, and centered finite-difference approximations of first and second derivatives.
- Recognizing that efforts to minimize truncation errors can sometimes increase roundoff errors.

YOU'VE GOT A PROBLEM

In Chap. 1, you developed a numerical model for the velocity of a bungee jumper. To solve the problem with a computer, you had to approximate the derivative of velocity with a finite difference.

$$\frac{dv}{dt} \cong \frac{\Delta v}{\Delta t} = \frac{v(t_{i+1}) - v(t_i)}{t_{i+1} - t_i}$$

Thus, the resulting solution is not exact—that is, it has error.

In addition, the computer you use to obtain the solution is also an imperfect tool. Because it is a digital device, the computer is limited in its ability to represent the magnitudes and precision of numbers. Consequently, the machine itself yields results that contain error.

So both your mathematical approximation and your digital computer cause your resulting model prediction to be uncertain. Your problem is: How do you deal with such uncertainty? This chapter introduces you to some approaches and concepts that engineers and scientists use to deal with this dilemma.

4.1 ERRORS

Engineers and scientists constantly find themselves having to accomplish objectives based on uncertain information. Although perfection is a laudable goal, it is rarely if ever attained. For example, despite the fact that the model developed from Newton's second law is an excellent approximation, it would never in practice exactly predict the jumper's fall. A variety of factors such as winds and slight variations in air resistance would result in deviations from the prediction. If these deviations are systematically high or low, then we might need to develop a new model. However, if they are randomly distributed and tightly grouped around the prediction, then the deviations might be considered negligible and the model deemed adequate. Numerical approximations also introduce similar discrepancies into the analysis.

This chapter covers basic topics related to the identification, quantification, and minimization of these errors. General information concerned with the quantification of error is reviewed in this section. This is followed by Sections 4.2 and 4.3, dealing with the two major forms of numerical error: roundoff error (due to computer approximations) and truncation error (due to mathematical approximations). We also describe how strategies to reduce truncation error sometimes increase roundoff. Finally, we briefly discuss errors not directly connected with the numerical methods themselves. These include blunders, model errors, and data uncertainty.

4.1.1 Accuracy and Precision

The errors associated with both calculations and measurements can be characterized with regard to their accuracy and precision. *Accuracy* refers to how closely a computed or measured value agrees with the true value. *Precision* refers to how closely individual computed or measured values agree with each other.

These concepts can be illustrated graphically using an analogy from target practice. The bullet holes on each target in Fig. 4.1 can be thought of as the predictions of a numerical technique, whereas the bull's-eye represents the truth. *Inaccuracy* (also called *bias*) is defined as systematic deviation from the truth. Thus, although the shots in Fig. 4.1*c* are more tightly grouped than in Fig. 4.1*a*, the two cases are equally biased because they are both centered on the upper left quadrant of the target. *Imprecision* (also called *uncertainty*), on the other hand, refers to the magnitude of the scatter. Therefore, although Fig. 4.1*b* and *d* are equally accurate (i.e., centered on the bull's-eye), the latter is more precise because the shots are tightly grouped.

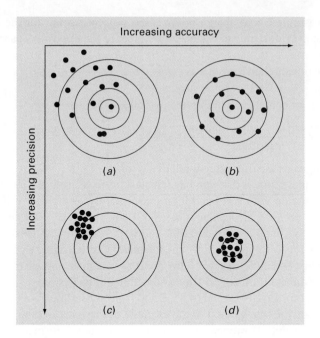

FIGURE 4.1
An example from marksmanship illustrating the concepts of accuracy and precision:
(a) inaccurate and imprecise, (b) accurate and imprecise, (c) inaccurate and precise,
and (d) accurate and precise.

Numerical methods should be sufficiently accurate or unbiased to meet the requirements of a particular problem. They also should be precise enough for adequate design. In this book, we will use the collective term *error* to represent both the inaccuracy and imprecision of our predictions.

4.1.2 Error Definitions

Numerical errors arise from the use of approximations to represent exact mathematical operations and quantities. For such errors, the relationship between the exact, or true, result and the approximation can be formulated as

True value = approximation + error   (4.1)

By rearranging Eq. (4.1), we find that the numerical error is equal to the discrepancy between the truth and the approximation, as in

E_t = true value − approximation   (4.2)

where E_t is used to designate the exact value of the error. The subscript t is included to designate that this is the "true" error. This is in contrast to other cases, as described shortly, where an "approximate" estimate of the error must be employed. Note that the true error is commonly expressed as an absolute value and referred to as the *absolute error*.

A shortcoming of this definition is that it takes no account of the order of magnitude of the value under examination. For example, an error of a centimeter is much more significant if we are measuring a rivet than a bridge. One way to account for the magnitudes of the quantities being evaluated is to normalize the error to the true value, as in

$$\text{True fractional relative error} = \frac{\text{true value} - \text{approximation}}{\text{true value}}$$

The relative error can also be multiplied by 100% to express it as

$$\varepsilon_t = \frac{\text{true value} - \text{approximation}}{\text{true value}} 100\% \qquad (4.3)$$

where ε_t designates the true percent relative error.

For example, suppose that you have the task of measuring the lengths of a bridge and a rivet and come up with 9999 and 9 cm, respectively. If the true values are 10,000 and 10 cm, respectively, the error in both cases is 1 cm. However, their percent relative errors can be computed using Eq. (4.3) as 0.01% and 10%, respectively. Thus, although both measurements have an absolute error of 1 cm, the relative error for the rivet is much greater. We would probably conclude that we have done an adequate job of measuring the bridge, whereas our estimate for the rivet leaves something to be desired.

Notice that for Eqs. (4.2) and (4.3), E and ε are subscripted with a t to signify that the error is based on the true value. For the example of the rivet and the bridge, we were provided with this value. However, in actual situations such information is rarely available. For numerical methods, the true value will only be known when we deal with functions that can be solved analytically. Such will typically be the case when we investigate the theoretical behavior of a particular technique for simple systems. However, in real-world applications, we will obviously not know the true answer *a priori*. For these situations, an alternative is to normalize the error using the best available estimate of the true value—that is, to the approximation itself, as in

$$\varepsilon_a = \frac{\text{approximate error}}{\text{approximation}} 100\% \qquad (4.4)$$

where the subscript a signifies that the error is normalized to an approximate value. Note also that for real-world applications, Eq. (4.2) cannot be used to calculate the error term in the numerator of Eq. (4.4). One of the challenges of numerical methods is to determine error estimates in the absence of knowledge regarding the true value. For example, certain numerical methods use *iteration* to compute answers. In such cases, a present approximation is made on the basis of a previous approximation. This process is performed repeatedly, or iteratively, to successively compute (hopefully) better and better approximations. For such cases, the error is often estimated as the difference between the previous and present approximations. Thus, percent relative error is determined according to

$$\varepsilon_a = \frac{\text{present approximation} - \text{previous approximation}}{\text{present approximation}} 100\% \qquad (4.5)$$

This and other approaches for expressing errors is elaborated on in subsequent chapters.

The signs of Eqs. (4.2) through (4.5) may be either positive or negative. If the approximation is greater than the true value (or the previous approximation is greater than the current approximation), the error is negative; if the approximation is less than the true value, the error is positive. Also, for Eqs. (4.3) to (4.5), the denominator may be less than zero, which can also lead to a negative error. Often, when performing computations, we may not be concerned with the sign of the error but are interested in whether the absolute value of the percent relative error is lower than a prespecified tolerance ε_s. Therefore, it is often useful to employ the absolute value of Eq. (4.5). For such cases, the computation is repeated until

$$|\varepsilon_a| < \varepsilon_s \tag{4.6}$$

This relationship is referred to as a *stopping criterion*. If it is satisfied, our result is assumed to be within the prespecified acceptable level ε_s. Note that for the remainder of this text, we almost always employ absolute values when using relative errors.

It is also convenient to relate these errors to the number of significant figures in the approximation. It can be shown (Scarborough, 1966) that if the following criterion is met, we can be assured that the result is correct to *at least n* significant figures.

$$\varepsilon_s = (0.5 \times 10^{2-n})\% \tag{4.7}$$

EXAMPLE 4.1   Error Estimates for Iterative Methods

Problem Statement.   In mathematics, functions can often be represented by infinite series. For example, the exponential function can be computed using

$$e^x = 1 + x + \frac{x^2}{2} + \frac{x^3}{3!} + \cdots + \frac{x^n}{n!} \tag{E4.1.1}$$

Thus, as more terms are added in sequence, the approximation becomes a better and better estimate of the true value of e^x. Equation (E4.1.1) is called a *Maclaurin series expansion*.

Starting with the simplest version, $e^x = 1$, add terms one at a time in order to estimate $e^{0.5}$. After each new term is added, compute the true and approximate percent relative errors with Eqs. (4.3) and (4.5), respectively. Note that the true value is $e^{0.5} = 1.648721\ldots$. Add terms until the absolute value of the approximate error estimate ε_a falls below a prespecified error criterion ε_s conforming to three significant figures.

Solution.   First, Eq. (4.7) can be employed to determine the error criterion that ensures a result that is correct to at least three significant figures:

$$\varepsilon_s = (0.5 \times 10^{2-3})\% = 0.05\%$$

Thus, we will add terms to the series until ε_a falls below this level.

The first estimate is simply equal to Eq. (E4.1.1) with a single term. Thus, the first estimate is equal to 1. The second estimate is then generated by adding the second term as in

$$e^x = 1 + x$$

or for $x = 0.5$

$$e^{0.5} = 1 + 0.5 = 1.5$$

This represents a true percent relative error of [Eq. (4.3)]

$$\varepsilon_t = \left| \frac{1.648721 - 1.5}{1.648721} \right| \times 100\% = 9.02\%$$

Equation (4.5) can be used to determine an approximate estimate of the error, as in

$$\varepsilon_a = \left| \frac{1.5 - 1}{1.5} \right| \times 100\% = 33.3\%$$

Because ε_a is not less than the required value of ε_s, we would continue the computation by adding another term, $x^2/2!$, and repeating the error calculations. The process is continued until $|\varepsilon_a| < \varepsilon_s$. The entire computation can be summarized as

Terms	Result	ε_t, %	ε_a, %
1	1	39.3	
2	1.5	9.02	33.3
3	1.625	1.44	7.69
4	1.645833333	0.175	1.27
5	1.648437500	0.0172	0.158
6	1.648697917	0.00142	0.0158

Thus, after six terms are included, the approximate error falls below $\varepsilon_s = 0.05\%$, and the computation is terminated. However, notice that, rather than three significant figures, the result is accurate to five! This is because, for this case, both Eqs. (4.5) and (4.7) are conservative. That is, they ensure that the result is at least as good as they specify. Although, this is not always the case for Eq. (4.5), it is true most of the time.

4.2 ROUNDOFF ERRORS

Roundoff errors arise because digital computers cannot represent some quantities exactly. They are important to engineering and scientific problem solving because they can lead to erroneous results. In certain cases, they can actually lead to a calculation going unstable and yielding obviously erroneous results. Such calculations are said to be *ill-conditioned*. Worse still, they can lead to subtler discrepancies that are difficult to detect.

There are two major facets of roundoff errors involved in numerical calculations:

1. Digital computers have size and precision limits on their ability to represent numbers.
2. Certain numerical manipulations are highly sensitive to roundoff errors. This can result from both mathematical considerations as well as from the way in which computers perform arithmetic operations.

4.2.1 Computer Number Representation

Numerical roundoff errors are directly related to the manner in which numbers are stored in a computer. The fundamental unit whereby information is represented is called a *word*. This is an entity that consists of a string of *binary digits*, or *bits*. Numbers are typically stored in one or more words. To understand how this is accomplished, we must first review some material related to number systems.

A *number system* is merely a convention for representing quantities. Because we have 10 fingers and 10 toes, the number system that we are most familiar with is the *decimal,* or *base-10,* number system. A base is the number used as the reference for constructing the system. The base-10 system uses the 10 digits—0, 1, 2, 3, 4, 5, 6, 7, 8, and 9—to represent numbers. By themselves, these digits are satisfactory for counting from 0 to 9.

For larger quantities, combinations of these basic digits are used, with the position or *place value* specifying the magnitude. The rightmost digit in a whole number represents a number from 0 to 9. The second digit from the right represents a multiple of 10. The third digit from the right represents a multiple of 100 and so on. For example, if we have the number 8642.9, then we have eight groups of 1000, six groups of 100, four groups of 10, two groups of 1, and nine groups of 0.1, or

$$(8 \times 10^3) + (6 \times 10^2) + (4 \times 10^1) + (2 \times 10^0) + (9 \times 10^{-1}) = 8642.9$$

This type of representation is called *positional notation.*

Now, because the decimal system is so familiar, it is not commonly realized that there are alternatives. For example, if human beings happened to have eight fingers and toes we would undoubtedly have developed an *octal,* or *base-8,* representation. In the same sense, our friend the computer is like a two-fingered animal who is limited to two states—either 0 or 1. This relates to the fact that the primary logic units of digital computers are on/off electronic components. Hence, numbers on the computer are represented with a *binary,* or *base-2,* system. Just as with the decimal system, quantities can be represented using positional notation. For example, the binary number 101.1 is equivalent to $(1 \times 2^2) + (0 \times 2^1) + (1 \times 2^0) + (1 \times 2^{-1}) = 4 + 0 + 1 + 0.5 = 5.5$ in the decimal system.

Integer Representation.   Now that we have reviewed how base-10 numbers can be represented in binary form, it is simple to conceive of how integers are represented on a computer. The most straightforward approach, called the *signed magnitude method,* employs the first bit of a word to indicate the sign, with a 0 for positive and a 1 for negative. The remaining bits are used to store the number. For example, the integer value of 173 is represented in binary as 10101101:

$$(10101101)_2 = 2^7 + 2^5 + 2^3 + 2^2 + 2^0 = 128 + 32 + 8 + 4 + 1 = (173)_{10}$$

Therefore, the binary equivalent of -173 would be stored on a 16-bit computer, as depicted in Fig. 4.2.

If such a scheme is employed, there clearly is a limited range of integers that can be represented. Again assuming a 16-bit word size, if one bit is used for the sign, the 15 remaining bits can represent binary integers from 0 to 111111111111111. The upper limit can

FIGURE 4.2
The binary representation of the decimal integer −173 on a 16-bit computer using the signed magnitude method.

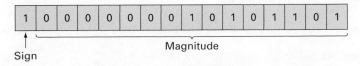

be converted to a decimal integer, as in $(1 \times 2^{14}) + (1 \times 2^{13}) + \cdots + (1 \times 2^{1}) + (1 \times 2^{0}) =$ 32,767. Note that this value can be simply evaluated as $2^{15} - 1$. Thus, a 16-bit computer word can store decimal integers ranging from $-32,767$ to $32,767$.

In addition, because zero is already defined as 0000000000000000, it is redundant to use the number 1000000000000000 to define a "minus zero." Therefore, it is conventionally employed to represent an additional negative number: $-32,768$, and the range is from $-32,768$ to $32,767$. For an n-bit word, the range would be from -2^{n} to $2^{n} - 1$. Thus, 32-bit integers would range from $-2,147,483,648$ to $+2,147,483,647$.

Note that, although it provides a nice way to illustrate our point, the signed magnitude method is not actually used to represent integers for conventional computers. A preferred approach called the *2s complement* technique directly incorporates the sign into the number's magnitude rather than providing a separate bit to represent plus or minus. Regardless, the range of numbers is still the same as for the signed magnitude method described above.

The foregoing serves to illustrate how all digital computers are limited in their capability to represent integers. That is, numbers above or below the range cannot be represented. A more serious limitation is encountered in the storage and manipulation of fractional quantities as described next.

Floating-Point Representation.   Fractional quantities are typically represented in computers using *floating-point format*. In this approach, which is very much like scientific notation, the number is expressed as

$$\pm s \times b^{e}$$

where s = the significand, b = the base of the number system being used, and e = the exponent.

Prior to being expressed in this form, the number is *normalized* by moving the decimal place over so that only one significant digit is to the left of the decimal point. This is done so computer memory is not wasted on storing useless nonsignificant zeros. For example, a value like 0.005678 could be represented in a wasteful manner as 0.005678×10^{0}. However, normalization would yield 5.678×10^{-3} which eliminates the useless zeroes.

Before describing the base-2 implementation used on computers, we will first explore the fundamental implications of such floating-point representation. In particular, what are the ramifications of the fact that in order to be stored in the computer, both the mantissa and the exponent must be limited to a finite number of bits? As in the next example, a nice way to do this is within the context of our more familiar base-10 decimal world.

EXAMPLE 4.2   Implications of Floating-Point Representation

Problem Statement.   Suppose that we had a hypothetical base-10 computer with a 5-digit word size. Assume that one digit is used for the sign, two for the exponent, and two for the mantissa. For simplicity, assume that one of the exponent digits is used for its sign, leaving a single digit for its magnitude.

Solution.   A general representation of the number following normalization would be

$$s_{1}d_{1}.d_{2} \times 10^{s_{0}d_{0}}$$

where s_0 and s_1 = the signs, d_0 = the magnitude of the exponent, and d_1 and d_2 = the magnitude of the significand digits.

Now, let's play with this system. First, what is the largest possible positive quantity that can be represented? Clearly, it would correspond to both signs being positive and all magnitude digits set to the largest possible value in base-10, that is, 9:

$$\text{Largest value} = +9.9 \times 10^{+9}$$

So the largest possible number would be a little less than 10 billion. Although this might seem like a big number, it's really not that big. For example, this computer would be incapable of representing a commonly used constant like Avogadro's number (6.022×10^{23}).

In the same sense, the smallest possible positive number would be

$$\text{Smallest value} = +1.0 \times 10^{-9}$$

Again, although this value might seem pretty small, you could not use it to represent a quantity like Planck's constant (6.626×10^{-34} J·s).

Similar negative values could also be developed. The resulting ranges are displayed in Fig. 4.3. Large positive and negative numbers that fall outside the range would cause an *overflow error*. In a similar sense, for very small quantities there is a "hole" at zero, and very small quantities would usually be converted to zero.

Recognize that the exponent overwhelmingly determines these range limitations. For example, if we increase the mantissa by one digit, the maximum value increases slightly to 9.99×10^9. In contrast, a one-digit increase in the exponent raises the maximum by 90 orders of magnitude to 9.9×10^{99}!

When it comes to precision, however, the situation is reversed. Whereas the significand plays a minor role in defining the range, it has a profound effect on specifying the precision. This is dramatically illustrated for this example where we have limited the significand to only 2 digits. As in Fig. 4.4, just as there is a "hole" at zero, there are also "holes" between values.

For example, a simple rational number with a finite number of digits like $2^{-5} = 0.03125$ would have to be stored as 3.1×10^{-2} or 0.031. Thus, a *roundoff error* is introduced. For this case, it represents a relative error of

$$\frac{0.03125 - 0.031}{0.03125} = 0.008$$

FIGURE 4.3
The number line showing the possible ranges corresponding to the hypothetical base-10 floating-point scheme described in Example 4.2.

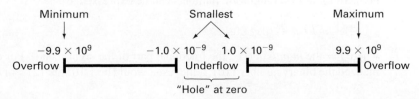

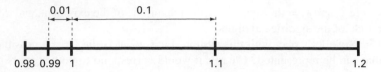

FIGURE 4.4
A small portion of the number line corresponding to the hypothetical base-10 floating-point scheme described in Example 4.2. The numbers indicate values that can be represented exactly. All other quantities falling in the "holes" between these values would exhibit some roundoff error.

While we could store a number like 0.03125 exactly by expanding the digits of the significand, quantities with infinite digits must always be approximated. For example, a commonly used constant such as π (= 3.14159...) would have to be represented as 3.1×10^0 or 3.1. For this case, it represented a relative error of

$$\frac{3.14159 - 3.1}{3.14159} = 0.0132$$

Although adding significand digits can improve the approximation, such quantities will always have some roundoff error when stored in a computer.

Another more subtle effect of floating-point representation is illustrated by Fig. 4.4. Notice how the interval between numbers increases as we move between orders of magnitude. For numbers with an exponent of -1 (that is, between 0.1 and 1), the spacing is 0.01. Once we cross over into the range from 1 to 10, the spacing increases to 0.1. This means that the roundoff error of a number will be proportional to its magnitude. In addition, it means that the relative error will have an upper bound. For this example, the maximum relative error would be 0.05. This value is called the *machine epsilon* (or machine precision).

As illustrated in Example 4.2, the fact that both the exponent and significand are finite means that there are both range and precision limits on floating-point representation. Now, let us examine how floating-point quantities are actually represented in a real computer using base-2 or binary numbers.

First, let's look at normalization. Since binary numbers consist exclusively of 0s and 1s, a bonus occurs when they are normalized. That is, the bit to the left of the binary point will always be one! This means that this leading bit does not have to be stored. Hence, nonzero binary floating-point numbers can be expressed as

$$\pm(1 + f) \times 2^e$$

where f = the *mantissa* (i.e., the fractional part of the significand). For example, if we normalized the binary number 1101.1, the result would be $1.1011 \times (2)^{-3}$ or $(1 + 0.1011) \times 2^{-3}$.

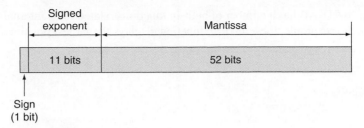

FIGURE 4.5
The manner in which a floating-point number is stored in an 8-byte word in IEEE double-precision format.

Thus, although the original number has five significant bits, we only have to store the four fractional bits: 0.1011.

By default, MATLAB has adopted the *IEEE double-precision format* in which eight bytes (64 bits) are used to represent floating-point numbers. As in Fig. 4.5, one bit is reserved for the number's sign. In a similar spirit to the way in which integers are stored, the exponent and its sign are stored in 11 bits. Finally, 52 bits are set aside for the mantissa. However, because of normalization, 53 significand bits can be stored.

Now, just as in Example 4.2, this means that the numbers will have a limited range and precision. However, because the IEEE format uses many more bits, the resulting number system can be used for practical purposes.

Range. In a fashion similar to the way in which integers are stored, the 11 bits used for the exponent translates into a range from -1022 to 1023. The largest positive number can be represented in binary as

$$\text{Largest value} = +1.1111\ldots1111 \times 2^{+1023}$$

where the 52 bits in the mantissa are all 1. Since the significand is approximately 2 (it is actually $2 - 2^{-52}$), the largest value is therefore $2^{1024} = 1.7977 \times 10^{308}$. In a similar fashion, the smallest positive number can be represented as

$$\text{Smallest value} = +1.0000\ldots0000 \times 2^{-1022}$$

This value can be translated into a base-10 value of $2^{-1022} = 2.2251 \times 10^{-308}$.

Precision. The 52 bits used for the mantissa correspond to about 15 to 16 base-10 digits. Thus, π would be expressed as

```
>> format long
>> pi

ans =
    3.14159265358979
```

Note that the machine epsilon is $2^{-52} = 2.2204 \times 10^{-16}$.

MATLAB has a number of built-in functions related to its internal number representation. For example, the `realmax` function displays the largest positive real number:

```
>> format long
>> realmax

ans =
    1.797693134862316e+308
```

Numbers occurring in computations that exceed this value create an overflow. In MATLAB they are set to infinity, `inf`. The `realmin` function displays the smallest positive real number:

```
>> realmin

ans =
    2.225073858507201e-308
```

Numbers that are smaller than this value create an *underflow* and, in MATLAB, are set to zero. Finally, the `eps` function displays the machine epsilon:

```
>> eps

ans =
    2.220446049250313e-016
```

4.2.2 Arithmetic Manipulations of Computer Numbers

Aside from the limitations of a computer's number system, the actual arithmetic manipulations involving these numbers can also result in roundoff error. To understand how this occurs, let's look at how the computer performs simple addition and subtraction.

Because of their familiarity, normalized base-10 numbers will be employed to illustrate the effect of roundoff errors on simple addition and subtraction. Other number bases would behave in a similar fashion. To simplify the discussion, we will employ a hypothetical decimal computer with a 4-digit mantissa and a 1-digit exponent.

When two floating-point numbers are added, the numbers are first expressed so that they have the same exponents. For example, if we want to add $1.557 + 0.04341$, the computer would express the numbers as $0.1557 \times 10^1 + 0.004341 \times 10^1$. Then the mantissas are added to give 0.160041×10^1. Now, because this hypothetical computer only carries a 4-digit mantissa, the excess number of digits get chopped off and the result is 0.1600×10^1. Notice how the last two digits of the second number (41) that were shifted to the right have essentially been lost from the computation.

Subtraction is performed identically to addition except that the sign of the subtrahend is reversed. For example, suppose that we are subtracting 26.86 from 36.41. That is,

$$
\begin{array}{r}
0.3641 \times 10^2 \\
-0.2686 \times 10^2 \\
\hline
0.0955 \times 10^2
\end{array}
$$

For this case the result must be normalized because the leading zero is unnecessary. So we must shift the decimal one place to the right to give $0.9550 \times 10^1 = 9.550$. Notice that the zero added to the end of the mantissa is not significant but is merely appended to fill the empty space created by the shift. Even more dramatic results would be obtained when the numbers are very close as in

$$
\begin{array}{r}
0.7642 \times 10^3 \\
-0.7641 \times 10^3 \\
\hline
0.0001 \times 10^3
\end{array}
$$

which would be converted to $0.1000 \times 10^0 = 0.1000$. Thus, for this case, three nonsignificant zeros are appended.

The subtracting of two nearly equal numbers is called *subtractive cancellation*. It is the classic example of how the manner in which computers handle mathematics can lead to numerical problems. Other calculations that can cause problems include:

Large Computations.   Certain methods require extremely large numbers of arithmetic manipulations to arrive at their final results. In addition, these computations are often interdependent. That is, the later calculations are dependent on the results of earlier ones. Consequently, even though an individual roundoff error could be small, the cumulative effect over the course of a large computation can be significant. A very simple case involves summing a round base-10 number that is not round in base-2. Suppose that the following M-file is constructed:

```
function sout = sumdemo()
s = 0;
for i = 1:10000
  s = s + 0.0001;
end
sout = s;
```

When this function is executed, the result is

```
>> format long
>> sumdemo

ans =
   0.99999999999991
```

The `format long` command lets us see the 15 significant-digit representation used by MATLAB. You would expect that sum would be equal to 1. However, although 0.0001 is a nice round number in base-10, it cannot be expressed exactly in base-2. Thus, the sum comes out to be slightly different than 1. We should note that MATLAB has features that are designed to minimize such errors. For example, suppose that you form a vector as in

```
>> format long
>> s = [0:0.0001:1];
```

For this case, rather than being equal to 0.99999999999991, the last entry will be exactly one as verified by

```
>> s(10001)

ans =
   1
```

Adding a Large and a Small Number. Suppose we add a small number, 0.0010, to a large number, 4000, using a hypothetical computer with the 4-digit mantissa and the 1-digit exponent. After modifying the smaller number so that its exponent matches the larger,

$$
\begin{array}{r}
0.4000 \times 10^4 \\
0.0000001 \times 10^4 \\
\hline
0.4000001 \times 10^4
\end{array}
$$

which is chopped to 0.4000×10^4. Thus, we might as well have not performed the addition! This type of error can occur in the computation of an infinite series. The initial terms in such series are often relatively large in comparison with the later terms. Thus, after a few terms have been added, we are in the situation of adding a small quantity to a large quantity. One way to mitigate this type of error is to sum the series in reverse order. In this way, each new term will be of comparable magnitude to the accumulated sum.

Smearing. Smearing occurs whenever the individual terms in a summation are larger than the summation itself. One case where this occurs is in a series of mixed signs.

Inner Products. As should be clear from the last sections, some infinite series are particularly prone to roundoff error. Fortunately, the calculation of series is not one of the more common operations in numerical methods. A far more ubiquitous manipulation is the calculation of inner products as in

$$
\sum_{i=1}^{n} x_i y_i = x_1 y_1 + x_2 y_2 + \cdots + x_n y_n
$$

This operation is very common, particularly in the solution of simultaneous linear algebraic equations. Such summations are prone to roundoff error. Consequently, it is often desirable to compute such summations in double precision as is done automatically in MATLAB.

4.3   TRUNCATION ERRORS

Truncation errors are those that result from using an approximation in place of an exact mathematical procedure. For example, in Chap. 1 we approximated the derivative of velocity of a bungee jumper by a finite-difference equation of the form [Eq. (1.11)]

$$
\frac{dv}{dt} \cong \frac{\Delta v}{\Delta t} = \frac{v(t_{i+1}) - v(t_i)}{t_{i+1} - t_i}
\tag{4.8}
$$

A truncation error was introduced into the numerical solution because the difference equation only approximates the true value of the derivative (recall Fig. 1.3). To gain insight into the properties of such errors, we now turn to a mathematical formulation that is used widely in numerical methods to express functions in an approximate fashion—the Taylor series.

4.3.1 The Taylor Series

Taylor's theorem and its associated formula, the Taylor series, is of great value in the study of numerical methods. In essence, the *Taylor theorem* states that any smooth function can be approximated as a polynomial. The *Taylor series* then provides a means to express this idea mathematically in a form that can be used to come up with practical results.

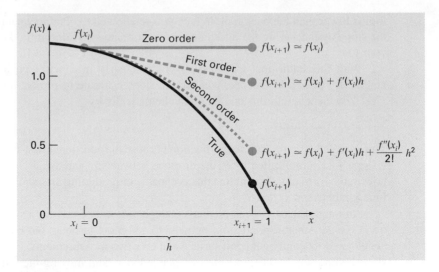

FIGURE 4.6

The approximation of $f(x) = -0.1x^4 - 0.15x^3 - 0.5x^2 - 0.25x + 1.2$ at $x = 1$ by zero-order, first-order, and second-order Taylor series expansions.

A useful way to gain insight into the Taylor series is to build it term by term. A good problem context for this exercise is to predict a function value at one point in terms of the function value and its derivatives at another point.

Suppose that you are blindfolded and taken to a location on the side of a hill facing downslope (Fig. 4.6). We'll call your horizontal location x_i and your vertical distance with respect to the base of the hill $f(x_i)$. You are given the task of predicting the height at a position x_{i+1}, which is a distance h away from you.

At first, you are placed on a platform that is completely horizontal so that you have no idea that the hill is sloping down away from you. At this point, what would be your best guess at the height at x_{i+1}? If you think about it (remember you have no idea whatsoever what's in front of you), the best guess would be the same height as where you're standing now! You could express this prediction mathematically as

$$f(x_{i+1}) \cong f(x_i) \tag{4.9}$$

This relationship, which is called the *zero-order approximation,* indicates that the value of f at the new point is the same as the value at the old point. This result makes intuitive sense because if x_i and x_{i+1} are close to each other, it is likely that the new value is probably similar to the old value.

Equation (4.9) provides a perfect estimate if the function being approximated is, in fact, a constant. For our problem, you would be right only if you happened to be standing on a perfectly flat plateau. However, if the function changes at all over the interval, additional terms of the Taylor series are required to provide a better estimate.

So now you are allowed to get off the platform and stand on the hill surface with one leg positioned in front of you and the other behind. You immediately sense that the front foot is lower than the back foot. In fact, you're allowed to obtain a quantitative estimate of the slope by measuring the difference in elevation and dividing it by the distance between your feet.

With this additional information, you're clearly in a better position to predict the height at $f(x_{i+1})$. In essence, you use the slope estimate to project a straight line out to x_{i+1}. You can express this prediction mathematically by

$$f(x_{i+1}) \cong f(x_i) + f'(x_i)h \tag{4.10}$$

This is called a *first-order approximation* because the additional first-order term consists of a slope $f'(x_i)$ multiplied by h, the distance between x_i and x_{i+1}. Thus, the expression is now in the form of a straight line that is capable of predicting an increase or decrease of the function between x_i and x_{i+1}.

Although Eq. (4.10) can predict a change, it is only exact for a straight-line, or *linear*, trend. To get a better prediction, we need to add more terms to our equation. So now you are allowed to stand on the hill surface and take two measurements. First, you measure the slope behind you by keeping one foot planted at x_i and moving the other one back a distance Δx. Let's call this slope $f_b'(x_i)$. Then you measure the slope in front of you by keeping one foot planted at x_i and moving the other one forward Δx. Let's call this slope $f_f'(x_i)$. You immediately recognize that the slope behind is milder than the one in front. Clearly the drop in height is "accelerating" in front of you. Thus, the odds are that $f(x_i)$ is even lower than your previous linear prediction.

As you might expect, you're now going to add a second-order term to your equation and make it into a parabola. The Taylor series provides the correct way to do this as in

$$f(x_{i+1}) \cong f(x_i) + f'(x_i)h + \frac{f''(x_i)}{2!}h^2 \tag{4.11}$$

To make use of this formula, you need an estimate of the second derivative. You can use the last two slopes you determined to estimate it as

$$f''(x_{i+1}) \cong \frac{f_f'(x_i) - f_b'(x_i)}{\Delta x} \tag{4.12}$$

Thus, the second derivative is merely a derivative of a derivative; in this case, the rate of change of the slope.

Before proceeding, let's look carefully at Eq. (4.11). Recognize that all the values subscripted i represent values that you have estimated. That is, they are numbers. Consequently, the only unknowns are the values at the prediction position x_{i+1}. Consequently, it is a quadratic equation of the form

$$f(h) \cong a_2 h^2 + a_1 h + a_0$$

Thus, we can see that the second-order Taylor series approximates the function with a second-order polynomial.

Clearly, we could keep adding more derivatives to capture more of the function's curvature. Thus, we arrive at the complete Taylor series expansion

$$f(x_{i+1}) = f(x_i) + f'(x_i)h + \frac{f''(x_i)}{2!}h^2 + \frac{f^{(3)}(x_i)}{3!}h^3 + \cdots + \frac{f^{(n)}(x_i)}{n!}h^n + R_n \tag{4.13}$$

Note that because Eq. (4.13) is an infinite series, an equal sign replaces the approximate sign that was used in Eqs. (4.9) through (4.11). A remainder term is also included to account for all terms from $n + 1$ to infinity:

$$R_n = \frac{f^{(n+1)}(\xi)}{(n + 1)!} h^{n+1} \tag{4.14}$$

where the subscript n connotes that this is the remainder for the nth-order approximation and ξ is a value of x that lies somewhere between x_i and x_{i+1}.

Thus, we can now see why the Taylor theorem states that any smooth function can be approximated as a polynomial and that the Taylor series provides a means to express this idea mathematically.

In general, the nth-order Taylor series expansion will be exact for an nth-order polynomial. For other differentiable and continuous functions, such as exponentials and sinusoids, a finite number of terms will not yield an exact estimate. Each additional term will contribute some improvement, however slight, to the approximation. This behavior will be demonstrated in Example 4.3. Only if an infinite number of terms are added will the series yield an exact result.

Although the foregoing is true, the practical value of Taylor series expansions is that, in most cases, the inclusion of only a few terms will result in an approximation that is close enough to the true value for practical purposes. The assessment of how many terms are required to get "close enough" is based on the remainder term of the expansion (Eq. 4.14). This relationship has two major drawbacks. First, ξ is not known exactly but merely lies somewhere between x_i and x_{i+1}. Second, to evaluate Eq. (4.14), we need to determine the $(n + 1)$th derivative of $f(x)$. To do this, we need to know $f(x)$. However, if we knew $f(x)$, there would be no need to perform the Taylor series expansion in the present context!

Despite this dilemma, Eq. (4.14) is still useful for gaining insight into truncation errors. This is because we *do* have control over the term h in the equation. In other words, we can choose how far away from x we want to evaluate $f(x)$, and we can control the number of terms we include in the expansion. Consequently, Eq. (4.14) is usually expressed as

$$R_n = O(h^{n+1})$$

where the nomenclature $O(h^{n+1})$ means that the truncation error is of the order of h^{n+1}. That is, the error is proportional to the step size h raised to the $(n + 1)$th power. Although this approximation implies nothing regarding the magnitude of the derivatives that multiply h^{n+1}, it is extremely useful in judging the comparative error of numerical methods based on Taylor series expansions. For example, if the error is $O(h)$, halving the step size will halve the error. On the other hand, if the error is $O(h^2)$, halving the step size will quarter the error.

In general, we can usually assume that the truncation error is decreased by the addition of terms to the Taylor series. In many cases, if h is sufficiently small, the first- and other lower-order terms usually account for a disproportionately high percent of the error. Thus, only a few terms are required to obtain an adequate approximation. This property is illustrated by the following example.

EXAMPLE 4.3   Approximation of a Function with a Taylor Series Expansion

Problem Statement. Use Taylor series expansions with $n = 0$ to 6 to approximate $f(x) = \cos x$ at $x_{i+1} = \pi/3$ on the basis of the value of $f(x)$ and its derivatives at $x_i = \pi/4$. Note that this means that $h = \pi/3 - \pi/4 = \pi/12$.

Solution. Our knowledge of the true function means that we can determine the correct value $f(\pi/3) = 0.5$. The zero-order approximation is [Eq. (4.9)]

$$f\left(\frac{\pi}{3}\right) \cong \cos\left(\frac{\pi}{4}\right) = 0.707106781$$

which represents a percent relative error of

$$\varepsilon_t = \left|\frac{0.5 - 0.707106781}{0.5}\right| 100\% = 41.4\%$$

For the first-order approximation, we add the first derivative term where $f'(x) = -\sin x$:

$$f\left(\frac{\pi}{3}\right) \cong \cos\left(\frac{\pi}{4}\right) - \sin\left(\frac{\pi}{4}\right)\left(\frac{\pi}{12}\right) = 0.521986659$$

which has $|\varepsilon_t| = 4.40\%$. For the second-order approximation, we add the second derivative term where $f''(x) = -\cos x$:

$$f\left(\frac{\pi}{3}\right) \cong \cos\left(\frac{\pi}{4}\right) - \sin\left(\frac{\pi}{4}\right)\left(\frac{\pi}{12}\right) - \frac{\cos(\pi/4)}{2}\left(\frac{\pi}{12}\right)^2 = 0.497754491$$

with $|\varepsilon_t| = 0.449\%$. Thus, the inclusion of additional terms results in an improved estimate. The process can be continued and the results listed as in

Order n	$f^{(n)}(x)$	$f(\pi/3)$	$|\varepsilon_t|$
0	$\cos x$	0.707106781	41.4
1	$-\sin x$	0.521986659	4.40
2	$-\cos x$	0.497754491	0.449
3	$\sin x$	0.499869147	2.62×10^{-2}
4	$\cos x$	0.500007551	1.51×10^{-3}
5	$-\sin x$	0.500000304	6.08×10^{-5}
6	$-\cos x$	0.499999988	2.44×10^{-6}

Notice that the derivatives never go to zero as would be the case for a polynomial. Therefore, each additional term results in some improvement in the estimate. However, also notice how most of the improvement comes with the initial terms. For this case, by the time we have added the third-order term, the error is reduced to 0.026%, which means that we have attained 99.974% of the true value. Consequently, although the addition of more terms will reduce the error further, the improvement becomes negligible.

4.3.2 The Remainder for the Taylor Series Expansion

Before demonstrating how the Taylor series is actually used to estimate numerical errors, we must explain why we included the argument ξ in Eq. (4.14). To do this, we will use a simple, visually based explanation.

Suppose that we truncated the Taylor series expansion [Eq. (4.13)] after the zero-order term to yield

$$f(x_{i+1}) \cong f(x_i)$$

A visual depiction of this zero-order prediction is shown in Fig. 4.7. The remainder, or error, of this prediction, which is also shown in the illustration, consists of the infinite series of terms that were truncated

$$R_0 = f'(x_i)h + \frac{f''(x_i)}{2!}h^2 + \frac{f^{(3)}(x_i)}{3!}h^3 + \cdots$$

It is obviously inconvenient to deal with the remainder in this infinite series format. One simplification might be to truncate the remainder itself, as in

$$R_0 \cong f'(x_i)h \tag{4.15}$$

Although, as stated in the previous section, lower-order derivatives usually account for a greater share of the remainder than the higher-order terms, this result is still inexact because of the neglected second- and higher-order terms. This "inexactness" is implied by the approximate equality symbol ($\cong$) employed in Eq. (4.15).

An alternative simplification that transforms the approximation into an equivalence is based on a graphical insight. As in Fig. 4.8, the *derivative mean-value theorem* states that

FIGURE 4.7
Graphical depiction of a zero-order Taylor series prediction and remainder.

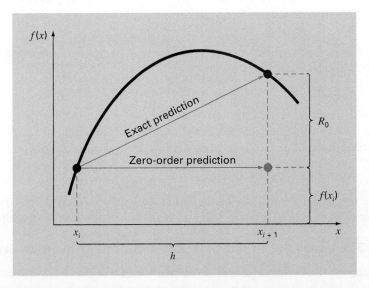

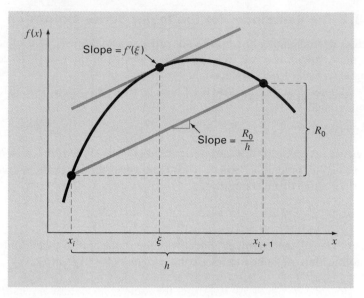

FIGURE 4.8
Graphical depiction of the derivative mean-value theorem.

if a function $f(x)$ and its first derivative are continuous over an interval from x_i to x_{i+1}, then there exists at least one point on the function that has a slope, designated by $f'(\xi)$, that is parallel to the line joining $f(x_i)$ and $f(x_{i+1})$. The parameter ξ marks the x value where this slope occurs (Fig. 4.8). A physical illustration of this theorem is that, if you travel between two points with an average velocity, there will be at least one moment during the course of the trip when you will be moving at that average velocity.

By invoking this theorem, it is simple to realize that, as illustrated in Fig. 4.8, the slope $f'(\xi)$ is equal to the rise R_0 divided by the run h, or

$$f'(\xi) = \frac{R_0}{h}$$

which can be rearranged to give

$$R_0 = f'(\xi)h \tag{4.16}$$

Thus, we have derived the zero-order version of Eq. (4.14). The higher-order versions are merely a logical extension of the reasoning used to derive Eq. (4.16). The first-order version is

$$R_1 = \frac{f''(\xi)}{2!}h^2 \tag{4.17}$$

For this case, the value of ξ conforms to the x value corresponding to the second derivative that makes Eq. (4.17) exact. Similar higher-order versions can be developed from Eq. (4.14).

4.3.3 Using the Taylor Series to Estimate Truncation Errors

Although the Taylor series will be extremely useful in estimating truncation errors throughout this book, it may not be clear to you how the expansion can actually be applied to numerical methods. In fact, we have already done so in our example of the bungee jumper. Recall that the objective of both Examples 1.1 and 1.2 was to predict velocity as a function of time. That is, we were interested in determining $v(t)$. As specified by Eq. (4.13), $v(t)$ can be expanded in a Taylor series:

$$v(t_{i+1}) = v(t_i) + v'(t_i)(t_{i+1} - t_i) + \frac{v''(t_i)}{2!}(t_{i+1} - t_i)^2 + \cdots + R_n$$

Now let us truncate the series after the first derivative term:

$$v(t_{i+1}) = v(t_i) + v'(t_i)(t_{i+1} - t_i) + R_1 \tag{4.18}$$

Equation (4.18) can be solved for

$$v'(t_i) = \underbrace{\frac{v(t_{i+1}) - v(t_i)}{t_{i+1} - t_i}}_{\substack{\text{First-order}\\\text{approximation}}} - \underbrace{\frac{R_1}{t_{i+1} - t_i}}_{\substack{\text{Truncation}\\\text{error}}} \tag{4.19}$$

The first part of Eq. (4.19) is exactly the same relationship that was used to approximate the derivative in Example 1.2 [Eq. (1.11)]. However, because of the Taylor series approach, we have now obtained an estimate of the truncation error associated with this approximation of the derivative. Using Eqs. (4.14) and (4.19) yields

$$\frac{R_1}{t_{i+1} - t_i} = \frac{v''(\xi)}{2!}(t_{i+1} - t_i)$$

or

$$\frac{R_1}{t_{i+1} - t_i} = O(t_{i+1} - t_i)$$

Thus, the estimate of the derivative [Eq. (1.11) or the first part of Eq. (4.19)] has a truncation error of order $t_{i+1} - t_i$. In other words, the error of our derivative approximation should be proportional to the step size. Consequently, if we halve the step size, we would expect to halve the error of the derivative.

4.3.4 Numerical Differentiation

Equation (4.19) is given a formal label in numerical methods—it is called a *finite difference*. It can be represented generally as

$$f'(x_i) = \frac{f(x_{i+1}) - f(x_i)}{x_{i+1} - x_i} + O(x_{i+1} - x_i) \tag{4.20}$$

or

$$f'(x_i) = \frac{f(x_{i+1}) - f(x_i)}{h} + O(h) \tag{4.21}$$

where h is called the step size—that is, the length of the interval over which the approximation is made, $x_{i+1} - x_i$. It is termed a "forward" difference because it utilizes data at i and $i + 1$ to estimate the derivative (Fig. 4.9a).

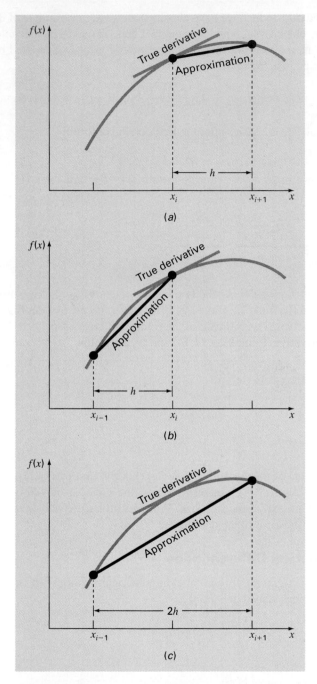

FIGURE 4.9
Graphical depiction of (a) forward, (b) backward, and (c) centered finite-difference approximations of the first derivative.

This forward difference is but one of many that can be developed from the Taylor series to approximate derivatives numerically. For example, backward and centered difference approximations of the first derivative can be developed in a fashion similar to the derivation of Eq. (4.19). The former utilizes values at x_{i-1} and x_i (Fig. 4.9b), whereas the latter uses values that are equally spaced around the point at which the derivative is estimated (Fig. 4.9c). More accurate approximations of the first derivative can be developed by including higher-order terms of the Taylor series. Finally, all the foregoing versions can also be developed for second, third, and higher derivatives. The following sections provide brief summaries illustrating how some of these cases are derived.

Backward Difference Approximation of the First Derivative. The Taylor series can be expanded backward to calculate a previous value on the basis of a present value, as in

$$f(x_{i-1}) = f(x_i) - f'(x_i)h + \frac{f''(x_i)}{2!}h^2 - \cdots \tag{4.22}$$

Truncating this equation after the first derivative and rearranging yields

$$f'(x_i) \cong \frac{f(x_i) - f(x_{i-1})}{h} \tag{4.23}$$

where the error is $O(h)$. See Fig. 4.9b for a graphical representation.

Centered Difference Approximation of the First Derivative. A third way to approximate the first derivative is to subtract Eq. (4.22) from the forward Taylor series expansion:

$$f(x_{i+1}) = f(x_i) + f'(x_i)h + \frac{f''(x_i)}{2!}h^2 + \cdots \tag{4.24}$$

to yield

$$f(x_{i+1}) = f(x_{i-1}) + 2f'(x_i)h + \frac{f^{(3)}(x_i)}{3!}h^3 + \cdots$$

which can be solved for

$$f'(x_i) = \frac{f(x_{i+1}) - f(x_{i-1})}{2h} - \frac{f^{(3)}(x_i)}{6}h^2 + \cdots$$

or

$$f'(x_i) = \frac{f(x_{i+1}) - f(x_{i-1})}{2h} - O(h^2) \tag{4.25}$$

Equation (4.25) is a *centered finite difference* representation of the first derivative. Notice that the truncation error is of the order of h^2 in contrast to the forward and backward approximations that were of the order of h. Consequently, the Taylor series analysis yields the practical information that the centered difference is a more accurate representation of the derivative (Fig. 4.9c). For example, if we halve the step size using a forward or backward difference, we would approximately halve the truncation error, whereas for the central difference, the error would be quartered.

EXAMPLE 4.4   Finite-Difference Approximations of Derivatives

Problem Statement. Use forward and backward difference approximations of $O(h)$ and a centered difference approximation of $O(h^2)$ to estimate the first derivative of

$$f(x) = -0.1x^4 - 0.15x^3 - 0.5x^2 - 0.25x + 1.2$$

at $x = 0.5$ using a step size $h = 0.5$. Repeat the computation using $h = 0.25$. Note that the derivative can be calculated directly as

$$f'(x) = -0.4x^3 - 0.45x^2 - 1.0x - 0.25$$

and can be used to compute the true value as $f'(0.5) = -0.9125$.

Solution. For $h = 0.5$, the function can be employed to determine

$$
\begin{aligned}
x_{i-1} &= 0 & f(x_{i-1}) &= 1.2 \\
x_i &= 0.5 & f(x_i) &= 0.925 \\
x_{i+1} &= 1.0 & f(x_{i+1}) &= 0.2
\end{aligned}
$$

These values can be used to compute the forward difference [Eq. (4.21)],

$$f'(0.5) \cong \frac{0.2 - 0.925}{0.5} = -1.45 \qquad |\varepsilon_t| = 58.9\%$$

the backward difference [Eq. (4.23)],

$$f'(0.5) \cong \frac{0.925 - 1.2}{0.5} = -0.55 \qquad |\varepsilon_t| = 39.7\%$$

and the centered difference [Eq. (4.25)],

$$f'(0.5) \cong \frac{0.2 - 1.2}{1.0} = -1.0 \qquad |\varepsilon_t| = 9.6\%$$

For $h = 0.25$,

$$
\begin{aligned}
x_{i-1} &= 0.25 & f(x_{i-1}) &= 1.10351563 \\
x_i &= 0.5 & f(x_i) &= 0.925 \\
x_{i+1} &= 0.75 & f(x_{i+1}) &= 0.63632813
\end{aligned}
$$

which can be used to compute the forward difference,

$$f'(0.5) \cong \frac{0.63632813 - 0.925}{0.25} = -1.155 \qquad |\varepsilon_t| = 26.5\%$$

the backward difference,

$$f'(0.5) \cong \frac{0.925 - 1.10351563}{0.25} = -0.714 \qquad |\varepsilon_t| = 21.7\%$$

and the centered difference,

$$f'(0.5) \cong \frac{0.63632813 - 1.10351563}{0.5} = -0.934 \qquad |\varepsilon_t| = 2.4\%$$

For both step sizes, the centered difference approximation is more accurate than forward or backward differences. Also, as predicted by the Taylor series analysis, halving the step size approximately halves the error of the backward and forward differences and quarters the error of the centered difference.

Finite-Difference Approximations of Higher Derivatives. Besides first derivatives, the Taylor series expansion can be used to derive numerical estimates of higher derivatives. To do this, we write a forward Taylor series expansion for $f(x_{i+2})$ in terms of $f(x_i)$:

$$f(x_{i+2}) = f(x_i) + f'(x_i)(2h) + \frac{f''(x_i)}{2!}(2h)^2 + \cdots \tag{4.26}$$

Equation (4.24) can be multiplied by 2 and subtracted from Eq. (4.26) to give

$$f(x_{i+2}) - 2f(x_{i+1}) = -f(x_i) + f''(x_i)h^2 + \cdots$$

which can be solved for

$$f''(x_i) = \frac{f(x_{i+2}) - 2f(x_{i+1}) + f(x_i)}{h^2} + O(h) \tag{4.27}$$

This relationship is called the *second forward finite difference*. Similar manipulations can be employed to derive a backward version

$$f''(x_i) = \frac{f(x_i) - 2f(x_{i-1}) + f(x_{i-2})}{h^2} + O(h)$$

and a centered version

$$f''(x_i) = \frac{f(x_{i+1}) - 2f(x_i) + f(x_{i-1})}{h^2} + O(h^2)$$

As was the case with the first-derivative approximations, the centered case is more accurate. Notice also that the centered version can be alternatively expressed as

$$f''(x_i) \cong \frac{\dfrac{f(x_{i+1}) - f(x_i)}{h} - \dfrac{f(x_i) - f(x_{i-1})}{h}}{h}$$

Thus, just as the second derivative is a derivative of a derivative, the second finite difference approximation is a difference of two first finite differences [recall Eq. (4.12)].

4.4 TOTAL NUMERICAL ERROR

The *total numerical error* is the summation of the truncation and roundoff errors. In general, the only way to minimize roundoff errors is to increase the number of significant figures of the computer. Further, we have noted that roundoff error may *increase* due to subtractive cancellation or due to an increase in the number of computations in an analysis. In contrast, Example 4.4 demonstrated that the truncation error can be reduced by decreasing the step size. Because a decrease in step size can lead to subtractive cancellation or to an increase in computations, the truncation errors are *decreased* as the roundoff errors are *increased*.

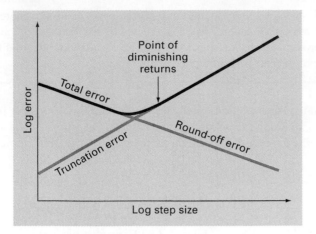

FIGURE 4.10
A graphical depiction of the trade-off between roundoff and truncation error that sometimes comes into play in the course of a numerical method. The point of diminishing returns is shown, where roundoff error begins to negate the benefits of step-size reduction.

Therefore, we are faced by the following dilemma: The strategy for decreasing one component of the total error leads to an increase of the other component. In a computation, we could conceivably decrease the step size to minimize truncation errors only to discover that in doing so, the roundoff error begins to dominate the solution and the total error grows! Thus, our remedy becomes our problem (Fig. 4.10). One challenge that we face is to determine an appropriate step size for a particular computation. We would like to choose a large step size to decrease the amount of calculations and roundoff errors without incurring the penalty of a large truncation error. If the total error is as shown in Fig. 4.10, the challenge is to identify the point of diminishing returns where roundoff error begins to negate the benefits of step-size reduction.

When using MATLAB, such situations are relatively uncommon because of its 15- to 16-digit precision. Nevertheless, they sometimes do occur and suggest a sort of "numerical uncertainty principle" that places an absolute limit on the accuracy that may be obtained using certain computerized numerical methods. We explore such a case in the following section.

4.4.1 Error Analysis of Numerical Differentiation

As described in Sec. 4.3.4, a centered difference approximation of the first derivative can be written as (Eq. 4.25)

$$f'(x_i) = \frac{f(x_{i+1}) - f(x_{i-1})}{2h} - \frac{f^{(3)}(\xi)}{6}h^2$$

$$(4.28)$$

True	Finite-difference	Truncation
value	approximation	error

Thus, if the two function values in the numerator of the finite-difference approximation have no roundoff error, the only error is due to truncation.

However, because we are using digital computers, the function values do include roundoff error as in

$$f(x_{i-1}) = \tilde{f}(x_{i-1}) + e_{i-1}$$

$$f(x_{i+1}) = \tilde{f}(x_{i+1}) + e_{i+1}$$

where the $\tilde{f}$'s are the rounded function values and the e's are the associated roundoff errors. Substituting these values into Eq. (4.28) gives

$$f'(x_i) = \frac{\tilde{f}(x_{i+1}) - \tilde{f}(x_{i-1})}{2h} + \frac{e_{i+1} - e_{i-1}}{2h} - \frac{f^{(3)}(\xi)}{6}h^2$$

| True | Finite-difference | Roundoff | Truncation |
| value | approximation | error | error |

We can see that the total error of the finite-difference approximation consists of a roundoff error that increases with step size and a truncation error that decreases with step size.

Assuming that the absolute value of each component of the roundoff error has an upper bound of ε, the maximum possible value of the difference $e_{i+1} - e_i$ will be 2ε. Further, assume that the third derivative has a maximum absolute value of M. An upper bound on the absolute value of the total error can therefore be represented as

$$\text{Total error} = \left| f'(x_i) - \frac{\tilde{f}(x_{i+1}) - \tilde{f}(x_{i-1})}{2h} \right| \leq \frac{\varepsilon}{h} + \frac{h^2 M}{6} \tag{4.29}$$

An optimal step size can be determined by differentiating Eq. (4.29), setting the result equal to zero and solving for

$$h_{opt} = \sqrt[3]{\frac{3\varepsilon}{M}} \tag{4.30}$$

EXAMPLE 4.5   Roundoff and Truncation Errors in Numerical Differentiation

Problem Statement.   In Example 4.4, we used a centered difference approximation of $O(h^2)$ to estimate the first derivative of the following function at $x = 0.5$,

$$f(x) = -0.1x^4 - 0.15x^3 - 0.5x^2 - 0.25x + 1.2$$

Perform the same computation starting with $h = 1$. Then progressively divide the step size by a factor of 10 to demonstrate how roundoff becomes dominant as the step size is reduced. Relate your results to Eq. (4.30). Recall that the true value of the derivative is -0.9125.

Solution.   We can develop the following M-file to perform the computations and plot the results. Notice that we pass both the function and its analytical derivative as arguments:

```
function diffex(func,dfunc,x,n)
format long
dftrue=dfunc(x);
h=1;
H(1)=h;
D(1)=(func(x+h)-func(x-h))/(2*h);
E(1)=abs(dftrue-D(1));
```

```
for i=2:n
  h=h/10;
  H(i)=h;
  D(i)=(func(x+h)-func(x-h))/(2*h);
  E(i)=abs(dftrue-D(i));
end
L=[H' D' E']';
fprintf('    step size    finite difference    true error\n');
fprintf('%14.10f %16.14f %16.13f\n',L);
loglog(H,E),xlabel('Step Size'),ylabel('Error')
title('Plot of Error Versus Step Size')
format short
```

The M-file can then be run using the following commands:

```
>> ff=@(x) -0.1*x^4-0.15*x^3-0.5*x^2-0.25*x+1.2;
>> df=@(x) -0.4*x^3-0.45*x^2-x-0.25;
>> diffex(ff,df,0.5,11)

   step size    finite difference        true error
   1.0000000000 -1.26250000000000    0.3500000000000
   0.1000000000 -0.91600000000000    0.0035000000000
   0.0100000000 -0.91253500000000    0.0000350000000
   0.0010000000 -0.91250035000001    0.0000003500000
   0.0001000000 -0.91250000349985    0.0000000034998
   0.0000100000 -0.91250000003318    0.0000000000332
   0.0000010000 -0.91250000000542    0.0000000000054
   0.0000001000 -0.91249999945031    0.0000000005497
   0.0000000100 -0.91250000333609    0.0000000033361
   0.0000000010 -0.91250001998944    0.0000000199894
   0.0000000001 -0.91250007550059    0.0000000755006
```

As depicted in Fig. 4.11, the results are as expected. At first, roundoff is minimal and the estimate is dominated by truncation error. Hence, as in Eq. (4.29), the total error drops by a factor of 100 each time we divide the step by 10. However, starting at $h = 0.0001$, we see roundoff error begin to creep in and erode the rate at which the error diminishes. A minimum error is reached at $h = 10^{-6}$. Beyond this point, the error increases as roundoff dominates.

Because we are dealing with an easily differentiable function, we can also investigate whether these results are consistent with Eq. (4.30). First, we can estimate M by evaluating the function's third derivative as

$$M = \left| f^{(3)}(0.5) \right| = |-2.4(0.5) - 0.9| = 2.1$$

Because MATLAB has a precision of about 15 to 16 base-10 digits, a rough estimate of the upper bound on roundoff would be about $\varepsilon = 0.5 \times 10^{-16}$. Substituting these values into Eq. (4.30) gives

$$h_{opt} = \sqrt[3]{\frac{3(0.5 \times 10^{-16})}{2.1}} = 4.3 \times 10^{-6}$$

which is on the same order as the result of 1×10^{-6} obtained with MATLAB.

FIGURE 4.11

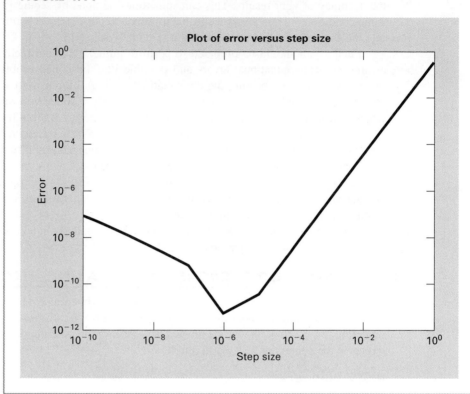

4.4.2 Control of Numerical Errors

For most practical cases, we do not know the exact error associated with numerical methods. The exception, of course, is when we know the exact solution, which makes our numerical approximations unnecessary. Therefore, for most engineering and scientific applications we must settle for some estimate of the error in our calculations.

There are no systematic and general approaches to evaluating numerical errors for all problems. In many cases error estimates are based on the experience and judgment of the engineer or scientist.

Although error analysis is to a certain extent an art, there are several practical programming guidelines we can suggest. First and foremost, avoid subtracting two nearly equal numbers. Loss of significance almost always occurs when this is done. Sometimes you can rearrange or reformulate the problem to avoid subtractive cancellation. If this is not possible, you may want to use extended-precision arithmetic. Furthermore, when adding and subtracting numbers, it is best to sort the numbers and work with the smallest numbers first. This avoids loss of significance.

Beyond these computational hints, one can attempt to predict total numerical errors using theoretical formulations. The Taylor series is our primary tool for analysis of such errors. Prediction of total numerical error is very complicated for even moderately sized problems and tends to be pessimistic. Therefore, it is usually attempted for only small-scale tasks.

The tendency is to push forward with the numerical computations and try to estimate the accuracy of your results. This can sometimes be done by seeing if the results satisfy some condition or equation as a check. Or it may be possible to substitute the results back into the original equation to check that it is actually satisfied.

Finally you should be prepared to perform numerical experiments to increase your awareness of computational errors and possible ill-conditioned problems. Such experiments may involve repeating the computations with a different step size or method and comparing the results. We may employ sensitivity analysis to see how our solution changes when we change model parameters or input values. We may want to try different numerical algorithms that have different theoretical foundations, are based on different computational strategies, or have different convergence properties and stability characteristics.

When the results of numerical computations are extremely critical and may involve loss of human life or have severe economic ramifications, it is appropriate to take special precautions. This may involve the use of two or more independent groups to solve the same problem so that their results can be compared.

The roles of errors will be a topic of concern and analysis in all sections of this book. We will leave these investigations to specific sections.

4.5 BLUNDERS, MODEL ERRORS, AND DATA UNCERTAINTY

Although the following sources of error are not directly connected with most of the numerical methods in this book, they can sometimes have great impact on the success of a modeling effort. Thus, they must always be kept in mind when applying numerical techniques in the context of real-world problems.

4.5.1 Blunders

We are all familiar with gross errors, or blunders. In the early years of computers, erroneous numerical results could sometimes be attributed to malfunctions of the computer itself. Today, this source of error is highly unlikely, and most blunders must be attributed to human imperfection.

Blunders can occur at any stage of the mathematical modeling process and can contribute to all the other components of error. They can be avoided only by sound knowledge of fundamental principles and by the care with which you approach and design your solution to a problem.

Blunders are usually disregarded in discussions of numerical methods. This is no doubt due to the fact that, try as we may, mistakes are to a certain extent unavoidable. However, we believe that there are a number of ways in which their occurrence can be minimized. In particular, the good programming habits that were outlined in Chap. 3 are extremely useful for mitigating programming blunders. In addition, there are usually simple ways to check whether a particular numerical method is working properly. Throughout this book, we discuss ways to check the results of numerical calculations.

4.5.2 Model Errors

Model errors relate to bias that can be ascribed to incomplete mathematical models. An example of a negligible model error is the fact that Newton's second law does not account for relativistic effects. This does not detract from the adequacy of the solution in Example 1.1

because these errors are minimal on the time and space scales associated with the bungee jumper problem.

However, suppose that air resistance is not proportional to the square of the fall velocity, as in Eq. (1.7), but is related to velocity and other factors in a different way. If such were the case, both the analytical and numerical solutions obtained in Chap. 1 would be erroneous because of model error. You should be cognizant of this type of error and realize that, if you are working with a poorly conceived model, no numerical method will provide adequate results.

4.5.3 Data Uncertainty

Errors sometimes enter into an analysis because of uncertainty in the physical data on which a model is based. For instance, suppose we wanted to test the bungee jumper model by having an individual make repeated jumps and then measuring his or her velocity after a specified time interval. Uncertainty would undoubtedly be associated with these measurements, as the parachutist would fall faster during some jumps than during others. These errors can exhibit both inaccuracy and imprecision. If our instruments consistently underestimate or overestimate the velocity, we are dealing with an inaccurate, or biased, device. On the other hand, if the measurements are randomly high and low, we are dealing with a question of precision.

Measurement errors can be quantified by summarizing the data with one or more well-chosen statistics that convey as much information as possible regarding specific characteristics of the data. These descriptive statistics are most often selected to represent (1) the location of the center of the distribution of the data and (2) the degree of spread of the data. As such, they provide a measure of the bias and imprecision, respectively. We will return to the topic of characterizing data uncertainty when we discuss regression in Part Four.

Although you must be cognizant of blunders, model errors, and uncertain data, the numerical methods used for building models can be studied, for the most part, independently of these errors. Therefore, for most of this book, we will assume that we have not made gross errors, we have a sound model, and we are dealing with error-free measurements. Under these conditions, we can study numerical errors without complicating factors.

PROBLEMS

4.1 Convert the following base-2 numbers to base 10: 1011001 and 110.00101.

4.2 Convert the following base-8 numbers to base 10: 71563 and 3.14.

4.3 For computers, the machine epsilon ε can also be thought of as the smallest number that when added to one gives a number greater than 1. An algorithm based on this idea can be developed as

Step 1: Set $\varepsilon = 1$.
Step 2: If $1 + \varepsilon$ is less than or equal to 1, then go to Step 5. Otherwise go to Step 3.
Step 3: $\varepsilon = \varepsilon/2$
Step 4: Return to Step 2
Step 5: $\varepsilon = 2 \times \varepsilon$

Write your own M-file based on this algorithm to determine the machine epsilon. Validate the result by comparing it with the value computed with the built-in function `eps`.

4.4 In a fashion similar to Prob. 4.3, develop your own M-file to determine the smallest positive real number used in MATLAB. Base your algorithm on the notion that your computer will be unable to reliably distinguish between zero and a quantity that is smaller than this number. Note that the result you obtain will differ from the value computed with `realmin`. Challenge question: Investigate the results by taking the base-2 logarithm of the number generated by your code and those obtained with `realmin`.

4.5 Although it is not commonly used, MATLAB allows numbers to be expressed in single precision. Each value is stored in 4 bytes with 1 bit for the sign, 23 bits for the mantissa, and 8 bits for the signed exponent. Determine the smallest and largest positive floating-point numbers as well as the machine epsilon for single precision representation. Note that the exponents range from -126 to 127.

4.6 For the hypothetical base-10 computer in Example 4.2, prove that the machine epsilon is 0.05.

4.7 The derivative of $f(x) = 1/(1 - 3x^2)^2$ is given by

$$\frac{6x}{(1 - 3x^2)^2}$$

Do you expect to have difficulties evaluating this function at $x = 0.577$? Try it using 3- and 4-digit arithmetic with chopping.

4.8 (a) Evaluate the polynomial

$$y = x^3 - 7x^2 + 8x - 0.35$$

at $x = 1.37$. Use 3-digit arithmetic with chopping. Evaluate the percent relative error.
(b) Repeat **(a)** but express y as

$$y = ((x - 7)x + 8)x - 0.35$$

Evaluate the error and compare with part **(a)**.

4.9 The following infinite series can be used to approximate e^x:

$$e^x = 1 + x + \frac{x^2}{2} + \frac{x^3}{3!} + \cdots + \frac{x^n}{n!}$$

(a) Prove that this Maclaurin series expansion is a special case of the Taylor series expansion (Eq. 4.13) with $x_i = 0$ and $h = x$.
(b) Use the Taylor series to estimate $f(x) = e^{-x}$ at $x_{i+1} = 1$ for $x_i = 0.2$. Employ the zero-, first-, second-, and third-order versions and compute the $|\varepsilon_t|$ for each case.

4.10 The Maclaurin series expansion for $\cos x$ is

$$\cos x = 1 - \frac{x^2}{2} + \frac{x^4}{4!} - \frac{x^6}{6!} + \frac{x^8}{8!} - \cdots$$

Starting with the simplest version, $\cos x = 1$, add terms one at a time to estimate $\cos(\pi/3)$. After each new term is added, compute the true and approximate percent relative errors. Use your pocket calculator or MATLAB to determine the true value. Add terms until the absolute value of the approximate error estimate falls below an error criterion conforming to two significant figures.

4.11 Perform the same computation as in Prob. 4.10, but use the Maclaurin series expansion for the $\sin x$ to estimate $\sin(\pi/3)$.

$$\sin x = x - \frac{x^3}{3!} + \frac{x^5}{5!} - \frac{x^7}{7!} + \cdots$$

4.12 Use zero- through third-order Taylor series expansions to predict $f(3)$ for

$$f(x) = 25x^3 - 6x^2 + 7x - 88$$

using a base point at $x = 1$. Compute the true percent relative error ε_t for each approximation.

4.13 Prove that Eq. (4.11) is exact for all values of x if $f(x) = ax^2 + bx + c$.

4.14 Use zero- through fourth-order Taylor series expansions to predict $f(2)$ for $f(x) = \ln x$ using a base point at $x = 1$. Compute the true percent relative error ε_t for each approximation. Discuss the meaning of the results.

4.15 Use forward and backward difference approximations of $O(h)$ and a centered difference approximation of $O(h^2)$ to estimate the first derivative of the function examined in Prob. 4.12. Evaluate the derivative at $x = 2$ using a step size of $h = 0.2$. Compare your results with the true value of the derivative. Interpret your results on the basis of the remainder term of the Taylor series expansion.

4.16 Use a centered difference approximation of $O(h^2)$ to estimate the second derivative of the function examined in Prob. 4.12. Perform the evaluation at $x = 2$ using step sizes of $h = 0.25$ and 0.125. Compare your estimates with the true value of the second derivative. Interpret your results on the basis of the remainder term of the Taylor series expansion.

4.17 If $|x| < 1$ it is known that

$$\frac{1}{1 - x} = 1 + x + x^2 + x^3 + \cdots$$

Repeat Prob. 4.10 for this series for $x = 0.1$.

4.18 To calculate a planet's space coordinates, we have to solve the function

$$f(x) = x - 1 - 0.5 \sin x$$

Let the base point be $a = x_i = \pi/2$ on the interval $[0, \pi]$. Determine the highest-order Taylor series expansion resulting in a maximum error of 0.015 on the specified interval. The error is equal to the absolute value of the difference between the given function and the specific Taylor series expansion. (Hint: Solve graphically.)

4.19 Consider the function $f(x) = x^3 - 2x + 4$ on the interval $[-2, 2]$ with $h = 0.25$. Use the forward, backward, and centered finite difference approximations for the first and second derivatives so as to graphically illustrate which approximation is most accurate. Graph all three first-derivative finite difference approximations along with the theoretical, and do the same for the second derivative as well.

4.20 Derive Eq. (4.30).

4.21 Repeat Example 4.5, but for $f(x) = \cos(x)$ at $x = \pi/6$.

4.22 Repeat Example 4.5, but for the forward divided difference (Eq. 4.21).

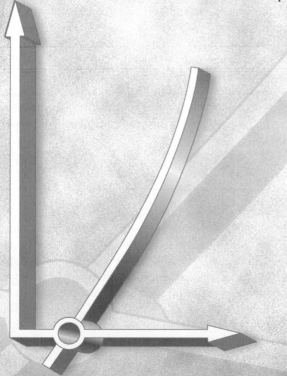

PART TWO

Roots and Optimization

2.1 OVERVIEW

Years ago, you learned to use the quadratic formula

$$x = \frac{-b \pm \sqrt{b^2 - 4ac}}{2a}$$

(PT2.1)

to solve

$$f(x) = ax^2 + bx + c = 0$$

(PT2.2)

The values calculated with Eq. (PT2.1) are called the "roots" of Eq. (PT2.2). They represent the values of x that make Eq. (PT2.2) equal to zero. For this reason, roots are sometimes called the *zeros* of the equation.

Although the quadratic formula is handy for solving Eq. (PT2.2), there are many other functions for which the root cannot be determined so easily. Before the advent of digital computers, there were a number of ways to solve for the roots of such equations. For some cases, the roots could be obtained by direct methods, as with Eq. (PT2.1). Although there were equations like this that could be solved directly, there were many more that could not. In such instances, the only alternative is an approximate solution technique.

One method to obtain an approximate solution is to plot the function and determine where it crosses the x axis. This point, which represents the x value for which $f(x) = 0$, is the root. Although graphical methods are useful for obtaining rough estimates of roots, they are limited because of their lack of precision. An alternative approach is to use *trial and error*. This "technique" consists of guessing a value of x and evaluating whether $f(x)$ is zero. If not (as is almost always the case), another guess is made, and $f(x)$ is again evaluated to determine whether the new value provides a better estimate of the root. The process is repeated until a guess results in an $f(x)$ that is close to zero.

111

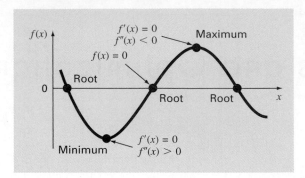

FIGURE PT2.1
A function of a single variable illustrating the difference between roots and optima.

Such haphazard methods are obviously inefficient and inadequate for the requirements of engineering practice. Numerical methods represent alternatives that are also approximate but employ systematic strategies to home in on the true root. As elaborated in the following pages, the combination of these systematic methods and computers makes the solution of most applied roots-of-equations problems a simple and efficient task.

Besides roots, another feature of functions of interest to engineers and scientists are its minimum and maximum values. The determination of such optimal values is referred to as *optimization*. As you learned in calculus, such solutions can be obtained analytically by determining the value at which the function is flat; that is, where its derivative is zero. Although such analytical solutions are sometimes feasible, most practical optimization problems require numerical, computer solutions. From a numerical standpoint, such numerical optimization methods are similar in spirit to the root location methods we just discussed. That is, both involve guessing and searching for a location on a function. The fundamental difference between the two types of problems is illustrated in Figure PT2.1. Root location involves searching for the location where the function equals zero. In contrast, optimization involves searching for the function's extreme points.

2.2   PART ORGANIZATION

The first two chapters in this part are devoted to root location. *Chapter 5* focuses on *bracketing methods* for finding roots. These methods start with guesses that bracket, or contain, the root and then systematically reduce the width of the bracket. Two specific methods are covered: *bisection* and *false position*. Graphical methods are used to provide visual insight into the techniques. Error formulations are developed to help you determine how much computational effort is required to estimate the root to a prespecified level of precision.

Chapter 6 covers *open methods*. These methods also involve systematic trial-and-error iterations but do not require that the initial guesses bracket the root. We will discover that these methods are usually more computationally efficient than bracketing methods but that they do not always work. We illustrate several open methods including the *fixed-point iteration, Newton-Raphson,* and *secant* methods.

Following the description of these individual open methods, we then discuss a hybrid approach called *Brent's root-finding method* that exhibits the reliability of the bracketing methods while exploiting the speed of the open methods. As such, it forms the basis for MATLAB's root-finding function, `fzero`. After illustrating how `fzero` can be used for engineering and scientific problems solving, Chap. 6 ends with a brief discussion of special methods devoted to finding the roots of *polynomials*. In particular, we describe MATLAB's excellent built-in capabilities for this task.

Chapter 7 deals with *optimization*. First, we describe two bracketing methods, *golden-section search* and *parabolic interpolation,* for finding the optima of a function of a single variable. Then, we discuss a robust, hybrid approach that combines golden-section search and quadratic interpolation. This approach, which again is attributed to Brent, forms the basis for MATLAB's one-dimensional root-finding function: `fminbnd`. After describing and illustrating `fminbnd`, the last part of the chapter provides a brief description of optimization of multidimensional functions. The emphasis is on describing and illustrating the use of MATLAB's capability in this area: the `fminsearch` function. Finally, the chapter ends with an example of how MATLAB can be employed to solve optimization problems in engineering and science.

5

Roots: Bracketing Methods

CHAPTER OBJECTIVES

The primary objective of this chapter is to acquaint you with bracketing methods for finding the root of a single nonlinear equation. Specific objectives and topics covered are

- Understanding what roots problems are and where they occur in engineering and science.
- Knowing how to determine a root graphically.
- Understanding the incremental search method and its shortcomings.
- Knowing how to solve a roots problem with the bisection method.
- Knowing how to estimate the error of bisection and why it differs from error estimates for other types of root location algorithms.
- Understanding false position and how it differs from bisection.

YOU'VE GOT A PROBLEM

Medical studies have established that a bungee jumper's chances of sustaining a significant vertebrae injury increase significantly if the free-fall velocity exceeds 36 m/s after 4 s of free fall. Your boss at the bungee-jumping company wants you to determine the mass at which this criterion is exceeded given a drag coefficient of 0.25 kg/m.

You know from your previous studies that the following analytical solution can be used to predict fall velocity as a function of time:

$$v(t) = \sqrt{\frac{gm}{c_d}} \tanh\left(\sqrt{\frac{gc_d}{m}}t\right) \tag{5.1}$$

Try as you might, you cannot manipulate this equation to explicitly solve for m—that is, you cannot isolate the mass on the left side of the equation.

An alternative way of looking at the problem involves subtracting $v(t)$ from both sides to give a new function:

$$f(m) = \sqrt{\frac{gm}{c_d}} \tanh\left(\sqrt{\frac{gc_d}{m}}\, t\right) - v(t) \qquad (5.2)$$

Now we can see that the answer to the problem is the value of m that makes the function equal to zero. Hence, we call this a "roots" problem. This chapter will introduce you to how the computer is used as a tool to obtain such solutions.

5.1 ROOTS IN ENGINEERING AND SCIENCE

Although they arise in other problem contexts, roots of equations frequently occur in the area of design. Table 5.1 lists a number of fundamental principles that are routinely used in design work. As introduced in Chap. 1, mathematical equations or models derived from these principles are employed to predict dependent variables as a function of independent variables, forcing functions, and parameters. Note that in each case, the dependent variables reflect the state or performance of the system, whereas the parameters represent its properties or composition.

An example of such a model is the equation for the bungee jumper's velocity. If the parameters are known, Eq. (5.1) can be used to predict the jumper's velocity. Such computations can be performed directly because v is expressed *explicitly* as a function of the model parameters. That is, it is isolated on one side of the equal sign.

However, as posed at the start of the chapter, suppose that we had to determine the mass for a jumper with a given drag coefficient to attain a prescribed velocity in a set time period. Although Eq. (5.1) provides a mathematical representation of the interrelationship among the model variables and parameters, it cannot be solved explicitly for mass. In such cases, m is said to be *implicit*.

TABLE 5.1 Fundamental principles used in design problems.

Fundamental Principle	Dependent Variable	Independent Variable	Parameters
Heat balance	Temperature	Time and position	Thermal properties of material, system geometry
Mass balance	Concentration or quantity of mass	Time and position	Chemical behavior of material, mass transfer, system geometry
Force balance	Magnitude and direction of forces	Time and position	Strength of material, structural properties, system geometry
Energy balance	Changes in kinetic and potential energy	Time and position	Thermal properties, mass of material, system geometry
Newton's laws of motion	Acceleration, velocity, or location	Time and position	Mass of material, system geometry, dissipative parameters
Kirchhoff's laws	Currents and voltages	Time	Electrical properties (resistance, capacitance, inductance)

This represents a real dilemma, because many design problems involve specifying the properties or composition of a system (as represented by its parameters) to ensure that it performs in a desired manner (as represented by its variables). Thus, these problems often require the determination of implicit parameters.

The solution to the dilemma is provided by numerical methods for roots of equations. To solve the problem using numerical methods, it is conventional to reexpress Eq. (5.1) by subtracting the dependent variable v from both sides of the equation to give Eq. (5.2). The value of m that makes $f(m) = 0$ is, therefore, the root of the equation. This value also represents the mass that solves the design problem.

The following pages deal with a variety of numerical and graphical methods for determining roots of relationships such as Eq. (5.2). These techniques can be applied to many other problems confronted routinely in engineering and science.

5.2 GRAPHICAL METHODS

A simple method for obtaining an estimate of the root of the equation $f(x) = 0$ is to make a plot of the function and observe where it crosses the x axis. This point, which represents the x value for which $f(x) = 0$, provides a rough approximation of the root.

EXAMPLE 5.1   The Graphical Approach

Problem Statement.   Use the graphical approach to determine the mass of the bungee jumper with a drag coefficient of 0.25 kg/m to have a velocity of 36 m/s after 4 s of free fall. Note: The acceleration of gravity is 9.81 m/s^2.

Solution.   The following MATLAB session sets up a plot of Eq. (5.2) versus mass:

```
>> cd = 0.25; g = 9.81; v = 36; t = 4;
>> mp = linspace(50,200);
>> fp = sqrt(g*mp/cd).*tanh(sqrt(g*cd./mp)*t)-v;
>> plot(mp,fp),grid
```

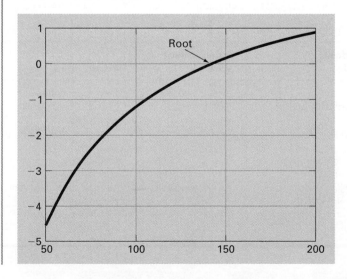

The function crosses the m axis between 140 and 150 kg. Visual inspection of the plot provides a rough estimate of the root of 145 kg (about 320 lb). The validity of the graphical estimate can be checked by substituting it into Eq. (5.2) to yield

```
>> sqrt(g*145/cd)*tanh(sqrt(g*cd/145)*t)-v

ans =
    0.0456
```

which is close to zero. It can also be checked by substituting it into Eq. (5.1) along with the parameter values from this example to give

```
>> sqrt(g*145/cd)*tanh(sqrt(g*cd/145)*t)

ans =
    36.0456
```

which is close to the desired fall velocity of 36 m/s.

Graphical techniques are of limited practical value because they are not very precise. However, graphical methods can be utilized to obtain rough estimates of roots. These estimates can be employed as starting guesses for numerical methods discussed in this chapter.

Aside from providing rough estimates of the root, graphical interpretations are useful for understanding the properties of the functions and anticipating the pitfalls of the numerical methods. For example, Fig. 5.1 shows a number of ways in which roots can occur (or be absent) in an interval prescribed by a lower bound x_l and an upper bound x_u. Figure 5.1b depicts the case where a single root is bracketed by negative and positive values of $f(x)$. However, Fig. 5.1d, where $f(x_l)$ and $f(x_u)$ are also on opposite sides of the x axis, shows three roots occurring within the interval. In general, if $f(x_l)$ and $f(x_u)$ have opposite signs, there are an odd number of roots in the interval. As indicated by Fig. 5.1a and c, if $f(x_l)$ and $f(x_u)$ have the same sign, there are either no roots or an even number of roots between the values.

Although these generalizations are usually true, there are cases where they do not hold. For example, functions that are tangential to the x axis (Fig. 5.2a) and discontinuous functions (Fig. 5.2b) can violate these principles. An example of a function that is tangential to the axis is the cubic equation $f(x) = (x-2)(x-2)(x-4)$. Notice that $x = 2$ makes two terms in this polynomial equal to zero. Mathematically, $x = 2$ is called a *multiple root*. Although they are beyond the scope of this book, there are special techniques that are expressly designed to locate multiple roots (Chapra and Canale, 2002).

The existence of cases of the type depicted in Fig. 5.2 makes it difficult to develop foolproof computer algorithms guaranteed to locate all the roots in an interval. However, when used in conjunction with graphical approaches, the methods described in the following sections are extremely useful for solving many problems confronted routinely by engineers, scientists, and applied mathematicians.

5.3 BRACKETING METHODS AND INITIAL GUESSES

If you had a roots problem in the days before computing, you'd often be told to use "trial and error" to come up with the root. That is, you'd repeatedly make guesses until the function was sufficiently close to zero. The process was greatly facilitated by the advent of software

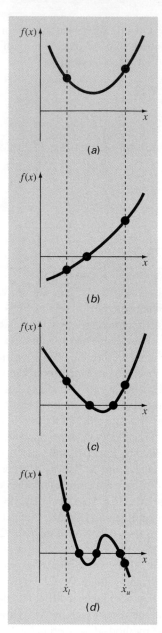

FIGURE 5.1
Illustration of a number of general ways that a root may occur in an interval prescribed by a lower bound x_l and an upper bound x_u. Parts (a) and (c) indicate that if both $f(x_l)$ and $f(x_u)$ have the same sign, either there will be no roots or there will be an even number of roots within the interval. Parts (b) and (d) indicate that if the function has different signs at the end points, there will be an odd number of roots in the interval.

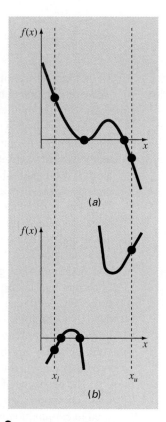

FIGURE 5.2
Illustration of some exceptions to the general cases depicted in Fig. 5.1. (a) Multiple roots that occur when the function is tangential to the x axis. For this case, although the end points are of opposite signs, there are an even number of axis interceptions for the interval. (b) Discontinuous functions where end points of opposite sign bracket an even number of roots. Special strategies are required for determining the roots for these cases.

tools such as spreadsheets. By allowing you to make many guesses rapidly, such tools can actually make the trial-and-error approach attractive for some problems.

But, for many other problems, it is preferable to have methods that come up with the correct answer automatically. Interestingly, as with trial and error, these approaches require an initial "guess" to get started. Then they systematically home in on the root in an iterative fashion.

The two major classes of methods available are distinguished by the type of initial guess. They are

- *Bracketing methods*. As the name implies, these are based on two initial guesses that "bracket" the root—that is, are on either side of the root.
- *Open methods*. These methods can involve one or more initial guesses, but there is no need for them to bracket the root.

For well-posed problems, the bracketing methods always work but converge slowly (i.e., they typically take more iterations to home in on the answer). In contrast, the open methods do not always work (i.e., they can diverge), but when they do they usually converge quicker.

In both cases, initial guesses are required. These may naturally arise from the physical context you are analyzing. However, in other cases, good initial guesses may not be obvious. In such cases, automated approaches to obtain guesses would be useful. The following section describes one such approach, the incremental search.

5.3.1 Incremental Search

When applying the graphical technique in Example 5.1, you observed that $f(x)$ changed sign on opposite sides of the root. In general, if $f(x)$ is real and continuous in the interval from x_l to x_u and $f(x_l)$ and $f(x_u)$ have opposite signs, that is,

$$f(x_l)f(x_u) < 0 \tag{5.3}$$

then there is at least one real root between x_l and x_u.

Incremental search methods capitalize on this observation by locating an interval where the function changes sign. A potential problem with an incremental search is the choice of the increment length. If the length is too small, the search can be very time consuming. On the other hand, if the length is too great, there is a possibility that closely spaced roots might be missed (Fig. 5.3). The problem is compounded by the possible existence of multiple roots.

An M-file can be developed[1] that implements an incremental search to locate the roots of a function func within the range from xmin to xmax (Fig. 5.4). An optional argument ns allows the user to specify the number of intervals within the range. If ns is omitted, it is automatically set to 50. A for loop is used to step through each interval. In the event that a sign change occurs, the upper and lower bounds are stored in an array xb.

---

[1] This function is a modified version of an M-file originally presented by Recktenwald (2000).

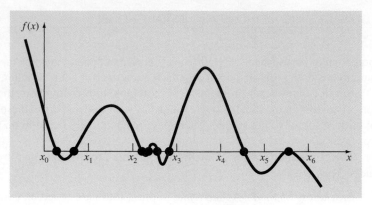

FIGURE 5.3
Cases where roots could be missed because the incremental length of the search procedure is too large. Note that the last root on the right is multiple and would be missed regardless of the increment length.

```
function xb = incsearch(func,xmin,xmax,ns)
% incsearch: incremental search root locator
%   xb = incsearch(func,xmin,xmax,ns):
%       finds brackets of x that contain sign changes
%       of a function on an interval
% input:
%   func = name of function
%   xmin, xmax = endpoints of interval
%   ns = number of subintervals (default = 50)
% output:
%   xb(k,1) is the lower bound of the kth sign change
%   xb(k,2) is the upper bound of the kth sign change
%   If no brackets found, xb = [].

if nargin < 4, ns = 50; end %if ns blank set to 50

% Incremental search
x = linspace(xmin,xmax,ns);
f = func(x);
nb = 0; xb = []; %xb is null unless sign change detected
for k = 1:length(x)-1
  if sign(f(k)) ~= sign(f(k+1)) %check for sign change
    nb = nb + 1;
    xb(nb,1) = x(k);
    xb(nb,2) = x(k+1);
  end
end
if isempty(xb)    %display that no brackets were found
  disp('no brackets found')
  disp('check interval or increase ns')
else
  disp('number of brackets:') %display number of brackets
  disp(nb)
end
```

FIGURE 5.4
An M-file to implement an incremental search.

EXAMPLE 5.2   Incremental Search

Problem Statement.   Use the M-file incsearch (Fig. 5.4) to identify brackets within the interval [3, 6] for the function:

$$f(x) = \sin(10x) + \cos(3x) \tag{5.4}$$

Solution.   The MATLAB session using the default number of intervals (50) is

```
>> incsearch(@x sin(10*x)+cos(3*x),3,6)
number of possible roots:
     5

ans =

     3.2449    3.3061
     3.3061    3.3673
     3.7347    3.7959
     4.6531    4.7143
     5.6327    5.6939
```

A plot of Eq. (5.4) along with the root locations is shown here.

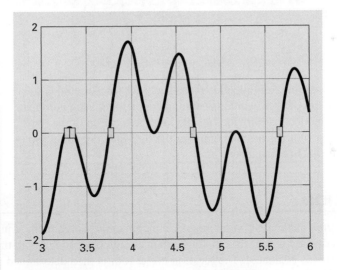

Although five sign changes are detected, because the subintervals are too wide, the function misses possible roots at $x \cong 4.25$ and 5.2. These possible roots look like they might be double roots. However, by using the zoom in tool, it is clear that each represents two real roots that are very close together. The function can be run again with more subintervals with the result that all nine sign changes are located

```
>> incsearch(@x sin(10*x)+cos(3*x),3,6,100)
number of possible roots:
     9

ans =
     3.2424    3.2727
     3.3636    3.3939
```

3.7273	3.7576
4.2121	4.2424
4.2424	4.2727
4.6970	4.7273
5.1515	5.1818
5.1818	5.2121
5.6667	5.6970

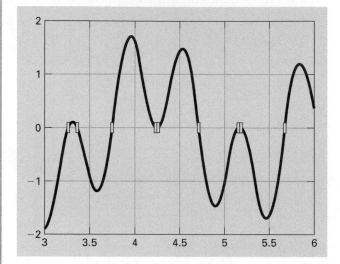

The foregoing example illustrates that brute-force methods such as incremental search are not foolproof. You would be wise to supplement such automatic techniques with any other information that provides insight into the location of the roots. Such information can be found by plotting the function and through understanding the physical problem from which the equation originated.

5.4 BISECTION

The *bisection method* is a variation of the incremental search method in which the interval is always divided in half. If a function changes sign over an interval, the function value at the midpoint is evaluated. The location of the root is then determined as lying within the subinterval where the sign change occurs. The subinterval then becomes the interval for the next iteration. The process is repeated until the root is known to the required precision. A graphical depiction of the method is provided in Fig. 5.5. The following example goes through the actual computations involved in the method.

EXAMPLE 5.3   The Bisection Method

Problem Statement.   Use bisection to solve the same problem approached graphically in Example 5.1.

Solution.   The first step in bisection is to guess two values of the unknown (in the present problem, m) that give values for $f(m)$ with different signs. From the graphical solution in

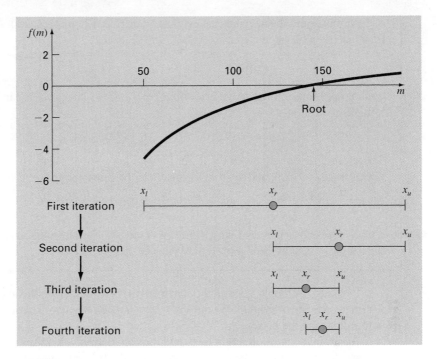

FIGURE 5.5
A graphical depiction of the bisection method. This plot corresponds to the first four iterations from Example 5.3.

Example 5.1, we can see that the function changes sign between values of 50 and 200. The plot obviously suggests better initial guesses, say 140 and 150, but for illustrative purposes let's assume we don't have the benefit of the plot and have made conservative guesses. Therefore, the initial estimate of the root x_r lies at the midpoint of the interval

$$x_r = \frac{50 + 200}{2} = 125$$

Note that the exact value of the root is 142.7376. This means that the value of 125 calculated here has a true percent relative error of

$$|\varepsilon_t| = \left| \frac{142.7376 - 125}{142.7376} \right| \times 100\% = 12.43\%$$

Next we compute the product of the function value at the lower bound and at the midpoint:

$$f(50)f(125) = -4.579(-0.409) = 1.871$$

which is greater than zero, and hence no sign change occurs between the lower bound and the midpoint. Consequently, the root must be located in the upper interval between 125 and 200. Therefore, we create a new interval by redefining the lower bound as 125.

At this point, the new interval extends from $x_l = 125$ to $x_u = 200$. A revised root estimate can then be calculated as

$$x_r = \frac{125 + 200}{2} = 162.5$$

which represents a true percent error of $|\varepsilon_t| = 13.85\%$. The process can be repeated to obtain refined estimates. For example,

$$f(125)f(162.5) = -0.409(0.359) = -0.147$$

Therefore, the root is now in the lower interval between 125 and 162.5. The upper bound is redefined as 162.5, and the root estimate for the third iteration is calculated as

$$x_r = \frac{125 + 162.5}{2} = 143.75$$

which represents a percent relative error of $\varepsilon_t = 0.709\%$. The method can be repeated until the result is accurate enough to satisfy your needs.

We ended Example 5.3 with the statement that the method could be continued to obtain a refined estimate of the root. We must now develop an objective criterion for deciding when to terminate the method.

An initial suggestion might be to end the calculation when the error falls below some prespecified level. For instance, in Example 5.3, the true relative error dropped from 12.43 to 0.709% during the course of the computation. We might decide that we should terminate when the error drops below, say, 0.5%. This strategy is flawed because the error estimates in the example were based on knowledge of the true root of the function. This would not be the case in an actual situation because there would be no point in using the method if we already knew the root.

Therefore, we require an error estimate that is not contingent on foreknowledge of the root. One way to do this is by estimating an approximate percent relative error as in [recall Eq. (4.5)]

$$|\varepsilon_a| = \left| \frac{x_r^{\text{new}} - x_r^{\text{old}}}{x_r^{\text{new}}} \right| 100\% \tag{5.5}$$

where x_r^{new} is the root for the present iteration and x_r^{old} is the root from the previous iteration. When ε_a becomes less than a prespecified stopping criterion ε_s, the computation is terminated.

EXAMPLE 5.4   Error Estimates for Bisection

Problem Statement.   Continue Example 5.3 until the approximate error falls below a stopping criterion of $\varepsilon_s = 0.5\%$. Use Eq. (5.5) to compute the errors.

Solution.   The results of the first two iterations for Example 5.3 were 125 and 162.5. Substituting these values into Eq. (5.5) yields

$$|\varepsilon_a| = \left| \frac{162.5 - 125}{162.5} \right| 100\% = 23.08\%$$

Recall that the true percent relative error for the root estimate of 162.5 was 13.85%. Therefore, $|\varepsilon_a|$ is greater than $|\varepsilon_t|$. This behavior is manifested for the other iterations:

Iteration	x_l	x_u	x_r	$|\varepsilon_a|$ (%)	$|\varepsilon_t|$ (%)
1	50	200	125		12.43
2	125	200	162.5	23.08	13.85
3	125	162.5	143.75	13.04	0.71
4	125	143.75	134.375	6.98	5.86
5	134.375	143.75	139.0625	3.37	2.58
6	139.0625	143.75	141.4063	1.66	0.93
7	141.4063	143.75	142.5781	0.82	0.11
8	142.5781	143.75	143.1641	0.41	0.30

Thus after eight iterations $|\varepsilon_a|$ finally falls below $\varepsilon_s = 0.5\%$, and the computation can be terminated.

These results are summarized in Fig. 5.6. The "ragged" nature of the true error is due to the fact that, for bisection, the true root can lie anywhere within the bracketing interval. The true and approximate errors are far apart when the interval happens to be centered on the true root. They are close when the true root falls at either end of the interval.

FIGURE 5.6
Errors for the bisection method. True and approximate errors are plotted versus the number of iterations.

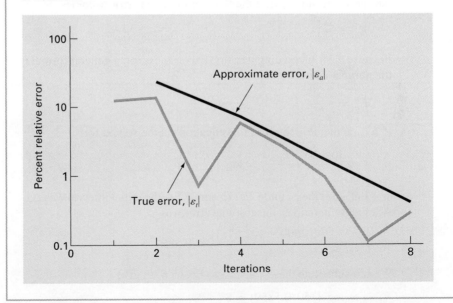

Although the approximate error does not provide an exact estimate of the true error, Fig. 5.6 suggests that $|\varepsilon_a|$ captures the general downward trend of $|\varepsilon_t|$. In addition, the plot exhibits the extremely attractive characteristic that $|\varepsilon_a|$ is always greater than $|\varepsilon_t|$. Thus, when $|\varepsilon_a|$ falls below ε_s, the computation could be terminated with confidence that the root is known to be at least as accurate as the prespecified acceptable level.

While it is dangerous to draw general conclusions from a single example, it can be demonstrated that $|\varepsilon_a|$ will always be greater than $|\varepsilon_t|$ for bisection. This is due to the fact that each time an approximate root is located using bisection as $x_r = (x_l + x_u)/2$, we know that the true root lies somewhere within an interval of $\Delta x = x_u - x_l$. Therefore, the root must lie within $\pm \Delta x/2$ of our estimate. For instance, when Example 5.4 was terminated, we could make the definitive statement that

$$x_r = 143.1641 \pm \frac{143.7500 - 142.5781}{2} = 143.1641 \pm 0.5859$$

In essence, Eq. (5.5) provides an upper bound on the true error. For this bound to be exceeded, the true root would have to fall outside the bracketing interval, which by definition could never occur for bisection. Other root-locating techniques do not always behave as nicely. Although bisection is generally slower than other methods, the neatness of its error analysis is a positive feature that makes it attractive for certain engineering and scientific applications.

Another benefit of the bisection method is that the number of iterations required to attain an absolute error can be computed *a priori*—that is, before starting the computation. This can be seen by recognizing that before starting the technique, the absolute error is

$$E_a^0 = x_u^0 - x_l^0 = \Delta x^0$$

where the superscript designates the iteration. Hence, before starting the method we are at the "zero iteration." After the first iteration, the error becomes

$$E_a^1 = \frac{\Delta x^0}{2}$$

Because each succeeding iteration halves the error, a general formula relating the error and the number of iterations n is

$$E_a^n = \frac{\Delta x^0}{2^n}$$

If $E_{a,d}$ is the desired error, this equation can be solved for[2]

$$n = \frac{\log(\Delta x^0/E_{a,d})}{\log 2} = \log_2\left(\frac{\Delta x^0}{E_{a,d}}\right) \tag{5.6}$$

Let's test the formula. For Example 5.4, the initial interval was $\Delta x_0 = 200 - 50 = 150$. After eight iterations, the absolute error was

$$E_a = \frac{|143.7500 - 142.5781|}{2} = 0.5859$$

We can substitute these values into Eq. (5.6) to give

$$n = \log_2(150/0.5859) = 8$$

---

[2] MATLAB provides the `log2` function to evaluate the base-2 logarithm directly. If the pocket calculator or computer language you are using does not include the base-2 logarithm as an intrinsic function, this equation shows a handy way to compute it. In general, $\log_b(x) = \log(x)/\log(b)$.

Thus, if we knew beforehand that an error of less than 0.5859 was acceptable, the formula tells us that eight iterations would yield the desired result.

Although we have emphasized the use of relative errors for obvious reasons, there will be cases where (usually through knowledge of the problem context) you will be able to specify an absolute error. For these cases, bisection along with Eq. (5.6) can provide a useful root location algorithm.

5.4.1 MATLAB M-file: bisect

An M-file to implement bisection is displayed in Fig. 5.7. It is passed the function (func) along with lower (xl) and upper (xu) guesses. In addition, an optional stopping criterion (es)

FIGURE 5.7
An M-file to implement the bisection method.

```
function [root,ea,iter]=bisect(func,xl,xu,es,maxit,varargin)
% bisect: root location zeroes
%    [root,ea,iter]=bisect(func,xl,xu,es,maxit,p1,p2,...):
%        uses bisection method to find the root of func
% input:
%    func = name of function
%    xl, xu = lower and upper guesses
%    es = desired relative error (default = 0.0001%)
%    maxit = maximum allowable iterations (default = 50)
%    p1,p2,... = additional parameters used by func
% output:
%    root = real root
%    ea = approximate relative error (%)
%    iter = number of iterations
if nargin<3,error('at least 3 input arguments required'),end
test = func(xl,varargin{:})*func(xu,varargin{:});
if test>0,error('no sign change'),end
if nargin<4|isempty(es), es=0.0001;end
if nargin<5|isempty(maxit), maxit=50;end
iter = 0; xr = xl;
while (1)
  xrold = xr;
  xr = (xl + xu)/2;
  iter = iter + 1;
  if xr ~= 0,ea = abs((xr - xrold)/xr) * 100;end
  test = func(xl,varargin{:})*func(xr,varargin{:});
  if test < 0
    xu = xr;
  elseif test > 0
    xl = xr;
  else
    ea = 0;
  end
  if ea <= es | iter >= maxit,break,end
end
root = xr;
```

and maximum iterations (`maxit`) can be entered. The function first checks whether there are sufficient arguments and if the initial guesses bracket a sign change. If not, an error message is displayed and the function is terminated. It also assigns default values if `maxit` and `es` are not supplied. Then a `while...break` loop is employed to implement the bisection algorithm until the approximate error falls below `es` or the iterations exceed `maxit`.

We can employ this function to solve the problem posed at the beginning of the chapter. Recall that you need to determine the mass at which a bungee jumper's free-fall velocity exceeds 36 m/s after 4 s of free fall given a drag coefficient of 0.25 kg/m. Thus, you have to find the root of

$$f(m) = \sqrt{\frac{9.81m}{0.25}} \tanh\left(\sqrt{\frac{9.81(0.25)}{m}} 4\right) - 36$$

In Example 5.1 we generated a plot of this function versus mass and estimated that the root fell between 140 and 150 kg. The `bisect` function from Fig. 5.7 can be used to determine the root as

```
>> fm=@(m) sqrt(9.81*m/0.25)*tanh(sqrt(9.81*0.25/m)*4)-36;
>> [mass ea iter]=bisect(fm,40,200)

mass =
   142.7377
ea =
   5.3450e-005
iter =
     21
```

Thus, a result of $m = 142.7377$ kg is obtained after 21 iterations with an approximate relative error of $\varepsilon_a = 0.00005345\%$. We can substitute the root back into the function to verify that it yields a value close to zero:

```
>> fm(mass)

ans =
   4.6089e-007
```

5.5 FALSE POSITION

False position (also called the linear interpolation method) is another well-known bracketing method. It is very similar to bisection with the exception that it uses a different strategy to come up with its new root estimate. Rather than bisecting the interval, it locates the root by joining $f(x_l)$ and $f(x_u)$ with a straight line (Fig. 5.8). The intersection of this line with the x axis represents an improved estimate of the root. Thus, the shape of the function influences the new root estimate. Using similar triangles, the intersection of the straight line with the x axis can be estimated as (see Chapra and Canale, 2002, for details),

$$x_r = x_u - \frac{f(x_u)(x_l - x_u)}{f(x_l) - f(x_u)} \tag{5.7}$$

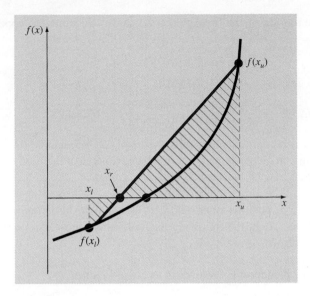

FIGURE 5.8
False position.

This is the *false-position formula*. The value of x_r computed with Eq. (5.7) then replaces whichever of the two initial guesses, x_l or x_u, yields a function value with the same sign as $f(x_r)$. In this way the values of x_l and x_u always bracket the true root. The process is repeated until the root is estimated adequately. The algorithm is identical to the one for bisection (Fig. 5.7) with the exception that Eq. (5.7) is used.

EXAMPLE 5.5   The False-Position Method

Problem Statement.   Use false position to solve the same problem approached graphically and with bisection in Examples 5.1 and 5.3.

Solution.   As in Example 5.3, initiate the computation with guesses of $x_l = 50$ and $x_u = 200$.

First iteration:

$$x_l = 50 \qquad f(x_l) = -4.579387$$
$$x_u = 200 \qquad f(x_u) = 0.860291$$
$$x_r = 200 - \frac{0.860291(50 - 200)}{-4.579387 - 0.860291} = 176.2773$$

which has a true relative error of 23.5%.

Second iteration:

$$f(x_l)f(x_r) = -2.592732$$

Therefore, the root lies in the first subinterval, and x_r becomes the upper limit for the next iteration, $x_u = 176.2773$.

$$x_l = 50 \qquad f(x_l) = -4.579387$$
$$x_u = 176.2773 \qquad f(x_u) = 0.566174$$
$$x_r = 176.2773 - \frac{0.566174(50 - 176.2773)}{-4.579387 - 0.566174} = 162.3828$$

which has true and approximate relative errors of 13.76% and 8.56%, respectively. Additional iterations can be performed to refine the estimates of the root.

Although false position often performs better than bisection, there are other cases where it does not. As in the following example, there are certain cases where bisection yields superior results.

EXAMPLE 5.6   A Case Where Bisection Is Preferable to False Position

Problem Statement.   Use bisection and false position to locate the root of

$$f(x) = x^{10} - 1$$

between $x = 0$ and 1.3.

Solution.   Using bisection, the results can be summarized as

Iteration	x_l	x_u	x_r	ε_a (%)	ε_t (%)
1	0	1.3	0.65	100.0	35
2	0.65	1.3	0.975	33.3	2.5
3	0.975	1.3	1.1375	14.3	13.8
4	0.975	1.1375	1.05625	7.7	5.6
5	0.975	1.05625	1.015625	4.0	1.6

Thus, after five iterations, the true error is reduced to less than 2%. For false position, a very different outcome is obtained:

Iteration	x_l	x_u	x_r	ε_a (%)	ε_t (%)
1	0	1.3	0.09430		90.6
2	0.09430	1.3	0.18176	48.1	81.8
3	0.18176	1.3	0.26287	30.9	73.7
4	0.26287	1.3	0.33811	22.3	66.2
5	0.33811	1.3	0.40788	17.1	59.2

After five iterations, the true error has only been reduced to about 59%. Insight into these results can be gained by examining a plot of the function. As in Fig. 5.9, the curve violates the premise on which false position was based—that is, if $f(x_l)$ is much closer to

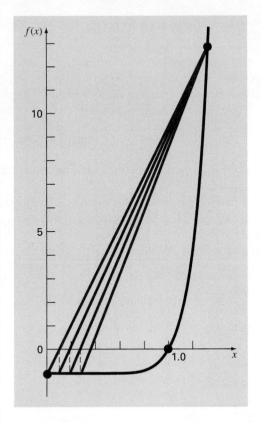

FIGURE 5.9
Plot of $f(x) = x^{10} - 1$, illustrating slow convergence of the false-position method.

zero than $f(x_u)$, then the root is closer to x_l than to x_u (recall Fig. 5.8). Because of the shape of the present function, the opposite is true.

The foregoing example illustrates that blanket generalizations regarding root-location methods are usually not possible. Although a method such as false position is often superior to bisection, there are invariably cases that violate this general conclusion. Therefore, in addition to using Eq. (5.5), the results should always be checked by substituting the root estimate into the original equation and determining whether the result is close to zero.

The example also illustrates a major weakness of the false-position method: its one-sidedness. That is, as iterations are proceeding, one of the bracketing points will tend to stay fixed. This can lead to poor convergence, particularly for functions with significant curvature. Possible remedies for this shortcoming are available elsewhere (Chapra and Canale, 2002).

5.6 CASE STUDY GREENHOUSE GASES AND RAINWATER

Background. It is well documented that the atmospheric levels of several so-called "greenhouse" gases have been increasing over the past 50 years. For example, Fig. 5.10 shows data for the partial pressure of carbon dioxide (CO_2) collected at Mauna Loa, Hawaii from 1958 through 2003. The trend in the data can be nicely fit with a quadratic polynomial,[3]

$$p_{CO_2} = 0.011825(t - 1980.5)^2 + 1.356975(t - 1980.5) + 339$$

where p_{CO_2} = CO_2 partial pressure (ppm). The data indicate that levels have increased a little over 19% over the period from 315 to 376 ppm.

One question that we can address is how this trend is affecting the pH of rainwater. Outside of urban and industrial areas, it is well documented that carbon dioxide is the primary determinant of the pH of the rain. pH is the measure of the activity of hydrogen ions and, therefore, its acidity or alkalinity. For dilute aqueous solutions, it can be computed as

$$pH = -\log_{10}[H^+] \tag{5.8}$$

where $[H^+]$ is the molar concentration of hydrogen ions.

The following five equations govern the chemistry of rainwater:

$$K_1 = 10^6 \frac{[H^+][HCO_3^-]}{K_H p_{CO_2}} \tag{5.9}$$

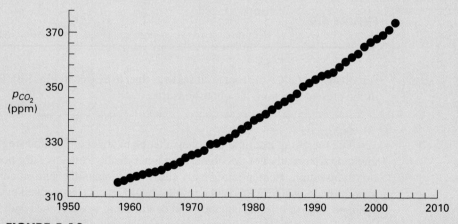

FIGURE 5.10
Average annual partial pressures of atmospheric carbon dioxide (ppm) measured at Mauna Loa, Hawaii.

[3] In Part Four, we will learn how to determine such polynomials.

$$K_2 = \frac{[H^+][CO_3^{-2}]}{[HCO_3^-]} \tag{5.10}$$

$$K_w = [H^+][OH^-] \tag{5.11}$$

$$c_T = \frac{K_H p_{CO_2}}{10^6} + [HCO_3^-] + [CO_3^{-2}] \tag{5.12}$$

$$0 = [HCO_3^-] + 2[CO_3^{-2}] + [OH^-] - [H^+] \tag{5.13}$$

where K_H = Henry's constant, and K_1, K_2, and K_w are equilibrium coefficients. The five unknowns are c_T = total inorganic carbon, $[HCO_3^-]$ = bicarbonate, $[CO_3^{-2}]$ = carbonate, $[H^+]$ = hydrogen ion, and $[OH^-]$ = hydroxyl ion. Notice how the partial pressure of CO_2 shows up in Eqs. (5.9) and (5.12).

Use these equations to compute the pH of rainwater given that $K_H = 10^{-1.46}$, $K_1 = 10^{-6.3}$, $K_2 = 10^{-10.3}$, and $K_w = 10^{-14}$. Compare the results in 1958 when the p_{CO_2} was 315 and in 2003 when it was 375 ppm. When selecting a numerical method for your computation, consider the following:

- You know with certainty that the pH of rain in pristine areas always falls between 2 and 12.
- You also know that pH can only be measured to two places of decimal precision.

Solution.   There are a variety of ways to solve this system of five equations. One way is to eliminate unknowns by combining them to produce a single function that only depends on $[H^+]$. To do this, first solve Eqs. (5.9) and (5.10) for

$$[HCO_3^-] = \frac{K_1}{10^6[H^+]} K_H p_{CO_2} \tag{5.14}$$

$$[CO_3^{-2}] = \frac{K_2[HCO_3^-]}{[H^+]} \tag{5.15}$$

Substitute Eq. (5.14) into (5.15)

$$[CO_3^{-2}] = \frac{K_2 K_1}{10^6[H^+]^2} K_H p_{CO_2} \tag{5.16}$$

Equations (5.14) and (5.16) can be substituted along with Eq. (5.11) into Eq. (5.13) to give

$$0 = \frac{K_1}{10^6[H^+]} K_H p_{CO_2} + 2\frac{K_2 K_1}{10^6[H^+]^2} K_H p_{CO_2} + \frac{K_w}{[H^+]} - [H^+] \tag{5.17}$$

Although it might not be immediately apparent, this result is a third-order polynomial in $[H^+]$. Thus, its root can be used to compute the pH of the rainwater.

Now we must decide which numerical method to employ to obtain the solution. There are two reasons why bisection would be a good choice. First, the fact that the pH always falls within the range from 2 to 12, provides us with two good initial guesses. Second, because the pH can only be measured to two decimal places of precision, we will be satisfied with an absolute error of $E_{a,d} = \pm 0.005$. Remember that given an initial bracket and the desired error, we can compute the number of iteration *a priori*. Substituting the present values into Eq. (5.6) gives

```
>> dx=12-2;
>> Ead=0.005;
>> n=log2(dx/Ead)

n =
    10.9658
```

Eleven iterations of bisection will produce the desired precision.

Before implementing bisection, we must first express Eq. (5.17) as a function. Because it is relatively complicated, we will store it as an M-file:

```
function f = fpH(pH,pCO2)
K1=10^-6.3;K2=10^-10.3;Kw=10^-14;
KH=10^-1.46;
H=10^-pH;
f=K1/(1e6*H)*KH*pCO2+2*K2*K1/(1e6*H)*KH*pCO2+Kw/H-H;
```

We can then use the M-file from Fig. 5.7 to obtain the solution. Notice how we have set the value of the desired relative error ($\varepsilon_a = 1 \times 10^{-8}$) at a very low level so that the iteration limit (`maxit`) is reached first so that exactly 11 iterations are implemented

```
>> [pH1958 ea iter]=bisect(@fpH,2,12,1e-8,11,315)

pH1958 =
    5.6279
ea =
    0.0868
iter =
    11
```

Thus, the pH is computed as 5.6279 with a relative error of 0.0868%. We can be confident that the rounded result of 5.63 is correct to two decimal places. This can be verified by performing another run with more iterations. For example, setting `maxit` to 50 yields

```
>> [pH1958 ea iter] = bisect(@fpH,2,12,1e-8,50,315)

pH1958 =
    5.6304
ea =
  5.1690e-009
iter =
    35
```

For 2003, the result is

```
>> [pH2003 ea iter]=bisect(@fpH,2,12,1e-8,11,375)

pH2003 =
    5.5889
```

5.6 CASE STUDY continued

```
ea =
    0.0874
iter =
    11
```

Interestingly, the results indicate that the 19% rise in atmospheric CO_2 levels has produced only a 0.67% drop in pH. Although this is certainly true, remember that the pH represents a logarithmic scale as defined by Eq. (5.8). Consequently, a unit drop in pH represents a 10-fold increase in the hydrogen ion. The concentration can be computed as $[H^+] = 10^{-pH}$ and its percent change can be calculated as.

```
>> ((10^-pH2003-10^-pH1958)/10^-pH1958)*100

ans =
    9.0930
```

Therefore, the hydrogen ion concentration has increased about 9%.

There is quite a lot of controversy related to the meaning of the greenhouse gas trends. Most of this debate focuses on whether the increases are contributing to global warming. However, regardless of the ultimate implications, it is sobering to realize that something as large as our atmosphere has changed so much over a relatively short time period. This case study illustrates how numerical methods and MATLAB can be employed to analyze and interpret such trends. Over the coming years, engineers and scientists can hopefully use such tools to gain increased understanding of such phenomena and help rationalize the debate over their ramifications.

PROBLEMS

5.1 Use bisection to determine the drag coefficient needed so that an 65-kg bungee jumper has a velocity of 35 m/s after 4.5 s of free fall. Note: The acceleration of gravity is 9.81 m/s². Start with initial guesses of $x_l = 0.2$ and $x_u = 0.3$ and iterate until the approximate relative error falls below 2%.

5.2 Develop your own M-file for bisection in a similar fashion to Fig. 5.7. However, rather than using the maximum iterations and Eq. (5.5), employ Eq. (5.6) as your stopping criterion. Make sure to round the result of Eq. (5.6) up to the next highest integer. Test your function by solving Prob. 5.1 using $E_{a,d} = 0.0001$.

5.3 Repeat Prob. 5.1, but use the false-position method to obtain your solution.

5.4 Develop an M-file for the false-position method. Test it by solving Prob. 5.1.

5.5 A beam is loaded as shown in Fig. P5.5. Use the bisection method to solve for the position inside the beam where there is no moment.

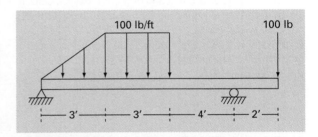

FIGURE P5.5

5.6 (a) Determine the roots of $f(x) = -14 - 20x + 19x^2 - 3x^3$ graphically. In addition, determine the first root of the function with **(b)** bisection and **(c)** false position. For **(b)** and **(c)** use initial guesses of $x_l = -1$ and $x_u = 0$ and a stopping criterion of 1%.

5.7 Locate the first nontrivial root of $\sin(x) = x^3$ where x is in radians. Use a graphical technique and bisection with the initial interval from 0.5 to 1. Perform the computation until ε_a is less than $\varepsilon_s = 2\%$.

5.8 Determine the positive real root of $\ln(x^4) = 0.7$ **(a)** graphically, **(b)** using three iterations of the bisection method, with initial guesses of $x_l = 0.5$ and $x_u = 2$, and **(c)** using three iterations of the false-position method, with the same initial guesses as in **(b)**.

5.9 The saturation concentration of dissolved oxygen in freshwater can be calculated with the equation

$$\ln o_{sf} = -139.34411 + \frac{1.575701 \times 10^5}{T_a}$$
$$- \frac{6.642308 \times 10^7}{T_a^2} + \frac{1.243800 \times 10^{10}}{T_a^3}$$
$$- \frac{8.621949 \times 10^{11}}{T_a^4}$$

where o_{sf} = the saturation concentration of dissolved oxygen in freshwater at 1 atm (mg L^{-1}); and T_a = absolute temperature (K). Remember that $T_a = T + 273.15$, where T = temperature (°C). According to this equation, saturation decreases with increasing temperature. For typical natural waters in temperate climates, the equation can be used to determine that oxygen concentration ranges from 14.621 mg/L at 0 °C to 6.413 mg/L at 40 °C. Given a value of oxygen concentration, this formula and the bisection method can be used to solve for temperature in °C.

(a) If the initial guesses are set as 0 and 40 °C, how many bisection iterations would be required to determine temperature to an absolute error of 0.05 °C?

(b) Based on **(a)**, develop and test a bisection M-file function to determine T as a function of a given oxygen concentration. Test your function for $o_{sf} = 8$, 10 and 12 mg/L. Check your results.

5.10 Water is flowing in a trapezoidal channel at a rate of $Q = 20$ m³/s. The critical depth y for such a channel must satisfy the equation

$$0 = 1 - \frac{Q^2}{g A_c^3} B$$

where $g = 9.81$ m/s², A_c = the cross-sectional area (m²), and B = the width of the channel at the surface (m). For this case, the width and the cross-sectional area can be related to depth y by

$$B = 3 + y$$

and

$$A_c = 3y + \frac{y^2}{2}$$

Solve for the critical depth using **(a)** the graphical method, **(b)** bisection, and **(c)** false position. For **(b)** and **(c)** use initial guesses of $x_l = 0.5$ and $x_u = 2.5$, and iterate until the approximate error falls below 1% or the number of iterations exceeds 10. Discuss your results.

5.11 The Michaelis-Menten model describes the kinetics of enzyme mediated reactions:

$$\frac{dS}{dt} = -v_m \frac{S}{k_s + S}$$

where S = substrate concentration (moles/L), v_m = maximum uptake rate (moles/L/d), and k_s = the half-saturation constant, which is the substrate level at which uptake is half of the maximum [moles/L]. If the initial substrate level at $t = 0$ is S_0, this differential equation can be solved for

$$S = S_0 - v_m t + k_s \ln(S/S_0)$$

Develop an M-file to generate a plot of S versus t for the case where $S_0 = 10$ moles/L, $v_m = 0.5$ moles/L/d, and $k_s = 2$ moles/L.

5.12 A reversible chemical reaction

$$2A + B \underset{\leftarrow}{\overset{\rightarrow}{}} C$$

can be characterized by the equilibrium relationship

$$K = \frac{c_c}{c_a^2 c_b}$$

where the nomenclature c_i represents the concentration of constituent i. Suppose that we define a variable x as representing the number of moles of C that are produced. Conservation of mass can be used to reformulate the equilibrium relationship as

$$K = \frac{(c_{c,0} + x)}{(c_{a,0} - 2x)^2 (c_{b,0} - x)}$$

where the subscript 0 designates the initial concentration of each constituent. If $K = 0.016$, $c_{a,0} = 42$, $c_{b,0} = 28$, and $c_{c,0} = 4$, determine the value of x.

(a) Obtain the solution graphically.

(b) On the basis of **(a)**, solve for the root with initial guesses of $x_l = 0$ and $x_u = 20$ to $\varepsilon_s = 0.5\%$. Choose either bisection or false position to obtain your solution. Justify your choice.

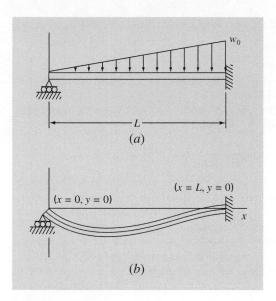

FIGURE P5.13

5.13 Figure P5.13*a* shows a uniform beam subject to a linearly increasing distributed load. The equation for the resulting elastic curve is (see Fig. P5.13*b*)

$$y = \frac{w_0}{120EIL}(-x^5 + 2L^2x^3 - L^4x) \tag{P5.13}$$

Use bisection to determine the point of maximum deflection (that is, the value of x where $dy/dx = 0$). Then substitute this value into Eq. (P5.13) to determine the value of the maximum deflection. Use the following parameter values in your computation: $L = 600$ cm, $E = 50,000$ kN/cm^2, $I = 30,000$ cm^4, and $w_0 = 2.5$ kN/cm.

5.14 You buy a $25,000 piece of equipment for nothing down at $5,500 per year for 6 years. What interest rate are you paying? The formula relating present worth P, annual payments A, number of years n, and interest rate i is

$$A = P\frac{i(1+i)^n}{(1+i)^n - 1}$$

5.15 Many fields of engineering require accurate population estimates. For example, transportation engineers might find it necessary to determine separately the population growth trends of a city and adjacent suburb. The population of the urban area is declining with time according to

$$P_u(t) = P_{u,\max}e^{-k_u t} + P_{u,\min}$$

while the suburban population is growing, as in

$$P_s(t) = \frac{P_{s,\max}}{1 + [P_{s,\max}/P_0 - 1]e^{-k_s t}}$$

where $P_{u,\max}$, k_u, $P_{s,\max}$, P_0, and $k_s =$ empirically derived parameters. Determine the time and corresponding values of $P_u(t)$ and $P_s(t)$ when the suburbs are 20% larger than the city. The parameter values are $P_{u,\max} = 75,000$, $k_u = 0.045$/yr, $P_{u,\min} = 100,000$ people, $P_{s,\max} = 300,000$ people, $P_0 = 10,000$ people, and $k_s = 0.08$/yr. To obtain your solutions, use **(a)** graphical, and **(b)** false-position methods.

5.16 The resistivity ρ of doped silicon is based on the charge q on an electron, the electron density n, and the electron mobility μ. The electron density is given in terms of the doping density N and the intrinsic carrier density n_i. The electron mobility is described by the temperature T, the reference temperature T_0, and the reference mobility μ_0. The equations required to compute the resistivity are

$$\rho = \frac{1}{qn\mu}$$

where

$$n = \frac{1}{2}\left(N + \sqrt{N^2 + 4n_i^2}\right) \quad \text{and} \quad \mu = \mu_0\left(\frac{T}{T_0}\right)^{-2.42}$$

Determine N, given $T_0 = 300$ K, $T = 1000$ K, $\mu_0 = 1350$ cm^2 (V s)$^{-1}$, $q = 1.7 \times 10^{-19}$ C, $n_i = 6.21 \times 10^9$ cm^{-3}, and a desired $\rho = 6.5 \times 10^6$ V s cm/C. Use **(a)** bisection and **(b)** the false position method.

5.17 A total charge Q is uniformly distributed around a ring-shaped conductor with radius a. A charge q is located at a distance x from the center of the ring (Fig. P5.17). The force exerted on the charge by the ring is given by

$$F = \frac{1}{4\pi e_0}\frac{qQx}{(x^2 + a^2)^{3/2}}$$

where $e_0 = 8.85 \times 10^{-12}$ C^2/(N m^2). Find the distance x where the force is 1.25 N if q and Q are 2×10^{-5} C for a ring with a radius of 0.9 m.

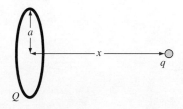

FIGURE P5.17

5.18 For fluid flow in pipes, friction is described by a dimensionless number, the *Fanning friction factor f*. The Fanning friction factor is dependent on a number of parameters related to the size of the pipe and the fluid, which can all be represented by another dimensionless quantity, the *Reynolds number* Re. A formula that predicts f given Re is the *von Karman equation:*

$$\frac{1}{\sqrt{f}} = 4 \log_{10} \left(\mathrm{Re} \sqrt{f} \right) - 0.4$$

Typical values for the Reynolds number for turbulent flow are 10,000 to 500,000 and for the Fanning friction factor are 0.001 to 0.01. Develop a function that uses bisection to solve for f given a user-supplied value of Re between 2,500 and 1,000,000. Design the function so that it ensures that the absolute error in the result is $E_{a,d} < 0.000005$.

5.19 Mechanical engineers, as well as most other engineers, use thermodynamics extensively in their work. The following polynomial can be used to relate the zero-pressure specific heat of dry air c_p kJ/(kg K) to temperature (K):

$$c_p = 0.99403 + 1.671 \times 10^{-4}T + 9.7215 \times 10^{-8}T^2$$
$$-9.5838 \times 10^{-11}T^3 + 1.9520 \times 10^{-14}T^4$$

Determine the temperature that corresponds to a specific heat of 1.1 kJ/(kg K).

5.20 The upward velocity of a rocket can be computed by the following formula:

$$v = u \ln \frac{m_0}{m_0 - qt} - gt$$

where v = upward velocity, u = the velocity at which fuel is expelled relative to the rocket, m_0 = the initial mass of the rocket at time $t = 0$, q = the fuel consumption rate, and g = the downward acceleration of gravity (assumed constant = 9.81 m/s²). If $u = 2000$ m/s, $m_0 = 150,000$ kg, and $q = 2700$ kg/s, compute the time at which $v = 750$ m/s. (Hint: t is somewhere between 10 and 50 s.) Determine your result so that it is within 1% of the true value. Check your answer.

6

Roots: Open Methods

CHAPTER OBJECTIVES

The primary objective of this chapter is to acquaint you with open methods for finding the root of a single nonlinear equation. Specific objectives and topics covered are

- Recognizing the difference between bracketing and open methods for root location.
- Understanding the fixed-point iteration method and how you can evaluate its convergence characteristics.
- Knowing how to solve a roots problem with the Newton-Raphson method and appreciating the concept of quadratic convergence.
- Knowing how to implement both the secant and the modified secant methods.
- Knowing how to use MATLAB's `fzero` function to estimate roots.
- Learning how to manipulate and determine the roots of polynomials with MATLAB.

For the bracketing methods in Chap. 5, the root is located within an interval prescribed by a lower and an upper bound. Repeated application of these methods always results in closer estimates of the true value of the root. Such methods are said to be *convergent* because they move closer to the truth as the computation progresses (Fig. 6.1a).

In contrast, the *open methods* described in this chapter require only a single starting value or two starting values that do not necessarily bracket the root. As such, they sometimes *diverge* or move away from the true root as the computation progresses (Fig. 6.1b). However, when the open methods converge (Fig. 6.1c) they usually do so much more quickly than the bracketing methods. We will begin our discussion of open techniques with a simple approach that is useful for illustrating their general form and also for demonstrating the concept of convergence.

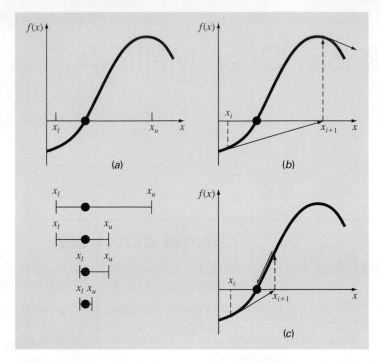

FIGURE 6.1
Graphical depiction of the fundamental difference between the (a) bracketing and (b) and (c) open methods for root location. In (a), which is bisection, the root is constrained within the interval prescribed by x_l and x_u. In contrast, for the open method depicted in (b) and (c), which is Newton-Raphson, a formula is used to project from x_i to x_{i+1} in an iterative fashion. Thus the method can either (b) diverge or (c) converge rapidly, depending on the shape of the function and the value of the initial guess.

6.1   SIMPLE FIXED-POINT ITERATION

As just mentioned, open methods employ a formula to predict the root. Such a formula can be developed for simple *fixed-point iteration* (or, as it is also called, one-point iteration or successive substitution) by rearranging the function $f(x) = 0$ so that x is on the left-hand side of the equation:

$$x = g(x) \tag{6.1}$$

This transformation can be accomplished either by algebraic manipulation or by simply adding x to both sides of the original equation.

The utility of Eq. (6.1) is that it provides a formula to predict a new value of x as a function of an old value of x. Thus, given an initial guess at the root x_i, Eq. (6.1) can be used to compute a new estimate x_{i+1} as expressed by the iterative formula

$$x_{i+1} = g(x_i) \tag{6.2}$$

As with many other iterative formulas in this book, the approximate error for this equation can be determined using the error estimator:

$$\varepsilon_a = \left| \frac{x_{i+1} - x_i}{x_{i+1}} \right| 100\% \tag{6.3}$$

EXAMPLE 6.1 Simple Fixed-Point Iteration

Problem Statement. Use simple fixed-point iteration to locate the root of $f(x) = e^{-x} - x$.

Solution. The function can be separated directly and expressed in the form of Eq. (6.2) as

$$x_{i+1} = e^{-x_i}$$

Starting with an initial guess of $x_0 = 0$, this iterative equation can be applied to compute:

i	x_i	$|\varepsilon_a|$, %	$|\varepsilon_t|$, %	$|\varepsilon_t|_i/|\varepsilon_t|_{i-1}$
0	0.0000		100.000	
1	1.0000	100.000	76.322	0.763
2	0.3679	171.828	35.135	0.460
3	0.6922	46.854	22.050	0.628
4	0.5005	38.309	11.755	0.533
5	0.6062	17.447	6.894	0.586
6	0.5454	11.157	3.835	0.556
7	0.5796	5.903	2.199	0.573
8	0.5601	3.481	1.239	0.564
9	0.5711	1.931	0.705	0.569
10	0.5649	1.109	0.399	0.566

Thus, each iteration brings the estimate closer to the true value of the root: 0.56714329.

Notice that the true percent relative error for each iteration of Example 6.1 is roughly proportional (by a factor of about 0.5 to 0.6) to the error from the previous iteration. This property, called *linear convergence,* is characteristic of fixed-point iteration.

Aside from the "rate" of convergence, we must comment at this point about the "possibility" of convergence. The concepts of convergence and divergence can be depicted graphically. Recall that in Section 5.2, we graphed a function to visualize its structure and behavior. Such an approach is employed in Fig. 6.2a for the function $f(x) = e^{-x} - x$. An alternative graphical approach is to separate the equation into two component parts, as in

$$f_1(x) = f_2(x)$$

Then the two equations

$$y_1 = f_1(x) \tag{6.4}$$

and

$$y_2 = f_2(x) \tag{6.5}$$

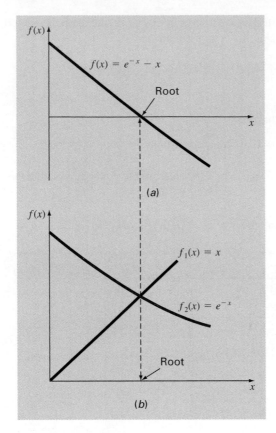

FIGURE 6.2
Two alternative graphical methods for determining the root of $f(x) = e^{-x} - x$. (a) Root at the point where it crosses the x axis; (b) root at the intersection of the component functions.

can be plotted separately (Fig. 6.2b). The x values corresponding to the intersections of these functions represent the roots of $f(x) = 0$.

The two-curve method can now be used to illustrate the convergence and divergence of fixed-point iteration. First, Eq. (6.1) can be reexpressed as a pair of equations $y_1 = x$ and $y_2 = g(x)$. These two equations can then be plotted separately. As was the case with Eqs. (6.4) and (6.5), the roots of $f(x) = 0$ correspond to the abscissa value at the intersection of the two curves. The function $y_1 = x$ and four different shapes for $y_2 = g(x)$ are plotted in Fig. 6.3.

For the first case (Fig. 6.3a), the initial guess of x_0 is used to determine the corresponding point on the y_2 curve $[x_0, g(x_0)]$. The point $[x_1, x_1]$ is located by moving left horizontally to the y_1 curve. These movements are equivalent to the first iteration of the fixed-point method:

$$x_1 = g(x_0)$$

Thus, in both the equation and in the plot, a starting value of x_0 is used to obtain an estimate of x_1. The next iteration consists of moving to $[x_1, g(x_1)]$ and then to $[x_2, x_2]$. This

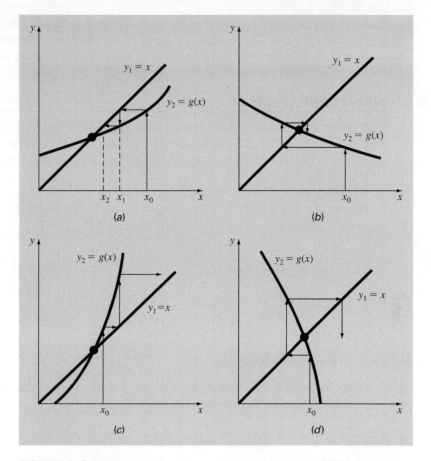

FIGURE 6.3
Graphical depiction of (a) and (b) convergence and (c) and (d) divergence of simple fixed-point iteration. Graphs (a) and (c) are called monotone patterns whereas (b) and (c) are called oscillating or spiral patterns. Note that convergence occurs when $|g'(x)| < 1$.

iteration is equivalent to the equation

$$x_2 = g(x_1)$$

The solution in Fig. 6.3*a* is *convergent* because the estimates of x move closer to the root with each iteration. The same is true for Fig. 6.3*b*. However, this is not the case for Fig. 6.3*c* and *d*, where the iterations diverge from the root.

A theoretical derivation can be used to gain insight into the process. As described in Chapra and Canale (2002), it can be shown that the error for any iteration is linearly proportional to the error from the previous iteration multiplied by the absolute value of the slope of g:

$$E_{i+1} = g'(\xi)E_i$$

Consequently, if $|g'| < 1$, the errors decrease with each iteration. For $|g'| > 1$ the errors grow. Notice also that if the derivative is positive, the errors will be positive, and hence the errors will have the same sign (Fig. 6.3*a* and *c*). If the derivative is negative, the errors will change sign on each iteration (Fig. 6.3*b* and *d*).

6.2 NEWTON-RAPHSON

Perhaps the most widely used of all root-locating formulas is the *Newton-Raphson method* (Fig. 6.4). If the initial guess at the root is x_i, a tangent can be extended from the point $[x_i, f(x_i)]$. The point where this tangent crosses the x axis usually represents an improved estimate of the root.

The Newton-Raphson method can be derived on the basis of this geometrical interpretation. As in Fig. 6.4, the first derivative at x is equivalent to the slope:

$$f'(x_i) = \frac{f(x_i) - 0}{x_i - x_{i+1}}$$

which can be rearranged to yield

$$x_{i+1} = x_i - \frac{f(x_i)}{f'(x_i)} \tag{6.6}$$

which is called the *Newton-Raphson formula*.

EXAMPLE 6.2   Newton-Raphson Method

Problem Statement.   Use the Newton-Raphson method to estimate the root of $f(x) = e^{-x} - x$ employing an initial guess of $x_0 = 0$.

Solution.   The first derivative of the function can be evaluated as

$$f'(x) = -e^{-x} - 1$$

which can be substituted along with the original function into Eq. (6.6) to give

$$x_{i+1} = x_i - \frac{e^{-x_i} - x_i}{-e^{-x_i} - 1}$$

Starting with an initial guess of $x_0 = 0$, this iterative equation can be applied to compute

i	x_i	$|\varepsilon_t|$, %
0	0	100
1	0.500000000	11.8
2	0.566311003	0.147
3	0.567143165	0.0000220
4	0.567143290	$<10^{-8}$

Thus, the approach rapidly converges on the true root. Notice that the true percent relative error at each iteration decreases much faster than it does in simple fixed-point iteration (compare with Example 6.1).

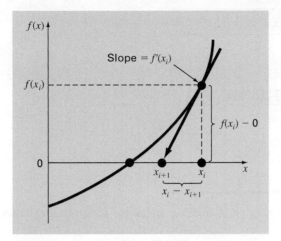

FIGURE 6.4
Graphical depiction of the Newton-Raphson method. A tangent to the function of x_i [that is, $f'(x)$] is extrapolated down to the x axis to provide an estimate of the root at x_{i+1}.

As with other root-location methods, Eq. (6.3) can be used as a termination criterion. In addition, a theoretical analysis (Chapra and Canale, 2002) provides insight regarding the rate of convergence as expressed by

$$E_{t,i+1} = \frac{-f''(x_r)}{2f'(x_r)} E_{t,i}^2 \qquad (6.7)$$

Thus, the error should be roughly proportional to the square of the previous error. In other words, the number of significant figures of accuracy approximately doubles with each iteration. This behavior is called *quadratic convergence* and is one of the major reasons for the popularity of the method.

Although the Newton-Raphson method is often very efficient, there are situations where it performs poorly. A special case—multiple roots—is discussed elsewhere (Chapra and Canale, 2002). However, even when dealing with simple roots, difficulties can also arise, as in the following example.

EXAMPLE 6.3 A Slowly Converging Function with Newton-Raphson

Problem Statement. Determine the positive root of $f(x) = x^{10} - 1$ using the Newton-Raphson method and an initial guess of $x = 0.5$.

Solution. The Newton-Raphson formula for this case is

$$x_{i+1} = x_i - \frac{x_i^{10} - 1}{10x_i^9}$$

which can be used to compute

i	x_i	$\lvert \varepsilon_a \rvert$, %
0	0.5	
1	51.65	99.032
2	46.485	11.111
3	41.8365	11.111
4	37.65285	11.111
⋮		
40	1.002316	2.130
41	1.000024	0.229
42	1	0.002

Thus, after the first poor prediction, the technique is converging on the true root of 1, but at a very slow rate.

Why does this happen? As shown in Fig. 6.5, a simple plot of the first few iterations is helpful in providing insight. Notice how the first guess is in a region where the slope is near zero. Thus, the first iteration flings the solution far away from the initial guess to a new value ($x = 51.65$) where $f(x)$ has an extremely high value. The solution then plods along for over 40 iterations until converging on the root with adequate accuracy.

FIGURE 6.5
Graphical depiction of the Newton-Raphson method for a case with slow convergence. The inset shows how a near-zero slope initially shoots the solution far from the root. Thereafter, the solution very slowly converges on the root.

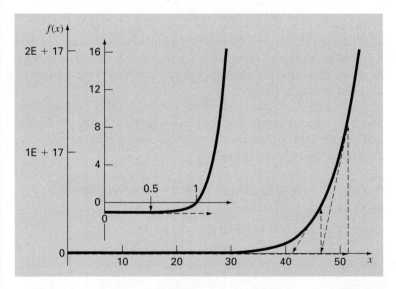

Aside from slow convergence due to the nature of the function, other difficulties can arise, as illustrated in Fig. 6.6. For example, Fig. 6.6a depicts the case where an inflection

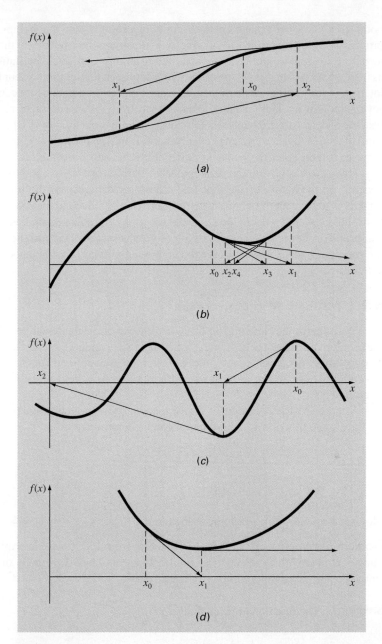

FIGURE 6.6
Four cases where the Newton-Raphson method exhibits poor convergence.

point (i.e., $f'(x) = 0$) occurs in the vicinity of a root. Notice that iterations beginning at x_0 progressively diverge from the root. Fig. 6.6b illustrates the tendency of the Newton-Raphson technique to oscillate around a local maximum or minimum. Such oscillations may persist, or, as in Fig. 6.6b, a near-zero slope is reached whereupon the solution is sent far from the area of interest. Figure 6.6c shows how an initial guess that is close to one root can jump to a location several roots away. This tendency to move away from the area of interest is due to the fact that near-zero slopes are encountered. Obviously, a zero slope $[f'(x) = 0]$ is a real disaster because it causes division by zero in the Newton-Raphson formula [Eq. (6.6)]. As in Fig. 6.6d, it means that the solution shoots off horizontally and never hits the x axis.

Thus, there is no general convergence criterion for Newton-Raphson. Its convergence depends on the nature of the function and on the accuracy of the initial guess. The only remedy is to have an initial guess that is "sufficiently" close to the root. And for some functions, no guess will work! Good guesses are usually predicated on knowledge of the physical problem setting or on devices such as graphs that provide insight into the behavior of the solution. It also suggests that good computer software should be designed to recognize slow convergence or divergence.

6.2.1 MATLAB M-file: `newtraph`

An algorithm for the Newton-Raphson method can be easily developed (Fig. 6.7). Note that the program must have access to the function (`func`) and its first derivative (`dfunc`). These can be simply accomplished by the inclusion of user-defined functions to compute these quantities. Alternatively, as in the algorithm in Fig. 6.7, they can be passed to the function as arguments.

After the M-file is entered and saved, it can be invoked to solve for root. For example, for the simple function $x^2 - 9$, the root can be determined as in

```
>> newtraph(@(x) x^2-9,@(x) 2*x,5)

ans =
     3
```

EXAMPLE 6.4   Newton-Raphson Bungee Jumper Problem

Problem Statement.   Use the M-file function from Fig. 6.7 to determine the mass of the bungee jumper with a drag coefficient of 0.25 kg/m to have a velocity of 36 m/s after 4 s of free fall. The acceleration of gravity is 9.81 m/s^2.

Solution.   The function to be evaluated is

$$f(m) = \sqrt{\frac{gm}{c_d}} \tanh\left(\sqrt{\frac{gc_d}{m}} t\right) - v(t) \tag{E6.4.1}$$

To apply the Newton-Raphson method, the derivative of this function must be evaluated with respect to the unknown, m:

$$\frac{df(m)}{dm} = \frac{1}{2}\sqrt{\frac{g}{mc_d}} \tanh\left(\sqrt{\frac{gc_d}{m}} t\right) - \frac{g}{2m} t \, \text{sech}^2\left(\sqrt{\frac{gc_d}{m}} t\right) \tag{E6.4.2}$$

```
function [root,ea,iter]=newtraph(func,dfunc,xr,es,maxit,varargin)
% newtraph: Newton-Raphson root location zeroes
%   [root,ea,iter]=newtraph(func,dfunc,xr,es,maxit,p1,p2,...):
%         uses Newton-Raphson method to find the root of func
% input:
%   func = name of function
%   dfunc = name of derivative of function
%   xr = initial guess
%   es = desired relative error (default = 0.0001%)
%   maxit = maximum allowable iterations (default = 50)
%   p1,p2,... = additional parameters used by function
% output:
%   root = real root
%   ea = approximate relative error (%)
%   iter = number of iterations

if nargin<3,error('at least 3 input arguments required'),end
if nargin<4|isempty(es),es=0.0001;end
if nargin<5|isempty(maxit),maxit=50;end
iter = 0;
while (1)
  xrold = xr;
  xr = xr - func(xr)/dfunc(xr);
  iter = iter + 1;
  if xr ~= 0, ea = abs((xr - xrold)/xr) * 100; end
  if ea <= es | iter >= maxit, break, end
end
root = xr;
```

FIGURE 6.7
An M-file to implement the Newton-Raphson method.

We should mention that although this derivative is not difficult to evaluate in principle, it involves a bit of concentration and effort to arrive at the final result.

The two formulas can now be used in conjunction with the function newtraph to evaluate the root:

```
>> y = @m sqrt(9.81*m/0.25)*tanh(sqrt(9.81*0.25/m)*4)-36;
>> dy = @m 1/2*sqrt(9.81/(m*0.25))*tanh((9.81*0.25/m) ...
        ^(1/2)*4)-9.81/(2*m)*sech(sqrt(9.81*0.25/m)*4)^2;

>> newtraph(y,dy,140,0.00001)

ans =
  142.7376
```

6.3 SECANT METHODS

As in Example 6.4, a potential problem in implementing the Newton-Raphson method is the evaluation of the derivative. Although this is not inconvenient for polynomials and many other functions, there are certain functions whose derivatives may be difficult or inconvenient to evaluate. For these cases, the derivative can be approximated by a backward finite divided difference:

$$f'(x_i) \cong \frac{f(x_{i-1}) - f(x_i)}{x_{i-1} - x_i}$$

This approximation can be substituted into Eq. (6.6) to yield the following iterative equation:

$$x_{i+1} = x_i - \frac{f(x_i)(x_{i-1} - x_i)}{f(x_{i-1}) - f(x_i)} \tag{6.8}$$

Equation (6.8) is the formula for the *secant method*. Notice that the approach requires two initial estimates of x. However, because $f(x)$ is not required to change signs between the estimates, it is not classified as a bracketing method.

Rather than using two arbitrary values to estimate the derivative, an alternative approach involves a fractional perturbation of the independent variable to estimate $f'(x)$,

$$f'(x_i) \cong \frac{f(x_i + \delta x_i) - f(x_i)}{\delta x_i} \text{ m}$$

where δ = a small perturbation fraction. This approximation can be substituted into Eq. (6.6) to yield the following iterative equation:

$$x_{i+1} = x_i - \frac{\delta x_i f(x_i)}{f(x_i + \delta x_i) - f(x_i)} \tag{6.9}$$

We call this the *modified secant method*. As in the following example, it provides a nice means to attain the efficiency of Newton-Raphson without having to compute derivatives.

EXAMPLE 6.5 Modified Secant Method

Problem Statement. Use the modified secant method to determine the mass of the bungee jumper with a drag coefficient of 0.25 kg/m to have a velocity of 36 m/s after 4 s of free fall. Note: The acceleration of gravity is 9.81 m/s^2. Use an initial guess of 50 kg and a value of 10^{-6} for the perturbation fraction.

Solution. Inserting the parameters into Eq. (6.9) yields

First iteration:

$x_0 = 50$ $\qquad\qquad\qquad f(x_0) = -4.57938708$

$x_0 + \delta x_0 = 50.00005$ $\qquad f(x_0 + \delta x_0) = -4.579381118$

$x_1 = 50 - \dfrac{10^{-6}(50)(-4.57938708)}{-4.579381118 - (-4.57938708)}$

$= 88.39931 (|\varepsilon_t| = 38.1\%; |\varepsilon_a| = 43.4\%)$

Second iteration:

$$x_1 = 88.39931 \qquad\qquad f(x_1) = -1.69220771$$

$$x_1 + \delta x_1 = 88.39940 \qquad f(x_1 + \delta x_1) = -1.692203516$$

$$x_2 = 88.39931 - \frac{10^{-6}(88.39931)(-1.69220771)}{-1.692203516 - (-1.69220771)}$$

$$= 124.08970(|\varepsilon_t| = 13.1\%; |\varepsilon_a| = 28.76\%)$$

The calculation can be continued to yield

i	x_i	$|\varepsilon_t|$, %	$|\varepsilon_a|$, %
0	50.0000	64.971	
1	88.3993	38.069	43.438
2	124.0897	13.064	28.762
3	140.5417	1.538	11.706
4	142.7072	0.021	1.517
5	142.7376	4.1×10^{-6}	0.021
6	142.7376	3.4×10^{-12}	4.1×10^{-6}

The choice of a proper value for δ is not automatic. If δ is too small, the method can be swamped by round-off error caused by subtractive cancellation in the denominator of Eq. (6.9). If it is too big, the technique can become inefficient and even divergent. However, if chosen correctly, it provides a nice alternative for cases where evaluating the derivative is difficult and developing two initial guesses is inconvenient.

Further, in its most general sense, a univariate function is merely an entity that returns a single value in return for values sent to it. Perceived in this sense, functions are not always simple formulas like the one-line equations solved in the preceding examples in this chapter. For example, a function might consist of many lines of code that could take a significant amount of execution time to evaluate. In some cases, the function might even represent an independent computer program. For such cases, the secant and modified secant methods are valuable.

6.4 MATLAB FUNCTION: fzero

The methods we have described to this point are either reliable but slow (bracketing) or fast but possibly unreliable (open). The MATLAB fzero function provides the best qualities of both. The fzero function is designed to find the real root of a single equation. A simple representation of its syntax is

```
fzero(function,x0)
```

where *function* is the name of the function being evaluated, and *x0* is the initial guess. Note that two guesses that bracket the root can be passed as a vector:

```
fzero(function,[x0 x1])
```

where *x0* and *x1* are guesses that bracket a sign change.

Here is a simple MATLAB session that solves for the root of a simple quadratic: $x^2 - 9$. Clearly two roots exist at -3 and 3. To find the negative root:

```
>> x = fzero(@(x) x^2-9,-4)

x =
    -3
```

If we want to find the positive root, use a guess that is near it:

```
>> x = fzero(@(x) x^2-9,4)

x =
    3
```

If we put in an initial guess of zero, it finds the negative root

```
>> x = fzero(@(x) x^2-9,0)

x =
    -3
```

If we wanted to ensure that we found the positive root, we could enter two guesses as in

```
>> x = fzero(@(x) x^2-9,[0 4])

x =
    3
```

Also, if a sign change does not occur between the two guesses, an error message is displayed

```
>> x = fzero(@(x) x^2-9,[-4 4])

??? Error using ==> fzero
The function values at the interval endpoints must ...
differ in sign.
```

The `fzero` function is a combination of the reliable bisection method with two faster algorithms: the secant method and inverse quadratic interpolation. *Inverse quadratic interpolation* is similar in spirit to the secant method. As in Fig. 6.8a, the secant method is based on computing a straight line that goes through two guesses. The intersection of this straight line with the x axis represents the new root estimate. The inverse quadratic interpolation uses a similar strategy but is based on computing a quadratic equation (i.e., a parabola) that goes through three points (Fig. 6.8b).

The `fzero` function works as follows. If a single initial guess is passed, it first performs a search to identify a sign change. This search differs from the incremental search described in Section 5.3.1, in that the search starts at the single initial guess and then takes increasingly bigger steps in both the positive and negative directions until a sign change is detected.

Thereafter, the fast methods (secant and inverse quadratic interpolation) are used unless an unacceptable result occurs (e.g., the root estimate falls outside the bracket). If a bad result happens, bisection is implemented until an acceptable root is obtained with one of

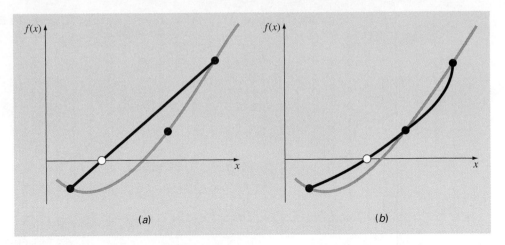

FIGURE 6.8
Comparison of (a) the secant method and (b) inverse quadratic interpolation. Note that the approach in (b) is called "inverse" because the quadratic function is written in y rather than in x.

the fast methods. As might be expected, bisection typically dominates at first but as the root is approached, the technique shifts to the faster methods.

A more complete representation of the fzero syntax can be written as

```
[x,fx] = fzero(function,x0,options,p1,p2,...)
```

where [xr, fx] = a vector containing the root x and the function evaluated at the root fx, options is a data structure created by the optimset function, and p1, p2... are any parameters that the function requires. Note that if you desire to pass in parameters but not use the options, pass an empty vector [] in its place.

The optimset function has the syntax

```
options = optimset('par₁',val₁,'par₂',val₂,...)
```

where the parameter par_i has the value val_i. A complete listing of all the possible parameters can be obtained by merely entering optimset at the command prompt. The parameters that are commonly used with the fzero function are

display: When set to 'iter' displays a detailed record of all the iterations.

tolx: A positive scalar that sets a termination tolerance on x.

EXAMPLE 6.6   The fzero and optimset Functions

Problem Statement.   Recall that in Example 6.3, we found the positive root of $f(x) = x^{10} - 1$ using the Newton-Raphson method with an initial guess of 0.5. Solve the same problem with optimset and fzero.

Solution.   An interactive MATLAB session can be implemented as follows:

```
>> options = optimset('display','iter');
>> [x,fx] = fzero(@(x) x^10-1,0.5,options)
```

```
 Func-count       x            f(x)        Procedure
     1            0.5        -0.999023      initial
     2         0.485858      -0.999267      search
     3         0.514142      -0.998709      search
     4          0.48         -0.999351      search
     5          0.52         -0.998554      search
     6         0.471716      -0.999454      search
                   .
                   .
                   .
    23         0.952548      -0.385007      search
    24          -0.14           -1          search
    25          1.14         2.70722        search
 Looking for a zero in the interval [-0.14, 1.14]
    26         0.205272         -1          interpolation
    27         0.672636      -0.981042      bisection
    28         0.906318      -0.626056      bisection
    29         1.02316       0.257278       bisection
    30         0.989128      -0.103551      interpolation
    31         0.998894     -0.0110017      interpolation
    32         1.00001    7.68385e-005      interpolation
    33            1      -3.83061e-007      interpolation
    34            1      -1.3245e-011       interpolation
    35            1            0            interpolation
 Zero found in the interval: [-0.14, 1.14].

 x =
      1

 fx =
      0
```

Thus, after 25 iterations of searching, `fzero` finds a sign change. It then uses interpolation and bisection until it gets close enough to the root so that interpolation takes over and rapidly converges on the root.

Suppose that we would like to use a less stringent tolerance. We can use the `optimset` function to set a low maximum tolerance and a less accurate estimate of the root results:

```
>> options = optimset ('tolx', 1e-3);
>> [x,fx] = fzero(@(x) x^10-1,0.5,options)

x =
   1.0009

fx =
   0.0090
```

6.5 POLYNOMIALS

Polynomials are a special type of nonlinear algebraic equation of the general form

$$f_n(x) = a_1 x^n + a_2 x^{n-1} + \cdots + a_{n-1} x^2 + a_n x + a_{n+1} \tag{6.10}$$

where n is the order of the polynomial, and the a's are constant coefficients. In many (but not all) cases, the coefficients will be real. For such cases, the roots can be real and/or complex. In general, an nth order polynomial will have n roots.

Polynomials have many applications in engineering and science. For example, they are used extensively in curve fitting. However, one of their most interesting and powerful applications is in characterizing dynamic systems—and, in particular, linear systems. Examples include reactors, mechanical devices, structures, and electrical circuits.

6.5.1 MATLAB Function: `roots`

If you are dealing with a problem where you must determine a single real root of a polynomial, the techniques such as bisection and the Newton-Raphson method can have utility. However, in many cases, engineers desire to determine all the roots, both real and complex. Unfortunately, simple techniques like bisection and Newton-Raphson are not available for determining all the roots of higher-order polynomials. However, MATLAB has an excellent built-in capability, the `roots` function, for this task.

The `roots` function has the syntax,

```
x = roots(c)
```

where x is a column vector containing the roots and c is a row vector containing the polynomial's coefficients.

So how does the `roots` function work? MATLAB is very good at finding the eigenvalues of a matrix. Consequently, the approach is to recast the root evaluation task as an eigenvalue problem. Because we will be describing eigenvalue problems later in the book, we will merely provide an overview here.

Suppose we have a polynomial

$$a_1 x^5 + a_2 x^4 + a_3 x^3 + a_4 x^2 + a_5 x + a_6 = 0 \tag{6.11}$$

Dividing by a_1 and rearranging yields

$$x^5 = -\frac{a_2}{a_1}x^4 - \frac{a_3}{a_1}x^3 - \frac{a_4}{a_1}x^2 - \frac{a_5}{a_1}x - \frac{a_6}{a_1} \tag{6.12}$$

A special matrix can be constructed by using the coefficients from the right-hand side as the first row and with 1's and 0's written for the other rows as shown:

$$\begin{bmatrix} -a_2/a_1 & -a_3/a_1 & -a_4/a_1 & -a_5/a_1 & -a_6/a_1 \\ 1 & 0 & 0 & 0 & 0 \\ 0 & 1 & 0 & 0 & 0 \\ 0 & 0 & 1 & 0 & 0 \\ 0 & 0 & 0 & 1 & 0 \end{bmatrix} \tag{6.13}$$

Equation (6.13) is called the polynomial's *companion matrix*. It has the useful property that its eigenvalues are the roots of the polynomial. Thus, the algorithm underlying the `roots` function consists of merely setting up the companion matrix and then using MATLAB's powerful eigenvalue evaluation function to determine the roots. Its application, along with some other related polynomial manipulation functions, are described in the following example.

We should note that `roots` has an inverse function called `poly`, which when passed the values of the roots, will return the polynomial's coefficients. Its syntax is

```
c = poly(r)
```

where r is a column vector containing the roots and c is a row vector containing the polynomial's coefficients.

EXAMPLE 6.7   Using MATLAB to Manipulate Polynomials and Determine Their Roots

Problem Statement.   Use the following equation to explore how MATLAB can be employed to manipulate polynomials:

$$f_5(x) = x^5 - 3.5x^4 + 2.75x^3 + 2.125x^2 - 3.875x + 1.25 \tag{E6.7.1}$$

Note that this polynomial has three real roots: 0.5, −1.0, and 2; and one pair of complex roots: $1 \pm 0.5i$.

Solution.   Polynomials are entered into MATLAB by storing the coefficients as a row vector. For example, entering the following line stores the coefficients in the vector `a`:

```
>> a = [1 -3.5 2.75 2.125 -3.875 1.25];
```

We can then proceed to manipulate the polynomial. For example we can evaluate it at $x = 1$, by typing

```
>> polyval(a,1)
```

with the result, $1(1)^5 - 3.5(1)^4 + 2.75(1)^3 + 2.125(1)^2 - 3.875(1) + 1.25 = -0.25$:

```
ans =
   -0.2500
```

We can create a quadratic polynomial that has roots corresponding to two of the original roots of Eq. (E6.7.1): 0.5 and −1. This quadratic is $(x - 0.5)(x + 1) = x^2 + 0.5x - 0.5$. It can be entered into MATLAB as the vector `b`:

```
>> b = [1 .5 -.5]

b =
   1.0000    0.5000    -0.5000
```

Note that the `poly` function can be used to perform the same task as in

```
>> b = poly([0.5 -1])

b =
   1.0000    0.5000    -0.5000
```

We can divide this polynomial into the original polynomial by

```
>> [q,r] = deconv(a,b)
```

with the result being a quotient (a third-order polynomial, `q`) and a remainder (`r`)

```
q =
   1.0000    -4.0000    5.2500    -2.5000
r =
      0       0       0       0       0       0
```

Because the polynomial is a perfect divisor, the remainder polynomial has zero coefficients. Now, the roots of the quotient polynomial can be determined as

```
>> x = roots(q)
```

with the expected result that the remaining roots of the original polynomial Eq. (E6.7.1) are found:

```
x =
2.0000
1.0000 + 0.5000i
1.0000 - 0.5000i
```

We can now multiply q by b to come up with the original polynomial:

```
>> a = conv(q,b)

a =
    1.0000   -3.5000    2.7500    2.1250   -3.8750    1.2500
```

We can then determine all the roots of the original polynomial by

```
>> x = roots(a)

x =
    2.0000
   -1.0000
    1.0000 + 0.5000i
    1.0000 - 0.5000i
    0.5000
```

Finally, we can return to the original polynomial again by using the `poly` function:

```
>> a = poly(x)

a =
    1.0000   -3.5000    2.7500    2.1250   -3.8750    1.2500
```

6.6 CASE STUDY   PIPE FRICTION

Background.   Determining fluid flow through pipes and tubes has great relevance in many areas of engineering and science. In engineering, typical applications include the flow of liquids and gases through pipelines and cooling systems. Scientists are interested in topics ranging from flow in blood vessels to nutrient transmission through a plant's vascular system.

The resistance to flow in such conduits is parameterized by a dimensionless number called the *friction factor*. For turbulent flow, the *Colebrook equation* provides a means to calculate the friction factor:

$$0 = \frac{1}{\sqrt{f}} + 2.0 \log\left(\frac{\varepsilon}{3.7D} + \frac{2.51}{\text{Re}\sqrt{f}}\right) \tag{6.14}$$

6.6 CASE STUDY   continued

where ε = the roughness (m), D = diameter (m), and Re = the *Reynolds number:*

$$\text{Re} = \frac{\rho V D}{\mu}$$

where ρ = the fluid's density (kg/m^3), V = its velocity (m/s), and μ = dynamic viscosity (N$\cdot$s/m^2). In addition to appearing in Eq. (6.14), the Reynolds number also serves as the criterion for whether flow is turbulent (Re > 4000).

In this case study, we will illustrate how the numerical methods covered in this part of the book can be employed to determine f for air flow through a smooth, thin tube. For this case, the parameters are $\rho = 1.23$ kg/m^3, $\mu = 1.79 \times 10^{-5}$ N$\cdot$s/m^2, $D = 0.005$ m, $V = 40$ m/s and $\varepsilon = 0.0015$ mm. Note that friction factors range from about 0.008 to 0.08. In addition, an explicit formulation called the *Swamee-Jain equation* provides an approximate estimate:

$$f = \frac{1.325}{\left[\ln\left(\dfrac{\varepsilon}{3.7D} + \dfrac{5.74}{\text{Re}^{0.9}}\right)\right]^2} \tag{6.15}$$

Solution.   The Reynolds number can be computed as

$$\text{Re} = \frac{\rho V D}{\mu} = \frac{1.23(40)0.005}{1.79 \times 10^{-5}} = 13{,}743$$

This value along with the other parameters can be substituted into Eq. (6.14) to give

$$g(f) = \frac{1}{\sqrt{f}} + 2.0 \log\left(\frac{0.0000015}{3.7(0.005)} + \frac{2.51}{13{,}743\sqrt{f}}\right)$$

Before determining the root, it is advisable to plot the function to estimate initial guesses and to anticipate possible difficulties. This can be done easily with MATLAB:

```
>> rho=1.23;mu=1.79e-5;D=0.005;V=40;e=0.0015/1000;
>> Re=rho*V*D/mu;
>> g=@(f) 1/sqrt(f)+2*log10(e/(3.7*D)+2.51/(Re*sqrt(f)));
>> fplot(g,[0.008 0.08]),grid,xlabel('f'),ylabel('g(f)')
```

As in Fig. 6.9, the root is located at about 0.03.

Because we are supplied initial guesses ($x_l = 0.008$ and $x_u = 0.08$), either of the bracketing methods from Chap. 5 could be used. For example, the `bisect` function developed in Fig. 5.7 gives a value of $f = 0.0289678$ with a percent relative error of error of 5.926×10^{-5} in 22 iterations. False position yields a result of similar precision in 26 iterations. Thus, although they produce the correct result, they are somewhat inefficient. This would not be important for a single application, but could become prohibitive if many evaluations were made.

6.6 CASE STUDY   continued

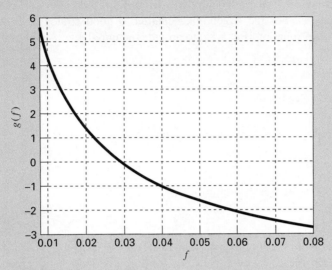

FIGURE 6.9

___

We could try to attain improved performance by turning to an open method. Because Eq. (6.14) is relatively straightforward to differentiate, the Newton-Raphson method is a good candidate. For example, using an initial guess at the lower end of the range ($x_0 = 0.008$), the `newtraph` function developed in Fig. 6.7 converges quickly:

```
>> dg=@(f) -2/log(10)*1.255/Re*f^(-3/2)/(e/D/3.7 ...
               +2.51/Re/sqrt(f))-0.5/f^(3/2);
>> [f ea iter]=newtraph(g,dg,0.008)

f =
    0.02896781017144
ea =
    6.870124190058040e-006
iter =
    6
```

However, when the initial guess is set at the upper end of the range ($x_0 = 0.08$), the routine diverges,

```
>> [f ea iter]=newtraph(g,dg,0.08)

f =
             NaN +            NaNi
```

As can be seen by inspecting Fig. 6.9, this occurs because the function's slope at the initial guess causes the first iteration to jump to a negative value. Further runs demonstrate that for this case, convergence only occurs when the initial guess is below about 0.066.

So we can see that although the Newton-Raphson is very efficient, it requires good initial guesses. For the Colebrook equation, a good strategy might be to employ the Swamee-Jain equation (Eq. 6.15) to provide the initial guess as in

```
>> fSJ=1.325/log(e/(3.7*D)+5.74/Re^0.9)^2

fSJ =
   0.02903099711265

>> [f ea iter]=newtraph(g,dg,fSJ)

f =
   0.02896781017144
ea =
   8.510189472800060e-010
iter =
   3
```

Aside from our homemade functions, we can also use MATLAB's built-in `fzero` function. However, just as with the Newton-Raphson method, divergence also occurs when `fzero` function is used with a single guess. However, in this case, guesses at the lower end of the range cause problems. For example,

```
>> fzero(g,0.008)

Exiting fzero: aborting search for an interval containing a sign
change because complex function value encountered ...
                                    during search.
(Function value at -0.0028 is -4.92028-20.2423i.)
Check function or try again with a different starting value.
ans =
   NaN
```

If the iterations are displayed using `optimset` (recall Example 6.6), it is revealed that a negative value occurs during the search phase before a sign change is detected and the routine aborts. However, for single initial guesses above about 0.016, the routine works nicely. For example, for the guess of 0.08 that caused problems for Newton-Raphson, `fzero` does just fine:

```
>> fzero(g,0.08)

ans =
   0.02896781017144
```

As a final note, let's see whether convergence is possible for simple fixed-point iteration. The easiest and most straightforward version involves solving for the first f in Eq. (6.14):

$$f_{i+1} = \frac{0.25}{\left(\log\left(\dfrac{\varepsilon}{3.7D} + \dfrac{2.51}{\text{Re}\sqrt{f_i}}\right)\right)^2} \tag{6.16}$$

6.6 CASE STUDY   continued

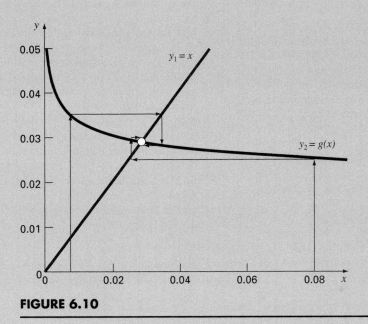

FIGURE 6.10

The two-curve display of this function depicted indicates a surprising result (Fig. 6.10). Recall that fixed-point iteration converges when the y_2 curve has a relatively flat slope (i.e., $|g'(\xi)| < 1$). As indicated by Fig. 6.10, the fact that the y_2 curve is quite flat in the range from $f = 0.008$ to 0.08 means that not only does fixed-point iteration converge, but it converges fairly rapidly! In fact, for initial guesses anywhere between 0.008 and 0.08, fixed-point iteration yields predictions with percent relative errors less than 0.008% in six or fewer iterations! Thus, this simple approach that requires only one guess and no derivative estimates performs really well for this particular case.

The take-home message from this case study is that even great, professionally developed software like MATLAB is not always foolproof. Further, there is usually no single method that works best for all problems. Sophisticated users understand the strengths and weaknesses of the available numerical techniques. In addition, they understand enough of the underlying theory so that they can effectively deal with situations where a method breaks down.

PROBLEMS

6.1 Employ fixed-point iteration to locate the root of

$$f(x) = 2\sin(\sqrt{x}) - x$$

Use an initial guess of $x_0 = 0.5$ and iterate until $\varepsilon_a \leq 0.01\%$.

6.2 Use **(a)** fixed-point iteration and **(b)** the Newton-Raphson method to determine a root of $f(x) = -x^2 + 1.8x + 2.5$ using $x_0 = 5$. Perform the computation until ε_a is less than $\varepsilon_s = 0.05\%$. Also check your final answer.

6.3 Determine the highest real root of $f(x) = 0.95x^3 - 5.9x^2 + 10.9x - 6$:

(a) Graphically.

(b) Using the Newton-Raphson method (three iterations, $x_i = 3.5$).

(c) Using the secant method (three iterations, $x_{i-1} = 2.5$ and $x_i = 3.5$).

(d) Using the modified secant method (three iterations, $x_i = 3.5$, $\delta = 0.01$).

(e) Determine all the roots with MATLAB.

6.4 Determine the lowest positive root of $f(x) = 8\sin(x)e^{-x} - 1$:

(a) Graphically.

(b) Using the Newton-Raphson method (three iterations, $x_i = 0.3$).

(c) Using the secant method (three iterations, $x_{i-1} = 0.5$ and $x_i = 0.4$).

(d) Using the modified secant method (five iterations, $x_i = 0.3$, $\delta = 0.01$).

6.5 Use **(a)** the Newton-Raphson method and **(b)** the modified secant method ($\delta = 0.05$) to determine a root of $f(x) = x^5 - 16.05x^4 + 88.75x^3 - 192.0375x^2 + 116.35x + 31.6875$ using an initial guess of $x = 0.5825$ and $\varepsilon_s = 0.01\%$. Explain your results.

6.6 Develop an M-file for the secant method. Along with the two initial guesses, pass the function as an argument. Test it by solving Prob. 6.3.

6.7 Develop an M-file for the modified secant method. Along with the initial guess and the perturbation fraction, pass the function as an argument. Test it by solving Prob. 6.3.

6.8 Differentiate Eq. (E6.4.1) to get Eq. (E6.4.2).

6.9 Employ the Newton-Raphson method to determine a real root for $f(x) = -1 + 6x - 4x^2 + 0.5x^3$, using an initial guess of **(a)** 4.5, and **(b)** 4.43. Discuss and use graphical and analytical methods to explain any peculiarities in your results.

6.10 The "divide and average" method, an old-time method for approximating the square root of any positive number a, can be formulated as

$$x_{i+1} = \frac{x_i + a/x_i}{2}$$

Prove that this formula is based on the Newton-Raphson algorithm.

6.11 **(a)** Apply the Newton-Raphson method to the function $f(x) = \tanh(x^2 - 9)$ to evaluate its known real root at $x = 3$. Use an initial guess of $x_0 = 3.1$ and take a minimum of three iterations. **(b)** Did the method exhibit convergence onto its real root? Sketch the plot with the results for each iteration labeled.

6.12 The polynomial $f(x) = 0.0074x^4 - 0.284x^3 + 3.355x^2 - 12.183x + 5$ has a real root between 15 and 20. Apply the Newton-Raphson method to this function using an initial guess of $x_0 = 16.15$. Explain your results.

6.13 In a chemical engineering process, water vapor (H_2O) is heated to sufficiently high temperatures that a significant portion of the water dissociates, or splits apart, to form oxygen (O_2) and hydrogen (H_2):

$$H_2O \rightleftarrows H_2 + \tfrac{1}{2}O_2$$

If it is assumed that this is the only reaction involved, the mole fraction x of H_2O that dissociates can be represented by

$$K = \frac{x}{1-x}\sqrt{\frac{2p_t}{2+x}} \qquad \text{(P6.13.1)}$$

where K is the reaction's equilibrium constant and p_t is the total pressure of the mixture. If $p_t = 3.5$ atm and $K = 0.04$, determine the value of x that satisfies Eq. (P6.13.1).

6.14 The Redlich-Kwong equation of state is given by

$$p = \frac{RT}{v-b} - \frac{a}{v(v+b)\sqrt{T}}$$

where R = the universal gas constant [= 0.518 kJ/(kg K)], T = absolute temperature (K), p = absolute pressure (kPa), and v = the volume of a kg of gas (m^3/kg). The parameters a and b are calculated by

$$a = 0.427\frac{R^2 T_c^{2.5}}{p_c} \qquad b = 0.0866R\frac{T_c}{p_c}$$

where $p_c = 4580$ kPa and $T_c = 191$ K. As a chemical engineer, you are asked to determine the amount of methane fuel that can be held in a 3-m^3 tank at a temperature of $-50\ ^\circ$C with a pressure of 65,000 kPa. Use a root locating method of your choice to calculate v and then determine the mass of methane contained in the tank.

6.15 The volume of liquid V in a hollow horizontal cylinder of radius r and length L is related to the depth of the liquid h by

$$V = \left[r^2\cos^{-1}\left(\frac{r-h}{r}\right) - (r-h)\sqrt{2rh - h^2}\right]L$$

Determine h given $r = 2$ m, $L = 5\ m^3$, and $V = 8.5\ m^3$.

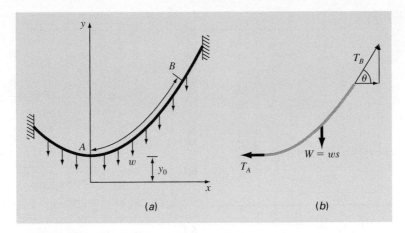

FIGURE P6.16

6.16 A catenary cable is one which is hung between two points not in the same vertical line. As depicted in Fig. P6.16a, it is subject to no loads other than its own weight. Thus, its weight acts as a uniform load per unit length along the cable w (N/m). A free-body diagram of a section AB is depicted in Fig. P6.16b, where T_A and T_B are the tension forces at the end. Based on horizontal and vertical force balances, the following differential equation model of the cable can be derived:

$$\frac{d^2y}{dx^2} = \frac{w}{T_A}\sqrt{1 + \left(\frac{dy}{dx}\right)^2}$$

Calculus can be employed to solve this equation for the height of the cable y as a function of distance x:

$$y = \frac{T_A}{w}\cosh\left(\frac{w}{T_A}x\right) + y_0 - \frac{T_A}{w}$$

(a) Use a numerical method to calculate a value for the parameter T_A given values for the parameters $w = 12$ and $y_0 = 6$, such that the cable has a height of $y = 15$ at $x = 50$.

(b) Develop a plot of y versus x for $x = -50$ to 100.

6.17 An oscillating current in an electric circuit is described by $I = 9e^{-t}\cos(2\pi t)$, where t is in seconds. Determine all values of t such that $I = 3$.

6.18 Figure P6.18 shows a circuit with a resistor, an inductor, and a capacitor in parallel. Kirchhoff's rules can be used to express the impedance of the system as

$$\frac{1}{Z} = \sqrt{\frac{1}{R^2} + \left(\omega C - \frac{1}{\omega L}\right)^2}$$

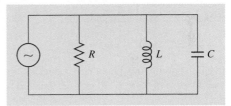

FIGURE P6.18

where Z = impedance (Ω), and ω is the angular frequency. Find the ω that results in an impedance of 75 Ω using the `fzero` function with initial guesses of 1 and 1000 for the following parameters: $R = 225\,\Omega$, $C = 0.6 \times 10^{-6}$ F, and $L = 0.5$ H.

6.19 Real mechanical systems may involve the deflection of nonlinear springs. In Fig. P6.19, a block of mass m is

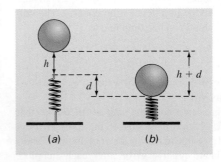

FIGURE P6.19

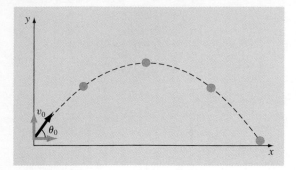

FIGURE P6.20

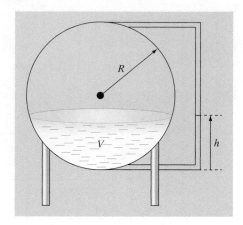

FIGURE P6.21

released a distance h above a nonlinear spring. The resistance force F of the spring is given by

$$F = -(k_1 d + k_2 d^{3/2})$$

Conservation of energy can be used to show that

$$0 = \frac{2k_2 d^{5/2}}{5} + \frac{1}{2}k_1 d^2 - mgd - mgh$$

Solve for d, given the following parameter values: $k_1 = 50,000$ g/s², $k_2 = 40$ g/(s² m⁵), $m = 90$ g, $g = 9.81$ m/s², and $h = 0.45$ m.

6.20 Aerospace engineers sometimes compute the trajectories of projectiles such as rockets. A related problem deals with the trajectory of a thrown ball. The trajectory of a ball thrown by a right fielder is defined by the (x, y) coordinates as displayed in Fig. P6.20. The trajectory can be modeled as

$$y = (\tan \theta_0)x - \frac{g}{2v_0^2 \cos^2 \theta_0}x^2 + y_0$$

Find the appropriate initial angle θ_0, if $v_0 = 20$ m/s, and the distance to the catcher is 35 m. Note that the throw leaves the right fielder's hand at an elevation of 2 m and the catcher receives it at 1 m.

6.21 You are designing a spherical tank (Fig. P6.21) to hold water for a small village in a developing country. The volume of liquid it can hold can be computed as

$$V = \pi h^2 \frac{[3R - h]}{3}$$

where V = volume [m³], h = depth of water in tank [m], and R = the tank radius [m].

If $R = 3$ m, what depth must the tank be filled to so that it holds 30 m³? Use three iterations of the most efficient numerical method possible to determine your answer. Determine the approximate relative error after each iteration. Also, provide justification for your choice of method. Extra

information: **(a)** For bracketing methods, initial guesses of 0 and R will bracket a single root for this example. **(b)** For open methods, an initial guess of R will always converge.
6.22 Perform the identical MATLAB operations as those in Example 6.7 to manipulate and find all the roots of the polynomial

$$f_5(x) = (x + 2)(x + 5)(x - 1)(x - 4)(x - 7)$$

6.23 In control systems analysis, transfer functions are developed that mathematically relate the dynamics of a system's input to its output. A transfer function for a robotic positioning system is given by

$$G(s) = \frac{C(s)}{N(s)} = \frac{s^3 + 12.5s^2 + 50.5s + 66}{s^4 + 19s^3 + 122s^2 + 296s + 192}$$

where $G(s)$ = system gain, $C(s)$ = system output, $N(s)$ = system input, and s = Laplace transform complex frequency. Use MATLAB to find the roots of the numerator and denominator and factor these into the form

$$G(s) = \frac{(s + a_1)(s + a_2)(s + a_3)}{(s + b_1)(s + b_2)(s + b_3)(s + b_4)}$$

where a_i and b_i = the roots of the numerator and denominator, respectively.
6.24 The Manning equation can be written for a rectangular open channel as

$$Q = \frac{\sqrt{S}(BH)^{5/3}}{n(B + 2H)^{2/3}}$$

where Q = flow (m³/s), S = slope (m/m), H = depth (m), and n = the Manning roughness coefficient. Develop a fixed-point iteration scheme to solve this equation for H given $Q = 5$, $S = 0.0002$, $B = 20$, and $n = 0.03$. Prove that your scheme converges for all initial guesses greater than or equal to zero.

6.25 See if you can develop a foolproof function to compute the friction factor based on the Colebrook equation as described in Sec. 6.6. Your function should return a precise result for Reynolds number ranging from 4000 to 10^7 and for ε/D ranging from 0.00001 to 0.05.

6.26 Use the Newton-Raphson method to find the root of

$$f(x) = e^{-0.5x}(4 - x) - 2$$

Employ initial guesses of **(a)** 2, **(b)** 6, and **(c)** 8. Explain your results.

6.27 Given

$$f(x) = -2x^6 - 1.5x^4 + 10x + 2$$

Use a root location technique to determine the maximum of this function. Perform iterations until the approximate relative error falls below 5%. If you use a bracketing method, use initial guesses of $x_l = 0$ and $x_u = 1$. If you use the Newton-Raphson or the modified secant method, use an initial guess of $x_i = 1$. If you use the secant method, use initial guesses of $x_{i-1} = 0$ and $x_i = 1$. Assuming that convergence is not an issue, choose the technique that is best suited to this problem. Justify your choice.

6.28 You must determine the root of the following easily differentiable function:

$$e^{0.5x} = 5 - 5x$$

Pick the best numerical technique, justify your choice, and then use that technique to determine the root. Note that it is known that for positive initial guesses, all techniques except fixed-point iteration will eventually converge. Perform iterations until the approximate relative error falls below 2%. If you use a bracketing method, use initial guesses of $x_l = 0$ and $x_u = 2$. If you use the Newton-Raphson or the modified secant method, use an initial guess of $x_i = 0.7$. If you use the secant method, use initial guesses of $x_{i-1} = 0$ and $x_i = 2$.

7

Optimization

CHAPTER OBJECTIVES

The primary objective of the present chapter is to introduce you to how optimization can be used to determine minima and maxima of both one-dimensional and multidimensional functions. Specific objectives and topics covered are

- Understanding why and where optimization occurs in engineering and scientific problem solving.
- Recognizing the difference between one-dimensional and multidimensional optimization.
- Distinguishing between global and local optima.
- Knowing how to recast a maximization problem so that it can be solved with a minimizing algorithm.
- Being able to define the golden ratio and understand why it makes one-dimensional optimization efficient.
- Locating the optimum of a single-variable function with the golden-section search.
- Locating the optimum of a single-variable function with parabolic interpolation.
- Knowing how to apply the `fminbnd` function to determine the minimum of a one-dimensional function.
- Being able to develop MATLAB contour and surface plots to visualize two-dimensional functions.
- Knowing how to apply the `fminsearch` function to determine the minimum of a multidimensional function.

YOU'VE GOT A PROBLEM

An object like a bungee jumper can be projected upward at a specified velocity. If it is subject to linear drag, its altitude as a function of time can be computed as

$$z = z_0 + \frac{m}{c}\left(v_0 + \frac{mg}{c}\right)\left(1 - e^{-(c/m)t}\right) - \frac{mg}{c}t \tag{7.1}$$

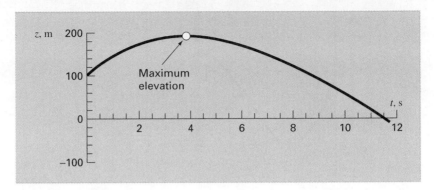

FIGURE 7.1
Elevation as a function of time for an object initially projected upward with an initial velocity.

where z = altitude (m) above the earth's surface (defined as $z = 0$), z_0 = the initial altitude (m), m = mass (kg), c = a linear drag coefficient (kg/s), v_0 = initial velocity (m/s), and t = time (s). Note that for this formulation, positive velocity is considered to be in the upward direction. Given the following parameter values: $g = 9.81$ m/s^2, $z_0 = 100$ m, $v_0 = 55$ m/s, $m = 80$ kg, and $c = 15$ kg/s, Eq. (7.1) can be used to calculate the jumper's altitude. As displayed in Fig. 7.1, the jumper rises to a peak elevation of about 190 m at about $t = 4$ s.

Suppose that you are given the job of determining the exact time of the peak elevation. The determination of such extreme values is referred to as optimization. This chapter will introduce you to how the computer is used to make such determinations.

7.1 INTRODUCTION AND BACKGROUND

In the most general sense, optimization is the process of creating something that is as effective as possible. As engineers, we must continuously design devices and products that perform tasks in an efficient fashion for the least cost. Thus, engineers are always confronting optimization problems that attempt to balance performance and limitations. In addition, scientists have interest in optimal phenomena ranging from the peak elevation of projectiles to the minimum free energy.

From a mathematical perspective, optimization deals with finding the maxima and minima of a function that depends on one or more variables. The goal is to determine the values of the variables that yield maxima or minima for the function. These can then be substituted back into the function to compute its optimal values.

Although these solutions can sometimes be obtained analytically, most practical optimization problems require numerical, computer solutions. From a numerical standpoint, optimization is similar in spirit to the root location methods we just covered in Chaps. 5 and 6. That is, both involve guessing and searching for a point on a function. The fundamental difference between the two types of problems is illustrated in Fig. 7.2. Root location involves searching for the location where the function equals zero. In contrast, optimization involves searching for the function's extreme points.

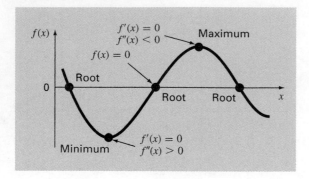

FIGURE 7.2
A function of a single variable illustrating the difference between roots and optima.

As can be seen in Fig. 7.2, the optimums are the points where the curve is flat. In mathematical terms, this corresponds to the x value where the derivative $f'(x)$ is equal to zero. Additionally, the second derivative, $f''(x)$, indicates whether the optimum is a minimum or a maximum: if $f''(x) < 0$, the point is a maximum; if $f''(x) > 0$, the point is a minimum.

Now, understanding the relationship between roots and optima would suggest a possible strategy for finding the latter. That is, you can differentiate the function and locate the root (that is, the zero) of the new function. In fact, some optimization methods do just this by solving the root problem: $f'(x) = 0$.

EXAMPLE 7.1   Determining the Optimum Analytically by Root Location

Problem Statement.   Determine the time and magnitude of the peak elevation based on Eq. (7.1). Use the following parameter values for your calculation: $g = 9.81$ m/s^2, $z_0 = 100$ m, $v_0 = 55$ m/s, $m = 80$ kg, and $c = 15$ kg/s.

Solution.   Equation (7.1) can be differentiated to give

$$\frac{dz}{dt} = v_0 e^{-(c/m)t} - \frac{mg}{c}\left(1 - e^{-(c/m)t}\right) \qquad \text{(E7.1.1)}$$

Note that because $v = dz/dt$, this is actually the equation for the velocity. The maximum elevation occurs at the value of t that drives this equation to zero. Thus, the problem amounts to determining the root. For this case, this can be accomplished by setting the derivative to zero and solving Eq. (E7.1.1) analytically for

$$t = \frac{m}{c}\ln\left(1 + \frac{cv_0}{mg}\right)$$

Substituting the parameters gives

$$t = \frac{80}{15}\ln\left(1 + \frac{15(55)}{80(9.81)}\right) = 3.83166 \text{ s}$$

This value along with the parameters can then be substituted into Eq. (7.1) to compute the maximum elevation as

$$z = 100 + \frac{80}{15}\left(50 + \frac{80(9.81)}{15}\right)\left(1 - e^{-(15/80)3.83166}\right) - \frac{80(9.81)}{15}(3.83166) = 192.8609 \text{ m}$$

We can verify that the result is a maximum by differentiating Eq. (E7.1.1) to obtain the second derivative

$$\frac{d^2z}{dt^2} = -\frac{c}{m}v_0 e^{-(c/m)t} - ge^{-(c/m)t} = -9.81\ \frac{\text{m}}{\text{s}^2}$$

The fact that the second derivative is negative tells us that we have a maximum. Further, the result makes physical sense since the acceleration should be solely equal to the force of gravity at the maximum when the vertical velocity (and hence drag) is zero.

Although an analytical solution was possible for this case, we could have obtained the same result using the root location methods described in Chaps. 5 and 6. This will be left as a homework exercise.

Although it is certainly possible to approach optimization as a roots problem, a variety of direct numerical optimization methods are available. These methods are available for both one-dimensional and multidimensional problems. As the name implies, one-dimensional problems involve functions that depend on a single dependent variable. As in Fig. 7.3a, the search then consists of climbing or descending one-dimensional peaks and valleys. Multidimensional problems involve functions that depend on two or more dependent variables.

FIGURE 7.3
(a) One-dimensional optimization. This figure also illustrates how minimization of $f(x)$ is equivalent to the maximization of $-f(x)$. (b) Two-dimensional optimization. Note that this figure can be taken to represent either a maximization (contours increase in elevation up to the maximum like a mountain) or a minimization (contours decrease in elevation down to the minimum like a valley).

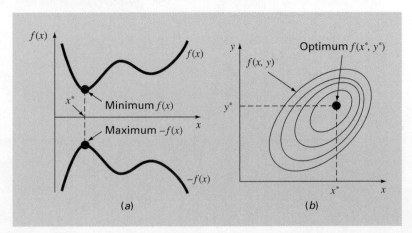

In the same spirit, a two-dimensional optimization can again be visualized as searching out peaks and valleys (Fig. 7.3*b*). However, just as in real hiking, we are not constrained to walk a single direction; instead the topography is examined to efficiently reach the goal.

Finally, the process of finding a maximum versus finding a minimum is essentially identical because the same value x^* both minimizes $f(x)$ and maximizes $-f(x)$. This equivalence is illustrated graphically for a one-dimensional function in Fig. 7.3*a*.

In the next section, we will describe some of the more common approaches for one-dimensional optimization. Then we will provide a brief description of how MATLAB can be employed to determine optima for multidimensional functions.

7.2 ONE-DIMENSIONAL OPTIMIZATION

This section will describe techniques to find the minimum or maximum of a function of a single variable $f(x)$. A useful image in this regard is the one-dimensional "roller coaster"–like function depicted in Fig. 7.4. Recall from Chaps. 5 and 6 that root location was complicated by the fact that several roots can occur for a single function. Similarly, both local and global optima can occur in optimization.

A *global optimum* represents the very best solution. A *local optimum,* though not the very best, is better than its immediate neighbors. Cases that include local optima are called *multimodal.* In such cases, we will almost always be interested in finding the global optimum. In addition, we must be concerned about mistaking a local result for the global optimum.

Just as in root location, optimization in one dimension can be divided into bracketing and open methods. As described in the next section, the golden-section search is an example of a bracketing method that is very similar in spirit to the bisection method for root location. This is followed by a somewhat more sophisticated bracketing approach—parabolic interpolation. We will then show how these two methods are combined and implemented with MATLAB's fminbnd function.

FIGURE 7.4
A function that asymptotically approaches zero at plus and minus ∞ and has two maximum and two minimum points in the vicinity of the origin. The two points to the right are local optima, whereas the two to the left are global.

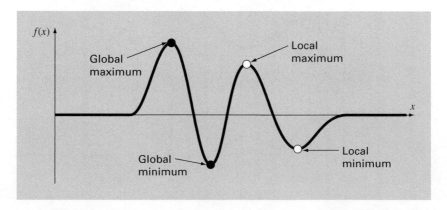

7.2.1 Golden-Section Search

In many cultures, certain numbers are ascribed magical qualities. For example, we in the West are all familiar with "lucky 7" and "Friday the 13th." Beyond such superstitious quantities, there are several well-known numbers that have such interesting and powerful mathematical properties that they could truly be called "magical." The most common of these are the ratio of a circle's circumference to its diameter π and the base of the natural logarithm e.

Although not as widely known, the *golden ratio* should surely be included in the pantheon of remarkable numbers. This quantity, which is typically represented by the Greek letter ϕ (pronounced: fee), was originally defined by Euclid (ca. 300 BCE) because of its role in the construction of the pentagram or five-pointed star. As depicted in Fig. 7.5, Euclid's definition reads: "A straight line is said to have been cut in extreme and mean ratio when, as the whole line is to the greater segment, so is the greater to the lesser."

The actual value of the golden ratio can be derived by expressing Euclid's definition as

$$\frac{\ell_1 + \ell_2}{\ell_1} = \frac{\ell_1}{\ell_2} \tag{7.2}$$

Multiplying by ℓ_1/ℓ_2 and collecting terms yields

$$\phi^2 - \phi - 1 = 0 \tag{7.3}$$

where $\phi = \ell_1/\ell_2$. The positive root of this equation is the golden ratio:

$$\phi = \frac{1 + \sqrt{5}}{2} = 1.61803398874989\ldots \tag{7.4}$$

The golden ratio has long been considered aesthetically pleasing in Western cultures. In addition, it arises in a variety of other contexts including biology. For our purposes, it provides the basis for the golden-section search, a simple, general-purpose method for determining the optimum of a single-variable function.

The golden-section search is similar in spirit to the bisection approach for locating roots in Chap. 5. Recall that bisection hinged on defining an interval, specified by a lower guess (x_l) and an upper guess (x_u) that bracketed a single root. The presence of a root between these bounds was verified by determining that $f(x_l)$ and $f(x_u)$ had different signs. The root was then estimated as the midpoint of this interval:

$$x_r = \frac{x_l + x_u}{2} \tag{7.5}$$

FIGURE 7.5
Euclid's definition of the golden ratio is based on dividing a line into two segments so that the ratio of the whole line to the larger segment is equal to the ratio of the larger segment to the smaller segment. This ratio is called the golden ratio.

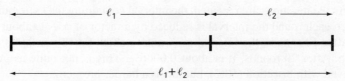

The final step in a bisection iteration involved determining a new smaller bracket. This was done by replacing whichever of the bounds x_l or x_u had a function value with the same sign as $f(x_r)$. A key advantage of this approach was that the new value x_r replaced one of the old bounds.

Now suppose that instead of a root, we were interested in determining the minimum of a one-dimensional function. As with bisection, we can start by defining an interval that contains a single answer. That is, the interval should contain a single minimum, and hence is called *unimodal*. We can adopt the same nomenclature as for bisection, where x_l and x_u defined the lower and upper bounds, respectively, of such an interval. However, in contrast to bisection, we need a new strategy for finding a minimum within the interval. Rather than using a single intermediate value (which is sufficient to detect a sign change, and hence a zero), we would need two intermediate function values to detect whether a minimum occurred.

The key to making this approach efficient is the wise choice of the intermediate points. As in bisection, the goal is to minimize function evaluations by replacing old values with new values. For bisection, this was accomplished by choosing the midpoint. For the golden-section search, the two intermediate points are chosen according to the golden ratio:

$$x_1 = x_l + d \tag{7.6}$$
$$x_2 = x_u - d \tag{7.7}$$

where

$$d = (\phi - 1)(x_u - x_l) \tag{7.8}$$

The function is evaluated at these two interior points. Two results can occur:

1. If, as in Fig. 7.6a, $f(x_1) < f(x_2)$, then $f(x_1)$ is the minimum, and the domain of x to the left of x_2, from x_l to x_2, can be eliminated because it does not contain the minimum. For this case, x_2 becomes the new x_l for the next round.
2. If $f(x_2) < f(x_1)$, then $f(x_2)$ is the minimum and the domain of x to the right of x_1, from x_1 to x_u would be eliminated. For this case, x_1 becomes the new x_u for the next round.

Now, here is the real benefit from the use of the golden ratio. Because the original x_1 and x_2 were chosen using the golden ratio, we do not have to recalculate all the function values for the next iteration. For example, for the case illustrated in Fig. 7.6, the old x_1 becomes the new x_2. This means that we already have the value for the new $f(x_2)$, since it is the same as the function value at the old x_1.

To complete the algorithm, we need only determine the new x_1. This is done with Eq. (7.6) with d computed with Eq. (7.8) based on the new values of x_l and x_u. A similar approach would be used for the alternate case where the optimum fell in the left subinterval. For this case, the new x_2 would be computed with Eq. (7.7).

As the iterations are repeated, the interval containing the extremum is reduced rapidly. In fact, each round the interval is reduced by a factor of $\phi - 1$ (about 61.8%). That means that after 10 rounds, the interval is shrunk to about 0.618^{10} or 0.008 or 0.8% of its initial length. After 20 rounds, it is about 0.0066%. This is not quite as good as the reduction achieved with bisection (50%), but this is a harder problem.

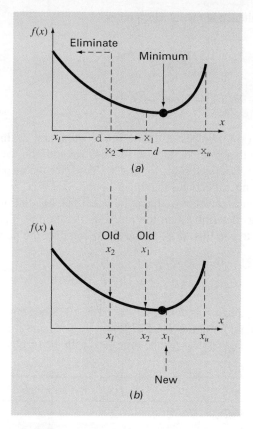

FIGURE 7.6
(a) The initial step of the golden-section search algorithm involves choosing two interior points according to the golden ratio. (b) The second step involves defining a new interval that encompasses the optimum.

EXAMPLE 7.2   Golden-Section Search

Problem Statement.   Use the golden-section search to find the minimum of

$$f(x) = \frac{x^2}{10} - 2\sin x$$

within the interval from $x_l = 0$ to $x_u = 4$.

Solution.   First, the golden ratio is used to create the two interior points:

$$d = 0.61803(4 - 0) = 2.4721$$
$$x_1 = 0 + 2.4721 = 2.4721$$
$$x_2 = 4 - 2.4721 = 1.5279$$

The function can be evaluated at the interior points:

$$f(x_2) = \frac{1.5279^2}{10} - 2\sin(1.5279) = -1.7647$$

$$f(x_1) = \frac{2.4721^2}{10} - 2\sin(2.4721) = -0.6300$$

Because $f(x_2) < f(x_1)$, our best estimate of the minimum at this point is that it is located at $x = 1.5279$ with a value of $f(x) = -1.7647$. In addition, we also know that the minimum is in the interval defined by x_l, x_2, and x_1. Thus, for the next iteration, the lower bound remains $x_l = 0$, and x_1 becomes the upper bound, that is, $x_u = 2.4721$. In addition, the former x_2 value becomes the new x_1, that is, $x_1 = 1.5279$. In addition, we do not have to recalculate $f(x_1)$, it was determined on the previous iteration as $f(1.5279) = -1.7647$.

All that remains is to use Eqs. (7.8) and (7.7) to compute the new value of d and x_2:

$$d = 0.61803(2.4721 - 0) = 1.5279$$

$$x_2 = 2.4721 - 1.5279 = 0.9443$$

The function evaluation at x_2 is $f(0.9943) = -1.5310$. Since this value is less than the function value at x_1, the minimum is $f(1.5279) = -1.7647$, and it is in the interval prescribed by x_2, x_1, and x_u. The process can be repeated, with the results tabulated here:

i	x_l	$f(x_l)$	x_2	$f(x_2)$	x_1	$f(x_1)$	x_u	$f(x_u)$	d
1	0	0	1.5279	−1.7647	2.4721	−0.6300	4.0000	3.1136	2.4721
2	0	0	0.9443	−1.5310	1.5279	−1.7647	2.4721	−0.6300	1.5279
3	0.9443	−1.5310	1.5279	−1.7647	1.8885	−1.5432	2.4721	−0.6300	0.9443
4	0.9443	−1.5310	1.3050	−1.7595	1.5279	−1.7647	1.8885	−1.5432	0.5836
5	1.3050	−1.7595	1.5279	−1.7647	1.6656	−1.7136	1.8885	−1.5432	0.3607
6	1.3050	−1.7595	1.4427	−1.7755	1.5279	−1.7647	1.6656	−1.7136	0.2229
7	1.3050	−1.7595	1.3901	−1.7742	1.4427	−1.7755	1.5279	−1.7647	0.1378
8	1.3901	−1.7742	1.4427	−1.7755	1.4752	−1.7732	1.5279	−1.7647	0.0851

Note that the current minimum is highlighted for every iteration. After the eighth iteration, the minimum occurs at $x = 1.4427$ with a function value of -1.7755. Thus, the result is converging on the true value of -1.7757 at $x = 1.4276$.

Recall that for bisection (Sec. 5.4), an exact upper bound for the error can be calculated at each iteration. Using similar reasoning, an upper bound for golden-section search can be derived as follows: Once an iteration is complete, the optimum will either fall in one of two intervals. If the optimum function value is at x_2, it will be in the lower interval (x_l, x_2, x_1). If optimum function value is at x_1, it will be in the upper interval (x_2, x_1, x_u). Because the interior points are symmetrical, either case can be used to define the error.

Looking at the upper interval (x_2, x_1, x_u), if the true value were at the far left, the maximum distance from the estimate would be

$$\Delta x_a = x_1 - x_2$$
$$= x_l + (\phi - 1)(x_u - x_l) - x_u + (\phi - 1)(x_u - x_l)$$
$$= (x_l - x_u) + 2(\phi - 1)(x_u - x_l)$$
$$= (2\phi - 3)(x_u - x_l)$$

or $0.2361 (x_u - x_l)$. If the true value were at the far right, the maximum distance from the estimate would be

$$\Delta x_b = x_u - x_1$$
$$= x_u - x_l - (\phi - 1)(x_u - x_l)$$
$$= (x_u - x_l) - (\phi - 1)(x_u - x_l)$$
$$= (2 - \phi)(x_u - x_l)$$

or $0.3820 (x_u - x_l)$. Therefore, this case would represent the maximum error. This result can then be normalized to the optimal value for that iteration x_{opt} to yield

$$\varepsilon_a = (2 - \phi) \left| \frac{x_u - x_l}{x_{\text{opt}}} \right| \times 100\% \tag{7.9}$$

This estimate provides a basis for terminating the iterations.

An M-file function for the golden-section search for minimization is presented in Fig. 7.7. The function returns the location of the minimum, the value of the function, the approximate error, and the number of iterations.

The M-file can be used to solve the problem from Example 7.1.

```
>> g=9.81;v0=55;m=80;c=15;z0=100;
>> z=@(t) -(z0+m/c*(v0+m*g/c)*(1-exp(-c/m*t))-m*g/c*t);
>> [xmin,fmin,ea,iter]=goldmin(z,0,8)

xmin =
      3.8317
fmin =
 -192.8609
ea =
    6.9356e-005
```

Notice how because this is a maximization, we have entered the negative of Eq. (7.1). Consequently, `fmin` corresponds to a maximum height of 192.8609.

You may be wondering why we have stressed the reduced function evaluations of the golden-section search. Of course, for solving a single optimization, the speed savings would be negligible. However, there are two important contexts where minimizing the number of function evaluations can be important. These are

1. **Many evaluations.** There are cases where the golden-section search algorithm may be a part of a much larger calculation. In such cases, it may be called many times. Therefore, keeping function evaluations to a minimum could pay great dividends for such cases.

```
function [x,fx,ea,iter]=goldmin(f,xl,xu,es,maxit,varargin)
% goldmin: minimization golden section search
%    [xopt,fopt,ea,iter]=goldmin(f,xl,xu,es,maxit,p1,p2,...):
%       uses golden section search to find the minimum of f
% input:
%    f = name of function
%    xl, xu = lower and upper guesses
%    es = desired relative error (default = 0.0001%)
%    maxit = maximum allowable iterations (default = 50)
%    p1,p2,... = additional parameters used by f
% output:
%    x = location of minimum
%    fx = minimum function value
%    ea = approximate relative error (%)
%    iter = number of iterations

if nargin<3,error('at least 3 input arguments required'),end
if nargin<4|isempty(es), es=0.0001;end
if nargin<5|isempty(maxit), maxit=50;end
phi=(1+sqrt(5))/2;
iter=0;
while(1)
  d = (phi-1)*(xu - xl);
  x1 = xl + d;
  x2 = xu - d;
  if f(x1,varargin{:}) < f(x2,varargin{:})
    xopt = x1;
    xl = x2;
  else
    xopt = x2;
    xu = x1;
  end
  iter=iter+1;
  if xopt~=0, ea = (2 - phi) * abs((xu - xl) / xopt) * 100;end
  if ea <= es | iter >= maxit,break,end
end
x=xopt;fx=f(xopt,varargin{:});
```

FIGURE 7.7
An M-file to determine the minimum of a function with the golden-section search.

2. Time-consuming evaluation. For pedagogical reasons, we use simple functions in most of our examples. You should understand that a function can be very complex and time-consuming to evaluate. For example, optimization can be used to estimate the parameters of a model consisting of a system of differential equations. For such cases, the "function" involves time-consuming model integration. Any method that minimizes such evaluations would be advantageous.

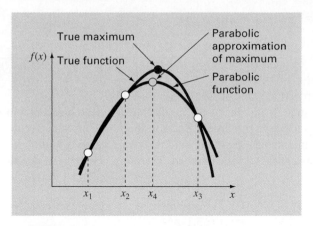

FIGURE 7.8
Graphical depiction of parabolic interpolation.

7.2.2 Parabolic Interpolation

Parabolic interpolation takes advantage of the fact that a second-order polynomial often provides a good approximation to the shape of $f(x)$ near an optimum (Fig. 7.8).

Just as there is only one straight line connecting two points, there is only one parabola connecting three points. Thus, if we have three points that jointly bracket an optimum, we can fit a parabola to the points. Then we can differentiate it, set the result equal to zero, and solve for an estimate of the optimal x. It can be shown through some algebraic manipulations that the result is

$$x_4 = x_2 - \frac{1}{2} \frac{(x_2 - x_1)^2 [f(x_2) - f(x_3)] - (x_2 - x_3)^2 [f(x_2) - f(x_1)]}{(x_2 - x_1) [f(x_2) - f(x_3)] - (x_2 - x_3) [f(x_2) - f(x_1)]} \tag{7.10}$$

where x_1, x_2, and x_3 are the initial guesses, and x_4 is the value of x that corresponds to the optimum value of the parabolic fit to the guesses.

EXAMPLE 7.3 | Parabolic Interpolation

Problem Statement.   Use parabolic interpolation to approximate the minimum of

$$f(x) = \frac{x^2}{10} - 2 \sin x$$

with initial guesses of $x_1 = 0$, $x_2 = 1$, and $x_3 = 4$.

Solution.   The function values at the three guesses can be evaluated:

$$
\begin{aligned}
x_1 &= 0 & f(x_1) &= 0 \\
x_2 &= 1 & f(x_2) &= -1.5829 \\
x_3 &= 4 & f(x_3) &= 3.1136
\end{aligned}
$$

and substituted into Eq. (7.10) to give

$$x_4 = 1 - \frac{1}{2} \frac{(1-0)^2 [-1.5829 - 3.1136] - (1-4)^2 [-1.5829 - 0]}{(1-0)[-1.5829 - 3.1136] - (1-4)[-1.5829 - 0]} = 1.5055$$

which has a function value of $f(1.5055) = -1.7691$.

Next, a strategy similar to the golden-section search can be employed to determine which point should be discarded. Because the function value for the new point is lower than for the intermediate point (x_2) and the new x value is to the right of the intermediate point, the lower guess (x_1) is discarded. Therefore, for the next iteration:

$$
\begin{aligned}
x_1 &= 1 & f(x_1) &= -1.5829 \\
x_2 &= 1.5055 & f(x_2) &= -1.7691 \\
x_3 &= 4 & f(x_3) &= 3.1136
\end{aligned}
$$

which can be substituted into Eq. (7.10) to give

$$
\begin{aligned}
x_4 &= 1.5055 - \frac{1}{2} \frac{(1.5055 - 1)^2 [-1.7691 - 3.1136] - (1.5055 - 4)^2 [-1.7691 - (-1.5829)]}{(1.5055 - 1)[-1.7691 - 3.1136] - (1.5055 - 4)[-1.7691 - (-1.5829)]} \\
&= 1.4903
\end{aligned}
$$

which has a function value of $f(1.4903) = -1.7714$. The process can be repeated, with the results tabulated here:

i	x_1	$f(x_1)$	x_2	$f(x_2)$	x_3	$f(x_3)$	x_4	$f(x_4)$
1	0.0000	0.0000	1.0000	−1.5829	4.0000	3.1136	1.5055	−1.7691
2	1.0000	−1.5829	1.5055	−1.7691	4.0000	3.1136	1.4903	−1.7714
3	1.0000	−1.5829	1.4903	−1.7714	1.5055	−1.7691	1.4256	−1.7757
4	1.0000	−1.5829	1.4256	−1.7757	1.4903	−1.7714	1.4266	−1.7757
5	1.4256	−1.7757	1.4266	−1.7757	1.4903	−1.7714	1.4275	−1.7757

Thus, within five iterations, the result is converging rapidly on the true value of -1.7757 at $x = 1.4276$.

7.2.3 MATLAB Function: `fminbnd`

Recall that in Sec. 6.4 we described the built-in MATLAB function `fzero`. This function combined several root-finding methods into a single algorithm that balanced reliability with efficiency.

The `fminbnd` function provides a similar approach for one-dimensional minimization. It combines the slow, dependable golden-section search with the faster, but possibly unreliable, parabolic interpolation. It first attempts parabolic interpolation and keeps applying it as long as acceptable results are obtained. If not, it uses the golden-section search to get matters in hand.

A simple expression of its syntax is

```
[xmin, fval] = fminbnd(function,x1,x2)
```

where x and fval are the location and value of the minimum, *function* is the name of the function being evaluated, and *x1* and *x2* are the bounds of the interval being searched.

Here is a simple MATLAB session that uses fminbnd to solve the problem from Example 7.1.

```
>> g=9.81;v0=55;m=80;c=15;z0=100;
>> z=@(t) -(z0+m/c*(v0+m*g/c)*(1-exp(-c/m*t))-m*g/c*t);
>> [x,f]=fminbnd(z,0,8)

x =
    3.8317
f =
 -192.8609
```

As with fzero, optional parameters can be specified using optimset. For example, we can display calculation details:

```
>> options = optimset('display','iter');
>> fminbnd(z,0,8,options)
```

Func-count	x	f(x)	Procedure
1	3.05573	-189.759	initial
2	4.94427	-187.19	golden
3	1.88854	-171.871	golden
4	3.87544	-192.851	parabolic
5	3.85836	-192.857	parabolic
6	3.83332	-192.861	parabolic
7	3.83162	-192.861	parabolic
8	3.83166	-192.861	parabolic
9	3.83169	-192.861	parabolic

```
Optimization terminated:
 the current x satisfies the termination criteria using
OPTIONS.TolX of 1.000000e-004

ans =
    3.8317
```

Thus, after three iterations, the method switches from golden to parabolic, and after eight iterations, the minimum is determined to a tolerance of 0.0001.

7.3 MULTIDIMENSIONAL OPTIMIZATION

Aside from one-dimensional functions, optimization also deals with multidimensional functions. Recall from Fig. 7.3a that our visual image of a one-dimensional search was like a roller coaster. For two-dimensional cases, the image becomes that of mountains and valleys (Fig. 7.3b). As in the following example, MATLAB's graphic capabilities provide a handy means to visualize such functions.

EXAMPLE 7.4   Visualizing a Two-Dimensional Function

Problem Statement. Use MATLAB's graphical capabilities to display the following function and visually estimate its minimum in the range $-2 \le x_1 \le 0$ and $0 \le x_2 \le 3$:

$$f(x_1, x_2) = 2 + x_1 - x_2 + 2x_1^2 + 2x_1 x_2 + x_2^2$$

Solution. The following script generates contour and mesh plots of the function:

```
x=linspace(-2,0,40);y=linspace(0,3,40);
[X,Y] = meshgrid(x,y);
Z=2+X-Y+2*X.^2+2*X.*Y+Y.^2;
subplot(1,2,1);
cs=contour(X,Y,Z);clabel(cs);
xlabel('x_1');ylabel('x_2');
title('(a) Contour plot');grid;
subplot(1,2,2);
cs=surfc(X,Y,Z);
zmin=floor(min(Z));
zmax=ceil(max(Z));
xlabel('x_1');ylabel('x_2');zlabel('f(x_1,x_2)');
title('(b) Mesh plot');
```

As displayed in Fig. 7.9, both plots indicate that function has a minimum value of about $f(x_1, x_2) = 0$ to 1 located at about $x_1 = -1$ and $x_2 = 1.5$.

FIGURE 7.9
(a) Contour and (b) mesh plots of a two-dimensional function.

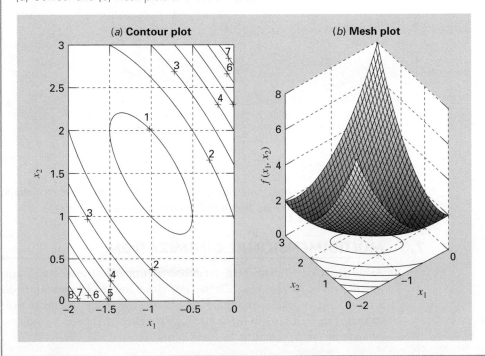

Techniques for multidimensional unconstrained optimization can be classified in a number of ways. For purposes of the present discussion, we will divide them depending on whether they require derivative evaluation. Those that require derivatives are called *gradient,* or *descent* (or ascent), methods. The approaches that do not require derivative evaluation are called *nongradient,* or *direct,* methods. As described next, the built-in MATLAB function `fminsearch` is a direct method.

7.3.1 MATLAB Function: `fminsearch`

Standard MATLAB has a function `fminsearch` that can be used to determine the minimum of a multidimensional function. It is based on the Nelder-Mead method, which is a direct-search method that uses only function values (does not require derivatives) and handles non-smooth objective functions. A simple expression of its syntax is

```
[xmin, fval] = fminsearch(function,x1,x2)
```

where *xmin* and *fval* are the location and value of the minimum, *function* is the name of the function being evaluated, and *x1* and *x2* are the bounds of the interval being searched.

Here is a simple MATLAB session that uses `fminsearch` to determine minimum for the function we just graphed in Example 7.4:

```
>> f=@(x) 2+x(1)-x(2)+2*x(1)^2+2*x(1)*x(2)+x(2)^2;
>> [x,fval]=fminsearch(f,[-0.5,0.5])

x =
   -1.0000    1.5000
fval =
    0.7500
```

7.4 CASE STUDY   EQUILIBRIUM AND MINIMUM POTENTIAL ENERGY

Background.   As in Fig. 7.10*a*, an unloaded spring can be attached to a wall mount. When a horizontal force is applied, the spring stretches. The displacement is related to the force by *Hookes law,* $F = kx$. The *potential energy* of the deformed state consists of the difference between the strain energy of the spring and the work done by the force:

$$PE(x) = 0.5kx^2 - Fx \tag{7.11}$$

FIGURE 7.10
(*a*) An unloaded spring attached to a wall mount. (*b*) Application of a horizontal force stretches the spring where the relationship between force and displacement is described by Hooke's law.

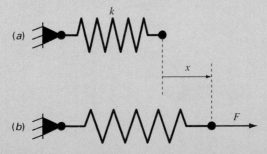

Equation (7.11) defines a parabola. Since the potential energy will be at a minimum at equilibrium, the solution for displacement can be viewed as a one-dimensional optimization problem. Because this equation is so easy to differentiate, we can solve for the displacement as $x = F/k$. For example, if $k = 2$ N/cm and $F = 5$ N, $x = 5N/(2$ N/cm$) = 2.5$ cm.

A more interesting two-dimensional case is shown in Fig. 7.11. In this system, there are two degrees of freedom in that the system can move both horizontally and vertically. In the same way that we approached the one-dimensional system, the equilibrium deformations are the values of x_1 and x_2 that minimize the potential energy:

$$PE(x_1, x_2) = 0.5k_a \left(\sqrt{x_1^2 + (L_a - x_2)^2} - L_a \right)^2$$

$$+ 0.5k_b \left(\sqrt{x_1^2 + (L_b + x_2)^2} - L_b \right)^2 - F_1 x_1 - F_2 x_2 \qquad (7.12)$$

If the parameters are $k_a = 9$ N/cm, $k_b = 2$ N/cm, $L_a = 10$ cm, $L_b = 10$ cm, $F_1 = 2$ N, and $F_2 = 4$ N, use MATLAB to solve for the displacements and the potential energy.

FIGURE 7.11
A two-spring system: (a) unloaded and (b) loaded.

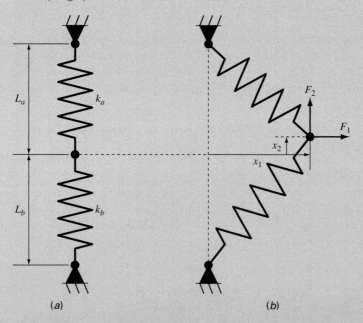

7.4 CASE STUDY continued

Solution.   An M-file can be developed to hold the potential energy function:

```
function p=PE(x,ka,kb,La,Lb,F1,F2)
PEa=0.5*ka*(sqrt(x(1)^2+(La-x(2))^2)-La)^2;
PEb=0.5*kb*(sqrt(x(1)^2+(Lb+x(2))^2)-Lb)^2;
W=F1*x(1)+F2*x(2);
p=PEa+PEb-W;
```

The solution can be obtained with the `fminsearch` function:

```
>> ka=9;kb=2;La=10;Lb=10;F1=2;F2=4;
>> [x,f]=fminsearch(@PE,[-0.5,0.5],[],ka,kb,La,Lb,F1,F2)

x =
     4.9523    1.2769
f =
    -9.6422
```

Thus, at equilibrium, the potential energy is -9.6422 N·cm. The connecting point is located 4.9523 cm to the right and 1.2759 cm above its original position.

PROBLEMS

7.1 Perform three iterations of the Newton-Raphson method to determine the root of Eq. (E7.1.1). Use the parameter values from Example 7.1 along with an initial guess of $t = 3$ s.

7.2 Given the formula

$$f(x) = -x^2 + 8x - 12$$

(a) Determine the maximum and the corresponding value of x for this function analytically (i.e., using differentiation).
(b) Verify that Eq. (7.10) yields the same results based on initial guesses of $x_1 = 0$, $x_2 = 2$, and $x_3 = 6$.

7.3 Consider the following function:

$$f(x) = 3 + 6x + 5x^2 + 3x^3 + 4x^4$$

Locate the minimum by finding the root of the derivative of this function. Use bisection with initial guesses of $x_l = -2$ and $x_u = 1$.

7.4 Given

$$f(x) = -1.5x^6 + 2x^4 + 12x$$

(a) Plot the function.
(b) Use analytical methods to prove that the function is concave for all values of x.

(c) Differentiate the function and then use a root-location method to solve for the maximum $f(x)$ and the corresponding value of x.

7.5 Solve for the value of x that maximizes $f(x)$ in Prob. 7.4 using the golden-section search. Employ initial guesses of $x_l = 0$ and $x_u = 2$, and perform three iterations.

7.6 Repeat Prob. 7.5, except use parabolic interpolation. Employ initial guesses of $x_1 = 0$, $x_2 = 1$, and $x_3 = 2$, and perform three iterations.

7.7 Employ the following methods to find the maximum of

$$f(x) = 4x - 1.8x^2 + 1.2x^3 - 0.3x^4$$

(a) Golden-section search ($x_l = -2$, $x_u = 4$, $\varepsilon_s = 1\%$).
(b) Parabolic interpolation ($x_1 = 1.75$, $x_2 = 2$, $x_3 = 2.5$, iterations = 5).

7.8 Consider the following function:

$$f(x) = x^4 + 2x^3 + 8x^2 + 5x$$

Use analytical and graphical methods to show the function has a minimum for some value of x in the range $-2 \le x \le 1$.

7.9 Employ the following methods to find the minimum of the function from Prob. 7.8:
(a) Golden-section search ($x_l = -2, x_u = 1, \varepsilon_s = 1\%$).
(b) Parabolic interpolation ($x_1 = -2, x_2 = -1, x_3 = 1$, iterations = 5).

7.10 Consider the following function:

$$f(x) = 2x + \frac{3}{x}$$

Perform 10 iterations of parabolic interpolation to locate the minimum. Comment on the convergence of your results ($x_1 = 0.1, x_2 = 0.5, x_3 = 5$)

7.11 Develop an M-file that is expressly designed to locate a maximum with the golden-section search. In other words, set if up so that it directly finds the maximum rather than finding the minimum of $-f(x)$. The subroutine should have the following features:

- Iterate until the relative error falls below a stopping criterion or exceeds a maximum number of iterations.
- Return both the optimal x and $f(x)$.

Test your program with the same problem as Example 7.1.
7.12 Develop an M-file to locate a minimum with the golden-section search. Rather than using the maximum iterations and Eq. (7.9) as the stopping criteria, determine the number of iterations needed to attain a desired tolerance. Test your function by solving Example 7.2 using $E_{a,d} = 0.0001$.
7.13 Develop an M-file to implement parabolic interpolation to locate a minimum. The subroutine should have the following features:

- Base it on two initial guesses, and have the program generate the third initial value at the midpoint of the interval.
- Check whether the guesses bracket a maximum. If not, the subroutine should not implement the algorithm, but should return an error message.
- Iterate until the relative error falls below a stopping criterion or exceeds a maximum number of iterations.
- Return both the optimal x and $f(x)$.

Test your program with the same problem as Example 7.3.
7.14 Pressure measurements are taken at certain points behind an airfoil over time. The data best fits the curve $y = 6 \cos x - 1.5 \sin x$ from $x = 0$ to 6 s. Use four iterations of the golden-search method to find the minimum pressure. Set $x_l = 2$ and $x_u = 4$.
7.15 The trajectory of a ball can be computed with

$$y = (\tan \theta_0)x - \frac{g}{2v_0^2 \cos^2 \theta_0}x^2 + y_0$$

where y = the height (m), θ_0 = the initial angle (radians), v_0 = the initial velocity (m/s), g = the gravitational

constant = 9.81 m/s², and y_0 = the initial height (m). Use the golden-section search to determine the maximum height given $y_0 = 1$ m, $v_0 = 25$ m/s, and $\theta_0 = 50°$. Iterate until the approximate error falls below $\varepsilon_s = 1\%$ using initial guesses of $x_l = 0$ and $x_u = 60$ m.
7.16 The deflection of a uniform beam subject to a linearly increasing distributed load can be computed as

$$y = \frac{w_0}{120EIL}(-x^5 + 2L^2x^3 - L^4x)$$

Given that $L = 600$ cm, $E = 50{,}000$ kN/cm², $I = 30{,}000$ cm⁴, and $w_0 = 2.5$ kN/cm, determine the point of maximum deflection **(a)** graphically, **(b)** using the golden-section search until the approximate error falls below $\varepsilon_s = 1\%$ with initial guesses of $x_l = 0$ and $x_u = L$.
7.17 A object with a mass of 100 kg is projected upward from the surface of the earth at a velocity of 50 m/s. If the object is subject to linear drag ($c = 15$ kg/s), use the golden-section search to determine the maximum height the object attains.
7.18 The normal distribution is a bell-shaped curve defined by

$$y = e^{-x^2}$$

Use the golden-section search to determine the location of the inflection point of this curve for positive x.
7.19 Use the `fminsearch` function to determine the minimum of

$$f(x, y) = 2y^2 - 2.25xy - 1.75y + 1.5x^2$$

7.20 Use the `fminsearch` function to determine the maximum of

$$f(x, y) = 4x + 2y + x^2 - 2x^4 + 2xy - 3y^2$$

7.21 Given the following function:

$$f(x, y) = -8x + x^2 + 12y + 4y^2 - 2xy$$

Determine the minimum **(a)** graphically, **(b)** numerically with the `fminsearch` function, and **(c)** substitute the result of **(b)** back into the function to determine the minimum $f(x, y)$.
7.22 The specific growth rate of a yeast that produces an antibiotic is a function of the food concentration c:

$$g = \frac{2c}{4 + 0.8c + c^2 + 0.2c^3}$$

As depicted in Fig. P7.22, growth goes to zero at very low concentrations due to food limitation. It also goes to zero at high concentrations due to toxicity effects. Find the value of c at which growth is a maximum.

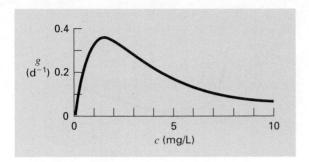

FIGURE P7.22
The specific growth rate of a yeast that produces an antibiotic versus the food concentration.

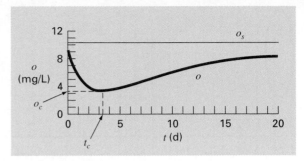

FIGURE P7.25
A dissolved oxygen "sag" below a point discharge of sewage into a river.

7.23 A compound A will be converted into B in a stirred tank reactor. The product B and unreacted A are purified in a separation unit. Unreacted A is recycled to the reactor. A process engineer has found that the initial cost of the system is a function of the conversion x_A. Find the conversion that will result in the lowest cost system. C is a proportionality constant.

$$\text{Cost} = C\left[\left(\frac{1}{(1-x_A)^2}\right)^{0.6} + 6\left(\frac{1}{x_A}\right)^{0.6}\right]$$

7.24 A finite-element model of a cantilever beam subject to loading and moments (Fig. P7.24) is given by optimizing

$$f(x, y) = 5x^2 - 5xy + 2.5y^2 - x - 1.5y$$

where $x =$ end displacement and $y =$ end moment. Find the values of x and y that minimize $f(x, y)$.

7.25 The *Streeter-Phelps model* can be used to compute the dissolved oxygen concentration in a river below a point discharge of sewage (Fig. P7.25),

$$o = o_s - \frac{k_d L_o}{k_d + k_s - k_a}\left(e^{-k_a t} - e^{-(k_d+k_s)t}\right)$$
$$-\frac{S_b}{k_a}(1 - e^{-k_a t})$$

(P7.25)

FIGURE P7.24
A cantilever beam.

where $o =$ dissolved oxygen concentration (mg/L), $o_s =$ oxygen saturation concentration (mg/L), $t =$ travel time (d), $L_o =$ biochemical oxygen demand (BOD) concentration at the mixing point (mg/L), $k_d =$ rate of decomposition of BOD (d^{-1}), $k_s =$ rate of settling of BOD (d^{-1}), $k_a =$ reaeration rate (d^{-1}), and $S_b =$ sediment oxygen demand (mg/L/d).

As indicated in Fig. P7.25, Eq. (P7.25) produces an oxygen "sag" that reaches a critical minimum level o_c, some travel time t_c below the point discharge. This point is called "critical" because it represents the location where biota that depend on oxygen (like fish) would be the most stressed. Determine the critical travel time and concentration, given the following values:

$o_s = 10$ mg/L	$k_d = 0.1$ d^{-1}	$k_a = 0.6$ d^{-1}
$k_s = 0.05$ d^{-1}	$L_o = 50$ mg/L	$S_b = 1$ mg/L/d

7.26 The two-dimensional distribution of pollutant concentration in a channel can be described by

$$c(x, y) = 7.9 + 0.13x + 0.21y - 0.05x^2$$
$$-0.016y^2 - 0.007xy$$

Determine the exact location of the peak concentration given the function and the knowledge that the peak lies within the bounds $-10 \le x \le 10$ and $0 \le y \le 20$.

7.27 A total charge Q is uniformly distributed around a ring-shaped conductor with radius a. A charge q is located at a distance x from the center of the ring (Fig. P7.27). The force exerted on the charge by the ring is given by

$$F = \frac{1}{4\pi e_0}\frac{qQx}{(x^2+a^2)^{3/2}}$$

where $e_0 = 8.85 \times 10^{-12}$ C^2/(N m^2), $q = Q = 2 \times 10^{-5}$ C, and $a = 0.9$ m. Determine the distance x where the force is a maximum.

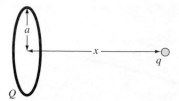

FIGURE P7.27

7.28 The torque transmitted to an induction motor is a function of the slip between the rotation of the stator field and the rotor speed s, where slip is defined as

$$s = \frac{n - n_R}{n}$$

where n = revolutions per second of rotating stator speed and n_R = rotor speed. Kirchhoff's laws can be used to show that the torque (expressed in dimensionless form) and slip are related by

$$T = \frac{15s(1-s)}{(1-s)(4s^2 - 3s + 4)}$$

Figure P7.28 shows this function. Use a numerical method to determine the slip at which the maximum torque occurs.

7.29 The total drag on an airfoil can be estimated by

$$D = \underbrace{0.01\sigma V^2}_{\text{Friction}} + \underbrace{\frac{0.95}{\sigma}\left(\frac{W}{V}\right)^2}_{\text{Lift}}$$

where D = drag, σ = ratio of air density between the flight altitude and sea level, W = weight, and V = velocity. As seen in Fig. P7.29, the two factors contributing to drag are affected differently as velocity increases. Whereas friction drag increases with velocity, the drag due to lift decreases.

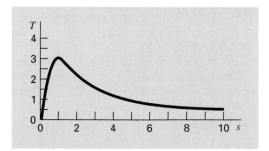

FIGURE P7.28
Torque transmitted to an inductor as a function of slip.

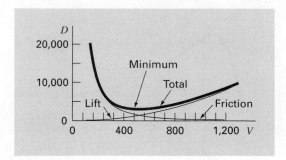

FIGURE P7.29
Plot of drag versus velocity for an airfoil.

The combination of the two factors leads to a minimum drag.
(a) If $\sigma = 0.6$ and $W = 16,000$, determine the minimum drag and the velocity at which it occurs.
(b) In addition, develop a sensitivity analysis to determine how this optimum varies in response to a range of $W = 12,000$ to $20,000$ with $\sigma = 0.6$.

7.30 Roller bearings are subject to fatigue failure caused by large contact loads F (Fig. P7.30). The problem of finding the location of the maximum stress along the x axis can be shown to be equivalent to maximizing the function:

$$f(x) = \frac{0.4}{\sqrt{1+x^2}} - \sqrt{1+x^2}\left(1 - \frac{0.4}{1+x^2}\right) + x$$

Find the x that maximizes $f(x)$.

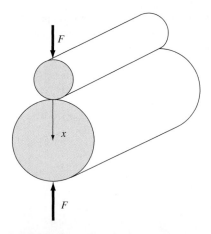

FIGURE P7.30
Roller bearings.

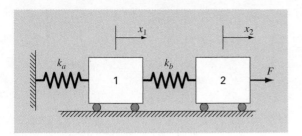

FIGURE P7.31
Two frictionless masses connected to a wall by a pair of linear elastic springs.

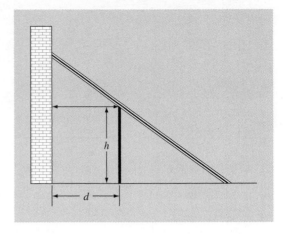

FIGURE P7.33
A ladder leaning against a fence and just touching a wall.

7.31 In a similar fashion to the case study described in Sec. 7.4, develop the potential energy function for the system depicted in Fig. P7.31. Develop contour and surface plots in MATLAB. Minimize the potential energy function to determine the equilibrium displacements x_1 and x_2 given the forcing function $F = 100$ N and the parameters $k_a = 20$ and $k_b = 15$ N/m.

7.32 As an agricultural engineer, you must design a trapezoidal open channel to carry irrigation water (Fig. P7.32). Determine the optimal dimensions to minimize the wetted perimeter for a cross-sectional area of 50 m². Are the relative dimensions universal?

7.33 Use the function `fminsearch` to determine the length of the shortest ladder that reaches from the ground over the fence to the building's wall (Fig. P7.33). Test it for the case where $h = d = 4$ m.

7.34 The length of the longest ladder that can negotiate the corner depicted in Fig. P7.34 can be determined by computing the value of θ that minimizes the following function:

$$L(\theta) = \frac{w_1}{\sin\theta} + \frac{w_2}{\sin(\pi - \alpha - \theta)}$$

For the case where $w_1 = w_2 = 2$ m, use a numerical method described in this chapter (including MATLAB's built-in capabilities) to develop a plot of L versus a range of α's from 45 to 135°.

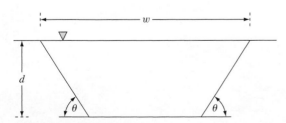

FIGURE P7.32

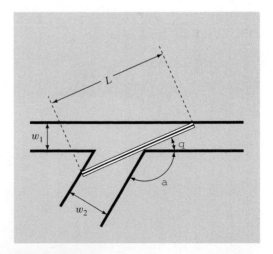

FIGURE P7.34
A ladder negotiating a corner formed by two hallways.

PART THREE

Linear Systems

3.1 OVERVIEW

What Are Linear Algebraic Equations?

In Part Two, we determined the value x that satisfied a single equation, $f(x) = 0$. Now, we deal with the case of determining the values $x_1, x_2, \ldots, x_n$ that simultaneously satisfy a set of equations:

$$
\begin{aligned}
f_1(x_1, x_2, \ldots, x_n) &= 0 \\
f_2(x_1, x_2, \ldots, x_n) &= 0 \\
&\vdots \\
f_n(x_1, x_2, \ldots, x_n) &= 0
\end{aligned}
$$

Such systems are either linear or nonlinear. In Part Three, we deal with *linear algebraic equations* that are of the general form

$$
\begin{aligned}
a_{11}x_1 + a_{12}x_2 + \cdots + a_{1n}x_n &= b_1 \\
a_{21}x_1 + a_{22}x_2 + \cdots + a_{2n}x_n &= b_2 \\
&\vdots \\
a_{n1}x_1 + a_{n2}x_2 + \cdots + a_{nn}x_n &= b_n
\end{aligned}
\qquad \text{(PT3.1)}
$$

where the a's are constant coefficients, the b's are constants, the x's are unknowns, and n is the number of equations. All other algebraic equations are nonlinear.

Linear Algebraic Equations in Engineering and Science

Many of the fundamental equations of engineering and science are based on conservation laws. Some familiar quantities that conform to such laws are mass, energy, and momentum. In mathematical terms, these principles lead to balance or continuity equations that relate system behavior as represented by the levels or

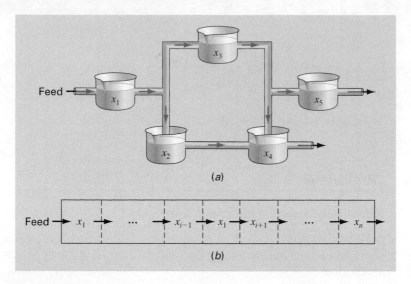

FIGURE PT3.1
Two types of systems that can be modeled using linear algebraic equations: (a) lumped variable system that involves coupled finite components and (b) distributed variable system that involves a continuum.

response of the quantity being modeled to the properties or characteristics of the system and the external stimuli or forcing functions acting on the system.

As an example, the principle of mass conservation can be used to formulate a model for a series of chemical reactors (Fig. PT3.1a). For this case, the quantity being modeled is the mass of the chemical in each reactor. The system properties are the reaction characteristics of the chemical and the reactors' sizes and flow rates. The forcing functions are the feed rates of the chemical into the system.

When we studied roots of equations, you saw how single-component systems result in a single equation that can be solved using root-location techniques. Multicomponent systems result in a coupled set of mathematical equations that must be solved simultaneously. The equations are coupled because the individual parts of the system are influenced by other parts. For example, in Fig. PT3.1a, reactor 4 receives chemical inputs from reactors 2 and 3. Consequently, its response is dependent on the quantity of chemical in these other reactors.

When these dependencies are expressed mathematically, the resulting equations are often of the linear algebraic form of Eq. (PT3.1). The x's are usually measures of the magnitudes of the responses of the individual components. Using Fig. PT3.1a as an example, x_1 might quantify the amount of chemical mass in the first reactor, x_2 might quantify the amount in the second, and so forth. The a's typically represent the properties and characteristics that bear on the interactions between components. For instance, the a's for Fig. PT3.1a might be reflective of the flow rates of mass between the reactors. Finally, the b's usually represent the forcing functions acting on the system, such as the feed rate.

Multicomponent problems of these types arise from both lumped (macro-) or distributed (micro-) variable mathematical models. *Lumped variable problems* involve coupled finite components. Examples include trusses, reactors, and electric circuits. The three bungee jumpers at the beginning of Chap. 8 are a lumped system.

Conversely, *distributed variable problems* attempt to describe the spatial detail of systems on a continuous or semicontinuous basis. The distribution of chemicals along the length of an elongated, rectangular reactor (Fig. PT3.1*b*) is an example of a continuous variable model. Differential equations derived from conservation laws specify the distribution of the dependent variable for such systems. These differential equations can be solved numerically by converting them to an equivalent system of simultaneous algebraic equations.

The solution of such sets of equations represents a major application area for the methods in the following chapters. These equations are coupled because the variables at one location are dependent on the variables in adjoining regions. For example, the concentration at the middle of the reactor in Fig. PT3.1*b* is a function of the concentration in adjoining regions. Similar examples could be developed for the spatial distribution of temperature, momentum, or electricity.

Aside from physical systems, simultaneous linear algebraic equations also arise in a variety of mathematical problem contexts. These result when mathematical functions are required to satisfy several conditions simultaneously. Each condition results in an equation that contains known coefficients and unknown variables. The techniques discussed in this part can be used to solve for the unknowns when the equations are linear and algebraic. Some widely used numerical techniques that employ simultaneous equations are regression analysis and spline interpolation.

3.2 PART ORGANIZATION

Due to its importance in formulating and solving linear algebraic equations, *Chap. 8* provides a brief overview of *matrix algebra*. Aside from covering the rudiments of matrix representation and manipulation, the chapter also describes how matrices are handled in MATLAB.

Chapter 9 is devoted to the most fundamental technique for solving linear algebraic systems: *Gauss elimination*. Before launching into a detailed discussion of this technique, a preliminary section deals with simple methods for solving small systems. These approaches are presented to provide you with visual insight and because one of the methods—the elimination of unknowns—represents the basis for Gauss elimination.

After this preliminary material, "naive" Gauss elimination is discussed. We start with this "stripped-down" version because it allows the fundamental technique to be elaborated on without complicating details. Then, in subsequent sections, we discuss potential problems of the naive approach and present a number of modifications to minimize and circumvent these problems. The focus of this discussion will be the process of switching rows, or *partial pivoting*. The chapter ends with a brief description of efficient methods for solving *tridiagonal matrices*.

Chapter 10 illustrates how Gauss elimination can be formulated as an *LU factorization*. Such solution techniques are valuable for cases where many right-hand-side vectors need to be evaluated. The chapter ends with a brief outline of how MATLAB solves linear systems.

Chapter 11 starts with a description of how *LU* factorization can be employed to efficiently calculate the *matrix inverse,* which has tremendous utility in analyzing stimulus-response relationships of physical systems. The remainder of the chapter is devoted to the important concept of matrix condition. The condition number is introduced as a measure of the roundoff errors that can result when solving ill-conditioned matrices.

Chapter 12 deals with iterative solution techniques, which are similar in spirit to the approximate methods for roots of equations discussed in Chap. 6. That is, they involve guessing a solution and then iterating to obtain a refined estimate. The emphasis is on the *Gauss-Seidel* method, although a description is provided of an alternative approach, the *Jacobi method*. The chapter ends with a brief description of how *nonlinear simultaneous equations* can be solved.

8

Linear Algebraic Equations and Matrices

CHAPTER OBJECTIVES

The primary objective of this chapter is to acquaint you with linear algebraic equations and their relationship to matrices and matrix algebra. Specific objectives and topics covered are

- Understanding matrix notation.
- Being able to identify the following types of matrices: identity, diagonal, symmetric, triangular, and tridiagonal.
- Knowing how to perform matrix multiplication and being able to assess when it is feasible.
- Knowing how to represent a system of linear algebraic equations in matrix form.
- Knowing how to solve linear algebraic equations with left division and matrix inversion in MATLAB.

YOU'VE GOT A PROBLEM

Suppose that three jumpers are connected by bungee cords. Figure 8.1a shows them being held in place vertically so that each cord is fully extended but unstretched. We can define three distances, x_1, x_2, and x_3, as measured downward from each of their unstretched positions. After they are released, gravity takes hold and the jumpers will eventually come to the equilibrium positions shown in Fig. 8.1b.

Suppose that you are asked to compute the displacement of each of the jumpers. If we assume that each cord behaves as a linear spring and follows Hooke's law, free-body diagrams can be developed for each jumper as depicted in Fig. 8.2.

Using Newton's second law, a steady-state force balance can be written for each jumper:

$$m_1 g + k_2(x_2 - x_1) - k_1 x_1 = 0$$
$$m_2 g + k_3(x_3 - x_2) - k_2(x_2 - x_1) = 0$$
$$m_3 g - k_3(x_3 - x_2) = 0$$

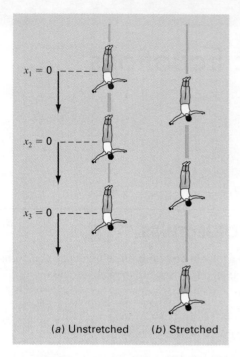

FIGURE 8.1
Three individuals connected by bungee cords.

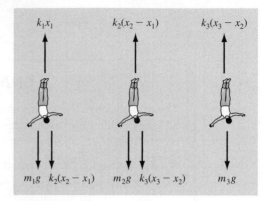

FIGURE 8.2
Free-body diagrams.

where m_i = the mass of jumper i (kg), k_j = the spring constant for cord j (N/m), x_i = the displacement of jumper i measured downward from the equilibrium position (m), and g = gravitational acceleration (9.81 m/s^2). Collecting terms gives

$$
\begin{aligned}
(k_1 + k_2)x_1 \quad\quad - k_2 x_2 \quad\quad\quad &= m_1 g \\
-k_2 x_1 + (k_2 + k_3)x_2 - k_3 x_3 &= m_2 g \\
-k_3 x_2 + k_3 x_3 &= m_3 g
\end{aligned}
\tag{8.1}
$$

Thus, the problem reduces to solving a system of three simultaneous equations for the three unknown displacements. Because we have used a linear law for the cords, these equations are linear algebraic equations. Chapters 8 through 12 will introduce you to how MATLAB is used to solve such systems of equations.

8.1   MATRIX ALGEBRA OVERVIEW

Knowledge of matrices is essential for understanding the solution of linear algebraic equations. The following sections outline how matrices provide a concise way to represent and manipulate linear algebraic equations.

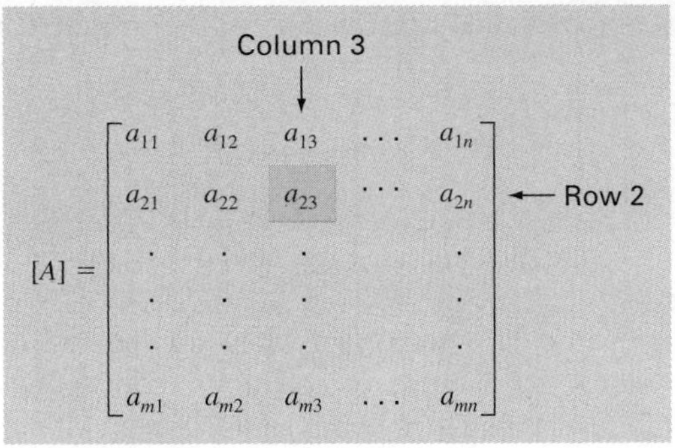

FIGURE 8.3
A matrix.

8.1.1 Matrix Notation

A *matrix* consists of a rectangular array of elements represented by a single symbol. As depicted in Fig. 8.3, $[A]$ is the shorthand notation for the matrix and a_{ij} designates an individual *element* of the matrix.

A horizontal set of elements is called a *row* and a vertical set is called a *column*. The first subscript i always designates the number of the row in which the element lies. The second subscript j designates the column. For example, element a_{23} is in row 2 and column 3.

The matrix in Fig. 8.3 has m rows and n columns and is said to have a dimension of m by n (or $m \times n$). It is referred to as an m by n matrix.

Matrices with row dimension $m = 1$, such as

$$[b] = [b_1 \quad b_2 \quad \cdots \quad b_n]$$

are called *row vectors*. Note that for simplicity, the first subscript of each element is dropped. Also, it should be mentioned that there are times when it is desirable to employ a special shorthand notation to distinguish a row matrix from other types of matrices. One way to accomplish this is to employ special open-topped brackets, as in $\lfloor b \rfloor$.[1]

Matrices with column dimension $n = 1$, such as

$$[c] = \begin{bmatrix} c_1 \\ c_2 \\ \vdots \\ c_m \end{bmatrix} \tag{8.2}$$

are referred to as *column vectors*. For simplicity, the second subscript is dropped. As with the row vector, there are occasions when it is desirable to employ a special shorthand notation to distinguish a column matrix from other types of matrices. One way to accomplish this is to employ special brackets, as in $\{c\}$.

---

[1] In addition to special brackets, we will use case to distinguish between vectors (lowercase) and matrices (uppercase).

Matrices where $m = n$ are called *square matrices*. For example, a 3×3 matrix is

$$[A] = \begin{bmatrix} a_{11} & a_{12} & a_{13} \\ a_{21} & a_{22} & a_{23} \\ a_{31} & a_{32} & a_{33} \end{bmatrix}$$

The diagonal consisting of the elements a_{11}, a_{22}, and a_{33} is termed the *principal* or *main diagonal* of the matrix.

Square matrices are particularly important when solving sets of simultaneous linear equations. For such systems, the number of equations (corresponding to rows) and the number of unknowns (corresponding to columns) must be equal for a unique solution to be possible. Consequently, square matrices of coefficients are encountered when dealing with such systems.

There are a number of special forms of square matrices that are important and should be noted:

A *symmetric matrix* is one where the rows equal the columns—that is, $a_{ij} = a_{ji}$ for all i's and j's. For example,

$$[A] = \begin{bmatrix} 5 & 1 & 2 \\ 1 & 3 & 7 \\ 2 & 7 & 8 \end{bmatrix}$$

is a 3×3 symmetric matrix.

A *diagonal matrix* is a square matrix where all elements off the main diagonal are equal to zero, as in

$$[A] = \begin{bmatrix} a_{11} & & \\ & a_{22} & \\ & & a_{33} \end{bmatrix}$$

Note that where large blocks of elements are zero, they are left blank.

An *identity matrix* is a diagonal matrix where all elements on the main diagonal are equal to 1, as in

$$[A] = \begin{bmatrix} 1 & & \\ & 1 & \\ & & 1 \end{bmatrix}$$

The symbol $[I]$ is used to denote the identity matrix. The identity matrix has properties similar to unity. That is,

$$[A][I] = [I][A] = [A]$$

An *upper triangular matrix* is one where all the elements below the main diagonal are zero, as in

$$[A] = \begin{bmatrix} a_{11} & a_{12} & a_{13} \\ & a_{22} & a_{23} \\ & & a_{33} \end{bmatrix}$$

A *lower triangular matrix* is one where all elements above the main diagonal are zero, as in

$$[A] = \begin{bmatrix} a_{11} & & \\ a_{21} & a_{22} & \\ a_{31} & a_{32} & a_{33} \end{bmatrix}$$

A *banded matrix* has all elements equal to zero, with the exception of a band centered on the main diagonal:

$$[A] = \begin{bmatrix} a_{11} & a_{12} & & \\ a_{21} & a_{22} & a_{23} & \\ & a_{32} & a_{33} & a_{34} \\ & & a_{43} & a_{44} \end{bmatrix}$$

The preceding matrix has a bandwidth of 3 and is given a special name—the *tridiagonal matrix*.

8.1.2 Matrix Operating Rules

Now that we have specified what we mean by a matrix, we can define some operating rules that govern its use. Two m by n matrices are equal if, and only if, every element in the first is equal to every element in the second—that is, $[A] = [B]$ if $a_{ij} = b_{ij}$ for all i and j.

Addition of two matrices, say, $[A]$ and $[B]$, is accomplished by adding corresponding terms in each matrix. The elements of the resulting matrix $[C]$ are computed as

$$c_{ij} = a_{ij} + b_{ij}$$

for $i = 1, 2, \ldots, m$ and $j = 1, 2, \ldots, n$. Similarly, the subtraction of two matrices, say, $[E]$ minus $[F]$, is obtained by subtracting corresponding terms, as in

$$d_{ij} = e_{ij} - f_{ij}$$

for $i = 1, 2, \ldots, m$ and $j = 1, 2, \ldots, n$. It follows directly from the preceding definitions that addition and subtraction can be performed only between matrices having the same dimensions.

Both addition and subtraction are commutative:

$$[A] + [B] = [B] + [A]$$

and associative:

$$([A] + [B]) + [C] = [A] + ([B] + [C])$$

The multiplication of a matrix $[A]$ by a scalar g is obtained by multiplying every element of $[A]$ by g. For example, for a 3×3 matrix:

$$[D] = g[A] = \begin{bmatrix} ga_{11} & ga_{12} & ga_{13} \\ ga_{21} & ga_{22} & ga_{23} \\ ga_{31} & ga_{32} & ga_{33} \end{bmatrix}$$

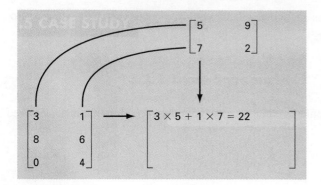

FIGURE 8.4
Visual depiction of how the rows and columns line up in matrix multiplication.

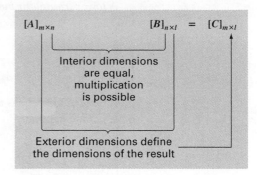

FIGURE 8.5
Matrix multiplication can be performed only if the inner dimensions are equal.

The product of two matrices is represented as $[C] = [A][B]$, where the elements of $[C]$ are defined as

$$c_{ij} = \sum_{k=1}^{n} a_{ik}b_{kj} \tag{8.3}$$

where $n =$ the column dimension of $[A]$ and the row dimension of $[B]$. That is, the c_{ij} element is obtained by adding the product of individual elements from the ith row of the first matrix, in this case $[A]$, by the jth column of the second matrix $[B]$. Figure 8.4 depicts how the rows and columns line up in matrix multiplication.

According to this definition, matrix multiplication can be performed only if the first matrix has as many columns as the number of rows in the second matrix. Thus, if $[A]$ is an m by n matrix, $[B]$ could be an n by l matrix. For this case, the resulting $[C]$ matrix would have the dimension of m by l. However, if $[B]$ were an m by l matrix, the multiplication could not be performed. Figure 8.5 provides an easy way to check whether two matrices can be multiplied.

If the dimensions of the matrices are suitable, matrix multiplication is *associative:*

$$([A][B])\,[C] = [A]([B][C])$$

and *distributive:*

$$[A]([B] + [C]) = [A][B] + [A][C]$$

or

$$([A] + [B])[C] = [A][C] + [B][C]$$

However, multiplication is not generally *commutative:*

$$[A][B] \neq [B][A]$$

That is, the order of multiplication is important.

Although multiplication is possible, matrix division is not a defined operation. However, if a matrix $[A]$ is square and nonsingular, there is another matrix $[A]^{-1}$, called the *inverse* of $[A]$, for which

$$[A][A]^{-1} = [A]^{-1}[A] = [I]$$

Thus, the multiplication of a matrix by the inverse is analogous to division, in the sense that a number divided by itself is equal to 1. That is, multiplication of a matrix by its inverse leads to the identity matrix.

The inverse of a 2×2 matrix can be represented simply by

$$[A]^{-1} = \frac{1}{a_{11}a_{22} - a_{12}a_{21}} \begin{bmatrix} a_{22} & -a_{12} \\ -a_{21} & a_{11} \end{bmatrix}$$

Similar formulas for higher-dimensional matrices are much more involved. Chapter 11 will deal with techniques for using numerical methods and the computer to calculate the inverse for such systems.

Two other matrix manipulations that will have utility in our discussion are the transpose and the augmentation of a matrix. The *transpose* of a matrix involves transforming its rows into columns and its columns into rows. For example, for the 3×3 matrix:

$$[A] = \begin{bmatrix} a_{11} & a_{12} & a_{13} \\ a_{21} & a_{22} & a_{23} \\ a_{31} & a_{32} & a_{33} \end{bmatrix}$$

the transpose, designated $[A]^T$, is defined as

$$[A]^T = \begin{bmatrix} a_{11} & a_{21} & a_{31} \\ a_{12} & a_{22} & a_{32} \\ a_{13} & a_{23} & a_{33} \end{bmatrix}$$

In other words, the element a_{ij} of the transpose is equal to the a_{ji} element of the original matrix.

The transpose has a variety of functions in matrix algebra. One simple advantage is that it allows a column vector to be written as a row, and vice versa. For example, if

$$\{c\} = \begin{Bmatrix} c_1 \\ c_1 \\ c_1 \end{Bmatrix}$$

then

$$\{c\}^T = \lfloor c_1 \quad c_2 \quad c_3 \rfloor$$

In addition, the transpose has numerous mathematical applications.

The final matrix manipulation that will have utility in our discussion is *augmentation*. A matrix is augmented by the addition of a column (or columns) to the original matrix. For example, suppose we have a 3×3 matrix of coefficients. We might wish to augment this matrix $[A]$ with a 3×3 identity matrix to yield a 3-by-6-dimensional matrix:

$$\begin{bmatrix} a_{11} & a_{11} & a_{11} & 1 & 0 & 0 \\ a_{21} & a_{21} & a_{21} & 0 & 1 & 0 \\ a_{31} & a_{31} & a_{31} & 0 & 0 & 1 \end{bmatrix}$$

Such an expression has utility when we must perform a set of identical operations on the rows of two matrices. Thus, we can perform the operations on the single augmented matrix rather than on the two individual matrices.

EXAMPLE 8.1   MATLAB Matrix Manipulations

Problem Statement.   The following example illustrates how a variety of matrix manipulations are implemented with MATLAB. It is best approached as a hands-on exercise on the computer.

Solution.   Create a 3 × 3 matrix:

```
>> A = [1 5 6;7 4 2;-3 6 7]

A =
        1       5       6
        7       4       2
       -3       6       7
```

The transpose of [A] can be obtained using the ' operator:

```
>> A'

ans =
        1       7      -3
        5       4       6
        6       2       7
```

Next we will create another 3 × 3 matrix on a row basis. First create three row vectors:

```
>> x = [8 6 9];
>> y = [-5 8 1];
>> z = [4 8 2];
```

Then we can combine these to form the matrix:

```
>> B = [x; y; z]

B =
        8       6       9
       -5       8       1
        4       8       2
```

We can add [A] and [B] together:

```
>> C = A+B

C =
        9      11      15
        2      12       3
        1      14       9
```

Further, we can subtract [*B*] from [*C*] to arrive back at [*A*]:

```
>> C = C-B

C =
      1      5      6
      7      4      2
     -3      6      7
```

Because their inner dimensions are equal, [*A*] and [*B*] can be multiplied

```
>> A*B

ans =
      7     94     26
     44     90     71
    -26     86     -7
```

Note that [*A*] and [*B*] can also be multiplied on an element-by-element basis by including a period with the multiplication operator as in

```
>> A.*B

ans =
      8     30     54
    -35     32      2
    -12     48     14
```

A 2 × 3 matrix can be set up

```
>> D = [1 4 3;5 8 1];
```

If [*A*] is multiplied times [*D*], an error message will occur

```
>> A*D

??? Error using ==> *
Inner matrix dimensions must agree.
```

However, if we reverse the order of multiplication so that the inner dimensions match, matrix multiplication works

```
>> D*A

ans =
     20     39     35
     58     63     53
```

The matrix inverse can be computed with the `inv` function:

```
>> AI = inv(A)

AI =
    0.2462     0.0154    -0.2154
   -0.8462     0.3846     0.6154
    0.8308    -0.3231    -0.4769
```

To test that this is the correct result, the inverse can be multiplied by the original matrix to give the identity matrix:

```
>> A*AI

ans =

    1.0000   -0.0000   -0.0000
    0.0000    1.0000   -0.0000
    0.0000   -0.0000    1.0000
```

The `eye` function can be used to generate an identity matrix:

```
>> I = eye(3)

I =

    1    0    0
    0    1    0
    0    0    1
```

Finally, matrices can be augmented simply as in

```
>> Aug = [A I]

Aug =

    1    5    6    1    0    0
    7    4    2    0    1    0
   -3    6    7    0    0    1
```

Note that the dimensions of a matrix can be determined by the `size` function:

```
>> [n,m] = size(Aug)

n =

    3

m =

    6
```

8.1.3 Representing Linear Algebraic Equations in Matrix Form

It should be clear that matrices provide a concise notation for representing simultaneous linear equations. For example, a 3×3 set of linear equations,

$$
\begin{aligned}
a_{11}x_1 + a_{12}x_2 + a_{13}x_3 &= b_1 \\
a_{21}x_1 + a_{22}x_2 + a_{23}x_3 &= b_2 \\
a_{31}x_1 + a_{32}x_2 + a_{33}x_3 &= b_3
\end{aligned}
\tag{8.4}
$$

can be expressed as

$$
[A]\{x\} = \{b\}
\tag{8.5}
$$

where $[A]$ is the matrix of coefficients:

$$[A] = \begin{bmatrix} a_{11} & a_{12} & a_{13} \\ a_{21} & a_{22} & a_{23} \\ a_{31} & a_{32} & a_{33} \end{bmatrix}$$

$\{b\}$ is the column vector of constants:

$$\{b\}^T = \lfloor b_1 \quad b_2 \quad b_3 \rfloor$$

and $\{x\}$ is the column vector of unknowns:

$$\{x\}^T = \lfloor x_1 \quad x_2 \quad x_3 \rfloor$$

Recall the definition of matrix multiplication [Eq. (8.3)] to convince yourself that Eqs. (8.4) and (8.5) are equivalent. Also, realize that Eq. (8.5) is a valid matrix multiplication because the number of columns n of the first matrix $[A]$ is equal to the number of rows n of the second matrix $\{x\}$.

This part of the book is devoted to solving Eq. (8.5) for $\{x\}$. A formal way to obtain a solution using matrix algebra is to multiply each side of the equation by the inverse of $[A]$ to yield

$$[A]^{-1}[A]\{x\} = [A]^{-1}\{b\}$$

Because $[A]^{-1}[A]$ equals the identity matrix, the equation becomes

$$\{x\} = [A]^{-1}\{b\} \tag{8.6}$$

Therefore, the equation has been solved for $\{x\}$. This is another example of how the inverse plays a role in matrix algebra that is similar to division. It should be noted that this is not a very efficient way to solve a system of equations. Thus, other approaches are employed in numerical algorithms. However, as discussed in Section 11.1.2, the matrix inverse itself has great value in the engineering analyses of such systems.

It should be noted that systems with more equations (rows) than unknowns (columns), $m > n$, are said to be *overdetermined*. A typical example is least-squares regression where an equation with n coefficients is fit to m data points (x, y). Conversely, systems with less equations than unknowns, $m < n$, are said to be *underdetermined*. A typical example of underdetermined systems is numerical optimization.

8.2 SOLVING LINEAR ALGEBRAIC EQUATIONS WITH MATLAB

MATLAB provides two direct ways to solve systems of linear algebraic equations. The most efficient way is to employ the backslash, or "left-division," operator as in

```
>> x = A\b
```

The second is to use matrix inversion:

```
>> x = inv(A)*b
```

As stated at the end of Section 8.1.3, the matrix inverse solution is less efficient than using the backslash. Both options are illustrated in the following example.

EXAMPLE 8.2   Solving the Bungee Jumper Problem with MATLAB

Problem Statement.   Use MATLAB to solve the bungee jumper problem described at the beginning of this chapter. The parameters for the problem are

Jumper	Mass (kg)	Spring Constant (N/m)	Unstretched Cord Length (m)
Top (1)	60	50	20
Middle (2)	70	100	20
Bottom (3)	80	50	20

Solution.   Substituting these parameter values into Eq. (8.1) gives

$$
\begin{bmatrix} 150 & -100 & 0 \\ -100 & 150 & -50 \\ 0 & -50 & 50 \end{bmatrix} \begin{Bmatrix} x_1 \\ x_2 \\ x_3 \end{Bmatrix} = \begin{Bmatrix} 588.6 \\ 686.7 \\ 784.8 \end{Bmatrix}
$$

Start up MATLAB and enter the coefficient matrix and the right-hand-side vector:

```
>> K = [150 -100 0;-100 150 -50;0 -50 50]

K =
   150   -100      0
  -100    150    -50
     0    -50     50

>> mg = [588.6; 686.7; 784.8]

mg =
  588.6000
  686.7000
  784.8000
```

Employing left division yields

```
>> x = K\mg

x =
   41.2020
   55.9170
   71.6130
```

Alternatively, multiplying the inverse of the coefficient matrix by the right-hand-side vector gives the same result:

```
>> x = inv(K)*mg

x =
    41.2020
    55.9170
    71.6130
```

Because the jumpers were connected by 20-m cords, their initial positions relative to the platform is

```
>> xi = [20;40;60];
```

Thus, their final positions can be calculated as

```
>> xf = x+xi

xf =
    61.2020
    95.9170
   131.6130
```

The results, which are displayed in Fig. 8.6, make sense. The first cord is extended the longest because it has a lower spring constant and is subject to the most weight (all three jumpers). Notice that the second and third cords are extended about the same amount. Because it is subject to the weight of two jumpers, one might expect the second cord to be extended longer than the third. However, because it is stiffer (i.e., it has a higher spring constant), it stretches less than expected based on the weight it carries.

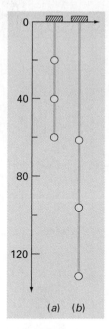

FIGURE 8.6
Positions of three individuals connected by bungee cords. (*a*) Unstretched and (*b*) stretched.

8.3 CASE STUDY   CURRENTS AND VOLTAGES IN CIRCUITS

Background.   Recall that in Chap. 1 (Table 1.1), we summarized some models and associated conservation laws that figure prominently in engineering. As in Fig. 8.7, each model represents a system of interacting elements. Consequently, steady-state balances derived from the conservation laws yield systems of simultaneous equations. In many cases, such systems are linear and hence can be expressed in matrix form. The present case study focuses on one such application: circuit analysis.

A common problem in electrical engineering involves determining the currents and voltages at various locations in resistor circuits. These problems are solved using *Kirchhoff's current* and *voltage rules*. The *current* (or point) *rule* states that the algebraic sum of all currents entering a node must be zero (Fig. 8.8*a*), or

$$\sum i = 0 \tag{8.7}$$

where all current entering the node is considered positive in sign. The current rule is an application of the principle of *conservation of charge* (recall Table 1.1).

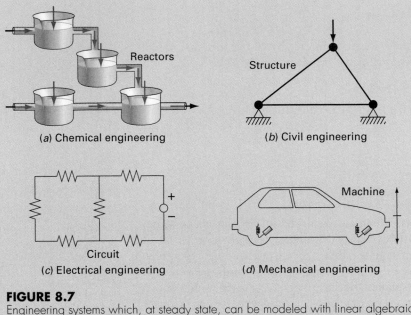

(a) Chemical engineering

(b) Civil engineering

(c) Electrical engineering

(d) Mechanical engineering

FIGURE 8.7
Engineering systems which, at steady state, can be modeled with linear algebraic equations.

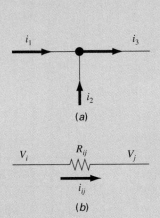

FIGURE 8.8
Schematic representations of (a) Kirchhoff's current rule and (b) Ohm's law.

The *voltage* (or loop) *rule* specifies that the algebraic sum of the potential differences (that is, voltage changes) in any loop must equal zero. For a resistor circuit, this is expressed as

$$\sum \xi - \sum iR = 0 \tag{8.8}$$

where ξ is the emf (electromotive force) of the voltage sources, and R is the resistance of any resistors on the loop. Note that the second term derives from *Ohm's law* (Fig. 8.8b), which states that the voltage drop across an ideal resistor is equal to the product of the current and the resistance. Kirchhoff's voltage rule is an expression of the *conservation of energy*.

Solution.   Application of these rules results in systems of simultaneous linear algebraic equations because the various loops within a circuit are interconnected. For example, consider the circuit shown in Fig. 8.9. The currents associated with this circuit are unknown both in magnitude and direction. This presents no great difficulty because one simply assumes a direction for each current. If the resultant solution from Kirchhoff's laws is negative, then the assumed direction was incorrect. For example, Fig. 8.10 shows some assumed currents.

8.3 CASE STUDY continued

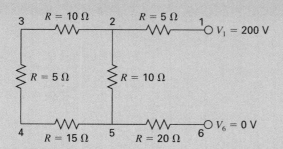

FIGURE 8.9
A resistor circuit to be solved using simultaneous
linear algebraic equations.

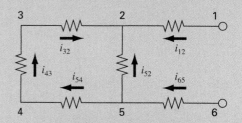

FIGURE 8.10
Assumed current directions.

Given these assumptions, Kirchhoff's current rule is applied at each node to yield

$$i_{12} + i_{52} + i_{32} = 0$$
$$i_{65} - i_{52} - i_{54} = 0$$
$$i_{43} - i_{32} = 0$$
$$i_{54} - i_{43} = 0$$

Application of the voltage rule to each of the two loops gives

$$i_{65} - i_{52} - i_{54} = 0$$
$$-i_{54}R_{54} - i_{43}R_{43} - i_{32}R_{32} + i_{52}R_{52} = 0$$
$$-i_{65}R_{65} - i_{52}R_{52} + i_{12}R_{12} - 200 = 0$$

or, substituting the resistances from Fig. 8.9 and bringing constants to the right-hand side,

$$-15i_{54} - 5i_{43} - 10i_{32} + 10i_{52} = 0$$
$$-20i_{65} - 10i_{52} + 5i_{12} = 200$$

Therefore, the problem amounts to solving six equations with six unknown currents. These equations can be expressed in matrix form as

$$
\begin{bmatrix}
1 & 1 & 1 & 0 & 0 & 0 \\
0 & -1 & 0 & 1 & -1 & 0 \\
0 & 0 & -1 & 0 & 0 & 1 \\
0 & 0 & 0 & 0 & 1 & -1 \\
0 & 10 & -10 & 0 & -15 & -5 \\
5 & -10 & 0 & -20 & 0 & 0
\end{bmatrix}
\begin{Bmatrix}
i_{12} \\
i_{52} \\
i_{32} \\
i_{65} \\
i_{54} \\
i_{43}
\end{Bmatrix}
=
\begin{Bmatrix}
0 \\
0 \\
0 \\
0 \\
0 \\
200
\end{Bmatrix}
$$

Although impractical to solve by hand, this system is easily handled by MATLAB. The solution is

```
>> A=[1 1 1 0 0 0
0 -1 0 1 -1 0
0 0 -1 0 0 1
0 0 0 0 1 -1
0 10 -10 0 -15 -5
5 -10 0 -20 0 0];
>> b=[0 0 0 0 0 200]';
>> current=A\b

current =
    6.1538
   -4.6154
   -1.5385
   -6.1538
   -1.5385
   -1.5385
```

Thus, with proper interpretation of the signs of the result, the circuit currents and voltages are as shown in Fig. 8.11. The advantages of using MATLAB for problems of this type should be evident.

FIGURE 8.11
The solution for currents and voltages obtained using MATLAB.

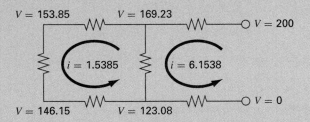

PROBLEMS

8.1 Given a square matrix $[A]$, write a single line MATLAB command that will create a new matrix $[Aug]$ that consists of the original matrix $[A]$ augmented by an identity matrix $[I]$.

8.2 A number of matrices are defined as

$$[A] = \begin{bmatrix} 4 & 7 \\ 1 & 2 \\ 5 & 6 \end{bmatrix} \qquad [B] = \begin{bmatrix} 4 & 3 & 7 \\ 1 & 2 & 7 \\ 2 & 0 & 4 \end{bmatrix}$$

$$\{C\} = \begin{Bmatrix} 3 \\ 6 \\ 1 \end{Bmatrix} \qquad [D] = \begin{bmatrix} 9 & 4 & 3 & -6 \\ 2 & -1 & 7 & 5 \end{bmatrix}$$

$$[E] = \begin{bmatrix} 1 & 5 & 8 \\ 7 & 2 & 3 \\ 4 & 0 & 6 \end{bmatrix}$$

$$[F] = \begin{bmatrix} 3 & 0 & 1 \\ 1 & 7 & 3 \end{bmatrix} \qquad \lfloor G \rfloor = \lfloor 7 \quad 6 \quad 4 \rfloor$$

Answer the following questions regarding these matrices:

(a) What are the dimensions of the matrices?
(b) Identify the square, column, and row matrices.
(c) What are the values of the elements: a_{12}, b_{23}, d_{32}, e_{22}, f_{12}, g_{12}?
(d) Perform the following operations:

(1) $[E] + [B]$ (2) $[A] + [F]$ (3) $[B] - [E]$
(4) $7 \times [B]$ (5) $[A] \times [B]$ (6) $\{C\}^T$
(7) $[B] \times [A]$ (8) $[D]^T$ (9) $[A] \times \{C\}$
(10) $[I] \times [B]$ (11) $[E]^T \times [E]$ (12) $\{C\}^T \times \{C\}$

8.3 Write the following set of equations in matrix form:

$$50 = 5x_3 - 7x_2$$
$$4x_2 + 7x_3 + 30 = 0$$
$$x_1 - 7x_3 = 40 - 3x_2 + 5x_1$$

Use MATLAB to solve for the unknowns. In addition, use it to compute the transpose and the inverse of the coefficient matrix.

8.4 Three matrices are defined as

$$[A] = \begin{bmatrix} 6 & -1 \\ 12 & 8 \\ -5 & 4 \end{bmatrix} \quad [B] = \begin{bmatrix} 4 & 0 \\ 0.5 & 2 \end{bmatrix} \quad [C] = \begin{bmatrix} 2 & -2 \\ -3 & 1 \end{bmatrix}$$

(a) Perform all possible multiplications that can be computed between pairs of these matrices.
(b) Justify why the remaining pairs cannot be multiplied.
(c) Use the results of (a) to illustrate why the order of multiplication is important.

8.5 Five reactors linked by pipes are shown in Fig. P8.5. The rate of mass flow through each pipe is computed as the product of flow (Q) and concentration (c). At steady state, the mass flow into and out of each reactor must be equal.

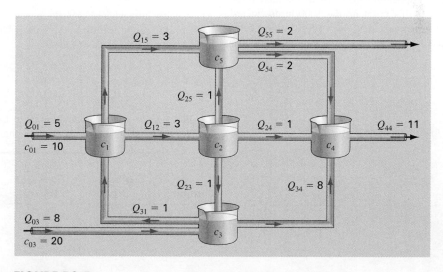

FIGURE P8.5

For example, for the first reactor, a *mass balance* can be written as

$$Q_{01}c_{01} + Q_{31}c_3 = Q_{15}c_1 + Q_{12}c_1$$

Write mass balances for the remaining reactors in Fig. P8.5 and express the equations in matrix form. Then use MATLAB to solve for the concentrations in each reactor.

8.6 An important problem in structural engineering is that of finding the forces in a statically determinate truss (Fig. P8.6). This type of structure can be described as a system of coupled linear algebraic equations derived from force balances. The sum of the forces in both horizontal and vertical directions must be zero at each node, because the system is at rest. Therefore, for node 1:

$$\sum F_H = 0 = -F_1 \cos 30° + F_3 \cos 60° + F_{1,h}$$
$$\sum F_V = 0 = -F_1 \sin 30° - F_3 \sin 60° + F_{1,v}$$

for node 2:

$$\sum F_H = 0 = F_2 + F_1 \cos 30° + F_{2,h} + H_2$$
$$\sum F_V = 0 = F_1 \sin 30° + F_{2,v} + V_2$$

for node 3:

$$\sum F_H = 0 = -F_2 - F_3 \cos 60° + F_{3,h}$$
$$\sum F_V = 0 = F_3 \sin 60° + F_{3,v} + V_3$$

where $F_{i,h}$ is the external horizontal force applied to node i (where a positive force is from left to right) and $F_{i,v}$ is the external vertical force applied to node i (where a positive force is upward). Thus, in this problem, the 1000-lb downward force on node 1 corresponds to $F_{i,v} = -1000$. For this case, all other $F_{i,v}$'s and $F_{i,h}$'s are zero. Express this set of linear

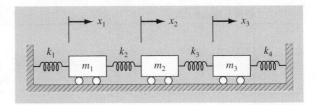

FIGURE P8.7

algebraic equations in matrix form and then use MATLAB to solve for the unknowns.

8.7 Consider the three mass-four spring system in Fig. P8.7. Determining the equations of motion from $\sum F_x = ma_x$ for each mass using its free-body diagram results in the following differential equations:

$$\ddot{x}_1 + \left(\frac{k_1 + k_2}{m_1}\right)x_1 - \left(\frac{k_2}{m_1}\right)x_2 = 0$$

$$\ddot{x}_2 - \left(\frac{k_2}{m_2}\right)x_1 + \left(\frac{k_2 + k_3}{m_2}\right)x_2 - \left(\frac{k_3}{m_2}\right)x_3 = 0$$

$$\ddot{x}_3 - \left(\frac{k_3}{m_3}\right)x_2 + \left(\frac{k_3 + k_4}{m_3}\right)x_3 = 0$$

where $k_1 = k_4 = 10$ N/m, $k_2 = k_3 = 30$ N/m, and $m_1 = m_2 = m_3 = m_4 = 1$ kg. The three equations can be written in matrix form:

$$0 = \{\text{Acceleration vector}\}$$
$$+ [k/m \text{ matrix}]\{\text{displacement vector } x\}$$

At a specific time where $x_1 = 0.05$ m, $x_2 = 0.04$ m, and $x_3 = 0.03$ m, this forms a tridiagonal matrix. Use MATLAB to solve for the acceleration of each mass.

8.8 Solve

$$\begin{bmatrix} 3 + 2i & 4 \\ -i & 1 \end{bmatrix} \begin{Bmatrix} z_1 \\ z_2 \end{Bmatrix} = \begin{Bmatrix} 2 + i \\ 3 \end{Bmatrix}$$

8.9 Perform the same computation as in Example 8.2, but use five parachutists with the following characteristics:

Jumper	Mass (kg)	Spring Constant (N/m)	Unstretched Cord Length (m)
1	55	80	10
2	75	50	10
3	60	70	10
4	75	100	10
5	90	20	10

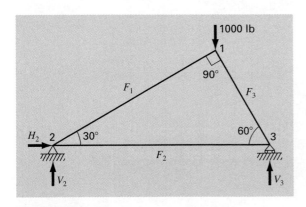

FIGURE P8.6

8.10 Three masses are suspended vertically by a series of identical springs where mass 1 is at the top and mass 3 is at the bottom. If $g = 9.81 \text{ m/s}^2$, $m_1 = 2$ kg, $m_2 = 3$ kg, $m_3 = 2.5$ kg, and the k's $= 10 \text{ kg/s}^2$, use MATLAB to solve for the displacements x.

8.11 Perform the same computation as in Sec. 8.3, but for the circuit in Fig. P8.11.

8.12 Perform the same computation as in Sec. 8.3, but for the circuit in Fig. P8.12.

8.13 Develop, debug, and test your own M-file to multiply two matrices—that is, $[X] = [Y][Z]$, where $[Y]$ is m by n and $[Z]$ is n by p. Employ `for...end` loops to implement the multiplication and include error traps to flag bad cases. Test the program using the matrices from Prob. 8.4.

8.14 Develop, debug, and test your own M-file to generate the transpose of a matrix. Employ `for...end` loops to implement the transpose. Test it on the matrices from Prob. 8.4.

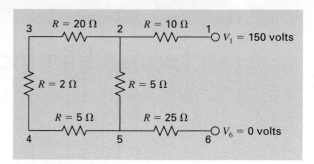

FIGURE P8.11

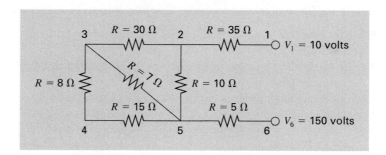

FIGURE P8.12

9

Gauss Elimination

CHAPTER OBJECTIVES

The primary objective of this chapter is to describe the Gauss elimination algorithm for solving linear algebraic equations. Specific objectives and topics covered are

- Knowing how to solve small sets of linear equations with the graphical method and Cramer's rule.
- Understanding how to implement forward elimination and back substitution as in Gauss elimination.
- Understanding how to count flops to evaluate the efficiency of an algorithm.
- Understanding the concepts of singularity and ill-condition.
- Understanding how partial pivoting is implemented and how it differs from complete pivoting.
- Recognizing how the banded structure of a tridiagonal system can be exploited to obtain extremely efficient solutions.

At the end of Chap. 8, we stated that MATLAB provides two simple and direct methods for solving systems of linear algebraic equations: left-division,

```
>> x = A\b
```

and matrix inversion,

```
>> x = inv(A)*b
```

Chapters 9 and 10 provide background on how such solutions are obtained. This material is included to provide insight into how MATLAB operates. In addition, it is intended to show how you can build your own solution algorithms in computational environments that do not have MATLAB's built-in capabilities.

The technique described in this chapter is called Gauss elimination because it involves combining equations to eliminate unknowns. Although it is one of the earliest methods for solving simultaneous equations, it remains among the most important algorithms in use today and is the basis for linear equation solving on many popular software packages including MATLAB.

9.1 SOLVING SMALL NUMBERS OF EQUATIONS

Before proceeding to Gauss elimination, we will describe several methods that are appropriate for solving small ($n \leq 3$) sets of simultaneous equations and that do not require a computer. These are the graphical method, Cramer's rule, and the elimination of unknowns.

9.1.1 The Graphical Method

A graphical solution is obtainable for two linear equations by plotting them on Cartesian coordinates with one axis corresponding to x_1 and the other to x_2. Because the equations are linear, each equation will plot as a straight line. For example, suppose that we have the following equations:

$$3x_1 + 2x_2 = 18$$
$$-x_1 + 2x_2 = 2$$

If we assume that x_1 is the abscissa, we can solve each of these equations for x_2:

$$x_2 = -\frac{3}{2}x_1 + 9$$

$$x_2 = \frac{1}{2}x_1 + 1$$

The equations are now in the form of straight lines—that is, $x_2 = $ (slope) $x_1 + $ intercept. When these equations are graphed, the values of x_1 and x_2 at the intersection of the lines represent the solution (Fig. 9.1). For this case, the solution is $x_1 = 4$ and $x_2 = 3$.

For three simultaneous equations, each equation would be represented by a plane in a three-dimensional coordinate system. The point where the three planes intersect would represent the solution. Beyond three equations, graphical methods break down and, consequently, have little practical value for solving simultaneous equations. However, they are useful in visualizing properties of the solutions.

For example, Fig. 9.2 depicts three cases that can pose problems when solving sets of linear equations. Fig. 9.2a shows the case where the two equations represent parallel lines. For such situations, there is no solution because the lines never cross. Figure 9.2b depicts the case where the two lines are coincident. For such situations there is an infinite number of solutions. Both types of systems are said to be *singular*.

In addition, systems that are very close to being singular (Fig. 9.2c) can also cause problems. These systems are said to be *ill-conditioned*. Graphically, this corresponds to the fact that it is difficult to identify the exact point at which the lines intersect. Ill-conditioned systems will also pose problems when they are encountered during the numerical solution of linear equations. This is because they will be extremely sensitive to roundoff error.

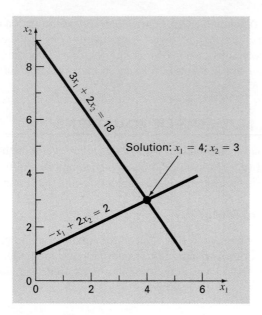

FIGURE 9.1
Graphical solution of a set of two simultaneous linear algebraic equations. The intersection of the lines represents the solution.

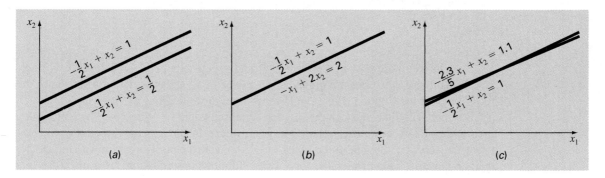

FIGURE 9.2
Graphical depiction of singular and ill-conditioned systems: (a) no solution, (b) infinite solutions, and (c) ill-conditioned system where the slopes are so close that the point of intersection is difficult to detect visually.

9.1.2 Determinants and Cramer's Rule

Cramer's rule is another solution technique that is best suited to small numbers of equations. Before describing this method, we will briefly review the concept of the determinant, which is used to implement Cramer's rule. In addition, the determinant has relevance to the evaluation of the ill-conditioning of a matrix.

Determinants. The determinant can be illustrated for a set of three equations:

$$[A]\{x\} = \{b\}$$

where $[A]$ is the coefficient matrix

$$[A] = \begin{bmatrix} a_{11} & a_{12} & a_{13} \\ a_{21} & a_{22} & a_{23} \\ a_{31} & a_{32} & a_{33} \end{bmatrix}$$

The *determinant* of this system is formed from the coefficients of $[A]$ and is represented as

$$D = \begin{vmatrix} a_{11} & a_{12} & a_{13} \\ a_{21} & a_{22} & a_{23} \\ a_{31} & a_{32} & a_{33} \end{vmatrix}$$

Although the determinant D and the coefficient matrix $[A]$ are composed of the same elements, they are completely different mathematical concepts. That is why they are distinguished visually by using brackets to enclose the matrix and straight lines to enclose the determinant. In contrast to a matrix, the determinant is a single number. For example, the value of the determinant for two simultaneous equations

$$D = \begin{vmatrix} a_{11} & a_{12} \\ a_{21} & a_{22} \end{vmatrix}$$

is calculated by

$$D = a_{11}a_{22} - a_{12}a_{21}$$

For the third-order case, the determinant can be computed as

$$D = a_{11}\begin{vmatrix} a_{22} & a_{23} \\ a_{32} & a_{33} \end{vmatrix} - a_{12}\begin{vmatrix} a_{21} & a_{23} \\ a_{31} & a_{33} \end{vmatrix} + a_{13}\begin{vmatrix} a_{21} & a_{22} \\ a_{31} & a_{32} \end{vmatrix} \tag{9.1}$$

where the 2 by 2 determinants are called *minors*.

EXAMPLE 9.1   Determinants

Problem Statement.   Compute values for the determinants of the systems represented in Figs. 9.1 and 9.2.

Solution.   For Fig. 9.1:

$$D = \begin{vmatrix} 3 & 2 \\ -1 & 2 \end{vmatrix} = 3(2) - 2(-1) = 8$$

For Fig. 9.2a:

$$D = \begin{vmatrix} -\frac{1}{2} & 1 \\ -\frac{1}{2} & 1 \end{vmatrix} = -\frac{1}{2}(1) - 1\left(\frac{-1}{2}\right) = 0$$

For Fig. 9.2b:

$$D = \begin{vmatrix} -\frac{1}{2} & 1 \\ -1 & 2 \end{vmatrix} = -\frac{1}{2}(2) - 1(-1) = 0$$

For Fig. 9.2c:

$$D = \begin{vmatrix} -\frac{1}{2} & 1 \\ -\frac{2.3}{5} & 1 \end{vmatrix} = -\frac{1}{2}(1) - 1\left(\frac{-2.3}{5}\right) = -0.04$$

In the foregoing example, the singular systems had zero determinants. Additionally, the results suggest that the system that is almost singular (Fig. 9.2c) has a determinant that is close to zero. These ideas will be pursued further in our subsequent discussion of ill-conditioning in Chap. 11.

Cramer's Rule. This rule states that each unknown in a system of linear algebraic equations may be expressed as a fraction of two determinants with denominator D and with the numerator obtained from D by replacing the column of coefficients of the unknown in question by the constants $b_1, b_2, \ldots, b_n$. For example, for three equations, x_1 would be computed as

$$x_1 = \frac{\begin{vmatrix} b_1 & a_{12} & a_{13} \\ b_2 & a_{22} & a_{23} \\ b_3 & a_{32} & a_{33} \end{vmatrix}}{D}$$

EXAMPLE 9.2   Cramer's Rule

Problem Statement. Use Cramer's rule to solve

$$0.3x_1 + 0.52x_2 + x_3 = -0.01$$
$$0.5x_1 + x_2 + 1.9x_3 = 0.67$$
$$0.1x_1 + 0.3\ x_2 + 0.5x_3 = -0.44$$

Solution. The determinant D can be evaluated as [Eq. (9.1)]:

$$D = 0.3\begin{vmatrix} 1 & 1.9 \\ 0.3 & 0.5 \end{vmatrix} - 0.52\begin{vmatrix} 0.5 & 1.9 \\ 0.1 & 0.5 \end{vmatrix} + 1\begin{vmatrix} 0.5 & 1 \\ 0.1 & 0.3 \end{vmatrix} = -0.0022$$

The solution can be calculated as

$$x_1 = \frac{\begin{vmatrix} -0.01 & 0.52 & 1 \\ 0.67 & 1 & 1.9 \\ -0.44 & 0.3 & 0.5 \end{vmatrix}}{-0.0022} = \frac{0.03278}{-0.0022} = -14.9$$

$$x_2 = \frac{\begin{vmatrix} 0.3 & -0.01 & 1 \\ 0.5 & 0.67 & 1.9 \\ 0.1 & -0.44 & 0.5 \end{vmatrix}}{-0.0022} = \frac{0.0649}{-0.0022} = -29.5$$

$$x_3 = \frac{\begin{vmatrix} 0.3 & 0.52 & -0.01 \\ 0.5 & 1 & 0.67 \\ 0.1 & 0.3 & -0.44 \end{vmatrix}}{-0.0022} = \frac{-0.04356}{-0.0022} = 19.8$$

The `det` **Function.** The determinant can be computed directly in MATLAB with the `det` function. For example, using the system from the previous example:

```
>> A=[0.3 0.52 1;0.5 1 1.9;0.1 0.3 0.5];
>> D=det(A)

D =
    -0.0022
```

Cramer's rule can be applied to compute x_1 as in

```
>> A(:,1)=[-0.01;0.67;-0.44]

A =
    -0.0100    0.5200    1.0000
     0.6700    1.0000    1.9000
    -0.4400    0.3000    0.5000

>> x1=det(A)/D

x1 =
   -14.9000
```

For more than three equations, Cramer's rule becomes impractical because, as the number of equations increases, the determinants are time consuming to evaluate by hand (or by computer). Consequently, more efficient alternatives are used. Some of these alternatives are based on the last noncomputer solution technique covered in Section 9.1.3—the elimination of unknowns.

9.1.3 Elimination of Unknowns

The elimination of unknowns by combining equations is an algebraic approach that can be illustrated for a set of two equations:

$$a_{11}x_1 + a_{12}x_2 = b_1 \tag{9.2}$$
$$a_{21}x_1 + a_{22}x_2 = b_2 \tag{9.3}$$

The basic strategy is to multiply the equations by constants so that one of the unknowns will be eliminated when the two equations are combined. The result is a single equation that can be solved for the remaining unknown. This value can then be substituted into either of the original equations to compute the other variable.

For example, Eq. (9.2) might be multiplied by a_{21} and Eq. (9.3) by a_{11} to give

$$a_{21}a_{11}x_1 + a_{21}a_{12}x_2 = a_{21}b_1 \tag{9.4}$$
$$a_{11}a_{21}x_1 + a_{11}a_{22}x_2 = a_{11}b_2 \tag{9.5}$$

Subtracting Eq. (9.4) from Eq. (9.5) will, therefore, eliminate the x_1 term from the equations to yield

$$a_{11}a_{22}x_2 - a_{21}a_{12}x_2 = a_{11}b_2 - a_{21}b_1$$

which can be solved for

$$x_2 = \frac{a_{11}b_2 - a_{21}b_1}{a_{11}a_{22} - a_{21}a_{12}} \tag{9.6}$$

Equation (9.6) can then be substituted into Eq. (9.2), which can be solved for

$$x_1 = \frac{a_{22}b_1 - a_{12}b_2}{a_{11}a_{22} - a_{21}a_{12}} \tag{9.7}$$

Notice that Eqs. (9.6) and (9.7) follow directly from Cramer's rule:

$$x_1 = \frac{\begin{vmatrix} b_1 & a_{12} \\ b_2 & a_{22} \end{vmatrix}}{\begin{vmatrix} a_{11} & a_{12} \\ a_{21} & a_{22} \end{vmatrix}} = \frac{a_{22}b_1 - a_{12}b_2}{a_{11}a_{22} - a_{21}a_{12}}$$

$$x_2 = \frac{\begin{vmatrix} a_{11} & b_1 \\ a_{21} & b_2 \end{vmatrix}}{\begin{vmatrix} a_{11} & a_{12} \\ a_{21} & a_{22} \end{vmatrix}} = \frac{a_{11}b_2 - a_{21}b_1}{a_{11}a_{22} - a_{21}a_{12}}$$

The elimination of unknowns can be extended to systems with more than two or three equations. However, the numerous calculations that are required for larger systems make the method extremely tedious to implement by hand. However, as described in Section 9.2, the technique can be formalized and readily programmed for the computer.

9.2 NAIVE GAUSS ELIMINATION

In Section 9.1.3, the elimination of unknowns was used to solve a pair of simultaneous equations. The procedure consisted of two steps (Fig. 9.3):

1. The equations were manipulated to eliminate one of the unknowns from the equations. The result of this elimination step was that we had one equation with one unknown.
2. Consequently, this equation could be solved directly and the result back-substituted into one of the original equations to solve for the remaining unknown.

This basic approach can be extended to large sets of equations by developing a systematic scheme or algorithm to eliminate unknowns and to back-substitute. Gauss elimination is the most basic of these schemes.

This section includes the systematic techniques for forward elimination and back substitution that comprise Gauss elimination. Although these techniques are ideally suited for implementation on computers, some modifications will be required to obtain a reliable algorithm. In particular, the computer program must avoid division by zero. The following method is called "naive" Gauss elimination because it does not avoid this problem. Section 9.3 will deal with the additional features required for an effective computer program.

The approach is designed to solve a general set of n equations:

$$a_{11}x_1 + a_{12}x_2 + a_{13}x_3 + \cdots + a_{1n}x_n = b_1 \tag{9.8a}$$

$$a_{21}x_1 + a_{22}x_2 + a_{23}x_3 + \cdots + a_{2n}x_n = b_2 \tag{9.8b}$$

$$\vdots \qquad \vdots$$

$$a_{n1}x_1 + a_{n2}x_2 + a_{n3}x_3 + \cdots + a_{nn}x_n = b_n \tag{9.8c}$$

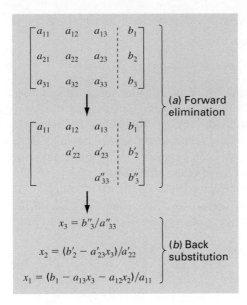

$$\begin{bmatrix} a_{11} & a_{12} & a_{13} & \vdots & b_1 \\ a_{21} & a_{22} & a_{23} & \vdots & b_2 \\ a_{31} & a_{32} & a_{33} & \vdots & b_3 \end{bmatrix}$$

$\downarrow$ (a) Forward elimination

$$\begin{bmatrix} a_{11} & a_{12} & a_{13} & \vdots & b_1 \\ & a'_{22} & a'_{23} & \vdots & b'_2 \\ & & a''_{33} & \vdots & b''_3 \end{bmatrix}$$

$\downarrow$

$$x_3 = b''_3 / a''_{33}$$

$$x_2 = (b'_2 - a'_{23}x_3)/a'_{22}$$

$$x_1 = (b_1 - a_{13}x_3 - a_{12}x_2)/a_{11}$$

(b) Back substitution

FIGURE 9.3
The two phases of Gauss elimination: (a) forward elimination and (b) back substitution.

As was the case with the solution of two equations, the technique for n equations consists of two phases: elimination of unknowns and solution through back substitution.

Forward Elimination of Unknowns. The first phase is designed to reduce the set of equations to an upper triangular system (Fig. 9.3a). The initial step will be to eliminate the first unknown x_1 from the second through the nth equations. To do this, multiply Eq. (9.8a) by a_{21}/a_{11} to give

$$a_{21}x_1 + \frac{a_{21}}{a_{11}}a_{12}x_2 + \frac{a_{21}}{a_{11}}a_{13}x_3 + \cdots + \frac{a_{21}}{a_{11}}a_{1n}x_n = \frac{a_{21}}{a_{11}}b_1 \qquad (9.9)$$

Now this equation can be subtracted from Eq. (9.8b) to give

$$\left(a_{22} - \frac{a_{21}}{a_{11}}a_{12}\right)x_2 + \cdots + \left(a_{2n} - \frac{a_{21}}{a_{11}}a_{1n}\right)x_n = b_2 - \frac{a_{21}}{a_{11}}b_1$$

or

$$a'_{22}x_2 + \cdots + a'_{2n}x_n = b'_2$$

where the prime indicates that the elements have been changed from their original values.
The procedure is then repeated for the remaining equations. For instance, Eq. (9.8a) can be multiplied by a_{31}/a_{11} and the result subtracted from the third equation. Repeating the procedure for the remaining equations results in the following modified system:

$$a_{11}x_1 + a_{12}x_2 + a_{13}x_3 + \cdots + a_{1n}x_n = b_1 \qquad (9.10a)$$

$$a'_{22}x_2 + a'_{23}x_3 + \cdots + a'_{2n}x_n = b'_2 \qquad (9.10b)$$

$$a'_{32}x_2 + a'_{33}x_3 + \cdots + a'_{3n}x_n = b'_3 \qquad (9.10c)$$

$$\vdots \qquad \qquad \vdots$$

$$a'_{n2}x_2 + a'_{n3}x_3 + \cdots + a'_{nn}x_n = b'_n \qquad (9.10d)$$

For the foregoing steps, Eq. (9.8a) is called the *pivot equation* and a_{11} is called the *pivot element*. Note that the process of multiplying the first row by a_{21}/a_{11} is equivalent to dividing it by a_{11} and multiplying it by a_{21}. Sometimes the division operation is referred to as *normalization*. We make this distinction because a zero pivot element can interfere with normalization by causing a division by zero. We will return to this important issue after we complete our description of naive Gauss elimination.

The next step is to eliminate x_2 from Eq. (9.10c) through (9.10d). To do this, multiply Eq. (9.10b) by a'_{32}/a'_{22} and subtract the result from Eq. (9.10c). Perform a similar elimination for the remaining equations to yield

$$a_{11}x_1 + a_{12}x_2 + a_{13}x_3 + \cdots + a_{1n}x_n = b_1$$

$$a'_{22}x_2 + a'_{23}x_3 + \cdots + a'_{2n}x_n = b'_2$$

$$a''_{33}x_3 + \cdots + a''_{3n}x_n = b''_3$$

$$\vdots \qquad \qquad \vdots$$

$$a''_{n3}x_3 + \cdots + a''_{nn}x_n = b''_n$$

where the double prime indicates that the elements have been modified twice.

The procedure can be continued using the remaining pivot equations. The final manipulation in the sequence is to use the $(n-1)$th equation to eliminate the x_{n-1} term from the nth equation. At this point, the system will have been transformed to an upper triangular system:

$$a_{11}x_1 + a_{12}x_2 + a_{13}x_3 + \cdots + a_{1n}x_n = b_1 \qquad (9.11a)$$

$$a'_{22}x_2 + a'_{23}x_3 + \cdots + a'_{2n}x_n = b'_2 \qquad (9.11b)$$

$$a''_{33}x_3 + \cdots + a''_{3n}x_n = b''_3 \qquad (9.11c)$$

$$\ddots \qquad \qquad \vdots$$

$$a_{nn}^{(n-1)}x_n = b_n^{(n-1)} \qquad (9.11d)$$

Back Substitution. Equation (9.11d) can now be solved for x_n:

$$x_n = \frac{b_n^{(n-1)}}{a_{nn}^{(n-1)}} \qquad (9.12)$$

This result can be back-substituted into the $(n-1)$th equation to solve for x_{n-1}. The procedure, which is repeated to evaluate the remaining x's, can be represented by the following formula:

$$x_i = \frac{b_i^{(i-1)} - \displaystyle\sum_{j=i+1}^{n} a_{ij}^{(i-1)}x_j}{a_{ii}^{(i-1)}} \qquad \text{for } i = n-1, n-2, \ldots, 1 \qquad (9.13)$$

EXAMPLE 9.3   Naive Gauss Elimination

Problem Statement.   Use Gauss elimination to solve

$$3x_1 - 0.1x_2 - 0.2x_3 = \quad 7.85 \tag{E9.3.1}$$
$$0.1x_1 + \quad 7x_2 - 0.3x_3 = -19.3 \tag{E9.3.2}$$
$$0.3x_1 - 0.2x_2 + 10x_3 = \quad 71.4 \tag{E9.3.3}$$

Solution.   The first part of the procedure is forward elimination. Multiply Eq. (E9.3.1) by $0.1/3$ and subtract the result from Eq. (E9.3.2) to give

$$7.00333x_2 - 0.293333x_3 = -19.5617$$

Then multiply Eq. (E9.3.1) by $0.3/3$ and subtract it from Eq. (E9.3.3). After these operations, the set of equations is

$$3x_1 - \quad 0.1x_2 - \quad 0.2x_3 = \quad 7.85 \tag{E9.3.4}$$
$$7.00333x_2 - 0.293333x_3 = -19.5617 \tag{E9.3.5}$$
$$- 0.190000x_2 + \quad 10.0200x_3 = \quad 70.6150 \tag{E9.3.6}$$

To complete the forward elimination, x_2 must be removed from Eq. (E9.3.6). To accomplish this, multiply Eq. (E9.3.5) by $-0.190000/7.00333$ and subtract the result from Eq. (E9.3.6). This eliminates x_2 from the third equation and reduces the system to an upper triangular form, as in

$$3x_1 - \quad 0.1x_2 - \quad 0.2x_3 = \quad 7.85 \tag{E9.3.7}$$
$$7.00333x_2 - 0.293333x_3 = -19.5617 \tag{E9.3.8}$$
$$10.0120x_3 = \quad 70.0843 \tag{E9.3.9}$$

We can now solve these equations by back substitution. First, Eq. (E9.3.9) can be solved for

$$x_3 = \frac{70.0843}{10.0120} = 7.00003$$

This result can be back-substituted into Eq. (E9.3.8), which can then be solved for

$$x_2 = \frac{-19.5617 + 0.293333(7.00003)}{7.00333} = -2.50000$$

Finally, $x_3 = 7.00003$ and $x_2 = -2.50000$ can be substituted back into Eq. (E9.3.7), which can be solved for

$$x_1 = \frac{7.85 + 0.1(-2.50000) + 0.2(7.00003)}{3} = 3.00000$$

Although there is a slight round-off error, the results are very close to the exact solution of $x_1 = 3$, $x_2 = -2.5$, and $x_3 = 7$. This can be verified by substituting the results into the original equation set:

$$3(3) - 0.1(-2.5) - 0.2(7.00003) = 7.84999 \cong 7.85$$
$$0.1(3) + 7(-2.5) - 0.3(7.00003) = -19.30000 = -19.3$$
$$0.3(3) - 0.2(-2.5) + 10(7.00003) = 71.4003 \cong 71.4$$

```
function x = GaussNaive(A,b)
% GaussNaive: naive Gauss elimination
%   x = GaussNaive(A,b): Gauss elimination without pivoting.
% input:
%   A = coefficient matrix
%   b = right hand side vector
% output:
%   x = solution vector

[m,n] = size(A);
if m~=n, error('Matrix A must be square'); end
nb = n+1;
Aug = [A b];
% forward elimination
for k = 1:n-1
  for i = k+1:n
    factor = Aug(i,k)/Aug(k,k);
    Aug(i,k:nb) = Aug(i,k:nb)-factor*Aug(k,k:nb);
  end
end
% back substitution
x = zeros(n,1);
x(n) = Aug(n,nb)/Aug(n,n);
for i = n-1:-1:1
  x(i) = (Aug(i,nb)-Aug(i,i+1:n)*x(i+1:n))/Aug(i,i);
end
```

FIGURE 9.4
An M-file to implement naive Gauss elimination.

9.2.1 MATLAB M-file: GaussNaive

An M-file that implements naive Gauss elimination is listed in Fig. 9.4. Notice that the coefficient matrix A and the right-hand-side vector b are combined in the augmented matrix Aug. Thus, the operations are performed on Aug rather than separately on A and b.

Two nested loops provide a concise representation of the forward elimination step. An outer loop moves down the matrix from one pivot row to the next. The inner loop moves below the pivot row to each of the subsequent rows where elimination is to take place. Finally, the actual elimination is represented by a single line that takes advantage of MATLAB's ability to perform matrix operations.

The back-substitution step follows directly from Eqs. (9.12) and (9.13). Again, MATLAB's ability to perform matrix operations allows Eq. (9.13) to be programmed as a single line.

9.2.2 Operation Counting

The execution time of Gauss elimination depends on the amount of *floating-point operations* (or *flops*) involved in the algorithm. On modern computers using math coprocessors, the time consumed to perform addition/subtraction and multiplication/division is about the same.

Therefore, totaling up these operations provides insight into which parts of the algorithm are most time consuming and how computation time increases as the system gets larger.

Before analyzing naive Gauss elimination, we will first define some quantities that facilitate operation counting:

$$\sum_{i=1}^{m} cf(i) = c \sum_{i=1}^{m} f(i) \qquad \sum_{i=1}^{m} f(i) + g(i) = \sum_{i=1}^{m} f(i) + \sum_{i=1}^{m} g(i) \qquad (9.14a,b)$$

$$\sum_{i=1}^{m} 1 = 1 + 1 + 1 + \cdots + 1 = m \qquad \sum_{i=k}^{m} 1 = m - k + 1 \qquad (9.14c,d)$$

$$\sum_{i=1}^{m} i = 1 + 2 + 3 + \cdots + m = \frac{m(m+1)}{2} = \frac{m^2}{2} + O(m) \qquad (9.14e)$$

$$\sum_{i=1}^{m} i^2 = 1^2 + 2^2 + 3^2 + \cdots + m^2 = \frac{m(m+1)(2m+1)}{6} = \frac{m^3}{3} + O(m^2) \qquad (9.14f)$$

where $O(m^n)$ means "terms of order m^n and lower."

Now let us examine the naive Gauss elimination algorithm (Fig. 9.4) in detail. We will first count the flops in the elimination stage. On the first pass through the outer loop, $k = 1$. Therefore, the limits on the inner loop are from $i = 2$ to n. According to Eq. (9.14d), this means that the number of iterations of the inner loop will be

$$\sum_{i=2}^{n} 1 = n - 2 + 1 = n - 1 \qquad (9.15)$$

For every one of these iterations, there is one division to calculate the factor. The next line then performs a multiplication and a subtraction for each column element from 2 to nb. Because $nb = n + 1$, going from 2 to nb results in n multiplications and n subtractions. Together with the single division, this amounts to $n + 1$ multiplications/divisions and n addition/subtractions for every iteration of the inner loop. The total for the first pass through the outer loop is therefore $(n - 1)(n + 1)$ multiplication/divisions and $(n - 1)(n)$ addition/subtractions.

Similar reasoning can be used to estimate the flops for the subsequent iterations of the outer loop. These can be summarized as

Outer Loop k	Inner Loop i	Addition/Subtraction Flops	Multiplication/Division Flops
1	2, n	$(n-1)(n)$	$(n-1)(n+1)$
2	3, n	$(n-2)(n-1)$	$(n-2)(n)$
⋮	⋮		
k	k+1, n	$(n-k)(n+1-k)$	$(n-k)(n+2-k)$
⋮	⋮		
n − 1	n, n	$(1)(2)$	$(1)(3)$

Therefore, the total addition/subtraction flops for elimination can be computed as

$$\sum_{k=1}^{n-1} (n-k)(n+1-k) = \sum_{k=1}^{n-1} [n(n+1) - k(2n+1) + k^2] \qquad (9.16)$$

or

$$n(n + 1) \sum_{k=1}^{n-1} 1 - (2n + 1) \sum_{k=1}^{n-1} k + \sum_{k=1}^{n-1} k^2 \tag{9.17}$$

Applying some of the relationships from Eq. (9.14) yields

$$[n^3 + O(n)] - [n^3 + O(n^2)] + \left[\frac{1}{3}n^3 + O(n^2)\right] = \frac{n^3}{3} + O(n) \tag{9.18}$$

A similar analysis for the multiplication/division flops yields

$$[n^3 + O(n^2)] - [n^3 + O(n)] + \left[\frac{1}{3}n^3 + O(n^2)\right] = \frac{n^3}{3} + O(n^2) \tag{9.19}$$

Summing these results gives

$$\frac{2n^3}{3} + O(n^2) \tag{9.20}$$

Thus, the total number of flops is equal to $2n^3/3$ plus an additional component proportional to terms of order n^2 and lower. The result is written in this way because as n gets large, the $O(n^2)$ and lower terms become negligible. We are therefore justified in concluding that for large n, the effort involved in forward elimination converges on $2n^3/3$.

Because only a single loop is used, back substitution is much simpler to evaluate. The number of addition/subtraction flops is equal to $n(n - 1)/2$. Because of the extra division prior to the loop, the number of multiplication/division flops is $n(n + 1)/2$. These can be added to arrive at a total of

$$n^2 + O(n) \tag{9.21}$$

Thus, the total effort in naive Gauss elimination can be represented as

$$\underbrace{\frac{2n^3}{3} + O(n^2)}_{\substack{\text{Forward} \\ \text{elimination}}} + \underbrace{n^2 + O(n)}_{\substack{\text{Back} \\ \text{substitution}}} \xrightarrow{\text{as } n \text{ increases}} \frac{2n^3}{3} + O(n^2) \tag{9.22}$$

Two useful general conclusions can be drawn from this analysis:

1. As the system gets larger, the computation time increases greatly. As in Table 9.1, the amount of flops increases nearly three orders of magnitude for every order of magnitude increase in the number of equations.

TABLE 9.1   Number of flops for naive Gauss elimination.

n	Elimination	Back Substitution	Total Flops	$2n^3/3$	Percent Due to Elimination
10	705	100	805	667	87.58%
100	671550	10000	681550	666667	98.53%
1000	6.67×10^8	1×10^6	6.68×10^8	6.67×10^8	99.85%

2. Most of the effort is incurred in the elimination step. Thus, efforts to make the method more efficient should probably focus on this step.

9.3 PIVOTING

The primary reason that the foregoing technique is called "naive" is that during both the elimination and the back-substitution phases, it is possible that a division by zero can occur. For example, if we use naive Gauss elimination to solve

$$2x_2 + 3x_3 = 8$$
$$4x_1 + 6x_2 + 7x_3 = -3$$
$$2x_1 - 3x_2 + 6x_3 = 5$$

the normalization of the first row would involve division by $a_{11} = 0$. Problems may also arise when the pivot element is close, rather than exactly equal, to zero because if the magnitude of the pivot element is small compared to the other elements, then round-off errors can be introduced.

Therefore, before each row is normalized, it is advantageous to determine the coefficient with the largest absolute value in the column below the pivot element. The rows can then be switched so that the largest element is the pivot element. This is called *partial pivoting*.

If columns as well as rows are searched for the largest element and then switched, the procedure is called *complete pivoting*. Complete pivoting is rarely used because switching columns changes the order of the x's and, consequently, adds significant and usually unjustified complexity to the computer program.

The following example illustrates the advantages of partial pivoting. Aside from avoiding division by zero, pivoting also minimizes round-off error. As such, it also serves as a partial remedy for ill-conditioning.

EXAMPLE 9.4   Partial Pivoting

Problem Statement.   Use Gauss elimination to solve

$$0.0003x_1 + 3.0000x_2 = 2.0001$$
$$1.0000x_1 + 1.0000x_2 = 1.0000$$

Note that in this form the first pivot element, $a_{11} = 0.0003$, is very close to zero. Then repeat the computation, but partial pivot by reversing the order of the equations. The exact solution is $x_1 = 1/3$ and $x_2 = 2/3$.

Solution.   Multiplying the first equation by $1/(0.0003)$ yields

$$x_1 + 10{,}000x_2 = 6667$$

which can be used to eliminate x_1 from the second equation:

$$-9999x_2 = -6666$$

which can be solved for $x_2 = 2/3$. This result can be substituted back into the first equation to evaluate x_1:

$$x_1 = \frac{2.0001 - 3(2/3)}{0.0003} \tag{E9.4.1}$$

Due to subtractive cancellation, the result is very sensitive to the number of significant figures carried in the computation:

Significant Figures	x_2	x_1	Absolute Value of Percent Relative Error for x_1
3	0.667	−3.33	1099
4	0.6667	0.0000	100
5	0.66667	0.30000	10
6	0.666667	0.330000	1
7	0.6666667	0.3330000	0.1

Note how the solution for x_1 is highly dependent on the number of significant figures. This is because in Eq. (E9.4.1), we are subtracting two almost-equal numbers.

On the other hand, if the equations are solved in reverse order, the row with the larger pivot element is normalized. The equations are

$$1.0000x_1 + 1.0000x_2 = 1.0000$$
$$0.0003x_1 + 3.0000x_2 = 2.0001$$

Elimination and substitution again yields $x_2 = 2/3$. For different numbers of significant figures, x_1 can be computed from the first equation, as in

$$x_1 = \frac{1 - (2/3)}{1}$$

This case is much less sensitive to the number of significant figures in the computation:

Significant Figures	x_2	x_1	Absolute Value of Percent Relative Error for x_1
3	0.667	0.333	0.1
4	0.6667	0.3333	0.01
5	0.66667	0.33333	0.001
6	0.666667	0.333333	0.0001
7	0.6666667	0.3333333	0.0000

Thus, a pivot strategy is much more satisfactory.

9.3.1 MATLAB M-file: GaussPivot

An M-file that implements Gauss elimination with partial pivoting is listed in Fig. 9.5. It is identical to the M-file for naive Gauss elimination presented previously in Section 9.2.1 with the exception of the bold portion that implements partial pivoting.

Notice how the built-in MATLAB function max is used to determine the largest available coefficient in the column below the pivot element. The max function has the syntax

```
[y,i] = max(x)
```

where y is the largest element in the vector x, and i is the index corresponding to that element.

```
function x = GaussPivot(A,b)
% GaussPivot: Gauss elimination pivoting
%   x = GaussPivot(A,b): Gauss elimination with pivoting.
% input:
%   A = coefficient matrix
%   b = right hand side vector
% output:
%   x = solution vector
[m,n]=size(A);
if m~=n, error('Matrix A must be square'); end
nb=n+1;
Aug=[A b];
% forward elimination
for k = 1:n-1
  % partial pivoting
  [big,i]=max(abs(Aug(k:n,k)));
  ipr=i+k-1;
  if ipr~=k
    Aug([k,ipr],:)=Aug([ipr,k],:);
  end
  for i = k+1:n
    factor=Aug(i,k)/Aug(k,k);
    Aug(i,k:nb)=Aug(i,k:nb)-factor*Aug(k,k:nb);
  end
end
% back substitution
x=zeros(n,1);
x(n)=Aug(n,nb)/Aug(n,n);
for i = n-1:-1:1
  x(i)=(Aug(i,nb)-Aug(i,i+1:n)*x(i+1:n))/Aug(i,i);
end
```

FIGURE 9.5
An M-file to implement the Gauss elimination with partial pivoting.

9.4 TRIDIAGONAL SYSTEMS

Certain matrices have a particular structure that can be exploited to develop efficient solution schemes. For example, a banded matrix is a square matrix that has all elements equal to zero, with the exception of a band centered on the main diagonal.

A *tridiagonal* system has a bandwidth of 3 and can be expressed generally as

$$\begin{bmatrix} f_1 & g_1 & & & & \\ e_2 & f_2 & g_2 & & & \\ & e_3 & f_3 & g_3 & & \\ & & \cdot & \cdot & \cdot & \\ & & & \cdot & \cdot & \cdot \\ & & & & \cdot & \cdot & \cdot \\ & & & & e_{n-1} & f_{n-1} & g_{n-1} \\ & & & & & e_n & f_n \end{bmatrix} \begin{Bmatrix} x_1 \\ x_2 \\ x_3 \\ \cdot \\ \cdot \\ \cdot \\ x_{n-1} \\ x_n \end{Bmatrix} = \begin{Bmatrix} r_1 \\ r_2 \\ r_3 \\ \cdot \\ \cdot \\ \cdot \\ r_{n-1} \\ r_n \end{Bmatrix} \qquad (9.23)$$

Notice that we have changed our notation for the coefficients from a's and b's to e's, f's, g's, and r's. This was done to avoid storing large numbers of useless zeros in the square matrix of a's. This space-saving modification is advantageous because the resulting algorithm requires less computer memory.

An algorithm to solve such systems can be directly patterned after Gauss elimination—that is, using forward elimination and back substitution. However, because most of the matrix elements are already zero, much less effort is expended than for a full matrix. This efficiency is illustrated in the following example.

EXAMPLE 9.5   Solution of a Tridiagonal System

Problem Statement.   Solve the following tridiagonal system:

$$\begin{bmatrix} 2.04 & -1 & & \\ -1 & 2.04 & -1 & \\ & -1 & 2.04 & -1 \\ & & -1 & 2.04 \end{bmatrix} \begin{Bmatrix} x_1 \\ x_2 \\ x_3 \\ x_4 \end{Bmatrix} = \begin{Bmatrix} 40.8 \\ 0.8 \\ 0.8 \\ 200.8 \end{Bmatrix}$$

Solution.   As with Gauss elimination, the first step involves transforming the matrix to upper triangular form. This is done by multiplying the first equation by the factor e_2/f_1 and subtracting the result from the second equation. This creates a zero in place of e_2 and transforms the other coefficients to new values,

$$f_2 = f_2 - \frac{e_2}{f_1} g_1 = 2.04 - \frac{-1}{2.04}(-1) = 1.550$$

$$r_2 = r_2 - \frac{e_2}{f_1} r_1 = 0.8 - \frac{-1}{2.04}(40.8) = 20.8$$

Notice that g_2 is unmodified because the element above it in the first row is zero.

After performing a similar calculation for the third and fourth rows, the system is transformed to the upper triangular form

$$\begin{bmatrix} 2.04 & -1 & & \\ & 1.550 & -1 & \\ & & 1.395 & -1 \\ & & & 1.323 \end{bmatrix} \begin{Bmatrix} x_1 \\ x_2 \\ x_3 \\ x_4 \end{Bmatrix} = \begin{Bmatrix} 40.8 \\ 20.8 \\ 14.221 \\ 210.996 \end{Bmatrix}$$

Now back substitution can be applied to generate the final solution:

$$x_4 = \frac{r_4}{f_4} = \frac{210.996}{1.323} = 159.480$$

$$x_3 = \frac{r_3 - g_3 x_4}{f_3} = \frac{14.221 - (-1)159.480}{1.395} = 124.538$$

$$x_2 = \frac{r_2 - g_2 x_3}{f_2} = \frac{20.800 - (-1)124.538}{1.550} = 93.778$$

$$x_1 = \frac{r_1 - g_1 x_2}{f_1} = \frac{40.800 - (-1)93.778}{2.040} = 65.970$$

```
function x = Tridiag(e,f,g,r)
% Tridiag: Tridiagonal equation solver banded system
%   x = Tridiag(e,f,g,r): Tridiagonal system solver.
% input:
%   e = subdiagonal vector
%   f = diagonal vector
%   g = superdiagonal vector
%   r = right hand side vector
% output:
%   x = solution vector
n=length(f);
% forward elimination
for k = 2:n
  factor = e(k)/f(k-1);
  f(k) = f(k) - factor*g(k-1);
  r(k) = r(k) - factor*r(k-1);
end
% back substitution
x(n) = r(n)/f(n);
for k = n-1:-1:1
  x(k) = (r(k)-g(k)*x(k+1))/f(k);
end
```

FIGURE 9.6
An M-file to solve a tridiagonal system.

9.4.1 MATLAB M-file: `Tridiag`

An M-file that solves a tridiagonal system of equations is listed in Fig. 9.6. Note that the algorithm does not include partial pivoting. Although pivoting is sometimes required, most tridiagonal systems routinely solved in engineering and science do not require pivoting.

Recall that the computational effort for Gauss elimination was proportional to n^3. Because of its sparseness, the effort involved in solving tridiagonal systems is proportional to n. Consequently, the algorithm in Fig. 9.6 executes much, much faster than Gauss elimination, particularly for large systems.

9.5 CASE STUDY   MODEL OF A HEATED ROD

Background.   Linear algebraic equations can arise when modeling distributed systems. For example, Fig. 9.7 shows a long, thin rod positioned between two walls that are held at constant temperatures. Heat flows through the rod as well as between the rod and the surrounding air. For the steady-state case, a differential equation based on heat conservation can be written for such a system as

$$\frac{d^2T}{dx^2} + h'(T_a - T) = 0 \tag{9.24}$$

9.5 CASE STUDY　continued

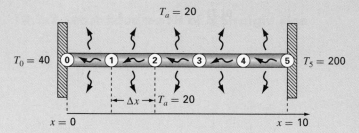

FIGURE 9.7
A noninsulated uniform rod positioned between two walls of constant but different temperature. The finite-difference representation employs four interior nodes.

where T = temperature (°C), x = distance along the rod (m), h' = a heat transfer coefficient between the rod and the surrounding air (m^{-2}), and T_a = the air temperature (°C).

Given values for the parameters, forcing functions, and boundary conditions, calculus can be used to develop an analytical solution. For example, if $h' = 0.01$, $T_a = 20$, $T(0) = 40$, and $T(10) = 200$, the solution is

$$T = 73.4523e^{0.1x} - 53.4523e^{-0.1x} + 20 \tag{9.25}$$

Although it provided a solution here, calculus does not work for all such problems. In such instances, numerical methods provide a valuable alternative. In this case study, we will use finite differences to transform this differential equation into a tridiagonal system of linear algebraic equations which can be readily solved using the numerical methods described in this chapter.

Solution.　Equation (9.24) can be transformed into a set of linear algebraic equations by conceptualizing the rod as consisting of a series of nodes. For example, the rod in Fig. 9.7 is divided into six equispaced nodes. Since the rod has a length of 10, the spacing between nodes is $\Delta x = 2$.

Calculus was necessary to solve Eq. (9.24) because it includes a second derivative. As we learned in Sec. 4.3.4, finite-difference approximations provide a means to transform derivatives into algebraic form. For example, the second derivative at each node can be approximated as

$$\frac{d^2 T}{dx^2} = \frac{T_{i+1} - 2T_i + T_{i-1}}{\Delta x^2}$$

where T_i designates the temperature at node i. This approximation can be substituted into Eq. (9.24) to give

$$\frac{T_{i+1} - 2T_i + T_{i-1}}{\Delta x^2} + h'(T_a - T_i) = 0$$

Collecting terms and substituting the parameters gives

$$-T_{i-1} + 2.04T_i - T_{i+1} = 0.8 \tag{9.26}$$

Thus, Eq. (9.24) has been transformed from a differential equation into an algebraic equation. Equation (9.26) can now be applied to each of the interior nodes:

$$
\begin{aligned}
-T_0 + 2.04T_1 - T_2 &= 0.8 \\
-T_1 + 2.04T_2 - T_3 &= 0.8 \\
-T_2 + 2.04T_3 - T_4 &= 0.8 \\
-T_3 + 2.04T_4 - T_5 &= 0.8
\end{aligned}
\tag{9.27}
$$

The values of the fixed end temperatures, $T_0 = 40$ and $T_5 = 200$, can be substituted and moved to the right-hand side. The results are four equations with four unknowns expressed in matrix form as

$$
\begin{bmatrix}
2.04 & -1 & 0 & 0 \\
-1 & 2.04 & -1 & 0 \\
0 & -1 & 2.04 & -1 \\
0 & 0 & -1 & 2.04
\end{bmatrix}
\begin{Bmatrix}
T_1 \\
T_2 \\
T_3 \\
T_4
\end{Bmatrix}
=
\begin{Bmatrix}
40.8 \\
0.8 \\
0.8 \\
200.8
\end{Bmatrix}
\tag{9.28}
$$

So our original differential equation has been converted into an equivalent system of linear algebraic equations. Consequently, we can use the techniques described in this chapter to solve for the temperatures. For example, using MATLAB

```
>> A=[2.04 -1 0 0
-1 2.04 -1 0
0 -1 2.04 -1
0 0 -1 2.04];
>> b=[40.8 0.8 0.8 200.8]';
>> T=(A\b)'

T =
   65.9698   93.7785  124.5382  159.4795
```

A plot can also be developed comparing these results with the analytical solution obtained with Eq. (9.25),

```
>> T=[40 T 200];
>> x=[0:2:10];
>> xanal=[0:10];
>> TT=@(x) 73.4523*exp(0.1*xanal)-53.4523*exp ...
       (-0.1*xanal)+20;
>> Tanal=TT(xanal);
>> plot(x,T,'o',xanal,Tanal)
```

As in Fig. 9.8, the numerical results are quite close to those obtained with calculus.

9.5 CASE STUDY   continued

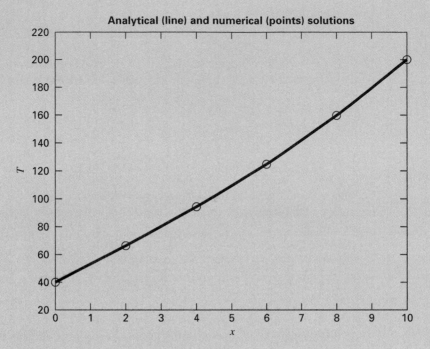

FIGURE 9.8
A plot of temperature versus distance along a heated rod. Both analytical (line) and numerical (points) solutions are displayed.

In addition to being a linear system, notice that Eq. (9.28) is also tridiagonal. We can use an efficient solution scheme like the M-file in Fig. 9.6 to obtain the solution:

```
>> e=[0 -1 -1 -1];
>> f=[2.04 2.04 2.04 2.04];
>> g=[-1 -1 -1 0];
>> r=[40.8 0.8 0.8 200.8];
>> Tridiag(e,f,g,r)

ans =
   65.9698   93.7785   124.5382   159.4795
```

The system is tridiagonal because each node depends only on its adjacent nodes. Because we numbered the nodes sequentially, the resulting equations are tridiagonal. Such cases often occur when solving differential equations based on conservation laws.

PROBLEMS

9.1 Determine the number of total flops as a function of the number of equations n for the tridiagonal algorithm (Fig. 9.6).

9.2 Use the graphical method to solve

$$4x_1 - 8x_2 = -24$$
$$x_1 + 6x_2 = 34$$

Check your results by substituting them back into the equations.

9.3 Given the system of equations

$$-1.1x_1 + 10x_2 = 120$$
$$-2x_1 + 17.4x_2 = 174$$

(a) Solve graphically and check your results by substituting them back into the equations.

(b) On the basis of the graphical solution, what do you expect regarding the condition of the system?

(c) Compute the determinant.

9.4 Given the system of equations

$$-3x_2 + 7x_3 = 2$$
$$x_1 + 2x_2 - x_3 = 3$$
$$5x_1 - 2x_2 = 2$$

(a) Compute the determinant.

(b) Use Cramer's rule to solve for the x's.

(c) Use Gauss elimination with partial pivoting to solve for the x's.

(d) Substitute your results back into the original equations to check your solution.

9.5 Given the equations

$$0.5x_1 - x_2 = -9.5$$
$$1.02x_1 - 2x_2 = -18.8$$

(a) Solve graphically.

(b) Compute the determinant.

(c) On the basis of **(a)** and **(b)**, what would you expect regarding the system's condition?

(d) Solve by the elimination of unknowns.

(e) Solve again, but with a_{11} modified slightly to 0.52. Interpret your results.

9.6 Given the equations

$$10x_1 + 2x_2 - x_3 = 27$$
$$-3x_1 - 6x_2 + 2x_3 = -61.5$$
$$x_1 + x_2 + 5x_3 = -21.5$$

(a) Solve by naive Gauss elimination. Show all steps of the computation.

(b) Substitute your results into the original equations to check your answers.

9.7 Given the equations

$$2x_1 - 6x_2 - x_3 = -38$$
$$-3x_1 - x_2 + 7x_3 = -34$$
$$-8x_1 + x_2 - 2x_3 = -20$$

(a) Solve by Gauss elimination with partial pivoting. Show all steps of the computation.

(b) Substitute your results into the original equations to check your answers.

9.8 Perform the same calculations as in Example 9.5, but for the tridiagonal system:

$$\begin{bmatrix} 0.8 & -0.4 & \\ -0.4 & 0.8 & -0.4 \\ & -0.4 & 0.8 \end{bmatrix} \begin{Bmatrix} x_1 \\ x_2 \\ x_3 \end{Bmatrix} = \begin{Bmatrix} 41 \\ 25 \\ 105 \end{Bmatrix}$$

9.9 Figure P9.9 shows three reactors linked by pipes. As indicated, the rate of transfer of chemicals through each pipe is equal to a flow rate (Q, with units of cubic meters per second) multiplied by the concentration of the reactor from which the flow originates (c, with units of milligrams per cubic meter). If the system is at a steady state, the transfer into each reactor will balance the transfer out. Develop mass-balance equations for the reactors and solve the three simultaneous linear algebraic equations for their concentrations.

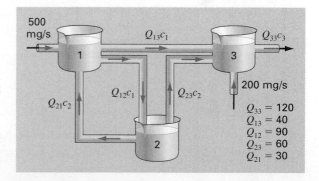

FIGURE P9.9
Three reactors linked by pipes. The rate of mass transfer through each pipe is equal to the product of flow Q and concentration c of the reactor from which the flow originates.

9.10 A civil engineer involved in construction requires 4800, 5800, and 5700 m³ of sand, fine gravel, and coarse gravel, respectively, for a building project. There are three pits from which these materials can be obtained. The composition of these pits is

	Sand %	Fine Gravel %	Coarse Gravel %
Pit1	55	30	15
Pit2	25	45	30
Pit3	25	20	55

How many cubic meters must be hauled from each pit in order to meet the engineer's needs?

9.11 An electrical engineer supervises the production of three types of electrical components. Three kinds of material—metal, plastic, and rubber—are required for production. The amounts needed to produce each component are

Component	Metal (g/ component)	Plastic (g/ component)	Rubber (g/ component)
1	15	0.30	1.0
2	17	0.40	1.2
3	19	0.55	1.5

If totals of 3.89, 0.095, and 0.282 kg of metal, plastic, and rubber, respectively, are available each day, how many components can be produced per day?

9.12 As described in Sec. 9.3, linear algebraic equations can arise in the solution of differential equations. For example,

the following differential equation results from a steady-state mass balance for a chemical in a one-dimensional canal:

$$0 = D\frac{d^2c}{dx^2} - U\frac{dc}{dx} - kc$$

where c = concentration, t = time, x = distance, D = diffusion coefficient, U = fluid velocity, and k = a first-order decay rate. Convert this differential equation to an equivalent system of simultaneous algebraic equations. Given $D = 2$, $U = 1$, $k = 0.2$, $c(0) = 80$ and $c(10) = 20$, solve these equations from $x = 0$ to 10 and develop a plot of concentration versus distance.

9.13 A stage extraction process is depicted in Fig. P9.13. In such systems, a stream containing a weight fraction y_{in} of a chemical enters from the left at a mass flow rate of F_1. Simultaneously, a solvent carrying a weight fraction x_{in} of the same chemical enters from the right at a flow rate of F_2. Thus, for stage i, a mass balance can be represented as

$$F_1 y_{i-1} + F_2 x_{i+1} = F_1 y_i + F_2 x_i \qquad (P9.13a)$$

At each stage, an equilibrium is assumed to be established between y_i and x_i as in

$$K = \frac{x_i}{y_i} \qquad (P9.13b)$$

where K is called a distribution coefficient. Equation (P9.13b) can be solved for x_i and substituted into Eq. (P9.13a) to yield

$$y_{i-1} - \left(1 + \frac{F_2}{F_1}K\right)y_i + \left(\frac{F_2}{F_1}K\right)y_{i+1} = 0 \qquad (P9.13c)$$

If $F_1 = 500$ kg/h, $y_{in} = 0.1$, $F_2 = 1000$ kg/h, $x_{in} = 0$, and $K = 4$, determine the values of y_{out} and x_{out} if a five-stage reactor is used. Note that Eq. (P9.13c) must be modified to account for the inflow weight fractions when applied to the first and last stages.

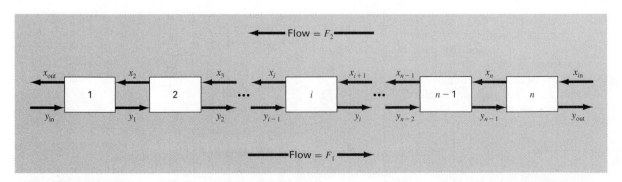

FIGURE P9.13
A stage extraction process.

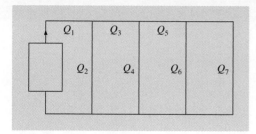

FIGURE P9.14

9.14 A peristaltic pump delivers a unit flow (Q_1) of a highly viscous fluid. The network is depicted in Fig. P9.14. Every pipe section has the same length and diameter. The mass and mechanical energy balance can be simplified to obtain the flows in every pipe. Solve the following system of equations to obtain the flow in every stream.

$$Q_3 + 2Q_4 - 2Q_2 = 0$$
$$Q_5 + 2Q_6 - 2Q_4 = 0$$
$$3Q_7 - 2Q_6 = 0$$
$$Q_1 = Q_2 + Q_3$$
$$Q_3 = Q_4 + Q_5$$
$$Q_5 = Q_6 + Q_7$$

9.15 A truss is loaded as shown in Fig. P9.15. Using the following set of equations, solve for the 10 unknowns, AB, BC, AD, BD, CD, DE, CE, A_x, A_y, and E_y.

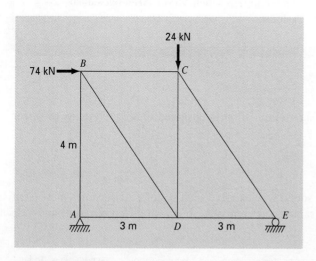

FIGURE P9.15

$$A_x + AD = 0$$
$$A_y + AB = 0$$
$$74 + BC + (3/5)BD = 0$$
$$-AB - (4/5)BD = 0$$
$$-BC + (3/5)CE = 0$$
$$-24 - CD - (4/5)CE = 0$$
$$-AD + DE - (3/5)BD = 0$$
$$CD + (4/5)BD = 0$$
$$-DE - (3/5)CE = 0$$
$$E_y + (4/5)CE = 0$$

9.16 A *pentadiagonal* system with a bandwidth of five can be expressed generally as

$$\begin{bmatrix} f_1 & g_1 & h_1 & & & \\ e_2 & f_2 & g_2 & h_2 & & \\ d_3 & e_3 & f_3 & g_3 & h_3 & \\ & \cdot & \cdot & \cdot & & \\ & & \cdot & \cdot & \cdot & \\ & & & \cdot & \cdot & \cdot \\ & & & d_{n-1} & e_{n-1} & f_{n-1} & g_{n-1} \\ & & & & d_n & e_n & f_n \end{bmatrix}$$

$$\times \begin{Bmatrix} x_1 \\ x_2 \\ x_3 \\ \cdot \\ \cdot \\ \cdot \\ x_{n-1} \\ x_n \end{Bmatrix} = \begin{Bmatrix} r_1 \\ r_2 \\ r_3 \\ \cdot \\ \cdot \\ \cdot \\ r_{n-1} \\ r_n \end{Bmatrix}$$

Develop an M-file to efficiently solve such systems without pivoting in a similar fashion to the algorithm used for tridiagonal matrices in Sec. 9.4.1. Test it for the following case:

$$\begin{bmatrix} 8 & -2 & -1 & 0 & 0 \\ -2 & 9 & -4 & -1 & 0 \\ -1 & -3 & 7 & -1 & -2 \\ 0 & -4 & -2 & 12 & -5 \\ 0 & 0 & -7 & -3 & 15 \end{bmatrix} \begin{Bmatrix} x_1 \\ x_2 \\ x_3 \\ x_4 \\ x_5 \end{Bmatrix} = \begin{Bmatrix} 5 \\ 2 \\ 1 \\ 1 \\ 5 \end{Bmatrix}$$

10

LU Factorization

CHAPTER OBJECTIVES

The primary objective of this chapter is to acquaint you with *LU* factorization[1].
Specific objectives and topics covered are

- Understanding that *LU* factorization involves decomposing the coefficient matrix into two triangular matrices that can then be used to efficiently evaluate different right-hand-side vectors.
- Knowing how to express Gauss elimination as an *LU* factorization.
- Given an *LU* factorization, knowing how to evaluate multiple right-hand-side vectors.
- Recognizing that Cholesky's method provides an efficient way to decompose a symmetric matrix and that the resulting triangular matrix and its transpose can be used to evaluate right-hand-side vectors efficiently.
- Understanding in general terms what happens when MATLAB's backslash operator is used to solve linear systems.

As described in Chap. 9, Gauss elimination is designed to solve systems of linear algebraic equations:

$$[A]\{x\} = \{b\} \tag{10.1}$$

Although it certainly represents a sound way to solve such systems, it becomes inefficient when solving equations with the same coefficients [A], but with different right-hand-side constants {b}.

[1] In the parlance of numerical methods, the terms "factorization" and "decomposition" are synonymous. To be consistent with the MATLAB documentation, we have chosen to employ the terminology *LU factorization* for the subject of this chapter. Note that *LU decomposition* is very commonly used to describe the same approach.

Recall that Gauss elimination involves two steps: forward elimination and back substitution (Fig. 9.3). As we learned in Section 9.2.2, the forward-elimination step comprises the bulk of the computational effort. This is particularly true for large systems of equations.

LU factorization methods separate the time-consuming elimination of the matrix [A] from the manipulations of the right-hand side {b}. Thus, once [A] has been "factored" or "decomposed," multiple right-hand-side vectors can be evaluated in an efficient manner.

Interestingly, Gauss elimination itself can be expressed as an *LU* factorization. Before showing how this can be done, let us first provide a mathematical overview of the factorization strategy.

10.1 OVERVIEW OF *LU* FACTORIZATION

Just as was the case with Gauss elimination, *LU* factorization requires pivoting to avoid division by zero. However, to simplify the following description, we will omit pivoting. In addition, the following explanation is limited to a set of three simultaneous equations. The results can be directly extended to n-dimensional systems.

Equation (10.1) can be rearranged to give

$$[A]\{x\} - \{b\} = 0 \tag{10.2}$$

Suppose that Eq. (10.2) could be expressed as an upper triangular system. For example, for a 3×3 system:

$$\begin{bmatrix} u_{11} & u_{12} & u_{13} \\ 0 & u_{22} & u_{23} \\ 0 & 0 & u_{33} \end{bmatrix} \begin{Bmatrix} x_1 \\ x_2 \\ x_3 \end{Bmatrix} = \begin{Bmatrix} d_1 \\ d_2 \\ d_3 \end{Bmatrix} \tag{10.3}$$

Recognize that this is similar to the manipulation that occurs in the first step of Gauss elimination. That is, elimination is used to reduce the system to upper triangular form. Equation (10.3) can also be expressed in matrix notation and rearranged to give

$$[U]\{x\} - \{d\} = 0 \tag{10.4}$$

Now assume that there is a lower diagonal matrix with 1's on the diagonal,

$$[L] = \begin{bmatrix} 1 & 0 & 0 \\ l_{21} & 1 & 0 \\ l_{31} & l_{32} & 1 \end{bmatrix} \tag{10.5}$$

that has the property that when Eq. (10.4) is premultiplied by it, Eq. (10.2) is the result. That is,

$$[L]\{[U]\{x\} - \{d\}\} = [A]\{x\} - \{b\} \tag{10.6}$$

If this equation holds, it follows from the rules for matrix multiplication that

$$[L][U] = [A] \tag{10.7}$$

and

$$[L]\{d\} = \{b\} \tag{10.8}$$

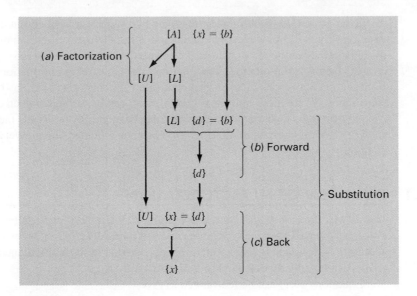

FIGURE 10.1
The steps in *LU* factorization.

A two-step strategy (see Fig. 10.1) for obtaining solutions can be based on Eqs. (10.3), (10.7), and (10.8):

1. *LU* factorization step. [A] is factored or "decomposed" into lower [L] and upper [U] triangular matrices.
2. Substitution step. [L] and [U] are used to determine a solution $\{x\}$ for a right-hand side $\{b\}$. This step itself consists of two steps. First, Eq. (10.8) is used to generate an intermediate vector $\{d\}$ by forward substitution. Then, the result is substituted into Eq. (10.3) which can be solved by back substitution for $\{x\}$.

Now let us show how Gauss elimination can be implemented in this way.

10.2 GAUSS ELIMINATION AS *LU* FACTORIZATION

Although it might appear at face value to be unrelated to *LU* factorization, Gauss elimination can be used to decompose [A] into [L] and [U]. This can be easily seen for [U], which is a direct product of the forward elimination. Recall that the forward-elimination step is intended to reduce the original coefficient matrix [A] to the form

$$[U] = \begin{bmatrix} a_{11} & a_{12} & a_{13} \\ 0 & a'_{22} & a'_{23} \\ 0 & 0 & a''_{33} \end{bmatrix} \tag{10.9}$$

which is in the desired upper triangular format.

Though it might not be as apparent, the matrix [L] is also produced during the step. This can be readily illustrated for a three-equation system,

$$\begin{bmatrix} a_{11} & a_{12} & a_{13} \\ a_{21} & a_{22} & a_{23} \\ a_{31} & a_{32} & a_{33} \end{bmatrix} \begin{Bmatrix} x_1 \\ x_2 \\ x_3 \end{Bmatrix} = \begin{Bmatrix} b_1 \\ b_2 \\ b_3 \end{Bmatrix}$$

The first step in Gauss elimination is to multiply row 1 by the factor [recall Eq. (9.9)]

$$f_{21} = \frac{a_{21}}{a_{11}}$$

and subtract the result from the second row to eliminate a_{21}. Similarly, row 1 is multiplied by

$$f_{31} = \frac{a_{31}}{a_{11}}$$

and the result subtracted from the third row to eliminate a_{31}. The final step is to multiply the modified second row by

$$f_{32} = \frac{a'_{32}}{a'_{22}}$$

and subtract the result from the third row to eliminate a_{32}.

Now suppose that we merely perform all these manipulations on the matrix [A]. Clearly, if we do not want to change the equations, we also have to do the same to the right-hand side {b}. But there is absolutely no reason that we have to perform the manipulations simultaneously. Thus, we could save the *f*'s and manipulate {b} later.

Where do we store the factors f_{21}, f_{31}, and f_{32}? Recall that the whole idea behind the elimination was to create zeros in a_{21}, a_{31}, and a_{32}. Thus, we can store f_{21} in a_{21}, f_{31} in a_{31}, and f_{32} in a_{32}. After elimination, the [A] matrix can therefore be written as

$$\begin{bmatrix} a_{11} & a_{12} & a_{13} \\ f_{21} & a'_{22} & a'_{23} \\ f_{31} & f_{32} & a''_{33} \end{bmatrix} \tag{10.10}$$

This matrix, in fact, represents an efficient storage of the *LU* factorization of [A],

$$[A] \rightarrow [L][U] \tag{10.11}$$

where

$$[U] = \begin{bmatrix} a_{11} & a_{12} & a_{13} \\ 0 & a'_{22} & a'_{23} \\ 0 & 0 & a''_{33} \end{bmatrix} \tag{10.12}$$

and

$$[L] = \begin{bmatrix} 1 & 0 & 0 \\ f_{21} & 1 & 0 \\ f_{31} & f_{32} & 1 \end{bmatrix} \tag{10.13}$$

The following example confirms that $[A] = [L][U]$.

EXAMPLE 10.1   *LU* Factorization with Gauss Elimination

Problem Statement. Derive an *LU* factorization based on the Gauss elimination performed previously in Example 9.3.

Solution. In Example 9.3, we used Gauss elimination to solve a set of linear algebraic equations that had the following coefficient matrix:

$$[A] = \begin{bmatrix} 3 & -0.1 & -0.2 \\ 0.1 & 7 & -0.3 \\ 0.3 & -0.2 & 10 \end{bmatrix}$$

After forward elimination, the following upper triangular matrix was obtained:

$$[U] = \begin{bmatrix} 3 & -0.1 & -0.2 \\ 0 & 7.00333 & -0.293333 \\ 0 & 0 & 10.0120 \end{bmatrix}$$

The factors employed to obtain the upper triangular matrix can be assembled into a lower triangular matrix. The elements a_{21} and a_{31} were eliminated by using the factors

$$f_{21} = \frac{0.1}{3} = 0.0333333 \qquad f_{31} = \frac{0.3}{3} = 0.1000000$$

and the element a_{32} was eliminated by using the factor

$$f_{32} = \frac{-0.19}{7.00333} = -0.0271300$$

Thus, the lower triangular matrix is

$$[L] = \begin{bmatrix} 1 & 0 & 0 \\ 0.0333333 & 1 & 0 \\ 0.100000 & -0.0271300 & 1 \end{bmatrix}$$

Consequently, the *LU* factorization is

$$[A] = [L][U] = \begin{bmatrix} 1 & 0 & 0 \\ 0.0333333 & 1 & 0 \\ 0.100000 & -0.0271300 & 1 \end{bmatrix} \begin{bmatrix} 3 & -0.1 & -0.2 \\ 0 & 7.00333 & -0.293333 \\ 0 & 0 & 10.0120 \end{bmatrix}$$

This result can be verified by performing the multiplication of $[L][U]$ to give

$$[L][U] = \begin{bmatrix} 3 & -0.1 & -0.2 \\ 0.0999999 & 7 & -0.3 \\ 0.3 & -0.2 & 9.99996 \end{bmatrix}$$

where the minor discrepancies are due to roundoff.

After the matrix is decomposed, a solution can be generated for a particular right-hand-side vector $\{b\}$. This is done in two steps. First, a forward-substitution step is executed by solving Eq. (10.8) for $\{d\}$. It is important to recognize that this merely amounts to performing the elimination manipulations on $\{b\}$. Thus, at the end of this step, the right-hand side will be in the same state that it would have been had we performed forward manipulation on [A] and {b} simultaneously.

The forward-substitution step can be represented concisely as

$$d_i = b_i - \sum_{j=1}^{i-1} l_{ij} b_j \qquad \text{for } i = 1, 2, \ldots, n$$

The second step then merely amounts to implementing back substitution to solve Eq. (10.3). Again, it is important to recognize that this is identical to the back-substitution phase of conventional Gauss elimination [compare with Eqs. (9.12) and (9.13)]:

$$x_n = d_n / a_{nn}$$

$$x_i = \frac{d_i - \sum_{j=i+1}^{n} u_{ij} x_j}{u_{ii}} \qquad \text{for } i = n-1, n-2, \ldots, 1$$

EXAMPLE 10.2   The Substitution Steps

Problem Statement.   Complete the problem initiated in Example 10.1 by generating the final solution with forward and back substitution.

Solution.   As just stated, the intent of forward substitution is to impose the elimination manipulations that we had formerly applied to [A] on the right-hand-side vector {b}. Recall that the system being solved is

$$\begin{bmatrix} 3 & -0.1 & -0.2 \\ 0.1 & 7 & -0.3 \\ 0.3 & -0.2 & 10 \end{bmatrix} \begin{Bmatrix} x_1 \\ x_2 \\ x_3 \end{Bmatrix} = \begin{Bmatrix} 7.85 \\ -19.3 \\ 71.4 \end{Bmatrix}$$

and that the forward-elimination phase of conventional Gauss elimination resulted in

$$\begin{bmatrix} 3 & -0.1 & -0.2 \\ 0 & 7.00333 & -0.293333 \\ 0 & 0 & 10.0120 \end{bmatrix} \begin{Bmatrix} x_1 \\ x_2 \\ x_3 \end{Bmatrix} = \begin{Bmatrix} 7.85 \\ -19.5617 \\ 70.0843 \end{Bmatrix}$$

The forward-substitution phase is implemented by applying Eq. (10.8):

$$\begin{bmatrix} 1 & 0 & 0 \\ 0.0333333 & 1 & 0 \\ 0.100000 & -0.0271300 & 1 \end{bmatrix} \begin{Bmatrix} d_1 \\ d_2 \\ d_3 \end{Bmatrix} = \begin{Bmatrix} 7.85 \\ -19.3 \\ 71.4 \end{Bmatrix}$$

or multiplying out the left-hand side:

$$\begin{aligned} d_1 &= 7.85 \\ 0.0333333 d_1 + d_2 &= -19.3 \\ 0.100000 d_1 - 0.0271300 d_2 + d_3 &= 71.4 \end{aligned}$$

We can solve the first equation for $d_1 = 7.85$, which can be substituted into the second equation to solve for

$$d_2 = -19.3 - 0.0333333(7.85) = -19.5617$$

Both d_1 and d_2 can be substituted into the third equation to give

$$d_3 = 71.4 - 0.1(7.85) + 0.02713(-19.5617) = 70.0843$$

Thus,

$$\{d\} = \left\{ \begin{array}{c} 7.85 \\ -19.5617 \\ 70.0843 \end{array} \right\}$$

This result can then be substituted into Eq. (10.3), $[U]\{x\} = \{d\}$:

$$\begin{bmatrix} 3 & -0.1 & -0.2 \\ 0 & 7.00333 & -0.293333 \\ 0 & 0 & 10.0120 \end{bmatrix} \left\{ \begin{array}{c} x_1 \\ x_2 \\ x_3 \end{array} \right\} = \left\{ \begin{array}{c} 7.85 \\ -19.5617 \\ 70.0843 \end{array} \right\}$$

which can be solved by back substitution (see Example 9.3 for details) for the final solution:

$$\{d\} = \left\{ \begin{array}{c} 3 \\ -2.5 \\ 7.00003 \end{array} \right\}$$

The *LU* factorization algorithm requires the same total flops as for Gauss elimination. The only difference is that a little less effort is expended in the factorization phase since the operations are not applied to the right-hand side. Conversely, the substitution phase takes a little more effort.

10.2.1 MATLAB Function: `lu`

MATLAB has a built-in function `lu` that generates the *LU* factorization. It has the general syntax:

```
[L,U] = lu(X)
```

where `L` and `U` are the lower triangular and upper triangular matrices, respectively, derived from the *LU* factorization of the matrix `X`. Note that this function uses partial pivoting to avoid division by zero. The following example shows how it can be employed to generate both the factorization and a solution for the same problem that was solved in Examples 10.1 and 10.2.

EXAMPLE 10.3   *LU* Factorization with MATLAB

Problem Statement. Use MATLAB to compute the *LU* factorization and find the solution for the same linear system analyzed in Examples 10.1 and 10.2:

$$\begin{bmatrix} 3 & -0.1 & -0.2 \\ 0.1 & 7 & -0.3 \\ 0.3 & -0.2 & 10 \end{bmatrix} \left\{ \begin{array}{c} x_1 \\ x_2 \\ x_3 \end{array} \right\} = \left\{ \begin{array}{c} 7.85 \\ -19.3 \\ 71.4 \end{array} \right\}$$

Solution.   The coefficient matrix and the right-hand-side vector can be entered in standard fashion as

```
>> A = [3 -.1 -.2;.1 7 -.3;.3 -.2 10];
>> b = [7.85; -19.3; 71.4];
```

Next, the *LU* factorization can be computed with

```
>> [L,U] = lu(A)

L =
    1.0000         0         0
    0.0333    1.0000         0
    0.1000   -0.0271    1.0000

U =
    3.0000   -0.1000   -0.2000
         0    7.0033   -0.2933
         0         0   10.0120
```

This is the same result that we obtained by hand in Example 10.1. We can test that it is correct by computing the original matrix as

```
>> L*U

ans =
    3.0000   -0.1000   -0.2000
    0.1000    7.0000   -0.3000
    0.3000   -0.2000   10.0000
```

To generate the solution, we first compute

```
>> d = L\b

d =
    7.8500
  -19.5617
   70.0843
```

And then use this result to compute the solution

```
>> x = U\d

x =
    3.0000
   -2.5000
    7.0000
```

These results conform to those obtained by hand in Example 10.2.

10.3 CHOLESKY FACTORIZATION

Recall from Chap. 8 that a symmetric matrix is one where $a_{ij} = a_{ji}$ for all i and j. In other words, $[A] = [A]^T$. Such systems occur commonly in both mathematical and engineering/science problem contexts.

Special solution techniques are available for such systems. They offer computational advantages because only half the storage is needed and only half the computation time is required for their solution.

One of the most popular approaches involves *Cholesky factorization* (also called Cholesky decomposition). This algorithm is based on the fact that a symmetric matrix can be decomposed, as in

$$[A] = [U]^T[U] \tag{10.14}$$

That is, the resulting triangular factors are the transpose of each other.

The terms of Eq. (10.14) can be multiplied out and set equal to each other. The factorization can be generated efficiently by recurrence relations. For the ith row:

$$u_{ii} = \sqrt{a_{ii} - \sum_{k=1}^{i-1} u_{ki}^2} \tag{10.15}$$

$$u_{ij} = \frac{a_{ij} - \sum_{k=1}^{i-1} u_{ki}u_{kj}}{u_{ii}} \quad \text{for } j = i+1, \ldots, n \tag{10.16}$$

EXAMPLE 10.4   Cholesky Factorization

Problem Statement.   Compute the Cholesky factorization for the symmetric matrix

$$[A] = \begin{bmatrix} 6 & 15 & 55 \\ 15 & 55 & 225 \\ 55 & 225 & 979 \end{bmatrix}$$

Solution.   For the first row ($i = 1$), Eq. (10.15) is employed to compute

$$u_{11} = \sqrt{a_{11}} = \sqrt{6} = 2.44949$$

Then, Eq. (10.16) can be used to determine

$$u_{12} = \frac{a_{12}}{u_{11}} = \frac{15}{2.44949} = 6.123724$$

$$u_{13} = \frac{a_{13}}{u_{11}} = \frac{55}{2.44949} = 22.45366$$

For the second row ($i = 2$):

$$u_{22} = \sqrt{a_{22} - u_{12}^2} = \sqrt{55 - (6.123724)^2} = 4.1833$$

$$u_{23} = \frac{a_{23} - u_{12}u_{13}}{u_{22}} = \frac{225 - 6.123724(22.45366)}{4.1833} = 20.9165$$

For the third row ($i = 3$):

$$u_{33} = \sqrt{a_{33} - u_{13}^2 - u_{23}^2} = \sqrt{979 - (22.45366)^2 - (20.9165)^2} = 6.110101$$

Thus, the Cholesky factorization yields

$$[U] = \begin{bmatrix} 2.44949 & 6.123724 & 22.45366 \\ & 4.1833 & 20.9165 \\ & & 6.110101 \end{bmatrix}$$

The validity of this factorization can be verified by substituting it and its transpose into Eq. (10.14) to see if their product yields the original matrix $[A]$. This is left for an exercise.

After obtaining the factorization, it can be used to determine a solution for a right-hand-side vector $\{b\}$ in a manner similar to LU factorization. First, an intermediate vector $\{d\}$ is created by solving

$$[U]^T\{d\} = \{b\} \tag{10.17}$$

Then, the final solution can be obtained by solving

$$[U]\{x\} = \{d\} \tag{10.18}$$

10.3.1 MATLAB Function: `chol`

MATLAB has a built-in function `chol` that generates the Cholesky factorization. It has the general syntax,

```
U = chol(X)
```

where U is an upper triangular matrix so that $U' * U = X$. The following example shows how it can be employed to generate both the factorization and a solution for the same matrix that we looked at in the previous example.

EXAMPLE 10.5   Cholesky Factorization with MATLAB

Problem Statement.   Use MATLAB to compute the Cholesky factorization for the same matrix we analyzed in Example 10.4.

$$[A] = \begin{bmatrix} 6 & 15 & 55 \\ 15 & 55 & 225 \\ 55 & 225 & 979 \end{bmatrix}$$

Also obtain a solution for a right-hand-side vector that is the sum of the rows of $[A]$. Note that for this case, the answer will be a vector of ones.

Solution.   The matrix is entered in standard fashion as

```
>> A = [6 15 55; 15 55 225; 55 225 979];
```

A right-hand-side vector that is the sum of the rows of [*A*] can be generated as

```
>> b = [sum(A(1,:)); sum(A(2,:)); sum(A(3,:))]

b =
          76
         295
        1259
```

Next, the Cholesky factorization can be computed with

```
>> U = chol(A)

U =
    2.4495     6.1237    22.4537
         0     4.1833    20.9165
         0          0     6.1101
```

We can test that this is correct by computing the original matrix as

```
>> U'*U

ans =
    6.0000    15.0000     55.0000
   15.0000    55.0000    225.0000
   55.0000   225.0000    979.0000
```

To generate the solution, we first compute

```
>> d = A'\b

d =
   31.0269
   25.0998
    6.1101
```

And then use this result to compute the solution

```
>> x = A\y

x =
    1.0000
    1.0000
    1.0000
```

10.4   MATLAB LEFT DIVISION

We previously introduced left division without any explanation of how it works. Now that we have some background on matrix solution techniques, we can provide a simplified description of its operation.

When we implement left division with the backslash operator, MATLAB invokes a highly sophisticated algorithm to obtain a solution. In essence, MATLAB examines the structure of the coefficient matrix and then implements an optimal method to obtain the solution. Although the details of the algorithm are beyond our scope, a simplified overview can be outlined.

First, MATLAB checks to see whether $[A]$ is in a format where a solution can be obtained without full Gauss elimination. These include systems that are (*a*) sparse and banded, (*b*) triangular (or easily transformed into triangular form), or (*c*) symmetric. If any of these cases are detected, the solution is obtained with the efficient techniques that are available for such systems. Some of the techniques include banded solvers, back and forward substitution, and Cholesky factorization.

If none of these simplified solutions are possible and the matrix is square,[2] a general triangular factorization is computed by Gauss elimination with partial pivoting and the solution obtained with substitution.

[2] It should be noted that in the event that $[A]$ is not square, a least-squares solution is obtained with an approach called *QR factorization.*

PROBLEMS

10.1 Determine the total flops as a function of the number of equations n for the (**a**) factorization, (**b**) forward substitution, and (**c**) back substitution phases of the *LU* factorization version of Gauss elimination.

10.2 Use the rules of matrix multiplication to prove that Eqs. (10.7) and (10.8) follow from Eq. (10.6).

10.3 Use naive Gauss elimination to factor the following system according to the description in Section 10.2:

$$10x_1 + 2x_2 - x_3 = 27$$
$$-3x_1 - 6x_2 + 2x_3 = -61.5$$
$$x_1 + x_2 + 5x_3 = -21.5$$

Then, multiply the resulting $[L]$ and $[U]$ matrices to determine that $[A]$ is produced.

10.4 Use *LU* factorization to solve the system of equations in Prob. 10.3. Show all the steps in the computation. Also solve the system for an alternative right-hand-side vector

$$\{b\}^T = \lfloor 12 \quad 18 \quad -6 \rfloor$$

10.5 Solve the following system of equations using *LU* factorization with partial pivoting:

$$2x_1 - 6x_2 - x_3 = -38$$
$$-3x_1 - x_2 + 7x_3 = -34$$
$$-8x_1 + x_2 - 2x_3 = -20$$

10.6 Develop your own M-file to determine the *LU* factorization of a square matrix without partial pivoting. That is, develop a function that is passed the square matrix and returns the triangular matrices $[L]$ and $[U]$. Test your function by using it to solve the system in Prob. 10.3. Confirm that your function is working properly by verifying that $[L][U] = [A]$ and by using the built-in function `lu`.

10.7 Confirm the validity of the Cholesky factorization of Example 10.4 by substituting the results into Eq. (10.14) to verify that the product of $[U]^T$ and $[U]$ yields $[A]$.

10.8 (**a**) Perform a Cholesky factorization of the following symmetric system by hand:

$$\begin{bmatrix} 8 & 20 & 15 \\ 20 & 80 & 50 \\ 15 & 50 & 60 \end{bmatrix} \begin{Bmatrix} x_1 \\ x_2 \\ x_3 \end{Bmatrix} = \begin{Bmatrix} 50 \\ 250 \\ 100 \end{Bmatrix}$$

(**b**) Verify your hand calculation with the built-in `chol` function. (**c**) Employ the results of the factorization $[U]$ to determine the solution for the right-hand-side vector.

10.9 Develop your own M-file to determine the Cholesky factorization of a symmetric matrix without pivoting. That is, develop a function that is passed the symmetric matrix and returns the matrix $[U]$. Test your function by using it to solve the system in Prob. 10.8 and use the built-in function `chol` to confirm that your function is working properly.

10.10 Solve the following set of equations with *LU* factorization:

$$3x_1 - 2x_2 + x_3 = -10$$
$$2x_1 + 6x_2 - 4x_3 = 44$$
$$-x_1 - 2x_2 + 5x_3 = -26$$

10.11 **(a)** Determine the *LU* factorization without pivoting by hand for the following matrix and check your results by validating that $[L][U] = [A]$.

$$\begin{bmatrix} 8 & 2 & 1 \\ 3 & 7 & 2 \\ 2 & 3 & 9 \end{bmatrix}$$

(b) Employ the result of **(a)** to compute the determinant.
(c) Repeat **(a)** and **(b)** using MATLAB.
10.12 Use the following *LU* factorization to **(a)** compute the determinant and **(b)** solve $[A]\{x\} = \{b\}$ with $\{b\}^T = \lfloor -10 \quad 44 \quad -26 \rfloor$.

$$[A] = [L][U] = \begin{bmatrix} 1 & & \\ 0.6667 & 1 & \\ -0.3333 & -0.3636 & 1 \end{bmatrix}$$

$$\times \begin{bmatrix} 3 & -2 & 1 \\ & 7.3333 & -4.6667 \\ & & 3.6364 \end{bmatrix}$$

10.13 Use Cholesky factorization to determine $[U]$ so that

$$[A] = [U]^T[U] = \begin{bmatrix} 2 & -1 & 0 \\ -1 & 2 & -1 \\ 0 & -1 & 2 \end{bmatrix}$$

10.14 Compute the Cholesky factorization of

$$[A] = \begin{bmatrix} 9 & 0 & 0 \\ 0 & 25 & 0 \\ 0 & 0 & 4 \end{bmatrix}$$

Do your results make sense in terms of Eqs. (10.15) and (10.16)?

11

Matrix Inverse and Condition

CHAPTER OBJECTIVES

The primary objective of this chapter is to show how to compute the matrix inverse and to illustrate how it can be used to analyze complex linear systems that occur in engineering and science. In addition, a method to assess a matrix solution's sensitivity to roundoff error is described. Specific objectives and topics covered are

- Knowing how to determine the matrix inverse in an efficient manner based on LU factorization.
- Understanding how the matrix inverse can be used to assess stimulus-response characteristics of engineering systems.
- Understanding the meaning of matrix and vector norms and how they are computed.
- Knowing how to use norms to compute the matrix condition number.
- Understanding how the magnitude of the condition number can be used to estimate the precision of solutions of linear algebraic equations.

11.1 THE MATRIX INVERSE

In our discussion of matrix operations (Section 8.1.2), we introduced the notion that if a matrix $[A]$ is square, there is another matrix $[A]^{-1}$, called the inverse of $[A]$, for which

$$[A][A]^{-1} = [A]^{-1}[A] = [I] \tag{11.1}$$

Now we will focus on how the inverse can be computed numerically. Then we will explore how it can be used for engineering analysis.

11.1.1 Calculating the Inverse

The inverse can be computed in a column-by-column fashion by generating solutions with unit vectors as the right-hand-side constants. For example, if the right-hand-side constant has a 1 in the first position and zeros elsewhere,

$$\{b\} = \begin{Bmatrix} 1 \\ 0 \\ 0 \end{Bmatrix} \qquad (11.2)$$

the resulting solution will be the first column of the matrix inverse. Similarly, if a unit vector with a 1 at the second row is used

$$\{b\} = \begin{Bmatrix} 0 \\ 1 \\ 0 \end{Bmatrix} \qquad (11.3)$$

the result will be the second column of the matrix inverse.

The best way to implement such a calculation is with *LU* factorization. Recall that one of the great strengths of *LU* factorization is that it provides a very efficient means to evaluate multiple right-hand-side vectors. Thus, it is ideal for evaluating the multiple unit vectors needed to compute the inverse.

EXAMPLE 11.1   Matrix Inversion

Problem Statement.   Employ *LU* factorization to determine the matrix inverse for the system from Example 10.1:

$$[A] = \begin{bmatrix} 3 & -0.1 & -0.2 \\ 0.1 & 7 & -0.3 \\ 0.3 & -0.2 & 10 \end{bmatrix}$$

Recall that the factorization resulted in the following lower and upper triangular matrices:

$$[U] = \begin{bmatrix} 3 & -0.1 & -0.2 \\ 0 & 7.00333 & -0.293333 \\ 0 & 0 & 10.0120 \end{bmatrix} \qquad [L] = \begin{bmatrix} 1 & 0 & 0 \\ 0.0333333 & 1 & 0 \\ 0.100000 & -0.0271300 & 1 \end{bmatrix}$$

Solution.   The first column of the matrix inverse can be determined by performing the forward-substitution solution procedure with a unit vector (with 1 in the first row) as the right-hand-side vector. Thus, the lower triangular system can be set up as (recall Eq. [10.8])

$$\begin{bmatrix} 1 & 0 & 0 \\ 0.0333333 & 1 & 0 \\ 0.100000 & -0.0271300 & 1 \end{bmatrix} \begin{Bmatrix} d_1 \\ d_2 \\ d_3 \end{Bmatrix} = \begin{Bmatrix} 1 \\ 0 \\ 0 \end{Bmatrix}$$

and solved with forward substitution for $\{d\}^T = \lfloor 1 \quad -0.03333 \quad -0.1009 \rfloor$. This vector can then be used as the right-hand side of the upper triangular system (recall Eq. [10.3]):

$$\begin{bmatrix} 3 & -0.1 & -0.2 \\ 0 & 7.00333 & -0.293333 \\ 0 & 0 & 10.0120 \end{bmatrix} \begin{Bmatrix} x_1 \\ x_2 \\ x_3 \end{Bmatrix} = \begin{Bmatrix} 1 \\ -0.03333 \\ -0.1009 \end{Bmatrix}$$

which can be solved by back substitution for $\{x\}^T = \lfloor 0.33249 \quad -0.00518 \quad -0.01008 \rfloor$, which is the first column of the matrix inverse:

$$[A]^{-1} = \begin{bmatrix} 0.33249 & 0 & 0 \\ -0.00518 & 0 & 0 \\ -0.01008 & 0 & 0 \end{bmatrix}$$

To determine the second column, Eq. (10.8) is formulated as

$$
\begin{bmatrix}
1 & 0 & 0 \\
0.0333333 & 1 & 0 \\
0.100000 & -0.0271300 & 1
\end{bmatrix}
\begin{Bmatrix}
d_1 \\
d_2 \\
d_3
\end{Bmatrix}
=
\begin{Bmatrix}
0 \\
1 \\
0
\end{Bmatrix}
$$

This can be solved for $\{d\}$, and the results are used with Eq. (10.3) to determine $\{x\}^T = \lfloor 0.004944 \quad 0.142903 \quad 0.00271 \rfloor$, which is the second column of the matrix inverse:

$$
[A]^{-1} =
\begin{bmatrix}
0.33249 & 0.004944 & 0 \\
-0.00518 & 0.142903 & 0 \\
-0.01008 & 0.002710 & 0
\end{bmatrix}
$$

Finally, the same procedures can be implemented with $\{b\}^T = \lfloor 0 \quad 0 \quad 1 \rfloor$ to solve for $\{x\}^T = \lfloor 0.006798 \quad 0.004183 \quad 0.09988 \rfloor$, which is the final column of the matrix inverse:

$$
[A]^{-1} =
\begin{bmatrix}
0.33249 & 0.004944 & 0.006798 \\
-0.00518 & 0.142903 & 0.004183 \\
-0.01008 & 0.002710 & 0.099880
\end{bmatrix}
$$

The validity of this result can be checked by verifying that $[A][A]^{-1} = [I]$.

11.1.2 Stimulus-Response Computations

As discussed in PT 3.1, many of the linear systems of equations arising in engineering and science are derived from conservation laws. The mathematical expression of these laws is some form of balance equation to ensure that a particular property—mass, force, heat, momentum, electrostatic potential—is conserved. For a force balance on a structure, the properties might be horizontal or vertical components of the forces acting on each node of the structure. For a mass balance, the properties might be the mass in each reactor of a chemical process. Other fields of engineering and science would yield similar examples.

A single balance equation can be written for each part of the system, resulting in a set of equations defining the behavior of the property for the entire system. These equations are interrelated, or coupled, in that each equation may include one or more of the variables from the other equations. For many cases, these systems are linear and, therefore, of the exact form dealt with in this chapter:

$$
[A]\{x\} = \{b\} \tag{11.4}
$$

Now, for balance equations, the terms of Eq. (11.4) have a definite physical interpretation. For example, the elements of $\{x\}$ are the levels of the property being balanced for each part of the system. In a force balance of a structure, they represent the horizontal and vertical forces in each member. For the mass balance, they are the mass of chemical in each reactor. In either case, they represent the system's *state* or *response*, which we are trying to determine.

The right-hand-side vector $\{b\}$ contains those elements of the balance that are independent of behavior of the system—that is, they are constants. In many problems, they represent the *forcing functions* or *external stimuli* that drive the system.

Finally, the matrix of coefficients [A] usually contains the *parameters* that express how the parts of the system *interact* or are coupled. Consequently, Eq. (11.4) might be reexpressed as

[Interactions]{response} = {stimuli}

As we know from previous chapters, there are a variety of ways to solve Eq. (11.4). However, using the matrix inverse yields a particularly interesting result. The formal solution can be expressed as

$$\{x\} = [A]^{-1}\{b\}$$

or (recalling our definition of matrix multiplication from Section 8.1.2)

$$x_1 = a_{11}^{-1}b_1 + a_{12}^{-1}b_2 + a_{13}^{-1}b_3$$
$$x_2 = a_{21}^{-1}b_1 + a_{22}^{-1}b_2 + a_{23}^{-1}b_3$$
$$x_3 = a_{31}^{-1}b_1 + a_{32}^{-1}b_2 + a_{33}^{-1}b_3$$

Thus, we find that the inverted matrix itself, aside from providing a solution, has extremely useful properties. That is, each of its elements represents the response of a single part of the system to a unit stimulus of any other part of the system.

Notice that these formulations are linear and, therefore, superposition and proportionality hold. *Superposition* means that if a system is subject to several different stimuli (the b's), the responses can be computed individually and the results summed to obtain a total response. *Proportionality* means that multiplying the stimuli by a quantity results in the response to those stimuli being multiplied by the same quantity. Thus, the coefficient a_{11}^{-1} is a proportionality constant that gives the value of x_1 due to a unit level of b_1. This result is independent of the effects of b_2 and b_3 on x_1, which are reflected in the coefficients a_{12}^{-1} and a_{13}^{-1}, respectively. Therefore, we can draw the general conclusion that the element a_{ij}^{-1} of the inverted matrix represents the value of x_i due to a unit quantity of b_j.

Using the example of the structure, element a_{ij}^{-1} of the matrix inverse would represent the force in member i due to a unit external force at node j. Even for small systems, such behavior of individual stimulus-response interactions would not be intuitively obvious. As such, the matrix inverse provides a powerful technique for understanding the interrelationships of component parts of complicated systems.

EXAMPLE 11.2   Analyzing the Bungee Jumper Problem

Problem Statement.   At the beginning of Chap. 8, we set up a problem involving three individuals suspended vertically connected by bungee cords. We derived a system of linear algebraic equations based on force balances for each jumper,

$$\begin{bmatrix} 150 & -100 & 0 \\ -100 & 150 & -50 \\ 0 & -50 & 50 \end{bmatrix} \begin{Bmatrix} x_1 \\ x_2 \\ x_3 \end{Bmatrix} = \begin{Bmatrix} 588.6 \\ 686.7 \\ 784.8 \end{Bmatrix}$$

In Example 8.2, we used MATLAB to solve this system for the vertical positions of the jumpers (the x's). In the present example, use MATLAB to compute the matrix inverse and interpret what it means.

Solution.   Start up MATLAB and enter the coefficient matrix:

```
>> K = [150 -100 0;-100 150 -50;0 -50 50];
```

The inverse can then be computed as

```
>> KI = inv(K)

KI =
    0.0200    0.0200    0.0200
    0.0200    0.0300    0.0300
    0.0200    0.0300    0.0500
```

Each element of the inverse, k_{ij}^{-1} of the inverted matrix represents the vertical change in position (in meters) of jumper i due to a unit change in force (in Newtons) applied to jumper j.

First, observe that the numbers in the first column ($j = 1$) indicate that the position of all three jumpers would increase by 0.02 m if the force on the first jumper was increased by 1 N. This makes sense, because the additional force would only elongate the first cord by that amount.

In contrast, the numbers in the second column ($j = 2$) indicate that applying a force of 1 N to the second jumper would move the first jumper down by 0.02 m, but the second and third by 0.03 m. The 0.02-m elongation of the first jumper makes sense because the first cord is subject to an extra 1 N regardless of whether the force is applied to the first or second jumper. However, for the second jumper the elongation is now 0.03 m because along with the first cord, the second cord also elongates due to the additional force. And of course, the third jumper shows the identical translation as the second jumper as there is no additional force on the third cord that connects them.

As expected, the third column ($j = 3$) indicates that applying a force of 1 N to the third jumper results in the first and second jumpers moving the same distances as occurred when the force was applied to the second jumper. However, now because of the additional elongation of the third cord, the third jumper is moved farther downward.

Superposition and proportionality can be demonstrated by using the inverse to determine how much farther the third jumper would move downward if additional forces of 10, 50, and 20 N were applied to the first, second, and third jumpers, respectively. This can be done simply by using the appropriate elements of the third row of the inverse to compute,

$$\Delta x_3 = k_{31}^{-1}\Delta F_1 + k_{32}^{-1}\Delta F_2 + k_{33}^{-1}\Delta F_3 = 0.02(10) + 0.03(50) + 0.05(20) = 2.7 \text{ m}$$

11.2   ERROR ANALYSIS AND SYSTEM CONDITION

Aside from its engineering and scientific applications, the inverse also provides a means to discern whether systems are ill-conditioned. Three direct methods can be devised for this purpose:

1.   Scale the matrix of coefficients [A] so that the largest element in each row is 1. Invert the scaled matrix and if there are elements of [A]$^{-1}$ that are several orders of magnitude greater than one, it is likely that the system is ill-conditioned.

2. Multiply the inverse by the original coefficient matrix and assess whether the result is close to the identity matrix. If not, it indicates ill-conditioning.
3. Invert the inverted matrix and assess whether the result is sufficiently close to the original coefficient matrix. If not, it again indicates that the system is ill-conditioned.

Although these methods can indicate ill-conditioning, it would be preferable to obtain a single number that could serve as an indicator of the problem. Attempts to formulate such a matrix condition number are based on the mathematical concept of the norm.

11.2.1 Vector and Matrix Norms

A *norm* is a real-valued function that provides a measure of the size or "length" of multi-component mathematical entities such as vectors and matrices.

A simple example is a vector in three-dimensional Euclidean space (Fig. 11.1) that can be represented as

$$\lfloor F \rfloor = \lfloor a \quad b \quad c \rfloor$$

where a, b, and c are the distances along the x, y, and z axes, respectively. The length of this vector—that is, the distance from the coordinate $(0, 0, 0)$ to (a, b, c)—can be simply computed as

$$\|F\|_e = \sqrt{a^2 + b^2 + c^2}$$

where the nomenclature $\|F\|_e$ indicates that this length is referred to as the *Euclidean norm* of $[F]$.

Similarly, for an n-dimensional vector $\lfloor X \rfloor = \lfloor x_1 \quad x_2 \quad \cdots \quad x_n \rfloor$, a Euclidean norm would be computed as

$$\|X\|_e = \sqrt{\sum_{i=1}^{n} x_i^2}$$

FIGURE 11.1
Graphical depiction of a vector in Euclidean space.

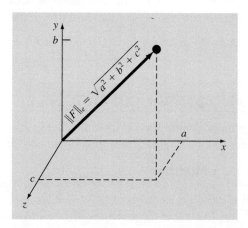

The concept can be extended further to a matrix $[A]$, as in

$$\|A\|_f = \sqrt{\sum_{i=1}^{n} \sum_{j=1}^{n} a_{i,j}^2} \tag{11.5}$$

which is given a special name—the *Frobenius norm*. As with the other vector norms, it provides a single value to quantify the "size" of $[A]$.

It should be noted that there are alternatives to the Euclidean and Frobenius norms. For vectors, there are alternatives called p norms that can be represented generally by

$$\|X\|_p = \left(\sum_{i=1}^{n} |x_i|^p \right)^{1/p}$$

We can see that the Euclidean norm and the 2 norm, $\|X\|_2$, are identical for vectors.

Other important examples are ($p = 1$)

$$\|X\|_1 = \sum_{i=1}^{n} |x_i|$$

which represents the norm as the sum of the absolute values of the elements. Another is the maximum-magnitude or uniform-vector norm ($p = \infty$),

$$\|X\|_\infty = \max_{1 \le i \le n} |x_i|$$

which defines the norm as the element with the largest absolute value.

Using a similar approach, norms can be developed for matrices. For example,

$$\|A\|_1 = \max_{1 \le j \le n} \sum_{i=1}^{n} |a_{ij}|$$

That is, a summation of the absolute values of the coefficients is performed for each column, and the largest of these summations is taken as the norm. This is called the *column-sum norm*.

A similar determination can be made for the rows, resulting in a uniform-matrix or *row-sum norm:*

$$\|A\|_\infty = \max_{1 \le i \le n} \sum_{j=1}^{n} |a_{ij}|$$

It should be noted that, in contrast to vectors, the 2 norm and the Frobenius norm for a matrix are not the same. Whereas the Frobenius norm $\|A\|_f$ can be easily determined by Eq. (11.5), the matrix 2 norm $\|A\|_2$ is calculated as

$$\|A\|_2 = (\mu_{\max})^{1/2}$$

where $\mu_{\max}$ is the largest eigenvalue of $[A]^T[A]$. In Appendix A, we will learn more about eigenvalues. For the time being, the important point is that the $\|A\|_2$, or *spectral norm,* is the minimum norm and, therefore, provides the tightest measure of size (Ortega, 1972).

11.2.2 Matrix Condition Number

Now that we have introduced the concept of the norm, we can use it to define

$$\text{Cond}[A] = \|A\| \cdot \|A^{-1}\|$$

where Cond[A] is called the *matrix condition number.* Note that for a matrix [A], this number will be greater than or equal to 1. It can be shown (Ralston and Rabinowitz, 1978; Gerald and Wheatley, 1989) that

$$\frac{\|\Delta X\|}{\|X\|} \leq \text{Cond}[A]\frac{\|\Delta A\|}{\|A\|}$$

That is, the relative error of the norm of the computed solution can be as large as the relative error of the norm of the coefficients of [A] multiplied by the condition number. For example, if the coefficients of [A] are known to t-digit precision (i.e., rounding errors are on the order of 10^{-t}) and Cond[A] = 10^c, the solution [X] may be valid to only $t - c$ digits (rounding errors $\approx 10^{c-t}$).

EXAMPLE 11.3 Matrix Condition Evaluation

Problem Statement. The Hilbert matrix, which is notoriously ill-conditioned, can be represented generally as

$$\begin{bmatrix} 1 & \frac{1}{2} & \frac{1}{3} & \cdots & \frac{1}{n} \\ \frac{1}{2} & \frac{1}{3} & \frac{1}{4} & \cdots & \frac{1}{n+1} \\ \vdots & \vdots & \vdots & & \vdots \\ \frac{1}{n} & \frac{1}{n+1} & \frac{1}{n+2} & \cdots & \frac{1}{2n-1} \end{bmatrix}$$

Use the row-sum norm to estimate the matrix condition number for the 3×3 Hilbert matrix:

$$[A] = \begin{bmatrix} 1 & \frac{1}{2} & \frac{1}{3} \\ \frac{1}{2} & \frac{1}{3} & \frac{1}{4} \\ \frac{1}{3} & \frac{1}{4} & \frac{1}{5} \end{bmatrix}$$

Solution. First, the matrix can be normalized so that the maximum element in each row is 1:

$$[A] = \begin{bmatrix} 1 & \frac{1}{2} & \frac{1}{3} \\ 1 & \frac{2}{3} & \frac{1}{2} \\ 1 & \frac{3}{4} & \frac{3}{5} \end{bmatrix}$$

Summing each of the rows gives 1.833, 2.1667, and 2.35. Thus, the third row has the largest sum and the row-sum norm is

$$\|A\|_\infty = 1 + \frac{3}{4} + \frac{3}{5} = 2.35$$

The inverse of the scaled matrix can be computed as

$$[A]^{-1} = \begin{bmatrix} 9 & -18 & 10 \\ -36 & 96 & -60 \\ 30 & -90 & 60 \end{bmatrix}$$

Note that the elements of this matrix are larger than the original matrix. This is also reflected in its row-sum norm, which is computed as

$$\|A^{-1}\|_\infty = |-36| + |96| + |-60| = 192$$

Thus, the condition number can be calculated as

$$\text{Cond}[A] = 2.35(192) = 451.2$$

The fact that the condition number is much greater than unity suggests that the system is ill-conditioned. The extent of the ill-conditioning can be quantified by calculating $c = \log 451.2 = 2.65$. Hence, the last three significant digits of the solution could exhibit rounding errors. Note that such estimates almost always overpredict the actual error. However, they are useful in alerting you to the possibility that roundoff errors may be significant.

11.2.3 Norms and Condition Number in MATLAB

MATLAB has built-in functions to compute both norms and condition numbers:

```
>> norm(X,p)
```

and

```
>> cond(X,p)
```

where X is the vector or matrix and p designates the type of norm or condition number (`1`, `2`, `inf`, or `'fro'`). Note that the `cond` function is equivalent to

```
>> norm(X,p) * norm(inv(X),p)
```

Also, note that if p is omitted, it is automatically set to 2.

EXAMPLE 11.4   Matrix Condition Evaluation with MATLAB

Problem Statement.   Use MATLAB to evaluate both the norms and condition numbers for the scaled Hilbert matrix previously analyzed in Example 11.3:

$$[A] = \begin{bmatrix} 1 & \frac{1}{2} & \frac{1}{3} \\ 1 & \frac{2}{3} & \frac{1}{2} \\ 1 & \frac{3}{4} & \frac{3}{5} \end{bmatrix}$$

(*a*) As in Example 11.3, first compute the row-sum versions ($p = \text{inf}$). (*b*) Also compute the Frobenius ($p = \text{'fro'}$) and the spectral ($p = 2$) condition numbers.

Solution: (*a*) First, enter the matrix:

```
>> A = [1 1/2 1/3;1 2/3 1/2;1 3/4 3/5];
```

Then, the row-sum norm and condition number can be computed as

```
>> norm(A,inf)

ans =
    2.3500

>> cond(A,inf)

ans =
  451.2000
```

These results correspond to those that were calculated by hand in Example 11.3.

(*b*) The condition numbers based on the Frobenius and spectral norms are

```
>> cond(A,'fro')

ans =
   368.0866

>> cond(A)

ans =
   366.3503
```

11.3 CASE STUDY   INDOOR AIR POLLUTION

Background.   As the name implies, indoor air pollution deals with air contamination in enclosed spaces such as homes, offices, and work areas. Suppose that you are studying the ventilation system for Bubba's Gas 'N Guzzle, a truck-stop restaurant located adjacent to an eight-lane freeway.

As depicted in Fig. 11.2, the restaurant serving area consists of two rooms for smokers and kids and one elongated room. Room 1 and section 3 have sources of carbon monoxide from smokers and a faulty grill, respectively. In addition, rooms 1 and 2 gain carbon monoxide from air intakes that unfortunately are positioned alongside the freeway.

FIGURE 11.2
Overhead view of rooms in a restaurant. The one-way arrows represent volumetric airflows, whereas the two-way arrows represent diffusive mixing. The smoker and grill loads add carbon monoxide mass to the system but negligible airflow.

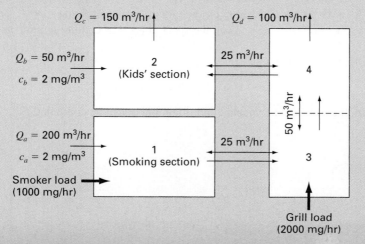

11.3 CASE STUDY   continued

Write steady-state mass balances for each room and solve the resulting linear algebraic equations for the concentration of carbon monoxide in each room. In addition, generate the matrix inverse and use it to analyze how the various sources affect the kids' room. For example, determine what percent of the carbon monoxide in the kids' section is due to (1) the smokers, (2) the grill, and (3) the intake vents. In addition, compute the improvement in the kids' section concentration if the carbon monoxide load is decreased by banning smoking and fixing the grill. Finally, analyze how the concentration in the kids' area would change if a screen is constructed so that the mixing between areas 2 and 4 is decreased to 5 m³/hr.

Solution.   Steady-state mass balances can be written for each room. For example, the balance for the smoking section (room 1) is

$$0 = W_{smoker} + \quad Q_a c_a - \quad Q_a c_1 \quad + E_{13}(c_3 - c_1)$$
$$\text{(Load)} + \text{(Inflow)} - \text{(Outflow)} + \quad \text{(Mixing)}$$

Similar balances can be written for the other rooms:

$$0 = Q_b c_b + (Q_a - Q_d)c_4 - Q_c c_2 + E_{24}(c_4 - c_2)$$

$$0 = W_{grill} + Q_a c_1 + E_{13}(c_1 - c_3) + E_{34}(c_4 - c_3) - Q_a c_3$$

$$0 = Q_a c_3 + E_{34}(c_3 - c_4) + E_{24}(c_2 - c_4) - Q_a c_4$$

Substituting the parameters yields the final system of equation:

$$\begin{bmatrix} 225 & 0 & -25 & 0 \\ 0 & 175 & 0 & -125 \\ -225 & 0 & 275 & -50 \\ 0 & -25 & -250 & 275 \end{bmatrix} \begin{Bmatrix} c_1 \\ c_2 \\ c_3 \\ c_4 \end{Bmatrix} = \begin{Bmatrix} 1400 \\ 100 \\ 2000 \\ 0 \end{Bmatrix}$$

MATLAB can be used to generate the solution. First, we can compute the inverse. Note that we use the "short g" format in order to obtain five significant digits of precision:

```
>> format short g
>> A=[225 0 -25 0
0 175 0 -125
-225 0 275 -50
0 -25 -250 275];
>> AI=inv(A)

AI =
      0.0049962     1.5326e-005     0.00055172     0.00010728
      0.0034483     0.0062069       0.0034483      0.0034483
      0.0049655     0.00013793      0.0049655      0.00096552
      0.0048276     0.00068966      0.0048276      0.0048276
```

The solution can then be generated as

```
>> b=[1400 100 2000 0]';
>> c=AI*b

c =
        8.0996
       12.345
       16.897
       16.483
```

Thus, we get the surprising result that the smoking section has the lowest carbon monoxide levels! The highest concentrations occur in rooms 3 and 4 with section 2 having an intermediate level. These results take place because (a) carbon monoxide is conservative and (b) the only air exhausts are out of sections 2 and 4 (Q_c and Q_d). Room 3 is so bad because not only does it get the load from the faulty grill, but it also receives the effluent from room 1.

Although the foregoing is interesting, the real power of linear systems comes from using the elements of the matrix inverse to understand how the parts of the system interact. For example, the elements of the matrix inverse can be used to determine the percent of the carbon monoxide in the kids' section due to each source:

The smokers:

$$c_{2,\text{smokers}} = a_{21}^{-1} W_{\text{smokers}} = 0.0034483(1000) = 3.4483$$

$$\%_{\text{smokers}} = \frac{3.4483}{12.345} \times 100\% = 27.93\%$$

The grill:

$$c_{2,\text{grill}} = a_{23}^{-1} W_{\text{grill}} = 0.0034483(2000) = 6.897$$

$$\%_{\text{grill}} = \frac{6.897}{12.345} \times 100\% = 55.87\%$$

The intakes:

$$c_{2,\text{intakes}} = a_{21}^{-1} Q_a c_a + a_{22}^{-1} Q_b c_b = 0.0034483(200)2 + 0.0062069(50)2$$
$$= 1.37931 + 0.62069 = 2$$

$$\%_{\text{grill}} = \frac{2}{12.345} \times 100\% = 16.20\%$$

The faulty grill is clearly the most significant source.

The inverse can also be employed to determine the impact of proposed remedies such as banning smoking and fixing the grill. Because the model is linear, superposition holds and the results can be determined individually and summed:

$$\Delta c_2 = a_{21}^{-1} \Delta W_{\text{smoker}} + a_{23}^{-1} \Delta W_{\text{grill}} = 0.0034483(-1000) + 0.0034483(-2000)$$
$$= -3.4483 - 6.8966 = -10.345$$

11.3 CASE STUDY continued

Note that the same computation would be made in MATLAB as

```
>> AI(2,1)*(-1000)+AI(2,3)*(-2000)

ans =
    -10.345
```

Implementing both remedies would reduce the concentration by 10.345 mg/m^3. The result would bring the kids' room concentration to $12.345 - 10.345 = 2$ mg/m^3. This makes sense, because in the absence of the smoker and grill loads, the only sources are the air intakes which are at 2 mg/m^3.

Because all the foregoing calculations involved changing the forcing functions, it was not necessary to recompute the solution. However, if the mixing between the kids' area and zone 4 is decreased, the matrix is changed

$$\begin{bmatrix} 225 & 0 & -25 & 0 \\ 0 & 155 & 0 & -105 \\ -225 & 0 & 275 & -50 \\ 0 & -5 & -250 & 255 \end{bmatrix} \begin{Bmatrix} c_1 \\ c_2 \\ c_3 \\ c_4 \end{Bmatrix} = \begin{Bmatrix} 1400 \\ 100 \\ 2000 \\ 0 \end{Bmatrix}$$

The results for this case involve a new solution. Using MATLAB, the result is

$$\begin{Bmatrix} c_1 \\ c_2 \\ c_3 \\ c_4 \end{Bmatrix} = \begin{Bmatrix} 8.1084 \\ 12.0800 \\ 16.9760 \\ 16.8800 \end{Bmatrix}$$

Therefore, this remedy would only improve the kids' area concentration by a paltry 0.265 mg/m^3.

PROBLEMS

11.1 Determine the matrix inverse for the following system:

$$10x_1 + 2x_2 - x_3 = 27$$
$$-3x_1 - 6x_2 + 2x_3 = -61.5$$
$$x_1 + x_2 + 5x_3 = -21.5$$

Check your results by verifying that $[A][A]^{-1} = [I]$. Do not use a pivoting strategy.

11.2 Determine the matrix inverse for the following system:

$$-8x_1 + x_2 - 2x_3 = -20$$
$$2x_1 - 6x_2 - x_3 = -38$$
$$-3x_1 - x_2 + 7x_3 = -34$$

11.3 The following system of equations is designed to determine concentrations (the c's in g/m^3) in a series of coupled reactors as a function of the amount of mass input to each reactor (the right-hand sides in g/day):

$$15c_1 - 3c_2 - c_3 = 3800$$
$$-3c_1 + 18c_2 - 6c_3 = 1200$$
$$-4c_1 - c_2 + 12c_3 = 2350$$

(a) Determine the matrix inverse.
(b) Use the inverse to determine the solution.
(c) Determine how much the rate of mass input to reactor 3 must be increased to induce a 10 g/m^3 rise in the concentration of reactor 1.
(d) How much will the concentration in reactor 3 be reduced if the rate of mass input to reactors 1 and 2 is reduced by 500 and 250 g/day, respectively?

11.4 Determine the matrix inverse for the system described in Prob. 8.5. Use the matrix inverse to determine the concentration in reactor 5 if the inflow concentrations are changed to $c_{01} = 20$ and $c_{03} = 50$.

11.5 Determine the matrix inverse for the system described in Prob. 8.6. Use the matrix inverse to determine the force in the three members (F_1, F_2 and F_3) if the vertical load at node 1 is doubled to $F_{1,v} = -2000$ lb and a horizontal load of $F_{3,h} = -500$ lb is applied to node 3.

11.6 Determine $\|A\|_f$, $\|A\|_1$, and $\|A\|_\infty$ for

$$[A] = \begin{bmatrix} 8 & 2 & -10 \\ -9 & 1 & 3 \\ 15 & -1 & 6 \end{bmatrix}$$

Before determining the norms, scale the matrix by making the maximum element in each row equal to one.

11.7 Determine the Frobenius and row-sum norms for the systems in Probs. 11.2 and 11.3.

11.8 Use MATLAB to determine the spectral condition number for the following system. Do not normalize the system:

$$\begin{bmatrix} 1 & 4 & 9 & 16 & 25 \\ 4 & 9 & 16 & 25 & 36 \\ 9 & 16 & 25 & 36 & 49 \\ 16 & 25 & 36 & 49 & 64 \\ 25 & 36 & 49 & 64 & 81 \end{bmatrix}$$

Compute the condition number based on the row-sum norm.

11.9 Besides the Hilbert matrix, there are other matrices that are inherently ill-conditioned. One such case is the *Vandermonde matrix*, which has the following form:

$$\begin{bmatrix} x_1^2 & x_1 & 1 \\ x_2^2 & x_2 & 1 \\ x_3^2 & x_3 & 1 \end{bmatrix}$$

(a) Determine the condition number based on the row-sum norm for the case where $x_1 = 4$, $x_2 = 2$, and $x_3 = 7$.
(b) Use MATLAB to compute the spectral and Frobenius condition numbers.

11.10 Use MATLAB to determine the spectral condition number for a 10-dimensional Hilbert matrix. How many digits of precision are expected to be lost due to ill-conditioning? Determine the solution for this system for the case where each element of the right-hand-side vector $\{b\}$ consists of the summation of the coefficients in its row. In other words, solve for the case where all the unknowns should be exactly one. Compare the resulting errors with those expected based on the condition number.

11.11 Repeat Prob. 11.10, but for the case of a six-dimensional Vandermonde matrix (see Prob. 11.9) where $x_1 = 4$, $x_2 = 2$, $x_3 = 7$, $x_4 = 10$, $x_5 = 3$, and $x_6 = 5$.

11.12 The Lower Colorado River consists of a series of four reservoirs as shown in Fig. P11.12.

Mass balances can be written for each reservoir, and the following set of simultaneous linear algebraic equations results:

$$\begin{bmatrix} 13.422 & 0 & 0 & 0 \\ -13.422 & 12.252 & 0 & 0 \\ 0 & -12.252 & 12.377 & 0 \\ 0 & 0 & -12.377 & 11.797 \end{bmatrix}$$

$$\times \begin{Bmatrix} c_1 \\ c_2 \\ c_3 \\ c_4 \end{Bmatrix} = \begin{Bmatrix} 750.5 \\ 300 \\ 102 \\ 30 \end{Bmatrix}$$

where the right-hand-side vector consists of the loadings of chloride to each of the four lakes and c_1, c_2, c_3, and $c_4 =$ the resulting chloride concentrations for Lakes Powell, Mead, Mohave, and Havasu, respectively.

(a) Use the matrix inverse to solve for the concentrations in each of the four lakes.
(b) How much must the loading to Lake Powell be reduced for the chloride concentration of Lake Havasu to be 75?
(c) Using the column-sum norm, compute the condition number and how many suspect digits would be generated by solving this system.

11.13 (a) Determine the matrix inverse and condition number for the following matrix:

$$\begin{bmatrix} 1 & 2 & 3 \\ 4 & 5 & 6 \\ 7 & 8 & 9 \end{bmatrix}$$

(b) Repeat (a) but change a_{33} slightly to 9.1.

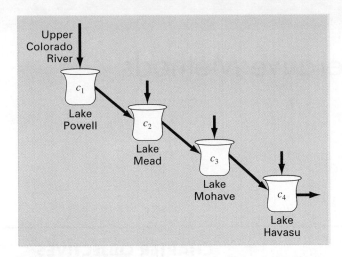

FIGURE P11.12
The Lower Colorado River.

11.14 Polynomial interpolation consists of determining the unique $(n-1)$th-order polynomial that fits n data points. Such polynomials have the general form,

$$f(x) = p_1 x^{n-1} + p_2 x^{n-2} + \cdots + p_{n-1} x + p_n \qquad \text{(P11.14)}$$

where the p's are constant coefficients. A straightforward way for computing the coefficients is to generate n linear algebraic equations that we can solve simultaneously for the coefficients. Suppose that we want to determine the coefficients of the fourth-order polynomial $f(x) = p_1 x^4 + p_2 x^3 + p_3 x^2 + p_4 x + p_5$ that passes through the following five points: (200, 0.746), (250, 0.675), (300, 0.616), (400, 0.525), and (500, 0.457). Each of these pairs can be substituted into Eq. (P11.14) to yield a system of five equations with five unknowns (the p's). Use this approach to solve for the coefficients. In addition, determine and interpret the condition number.

12

Iterative Methods

terative or approximate methods provide an alternative to the elimination methods described to this point. Such approaches are similar to the techniques we developed to obtain the roots of a single equation in Chaps. 5 and 6. Those approaches consisted of guessing a value and then using a systematic method to obtain a refined estimate of the root. Because the present part of the book deals with a similar problem—obtaining the values that simultaneously satisfy a set of equations—we might suspect that such approximate methods could be useful in this context. In this chapter, we will present approaches for solving both linear and nonlinear simultaneous equations.

12.1 LINEAR SYSTEMS: GAUSS-SEIDEL

The *Gauss-Seidel method* is the most commonly used iterative method for solving linear algebraic equations. Assume that we are given a set of n equations:

$$[A]\{x\} = \{b\}$$

Suppose that for conciseness we limit ourselves to a 3×3 set of equations. If the diagonal elements are all nonzero, the first equation can be solved for x_1, the second for x_2, and the third for x_3 to yield

$$x_1^j = \frac{b_1 - a_{12}x_2^{j-1} - a_{13}x_3^{j-1}}{a_{11}} \tag{12.1a}$$

$$x_2^j = \frac{b_2 - a_{21}x_1^j - a_{23}x_3^{j-1}}{a_{22}} \tag{12.1b}$$

$$x_3^j = \frac{b_3 - a_{31}x_1^j - a_{32}x_2^j}{a_{33}} \tag{12.1c}$$

where j and $j - 1$ are the present and previous iterations.

To start the solution process, initial guesses must be made for the x's. A simple approach is to assume that they are all zero. These zeros can be substituted into Eq. (12.1a), which can be used to calculate a new value for $x_1 = b_1/a_{11}$. Then we substitute this new value of x_1 along with the previous guess of zero for x_3 into Eq. (12.1b) to compute a new value for x_2. The process is repeated for Eq. (12.1c) to calculate a new estimate for x_3. Then we return to the first equation and repeat the entire procedure until our solution converges closely enough to the true values. Convergence can be checked using the criterion that for all i,

$$\varepsilon_{a,i} = \left| \frac{x_i^j - x_i^{j-1}}{x_i^j} \right| \times 100\% \leq \varepsilon_s \tag{12.2}$$

EXAMPLE 12.1 Gauss-Seidel Method

Problem Statement.   Use the Gauss-Seidel method to obtain the solution for

$$3x_1 - 0.1x_2 - 0.2x_3 = 7.85$$
$$0.1x_1 + 7x_2 - 0.3x_3 = -19.3$$
$$0.3x_1 - 0.2x_2 + 10x_3 = 71.4$$

Note that the solution is $\{x\}^T = \lfloor 3 \quad -2.5 \quad 7 \rfloor$.

Solution.   First, solve each of the equations for its unknown on the diagonal:

$$x_1 = \frac{7.85 + 0.1x_2 + 0.2x_3}{3} \tag{E12.1.1}$$

$$x_2 = \frac{-19.3 - 0.1x_1 + 0.3x_3}{7} \tag{E12.1.2}$$

$$x_3 = \frac{71.4 - 0.3x_1 + 0.2x_2}{10} \tag{E12.1.3}$$

By assuming that x_2 and x_3 are zero, Eq. (E12.1.1) can be used to compute

$$x_1 = \frac{7.85 + 0.1(0) + 0.2(0)}{3} = 2.616667$$

This value, along with the assumed value of $x_3 = 0$, can be substituted into Eq. (E12.1.2) to calculate

$$x_2 = \frac{-19.3 - 0.1(2.616667) + 0.3(0)}{7} = -2.794524$$

The first iteration is completed by substituting the calculated values for x_1 and x_2 into Eq. (E12.1.3) to yield

$$x_3 = \frac{71.4 - 0.3(2.616667) + 0.2(-2.794524)}{10} = 7.005610$$

For the second iteration, the same process is repeated to compute

$$x_1 = \frac{7.85 + 0.1(-2.794524) + 0.2(7.005610)}{3} = 2.990557$$

$$x_2 = \frac{-19.3 - 0.1(2.990557) + 0.3(7.005610)}{7} = -2.499625$$

$$x_3 = \frac{71.4 - 0.3(2.990557) + 0.2(-2.499625)}{10} = 7.000291$$

The method is, therefore, converging on the true solution. Additional iterations could be applied to improve the answers. However, in an actual problem, we would not know the true answer *a priori*. Consequently, Eq. (12.2) provides a means to estimate the error. For example, for x_1:

$$\varepsilon_{a,1} = \left| \frac{2.990557 - 2.616667}{2.990557} \right| \times 100\% = 12.5\%$$

For x_2 and x_3, the error estimates are $\varepsilon_{a,2} = 11.8\%$ and $\varepsilon_{a,3} = 0.076\%$. Note that, as was the case when determining roots of a single equation, formulations such as Eq. (12.2) usually provide a conservative appraisal of convergence. Thus, when they are met, they ensure that the result is known to at least the tolerance specified by ε_s.

As each new x value is computed for the Gauss-Seidel method, it is immediately used in the next equation to determine another x value. Thus, if the solution is converging, the best available estimates will be employed. An alternative approach, called *Jacobi iteration*, utilizes a somewhat different tactic. Rather than using the latest available x's, this technique uses Eq. (12.1) to compute a set of new x's on the basis of a set of old x's. Thus, as new values are generated, they are not immediately used but rather are retained for the next iteration.

The difference between the Gauss-Seidel method and Jacobi iteration is depicted in Fig. 12.1. Although there are certain cases where the Jacobi method is useful, Gauss-Seidel's utilization of the best available estimates usually makes it the method of preference.

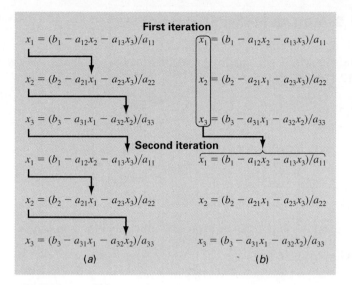

FIGURE 12.1
Graphical depiction of the difference between (*a*) the Gauss-Seidel and (*b*) the Jacobi iterative methods for solving simultaneous linear algebraic equations.

12.1.1 Convergence and Diagonal Dominance

Note that the Gauss-Seidel method is similar in spirit to the technique of simple fixed-point iteration that was used in Section 6.1 to solve for the roots of a single equation. Recall that simple fixed-point iteration was sometimes nonconvergent. That is, as the iterations progressed, the answer moved farther and farther from the correct result.

Although the Gauss-Seidel method can also diverge, because it is designed for linear systems, its ability to converge is much more predictable than for fixed-point iteration of nonlinear equations. It can be shown that if the following condition holds, Gauss-Seidel will converge:

$$|a_{ii}| > \sum_{\substack{j=1 \\ j \neq i}}^{n} |a_{ij}| \tag{12.3}$$

That is, the absolute value of the diagonal coefficient in each of the equations must be larger than the sum of the absolute values of the other coefficients in the equation. Such systems are said to be *diagonally dominant*. This criterion is sufficient but not necessary for convergence. That is, although the method may sometimes work if Eq. (12.3) is not met, convergence is guaranteed if the condition is satisfied. Fortunately, many engineering and scientific problems of practical importance fulfill this requirement. Therefore, Gauss-Seidel represents a feasible approach to solve many problems in engineering and science.

12.1.2 MATLAB M-file: GaussSeidel

Before developing an algorithm, let us first recast Gauss-Seidel in a form that is compatible with MATLAB's ability to perform matrix operations. This is done by expressing Eq. (12.1) as

$$x_1^{new} = \frac{b_1}{a_{11}} \qquad\qquad -\frac{a_{12}}{a_{11}}x_2^{old} - \frac{a_{13}}{a_{11}}x_3^{old}$$

$$x_2^{new} = \frac{b_2}{a_{22}} - \frac{a_{21}}{a_{22}}x_1^{new} \qquad\qquad -\frac{a_{23}}{a_{22}}x_3^{old}$$

$$x_3^{new} = \frac{b_3}{a_{33}} - \frac{a_{31}}{a_{33}}x_1^{new} - \frac{a_{32}}{a_{33}}x_2^{new}$$

Notice that the solution can be expressed concisely in matrix form as

$$\{x\} = \{d\} - [C]\{x\} \tag{12.4}$$

where

$$\{d\} = \begin{Bmatrix} b_1/a_{11} \\ b_2/a_{22} \\ b_3/a_{33} \end{Bmatrix}$$

and

$$[C] = \begin{bmatrix} 0 & a_{12}/a_{11} & a_{13}/a_{11} \\ a_{21}/a_{22} & 0 & a_{23}/a_{22} \\ a_{31}/a_{33} & a_{32}/a_{33} & 0 \end{bmatrix}$$

An M-file to implement Eq. (12.4) is listed in Fig. 12.2.

12.1.3 Relaxation

Relaxation represents a slight modification of the Gauss-Seidel method that is designed to enhance convergence. After each new value of x is computed using Eq. (12.1), that value is modified by a weighted average of the results of the previous and the present iterations:

$$x_i^{new} = \lambda x_i^{new} + (1 - \lambda)x_i^{old} \tag{12.5}$$

where λ is a weighting factor that is assigned a value between 0 and 2.

If $\lambda = 1$, $(1 - \lambda)$ is equal to 0 and the result is unmodified. However, if λ is set at a value between 0 and 1, the result is a weighted average of the present and the previous results. This type of modification is called *underrelaxation*. It is typically employed to make a nonconvergent system converge or to hasten convergence by dampening out oscillations.

For values of λ from 1 to 2, extra weight is placed on the present value. In this instance, there is an implicit assumption that the new value is moving in the correct direction toward the true solution but at too slow a rate. Thus, the added weight of λ is intended to improve the estimate by pushing it closer to the truth. Hence, this type of modification, which is called *overrelaxation,* is designed to accelerate the convergence of an already convergent system. The approach is also called *successive overrelaxation,* or SOR.

```
function x = GaussSeidel(A,b,es,maxit)
% GaussSeidel: Gauss Seidel method
%   x = GaussSeidel(A,b): Gauss Seidel without relaxation
% input:
%   A = coefficient matrix
%   b = right hand side vector
%   es = stop criterion (default = 0.00001%)
%   maxit = max iterations (default = 50)
% output:
%   x = solution vector

if nargin<2,error('at least 2 input arguments required'),end
if nargin<4|isempty(maxit),maxit=50;end
if nargin<3|isempty(es),es=0.00001;end
[m,n] = size(A);
if m~=n, error('Matrix A must be square'); end
C = A;
for i = 1:n
  C(i,i) = 0;
  x(i) = 0;
end
x = x';
for i = 1:n
  C(i,1:n) = C(i,1:n)/A(i,i);
end
for i = 1:n
  d(i) = b(i)/A(i,i);
end
iter = 0;
while (1)
  xold = x;
  for i = 1:n
    x(i) = d(i)-C(i,:)*x;
    if x(i) ~= 0
      ea(i) = abs((x(i) - xold(i))/x(i)) * 100;
    end
  end
  iter = iter+1;
  if max(ea)<=es | iter >= maxit, break, end
end
```

FIGURE 12.2
MATLAB M-file to implement Gauss-Seidel.

The choice of a proper value for λ is highly problem-specific and is often determined empirically. For a single solution of a set of equations it is often unnecessary. However, if the system under study is to be solved repeatedly, the efficiency introduced by a wise choice of λ can be extremely important. Good examples are the very large systems of linear algebraic equations that can occur when solving partial differential equations in a variety of engineering and scientific problem contexts.

12.2 NONLINEAR SYSTEMS

The following is a set of two simultaneous nonlinear equations with two unknowns:

$$x_1^2 + x_1 x_2 = 10 \tag{12.6a}$$

$$x_2 + 3x_1 x_2^2 = 57 \tag{12.6b}$$

In contrast to linear systems which plot as straight lines (recall Fig. 9.1), these equations plot as curves on an x_2 versus x_1 graph. As in Fig. 12.3, the solution is the intersection of the curves.

Just as we did when we determined roots for single nonlinear equations, such systems of equations can be expressed generally as

$$\begin{aligned}
f_1(x_1, x_2, \ldots, x_n) &= 0 \\
f_2(x_1, x_2, \ldots, x_n) &= 0 \\
&\vdots \\
f_n(x_1, x_2, \ldots, x_n) &= 0
\end{aligned} \tag{12.7}$$

Therefore, the solution are the values of the x's that make the equations equal to zero.

12.2.1 Successive Substitution

A simple approach for solving Eq. (12.7) is to use the same strategy that was employed for fixed-point iteration and the Gauss-Seidel method. That is, each one of the nonlinear equations can be solved for one of the unknowns. These equations can then be implemented iteratively to compute new values which (hopefully) will converge on the solutions. This approach, which is called *successive substitution,* is illustrated in the following example.

FIGURE 12.3
Graphical depiction of the solution of two simultaneous nonlinear equations.

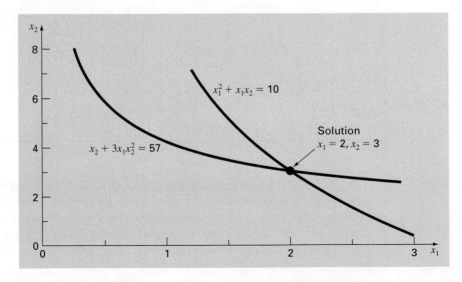

EXAMPLE 12.2   Successive Substitution for a Nonlinear System

Problem Statement.   Use successive substitution to determine the roots of Eq. (12.6). Note that a correct pair of roots is $x_1 = 2$ and $x_2 = 3$. Initiate the computation with guesses of $x_1 = 1.5$ and $x_2 = 3.5$.

Solution.   Equation (12.6a) can be solved for

$$x_1 = \frac{10 - x_1^2}{x_2} \qquad\qquad\qquad\qquad\qquad (\text{E}12.2.1)$$

and Eq. (12.6b) can be solved for

$$x_2 = 57 - 3x_1 x_2^2 \qquad\qquad\qquad\qquad\qquad (\text{E}12.2.2)$$

On the basis of the initial guesses, Eq. (E12.2.1) can be used to determine a new value of x_1:

$$x_1 = \frac{10 - (1.5)^2}{3.5} = 2.21429$$

This result and the initial value of $x_2 = 3.5$ can be substituted into Eq. (E12.2.2) to determine a new value of x_2:

$$x_2 = 57 - 3(2.21429)(3.5)^2 = -24.37516$$

Thus, the approach seems to be diverging. This behavior is even more pronounced on the second iteration:

$$x_1 = \frac{10 - (2.21429)^2}{-24.37516} = -0.20910$$

$$x_2 = 57 - 3(-0.20910)(-24.37516)^2 = 429.709$$

Obviously, the approach is deteriorating.

Now we will repeat the computation but with the original equations set up in a different format. For example, an alternative solution of Eq. (12.6a) is

$$x_1 = \sqrt{10 - x_1 x_2}$$

and of Eq. (12.6b) is

$$x_2 = \sqrt{\frac{57 - x_2}{3x_1}}$$

Now the results are more satisfactory:

$$x_1 = \sqrt{10 - 1.5(3.5)} = 2.17945$$

$$x_2 = \sqrt{\frac{57 - 3.5}{3(2.17945)}} = 2.86051$$

$$x_1 = \sqrt{10 - 2.17945(2.86051)} = 1.94053$$

$$x_2 = \sqrt{\frac{57 - 2.86051}{3(1.94053)}} = 3.04955$$

Thus, the approach is converging on the true values of $x_1 = 2$ and $x_2 = 3$.

The previous example illustrates the most serious shortcoming of successive substitution—that is, convergence often depends on the manner in which the equations are formulated. Additionally, even in those instances where convergence is possible, divergence can occur if the initial guesses are insufficiently close to the true solution. These criteria are so restrictive that fixed-point iteration has limited utility for solving nonlinear systems.

12.2.2 Newton-Raphson

Just as fixed-point iteration can be used to solve systems of nonlinear equations, other open root location methods such as the Newton-Raphson method can be used for the same purpose. Recall that the Newton-Raphson method was predicated on employing the derivative (i.e., the slope) of a function to estimate its intercept with the axis of the independent variable—that is, the root. In Chap. 6, we used a graphical derivation to compute this estimate. An alternative is to derive it from a first-order Taylor series expansion:

$$f(x_{i+1}) = f(x_i) + (x_{i+1} - x_i)f'(x_i) \tag{12.8}$$

where x_i is the initial guess at the root and x_{i+1} is the point at which the slope intercepts the x axis. At this intercept, $f(x_{i+1})$ by definition equals zero and Eq. (12.8) can be rearranged to yield

$$x_{i+1} = x_i - \frac{f(x_i)}{f'(x_i)} \tag{12.9}$$

which is the single-equation form of the Newton-Raphson method.

The multiequation form is derived in an identical fashion. However, a multivariable Taylor series must be used to account for the fact that more than one independent variable contributes to the determination of the root. For the two-variable case, a first-order Taylor series can be written for each nonlinear equation as

$$f_{1,i+1} = f_{1,i} + (x_{1,i+1} - x_{1,i})\frac{\partial f_{1,i}}{\partial x_1} + (x_{2,i+1} - x_{2,i})\frac{\partial f_{1,i}}{\partial x_2} \tag{12.10a}$$

$$f_{2,i+1} = f_{2,i} + (x_{1,i+1} - x_{1,i})\frac{\partial f_{2,i}}{\partial x_1} + (x_{2,i+1} - x_{2,i})\frac{\partial f_{2,i}}{\partial x_2} \tag{12.10b}$$

Just as for the single-equation version, the root estimate corresponds to the values of x_1 and x_2, where $f_{1,i+1}$ and $f_{2,i+1}$ equal zero. For this situation, Eq. (12.10) can be rearranged to give

$$\frac{\partial f_{1,i}}{\partial x_1}x_{1,i+1} + \frac{\partial f_{1,i}}{\partial x_2}x_{2,i+1} = -f_{1,i} + x_{1,i}\frac{\partial f_{1,i}}{\partial x_1} + x_{2,i}\frac{\partial f_{1,i}}{\partial x_2} \tag{12.11a}$$

$$\frac{\partial f_{2,i}}{\partial x_1}x_{1,i+1} + \frac{\partial f_{2,i}}{\partial x_2}x_{2,i+1} = -f_{2,i} + x_{1,i}\frac{\partial f_{2,i}}{\partial x_1} + x_{2,i}\frac{\partial f_{2,i}}{\partial x_2} \tag{12.11b}$$

Because all values subscripted with i's are known (they correspond to the latest guess or approximation), the only unknowns are $x_{1,i+1}$ and $x_{2,i+1}$. Thus, Eq. (12.11) is a set of two linear equations with two unknowns. Consequently, algebraic manipulations (e.g., Cramer's rule) can be employed to solve for

$$x_{1,i+1} = x_{1,i} - \frac{f_{1,i}\dfrac{\partial f_{2,i}}{\partial x_2} - f_{2,i}\dfrac{\partial f_{1,i}}{\partial x_2}}{\dfrac{\partial f_{1,i}}{\partial x_1}\dfrac{\partial f_{2,i}}{\partial x_2} - \dfrac{\partial f_{1,i}}{\partial x_2}\dfrac{\partial f_{2,i}}{\partial x_1}} \tag{12.12a}$$

$$x_{2,i+1} = x_{2,i} - \frac{f_{2,i}\dfrac{\partial f_{1,i}}{\partial x_1} - f_{1,i}\dfrac{\partial f_{2,i}}{\partial x_1}}{\dfrac{\partial f_{1,i}}{\partial x_1}\dfrac{\partial f_{2,i}}{\partial x_2} - \dfrac{\partial f_{1,i}}{\partial x_2}\dfrac{\partial f_{2,i}}{\partial x_1}} \tag{12.12b}$$

The denominator of each of these equations is formally referred to as the determinant of the *Jacobian* of the system.

Equation (12.12) is the two-equation version of the Newton-Raphson method. As in the following example, it can be employed iteratively to home in on the roots of two simultaneous equations.

EXAMPLE 12.3   Newton-Raphson for a Nonlinear System

Problem Statement.   Use the multiple-equation Newton-Raphson method to determine roots of Eq. (12.6). Initiate the computation with guesses of $x_1 = 1.5$ and $x_2 = 3.5$.

Solution.   First compute the partial derivatives and evaluate them at the initial guesses of x and y:

$$\frac{\partial f_{1,0}}{\partial x_1} = 2x_1 + x_2 = 2(1.5) + 3.5 = 6.5 \qquad \frac{\partial f_{1,0}}{\partial x_2} = x_1 = 1.5$$

$$\frac{\partial f_{2,0}}{\partial x_1} = 3x_2^2 = 3(3.5)^2 = 36.75 \qquad \frac{\partial f_{2,0}}{\partial x_2} = 1 + 6x_1x_2 = 1 + 6(1.5)(3.5) = 32.5$$

Thus, the determinant of the Jacobian for the first iteration is

$$6.5(32.5) - 1.5(36.75) = 156.125$$

The values of the functions can be evaluated at the initial guesses as

$$f_{1,0} = (1.5)^2 + 1.5(3.5) - 10 = -2.5$$
$$f_{2,0} = 3.5 + 3(1.5)(3.5)^2 - 57 = 1.625$$

These values can be substituted into Eq. (12.12) to give

$$x_1 = 1.5 - \frac{-2.5(32.5) - 1.625(1.5)}{156.125} = 2.03603$$

$$x_2 = 3.5 - \frac{1.625(6.5) - (-2.5)(36.75)}{156.125} = 2.84388$$

Thus, the results are converging to the true values of $x_1 = 2$ and $x_2 = 3$. The computation can be repeated until an acceptable accuracy is obtained.

When the multiequation Newton-Raphson works, it exhibits the same speedy quadratic convergence as the single-equation version. However, just as with successive substitution, it can diverge if the initial guesses are not sufficiently close to the true roots. Whereas graphical methods could be employed to derive good guesses for the single-equation case, no such simple procedure is available for the multiequation version. Although there are some advanced approaches for obtaining acceptable first estimates, often the initial guesses must be obtained on the basis of trial and error and knowledge of the physical system being modeled.

The two-equation Newton-Raphson approach can be generalized to solve n simultaneous equations. To do this, Eq. (12.11) can be written for the kth equation as

$$\frac{\partial f_{k,i}}{\partial x_1}x_{1,i+1} + \frac{\partial f_{k,i}}{\partial x_2}x_{2,i+1} + \cdots + \frac{\partial f_{k,i}}{\partial x_n}x_{n,i+1} = -f_{k,i} + x_{1,i}\frac{\partial f_{k,i}}{\partial x_1} + x_{2,i}\frac{\partial f_{k,i}}{\partial x_2}$$

$$+ \cdots + x_{n,i}\frac{\partial f_{k,i}}{\partial x_n} \tag{12.13}$$

where the first subscript k represents the equation or unknown and the second subscript denotes whether the value or function in question is at the present value (i) or at the next value ($i + 1$). Notice that the only unknowns in Eq. (12.13) are the $x_{k,i+1}$ terms on the left-hand side. All other quantities are located at the present value (i) and, thus, are known at any iteration. Consequently, the set of equations generally represented by Eq. (12.13) (i.e., with $k = 1, 2, \ldots, n$) constitutes a set of linear simultaneous equations that can be solved numerically by the elimination methods elaborated in previous chapters.

Matrix notation can be employed to express Eq. (12.13) concisely as

$$[J]\{x_{i+1}\} = -\{f\} + [J]\{x_i\} \tag{12.14}$$

where the partial derivatives evaluated at i are written as the *Jacobian matrix* consisting of the partial derivatives:

$$[J] = \begin{bmatrix} \dfrac{\partial f_{1,i}}{\partial x_1} & \dfrac{\partial f_{1,i}}{\partial x_2} & \cdots & \dfrac{\partial f_{1,i}}{\partial x_n} \\[2mm] \dfrac{\partial f_{2,i}}{\partial x_1} & \dfrac{\partial f_{2,i}}{\partial x_2} & \cdots & \dfrac{\partial f_{2,i}}{\partial x_n} \\[2mm] \vdots & \vdots & & \vdots \\[2mm] \dfrac{\partial f_{n,i}}{\partial x_1} & \dfrac{\partial f_{n,i}}{\partial x_2} & \cdots & \dfrac{\partial f_{n,i}}{\partial x_n} \end{bmatrix} \tag{12.15}$$

The initial and final values are expressed in vector form as

$$\{x_i\}^T = \lfloor x_{1,i} \quad x_{2,i} \quad \cdots \quad x_{n,i} \rfloor$$

and

$$\{x_{i+1}\}^T = \lfloor x_{1,i+1} \quad x_{2,i+1} \quad \cdots \quad x_{n,i+1} \rfloor$$

Finally, the function values at i can be expressed as

$$\{f\}^T = \lfloor f_{1,i} \quad f_{2,i} \quad \cdots \quad f_{n,i} \rfloor$$

Equation (12.14) can be solved using a technique such as Gauss elimination. This process can be repeated iteratively to obtain refined estimates in a fashion similar to the two-equation case in Example 12.3.

Insight into the solution can be obtained by solving Eq. (12.14) with matrix inversion. Recall that the single-equation version of the Newton-Raphson method is

$$x_{i+1} = x_i - \frac{f(x_i)}{f'(x_i)} \tag{12.16}$$

If Eq. (12.14) is solved by multiplying it by the inverse of the Jacobian, the result is

$$\{x_{i+1}\} = \{x_i\} - [J]^{-1}\{f\} \tag{12.17}$$

Comparison of Eqs. (12.16) and (12.17) clearly illustrates the parallels between the two equations. In essence, the Jacobian is analogous to the derivative of a multivariate function.

Such matrix calculations can be implemented very efficiently in MATLAB. We can illustrate this by using MATLAB to duplicate the calculations from Example 12.3. After defining the initial guesses, we can compute the Jacobian and the function values as

```
>> x=[1.5;3.5];
>> J=[2*x(1)+x(2) x(1);3*x(2)^2 1+6*x(1)*x(2)]

J =
    6.5000    1.5000
   36.7500   32.5000

>> f=[x(1)^2+x(1)*x(2)-10;x(2)+3*x(1)*x(2)^2-57]

f =
   -2.5000
    1.6250
```

Then, we can implement Eq. (12.17) to yield the improved estimates

```
>> x=x-J\f

x =
    2.0360
    2.8439
```

Although we could continue the iterations in the command mode, a nicer alternative is to express the algorithm as an M-file. As in Figure 12.4, this routine is passed an M-file that computes the function values and the Jacobian at a given value of x. It then calls this function and implements Eq. (12.17) in an iterative fashion. The routine iterates until an upper limit of iterations (`maxit`) or a specified percent relative error (`es`) is reached.

We should note that there are two shortcomings to the foregoing approach. First, Eq. (12.15) is sometimes inconvenient to evaluate. Therefore, variations of the Newton-Raphson approach have been developed to circumvent this dilemma. As might be expected, most are based on using finite-difference approximations for the partial derivatives that comprise $[J]$. The second shortcoming of the multiequation Newton-Raphson method is that excellent initial guesses are usually required to ensure convergence. Because these are sometimes difficult or inconvenient to obtain, alternative approaches that are slower

```
function [x,f,ea,iter]=newtmult(func,x0,es,maxit,varargin)
% newtmult: Newton-Raphson root zeroes nonlinear systems
%   [x,f,ea,iter]=newtmult(func,x0,es,maxit,p1,p2,...):
%     uses the Newton-Raphson method to find the roots of
%     a system of nonlinear equations
% input:
%   func = name of function that returns f and J
%   x0 = initial guess
%   es = desired percent relative error (default = 0.0001%)
%   maxit = maximum allowable iterations (default = 50)
%   p1,p2,... = additional parameters used by function
% output:
%   x = vector of roots
%   f = vector of functions evaluated at roots
%   ea = approximate percent relative error (%)
%   iter = number of iterations

if nargin<2,error('at least 2 input arguments required'),end
if nargin<3|isempty(es),es=0.0001;end
if nargin<4|isempty(maxit),maxit=50;end
iter = 0;
x=x0;
while (1)
  [J,f]=func(x,varargin{:});
  dx=J\f;
  x=x-dx;
  iter = iter + 1;
  ea=100*max(abs(dx./x));
  if iter>=maxit|ea<=es, break, end
end
```

FIGURE 12.4
MATLAB M-file to implement Newton-Raphson method for nonlinear systems of equations.

than Newton-Raphson but which have better convergence behavior have been developed. One approach is to reformulate the nonlinear system as a single function:

$$F(x) = \sum_{i=1}^{n} [f_i(x_1, x_2, \ldots, x_n)]^2$$

where $f_i(x_1, x_2, \ldots, x_n)$ is the ith member of the original system of Eq. (12.7). The values of x that minimize this function also represent the solution of the nonlinear system. Therefore, nonlinear optimization techniques can be employed to obtain solutions.

12.3 CASE STUDY CHEMICAL REACTIONS

Background. Nonlinear systems of equations occur frequently in the characterization of chemical reactions. For example, the following chemical reactions take place in a closed system:

$$2A + B \underset{\leftarrow}{\overset{\rightarrow}{}} C \tag{12.18}$$

$$A + D \underset{\leftarrow}{\overset{\rightarrow}{}} C \tag{12.19}$$

At equilibrium, they can be characterized by

$$K_1 = \frac{c_c}{c_a^2 c_b} \tag{12.20}$$

$$K_2 = \frac{c_c}{c_a c_d} \tag{12.21}$$

where the nomenclature c_i represents the concentration of constituent i. If x_1 and x_2 are the number of moles of C that are produced due to the first and second reactions, respectively, formulate the equilibrium relationships as a pair of two simultaneous nonlinear equations. If $K_1 = 4 \times 10^{-4}$, $K_2 = 3.7 \times 10^{-2}$, $c_{a,0} = 50$, $c_{b,0} = 20$, $c_{c,0} = 5$, and $c_{d,0} = 10$, employ the Newton-Raphson method to solve these equations.

Solution. Using the stoichiometry of Eqs. (12.18) and (12.19), the concentrations of each constituent can be represented in terms of x_1 and x_2 as

$$c_a = c_{a,0} - 2x_1 - x_2 \tag{12.22}$$

$$c_b = c_{b,0} - x_1 \tag{12.23}$$

$$c_c = c_{c,0} + x_1 + x_2 \tag{12.24}$$

$$c_d = c_{d,0} - x_2 \tag{12.25}$$

where the subscript 0 designates the initial concentration of each constituent. These values can be substituted into Eqs. (12.20) and (12.21) to give

$$K_1 = \frac{(c_{c,0} + x_1 + x_2)}{(c_{a,0} - 2x_1 - x_2)^2 (c_{b,0} - x_1)}$$

$$K_2 = \frac{(c_{c,0} + x_1 + x_2)}{(c_{a,0} - 2x_1 - x_2)(c_{d,0} - x_2)}$$

Given the parameter values, these are two nonlinear equations with two unknowns. Thus, the solution to this problem involves determining the roots of

$$f_1(x_1, x_2) = \frac{5 + x_1 + x_2}{(50 - 2x_1 - x_2)^2 (20 - x_1)} - 4 \times 10^{-4} \tag{12.26}$$

$$f_2(x_1, x_2) = \frac{(5 + x_1 + x_2)}{(50 - 2x_1 - x_2)(10 - x_2)} - 3.7 \times 10^{-2} \tag{12.27}$$

In order to use Newton-Raphson, we must determine the Jacobian by taking the partial derivatives of Eqs. (12.26) and (12.27). Although this is certainly possible, evaluating the derivatives is time consuming. An alternative is to represent them by finite differences in a fashion similar to the approach used for the modified secant method in Sec. 6.3. For example, the partial derivatives comprising the Jacobian can be evaluated as

$$\frac{\partial f_1}{\partial x_1} = \frac{f_1(x_1 + \delta x_1, x_2) - f_1(x_1, x_2)}{\delta x_1} \qquad \frac{\partial f_1}{\partial x_2} = \frac{f_1(x_1, x_2 + \delta x_2) - f_1(x_1, x_2)}{\delta x_2}$$

$$\frac{\partial f_2}{\partial x_1} = \frac{f_2(x_1 + \delta x_1, x_2) - f_2(x_1, x_2)}{\delta x_1} \qquad \frac{\partial f_2}{\partial x_2} = \frac{f_2(x_1, x_2 + \delta x_2) - f_2(x_1, x_2)}{\delta x_2}$$

These relationships can then be expressed as an M-file to compute both the function values and the Jacobian as

```
function [J,f]=jfreact(x,varargin)
del=0.000001;
df1dx1=(u(x(1)+del*x(1),x(2))-u(x(1),x(2)))/(del*x(1));
df1dx2=(u(x(1),x(2)+del*x(2))-u(x(1),x(2)))/(del*x(2));
df2dx1=(v(x(1)+del*x(1),x(2))-v(x(1),x(2)))/(del*x(1));
df2dx2=(v(x(1),x(2)+del*x(2))-v(x(1),x(2)))/(del*x(2));
J=[df1dx1 df1dx2;df2dx1 df2dx2];
f1=u(x(1),x(2));
f2=v(x(1),x(2));
f=[f1;f2];

function f=u(x,y)
f = (5 + x + y) / (50 - 2 * x - y) ^ 2 / (20 - x) - 0.0004;

function f=v(x,y)
f = (5 + x + y) / (50 - 2 * x - y) / (10 - y) - 0.037;
```

The function `newtmult` (Fig. 12.4) can then be employed to determine the roots given initial guesses of $x_1 = x_2 = 3$:

```
>> format short e
>> [x,f,ea,iter]=newtmult(@jfreact,x0)

x =
  3.3366e+000
  2.6772e+000

f =
 -7.1286e-017
  8.5973e-014

ea =
  5.2237e-010

iter =
    4
```

12.3 CASE STUDY continued

After four iterations, a solution of $x_1 = 3.3366$ and $x_2 = 2.6772$ is obtained. These values can then be substituted into Eq. (12.22) through (12.25) to compute the equilibrium concentrations of the four constituents:

$$c_a = 50 - 2(3.3366) - 2.6772 = 40.6496$$

$$c_b = 20 - 3.3366 = 16.6634$$

$$c_c = 5 + 3.3366 + 2.6772 = 11.0138$$

$$c_d = 10 - 2.6772 = 7.3228$$

PROBLEMS

12.1 **(a)** Use the Gauss-Seidel method to solve the following system until the percent relative error falls below $\varepsilon_s = 5\%$:

$$\begin{bmatrix} 0.8 & -0.4 & \\ -0.4 & 0.8 & -0.4 \\ & -0.4 & 0.8 \end{bmatrix} \begin{Bmatrix} x_1 \\ x_2 \\ x_3 \end{Bmatrix} = \begin{Bmatrix} 41 \\ 25 \\ 105 \end{Bmatrix}$$

(b) Repeat **(a)** but use overrelaxation with $\lambda = 1.2$.

12.2 Use the Gauss-Seidel method to solve the following system until the percent relative error falls below $\varepsilon_s = 5\%$:

$$10x_1 + 2x_2 - x_3 = 27$$
$$-3x_1 - 6x_2 + 2x_3 = -61.5$$
$$x_1 + x_2 + 5x_3 = -21.5$$

12.3 Repeat Prob. 12.2 but use Jacobi iteration.

12.4 The following system of equations is designed to determine concentrations (the c's in g/m^3) in a series of coupled reactors as a function of the amount of mass input to each reactor (the right-hand sides in g/day):

$$15c_1 - 3c_2 - c_3 = 3800$$
$$-3c_1 + 18c_2 - 6c_3 = 1200$$
$$-4c_1 - c_2 + 12c_3 = 2350$$

Solve this problem with the Gauss-Seidel method to $\varepsilon_s = 5\%$.

12.5 Use the Gauss-Seidel method **(a)** without relaxation and **(b)** with relaxation ($\lambda = 1.2$) to solve the following system to a tolerance of $\varepsilon_s = 5\%$. If necessary, rearrange the equations to achieve convergence.

$$2x_1 - 6x_2 - x_3 = -38$$
$$-3x_1 - x_2 + 7x_3 = -34$$
$$-8x_1 + x_2 - 2x_3 = -20$$

12.6 Of the following three sets of linear equations, identify the set(s) that you could not solve using an iterative method such as Gauss-Seidel. Show using any number of iterations that is necessary that your solution does not converge. Clearly state your convergence criteria (how you know it is not converging).

Set One	Set Two	Set Three
$9x + 3y + z = 13$	$x + y + 6z = 8$	$-3x + 4y + 5z = 6$
$-6x + 8z = 2$	$x + 5y - z = 5$	$-2x + 2y - 3z = -3$
$2x + 5y - z = 6$	$4x + 2y - 2z = 4$	$2y - z = 1$

12.7 Determine the solution of the simultaneous nonlinear equations

$$y = -x^2 + x + 0.75$$
$$y + 5xy = x^2$$

Use the Newton-Raphson method and employ initial guesses of $x = y = 1.2$.

12.8 Determine the solution of the simultaneous nonlinear equations:

$$x^2 = 5 - y^2$$
$$y + 1 = x^2$$

(a) Graphically.
(b) Successive substitution using initial guesses of $x = y = 1.5$.
(c) Newton-Raphson using initial guesses of $x = y = 1.5$.

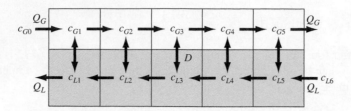

FIGURE P12.9

12.9 Figure P12.9 depicts a chemical exchange process consisting of a series of reactors in which a gas flowing from left to right is passed over a liquid flowing from right to left. The transfer of a chemical from the gas into the liquid occurs at a rate that is proportional to the difference between the gas and liquid concentrations in each reactor. At steady state, a mass balance for the first reactor can be written for the gas as

$$Q_G c_{G0} - Q_G c_{G1} + D(c_{L1} - c_{G1}) = 0$$

and for the liquid as

$$Q_L c_{L2} - Q_L c_{L1} + D(c_{G1} - c_{L1}) = 0$$

where Q_G and Q_L are the gas and liquid flow rates, respectively, and D = the gas-liquid exchange rate. Similar balances can be written for the other reactors. Solve for the concentrations given the following values: $Q_G = 2$, $Q_L = 1$, $D = 0.8$, $c_{G0} = 100$, $c_{L6} = 10$.

12.10 The steady-state distribution of temperature on a heated plate can be modeled by the *Laplace equation*:

$$0 = \frac{\partial^2 T}{\partial x^2} + \frac{\partial^2 T}{\partial y^2}$$

If the plate is represented by a series of nodes (Fig. P12.10), centered finite differences can be substituted for the second derivatives, which result in a system of linear algebraic equations. Use the Gauss-Seidel method to solve for the temperatures of the nodes in Fig. P12.10.

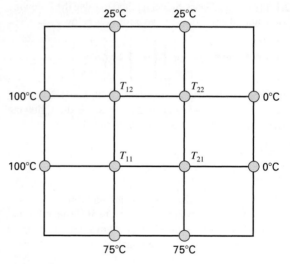

FIGURE P12.10

PART FOUR

Curve Fitting

4.1 OVERVIEW

What Is Curve Fitting?

Data is often given for discrete values along a continuum. However, you may require estimates at points between the discrete values. Chapters 13 through 16 describe techniques to fit curves to such data to obtain intermediate estimates. In addition, you may require a simplified version of a complicated function. One way to do this is to compute values of the function at a number of discrete values along the range of interest. Then, a simpler function may be derived to fit these values. Both of these applications are known as *curve fitting*.

There are two general approaches for curve fitting that are distinguished from each other on the basis of the amount of error associated with the data. First, where the data exhibits a significant degree of error or "scatter," the strategy is to derive a single curve that represents the general trend of the data. Because any individual data point may be incorrect, we make no effort to intersect every point. Rather, the curve is designed to follow the pattern of the points taken as a group. One approach of this nature is called *least-squares regression* (Fig. PT4.1*a*).

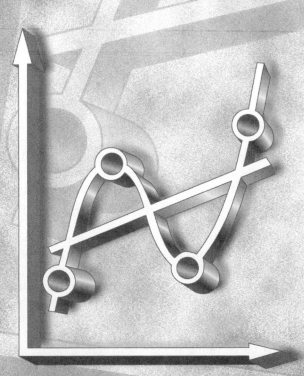

Second, where the data is known to be very precise, the basic approach is to fit a curve or a series of curves that pass directly through each of the points. Such data usually originates from tables. Examples are values for the density of water or for the heat capacity of gases as a function of temperature. The estimation of values between well-known discrete points is called *interpolation* (Fig. PT4.1*b* and *c*).

Curve Fitting and Engineering and Science. Your first exposure to curve fitting may have been to determine intermediate values from tabulated data—for instance, from interest tables for engineering economics or from steam tables for thermodynamics. Throughout the remainder of your career, you will have frequent occasion to estimate intermediate values from such tables.

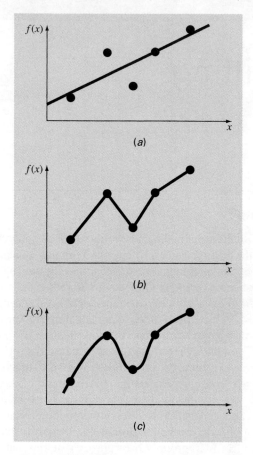

FIGURE PT4.1
Three attempts to fit a "best" curve through five data points: (a) least-squares regression, (b) linear interpolation, and (c) curvilinear interpolation.

Although many of the widely used engineering and scientific properties have been tabulated, there are a great many more that are not available in this convenient form. Special cases and new problem contexts often require that you measure your own data and develop your own predictive relationships. Two types of applications are generally encountered when fitting experimental data: trend analysis and hypothesis testing.

Trend analysis represents the process of using the pattern of the data to make predictions. For cases where the data is measured with high precision, you might utilize interpolating polynomials. Imprecise data is often analyzed with least-squares regression.

Trend analysis may be used to predict or forecast values of the dependent variable. This can involve extrapolation beyond the limits of the observed data or interpolation within the range of the data. All fields of engineering and science involve problems of this type.

A second application of experimental curve fitting is *hypothesis testing*. Here, an existing mathematical model is compared with measured data. If the model coefficients are unknown, it may be necessary to determine values that best fit the observed data. On the other hand, if estimates of the model coefficients are already available, it may be appropriate to compare predicted values of the model with observed values to test the adequacy of the model. Often, alternative models are compared and the "best" one is selected on the basis of empirical observations.

In addition to the foregoing engineering and scientific applications, curve fitting is important in other numerical methods such as integration and the approximate solution of differential equations. Finally, curve-fitting techniques can be used to derive simple functions to approximate complicated functions.

4.2 PART ORGANIZATION

After a brief review of statistics, *Chap. 13* focuses on *linear regression;* that is, how to determine the "best" straight line through a set of uncertain data points. Besides discussing how to calculate the slope and intercept of this straight line, we also present quantitative and visual methods for evaluating the validity of the results. In addition, we describe several approaches for the linearization of nonlinear equations.

Chapter 14 begins with brief discussions of polynomial and multiple linear regression. *Polynomial regression* deals with developing a best fit of parabolas, cubics, or higher-order polynomials. This is followed by a description of *multiple linear regression,* which is designed for the case where the dependent variable y is a linear function of two or more independent variables $x_1, x_2, \ldots, x_m$. This approach has special utility for evaluating experimental data where the variable of interest is dependent on a number of different factors.

After multiple regression, we illustrate how polynomial and multiple regression are both subsets of a *general linear least-squares model.* Among other things, this will allow us to introduce a concise matrix representation of regression and discuss its general statistical properties. Finally, the last sections of Chap. 14 are devoted to *nonlinear regression.* This approach is designed to compute a least-squares fit of a nonlinear equation to data.

In *Chap. 15,* the alternative curve-fitting technique called *interpolation* is described. As discussed previously, interpolation is used for estimating intermediate values between precise data points. In Chap. 15, polynomials are derived for this purpose. We introduce the basic concept of polynomial interpolation by using straight lines and parabolas to connect points. Then, we develop a generalized procedure for fitting an nth-order polynomial. Two formats are presented for expressing these polynomials in equation form. The first, called *Newton's interpolating polynomial,* is preferable when the appropriate order of the polynomial is unknown. The second, called the *Lagrange interpolating polynomial,* has advantages when the proper order is known beforehand.

Finally, *Chap. 16* presents an alternative technique for fitting precise data points. This technique, called *spline interpolation,* fits polynomials to data but in a piecewise fashion. As such, it is particularly well suited for fitting data that is generally smooth but exhibits abrupt local changes. The chapter ends with an overview of how piecewise interpolation is implemented in MATLAB.

Linear Regression

YOU'VE GOT A PROBLEM

In Chap. 1, we noted that a free-falling object such as a bungee jumper is subject to the upward force of air resistance. As a first approximation, we assumed that this force was proportional to the square of velocity as in

$$F_U = c_d v^2 \tag{13.1}$$

where F_U = the upward force of air resistance [N = kg m/s^2], c_d = a drag coefficient (kg/m), and v = velocity [m/s].

Expressions such as Eq. (13.1) come from the field of fluid mechanics. Although such relationships derive in part from theory, experiments play a critical role in their formulation. One such experiment is depicted in Fig. 13.1. An individual is suspended in a wind

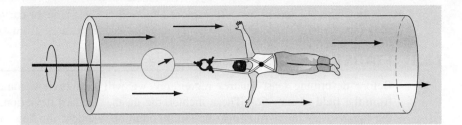

FIGURE 13.1
Wind tunnel experiment to measure how the force of air resistance depends on velocity.

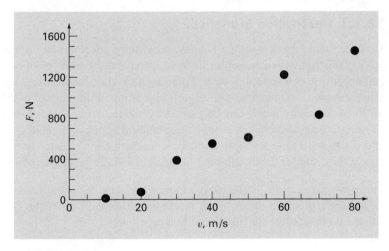

FIGURE 13.2
Plot of force versus wind velocity for an object suspended in a wind tunnel.

TABLE 13.1 Experimental data for force (N) and velocity (m/s) from a wind tunnel experiment.

v, m/s	10	20	30	40	50	60	70	80
F, N	25	70	380	550	610	1220	830	1450

tunnel (any volunteers?) and the force measured for various levels of wind velocity. The result might be as listed in Table 13.1.

The relationship can be visualized by plotting force versus velocity. As in Fig. 13.2, several features of the relationship bear mention. First, the points indicate that the force increases as velocity increases. Second, the points do not increase smoothly, but exhibit rather significant scatter, particularly at the higher velocities. Finally, although it may not be obvious, the relationship between force and velocity may not be linear. This conclusion becomes more apparent if we assume that force is zero for zero velocity.

In Chaps. 13 and 14, we will explore how to fit a "best" line or curve to such data. In so doing, we will illustrate how relationships like Eq. (13.1) arise from experimental data.

13.1 STATISTICS REVIEW

Before describing least-squares regression, we will first review some basic concepts from the field of statistics. These include the mean, standard deviation, residual sum of the squares, and the normal distribution. In addition, we describe how simple descriptive statistics and distributions can be generated in MATLAB. If you are familiar with these subjects, feel free to skip the following pages and proceed directly to Sec. 13.2. If you are unfamiliar with these concepts or are in need of a review, the following material is designed as a brief introduction.

13.1.1 Descriptive Statistics

Suppose that in the course of an engineering study, several measurements were made of a particular quantity. For example, Table 13.2 contains 24 readings of the coefficient of thermal expansion of a structural steel. Taken at face value, the data provides a limited amount of information—that is, that the values range from a minimum of 6.395 to a maximum of 6.775. Additional insight can be gained by summarizing the data in one or more well-chosen statistics that convey as much information as possible about specific characteristics of the data set. These descriptive statistics are most often selected to represent (1) the location of the center of the distribution of the data and (2) the degree of spread of the data set.

Measure of Location.   The most common measure of central tendency is the arithmetic mean. The *arithmetic mean* ($\bar{y}$) of a sample is defined as the sum of the individual data points (y_i) divided by the number of points (n), or

$$\bar{y} = \frac{\sum y_i}{n} \tag{13.2}$$

where the summation (and all the succeeding summations in this section) is from $i = 1$ through n.

There are several alternatives to the arithmetic mean. The *median* is the midpoint of a group of data. It is calculated by first putting the data in ascending order. If the number of measurements is odd, the median is the middle value. If the number is even, it is the arithmetic mean of the two middle values. The median is sometimes called the *50th percentile*.

The *mode* is the value that occurs most frequently. The concept usually has direct utility only when dealing with discrete or coarsely rounded data. For continuous variables such as the data in Table 13.2, the concept is not very practical. For example, there are actually

TABLE 13.2 Measurements of the coefficient of thermal expansion of structural steel [($\times 10^{-6}$) in/(in · °F)].

6.495	6.595	6.615	6.635	6.485	6.555
6.665	6.505	6.435	6.625	6.715	6.655
6.755	6.625	6.715	6.575	6.655	6.605
6.565	6.515	6.555	6.395	6.775	6.685

four modes for this data: 6.555, 6.625, 6.655, and 6.715, which all occur twice. If the numbers had not been rounded to 3 decimal digits, it would be unlikely that any of the values would even have repeated twice. However, if continuous data are grouped into equispaced intervals, it can be an informative statistic. We will return to the mode when we describe histograms later in this section.

Measures of Spread. The simplest measure of spread is the *range,* the difference between the largest and the smallest value. Although it is certainly easy to determine, it is not considered a very reliable measure because it is highly sensitive to the sample size and is very sensitive to extreme values.

The most common measure of spread for a sample is the *standard deviation* (s_y) about the mean:

$$s_y = \sqrt{\frac{S_t}{n-1}} \tag{13.3}$$

where S_t is the total sum of the squares of the residuals between the data points and the mean, or

$$S_t = \sum (y_i - \bar{y})^2 \tag{13.4}$$

Thus, if the individual measurements are spread out widely around the mean, S_t (and, consequently, s_y) will be large. If they are grouped tightly, the standard deviation will be small. The spread can also be represented by the square of the standard deviation, which is called the *variance:*

$$s_y^2 = \frac{\sum (y_i - \bar{y})^2}{n-1} \tag{13.5}$$

Note that the denominator in both Eqs. (13.3) and (13.5) is $n-1$. The quantity $n-1$ is referred to as the *degrees of freedom.* Hence S_t and s_y are said to be based on $n-1$ degrees of freedom. This nomenclature derives from the fact that the sum of the quantities upon which S_t is based (i.e., $\bar{y} - y_1, \bar{y} - y_2, \ldots, \bar{y} - y_n$) is zero. Consequently, if $\bar{y}$ is known and $n-1$ of the values are specified, the remaining value is fixed. Thus, only $n-1$ of the values are said to be freely determined. Another justification for dividing by $n-1$ is the fact that there is no such thing as the spread of a single data point. For the case where $n=1$, Eqs. (13.3) and (13.5) yield a meaningless result of infinity.

We should note that an alternative, more convenient formula is available to compute the variance:

$$s_y^2 = \frac{\sum y_i^2 - \left(\sum y_i\right)^2 / n}{n-1} \tag{13.6}$$

This version does not require precomputation of $\bar{y}$ and yields an identical result as Eq. (13.5).

A final statistic that has utility in quantifying the spread of data is the coefficient of variation (c.v.). This statistic is the ratio of the standard deviation to the mean. As such, it provides a normalized measure of the spread. It is often multiplied by 100 so that it can be expressed in the form of a percent:

$$\text{c.v.} = \frac{s_y}{\bar{y}} \times 100\% \tag{13.7}$$

EXAMPLE 13.1   Simple Statistics of a Sample

Problem Statement. Compute the mean, median, variance, standard deviation, and coefficient of variation for the data in Table 13.2.

Solution. The data can be assembled in tabular form and the necessary sums computed as in Table 13.3.

The mean can be computed as [Eq. (13.2)],

$$\bar{y} = \frac{158.4}{24} = 6.6$$

Because there are an even number of values, the median is computed as the arithmetic mean of the middle two values: $(6.605 + 6.615)/2 = 6.61$.

As in Table 13.3, the sum of the squares of the residuals is 0.217000, which can be used to compute the standard deviation [Eq. (13.3)]:

$$s_y = \sqrt{\frac{0.217000}{24 - 1}} = 0.097133$$

TABLE 13.3 Data and summations for computing simple descriptive statistics for the coefficients of thermal expansion from Table 13.2.

i	y_i	$(y_i - \bar{y})^2$	y_i^2
1	6.395	0.04203	40.896
2	6.435	0.02723	41.409
3	6.485	0.01323	42.055
4	6.495	0.01103	42.185
5	6.505	0.00903	42.315
6	6.515	0.00723	42.445
7	6.555	0.00203	42.968
8	6.555	0.00203	42.968
9	6.565	0.00123	43.099
10	6.575	0.00063	43.231
11	6.595	0.00003	43.494
12	6.605	0.00002	43.626
13	6.615	0.00022	43.758
14	6.625	0.00062	43.891
15	6.625	0.00062	43.891
16	6.635	0.00122	44.023
17	6.655	0.00302	44.289
18	6.655	0.00302	44.289
19	6.665	0.00422	44.422
20	6.685	0.00722	44.689
21	6.715	0.01322	45.091
22	6.715	0.01322	45.091
23	6.755	0.02402	45.630
24	6.775	0.03062	45.901
Σ	158.400	0.21700	1045.657

the variance [Eq. (13.5)]:

$$s_y^2 = (0.097133)^2 = 0.009435$$

and the coefficient of variation [Eq. (13.7)]:

$$c.v. = \frac{0.097133}{6.6} \times 100\% = 1.47\%$$

The validity of Eq. (13.6) can also be verified by computing

$$s_y^2 = \frac{1045.657 - (158.400)^2/24}{24 - 1} = 0.009435$$

13.1.2 The Normal Distribution

Another characteristic that bears on the present discussion is the data distribution—that is, the shape with which the data is spread around the mean. A histogram provides a simple visual representation of the distribution. A *histogram* is constructed by sorting the measurements into intervals, or *bins*. The units of measurement are plotted on the abscissa and the frequency of occurrence of each interval is plotted on the ordinate.

As an example, a histogram can be created for the data from Table 13.2. The result (Fig. 13.4) suggests that most of the data is grouped close to the mean value of 6.6. Notice also, that now that we have grouped the data, we can see that the bin with the most values is from 6.6 to 6.64. Although we could say that the mode is the midpoint of this bin, 6.62, it is more common to report the most frequent range as the *modal class interval*.

If we have a very large set of data, the histogram often can be approximated by a smooth curve. The symmetric, bell-shaped curve superimposed on Fig. 13.3 is one such characteristic shape—the *normal distribution*. Given enough additional measurements, the histogram for this particular case could eventually approach the normal distribution.

FIGURE 13.3
A histogram used to depict the distribution of data. As the number of data points increases, the histogram often approaches the smooth, bell-shaped curve called the normal distribution.

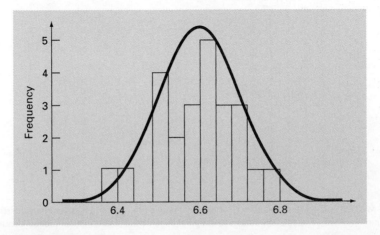

The concepts of the mean, standard deviation, residual sum of the squares, and normal distribution all have great relevance to engineering and science. A very simple example is their use to quantify the confidence that can be ascribed to a particular measurement. If a quantity is normally distributed, the range defined by $\bar{y} - s_y$ to $\bar{y} + s_y$ will encompass approximately 68% of the total measurements. Similarly, the range defined by $\bar{y} - 2s_y$ to $\bar{y} + 2s_y$ will encompass approximately 95%.

For example, for the data in Table 13.2, we calculated in Example 13.1 that $\bar{y} = 6.6$ and $s_y = 0.097133$. Based on our analysis, we can tentatively make the statement that approximately 95% of the readings should fall between 6.405734 and 6.794266. Because it is so far outside these bounds, if someone told us that they had measured a value of 7.35, we would suspect that the measurement might be erroneous.

13.1.3 Descriptive Statistics in MATLAB

Standard MATLAB has several functions to compute descriptive statistics.[1] For example, the arithmetic mean is computed as `mean(x)`. If x is a vector, the function returns the mean of the vector's values. If it is a matrix, it returns a row vector containing the arithmetic mean of each column of x. The following is the result of using mean and the other statistical functions to analyze a column vector s that holds the data from Table 13.2:

```
>> format short g
>> mean(s),median(s),mode(s)

ans =
          6.6
ans =
         6.61
ans =
        6.555
>> min(s),max(s)

ans =
        6.395
ans =
        6.775
>> range=max(s)-min(s)

range =
         0.38
>> var(s),std(s)

ans =
     0.0094348
ans =
     0.097133
```

[1] MATLAB also offers a Statistics Toolbox that provides a wide range of common statistical tasks, from random number generation, to curve fitting, to design of experiments and statistical process control.

These results are consistent with those obtained previously in Example 13.1. Note that although there are four values that occur twice, the `mode` function only returns one of the values: 6.555.

MATLAB can also be used to generate a histogram based on the `hist` function. The `hist` function has the syntax

```
[n, x] = hist(y, x)
```

where n = the number of elements in each bin, x = a vector specifying the midpoint of each bin, and y is the vector being analyzed. For the data from Table 13.2, the result is

```
>> hist(s)
>> [n,x] =hist(s)

n =
     1    1    3    1    4    3    5    2    2    2
x =
   6.414 6.452 6.49 6.528 6.566 6.604 6.642 6.68 6.718 6.756
```

The resulting histogram depicted in Fig. 13.4 is similar to the one we generated by hand in Fig. 13.3. Note that all the arguments and outputs with the exception of y are optional. For example, `hist(y)` without output arguments just produces a histogram bar plot with 10 bins determined automatically based on the range of values in y.

FIGURE 13.4
Histogram generated with the MATLAB `hist` function.

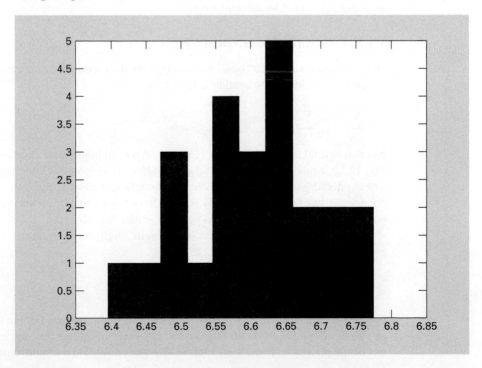

13.2 LINEAR LEAST-SQUARES REGRESSION

Where substantial error is associated with data, the best curve-fitting strategy is to derive an approximating function that fits the shape or general trend of the data without necessarily matching the individual points. One approach to do this is to visually inspect the plotted data and then sketch a "best" line through the points. Although such "eyeball" approaches have commonsense appeal and are valid for "back-of-the-envelope" calculations, they are deficient because they are arbitrary. That is, unless the points define a perfect straight line (in which case, interpolation would be appropriate), different analysts would draw different lines.

To remove this subjectivity, some criterion must be devised to establish a basis for the fit. One way to do this is to derive a curve that minimizes the discrepancy between the data points and the curve. To do this, we must first quantify the discrepancy. The simplest example is fitting a straight line to a set of paired observations: $(x_1, y_1), (x_2, y_2), \ldots, (x_n, y_n)$. The mathematical expression for the straight line is

$$y = a_0 + a_1 x + e \tag{13.8}$$

where a_0 and a_1 are coefficients representing the intercept and the slope, respectively, and e is the error, or *residual,* between the model and the observations, which can be represented by rearranging Eq. (13.8) as

$$e = y - a_0 - a_1 x \tag{13.9}$$

Thus, the residual is the discrepancy between the true value of y and the approximate value, $a_0 + a_1 x$, predicted by the linear equation.

13.2.1 Criteria for a "Best" Fit

One strategy for fitting a "best" line through the data would be to minimize the sum of the residual errors for all the available data, as in

$$\sum_{i=1}^{n} e_i = \sum_{i=1}^{n} (y_i - a_0 - a_1 x_i) \tag{13.10}$$

where n = total number of points. However, this is an inadequate criterion, as illustrated by Fig. 13.5a, which depicts the fit of a straight line to two points. Obviously, the best fit is the line connecting the points. However, any straight line passing through the midpoint of the connecting line (except a perfectly vertical line) results in a minimum value of Eq. (13.10) equal to zero because positive and negative errors cancel.

One way to remove the effect of the signs might be to minimize the sum of the absolute values of the discrepancies, as in

$$\sum_{i=1}^{n} |e_i| = \sum_{i=1}^{n} |y_i - a_0 - a_1 x_i| \tag{13.11}$$

Figure 13.5b demonstrates why this criterion is also inadequate. For the four points shown, any straight line falling within the dashed lines will minimize the sum of the absolute values of the residuals. Thus, this criterion also does not yield a unique best fit.

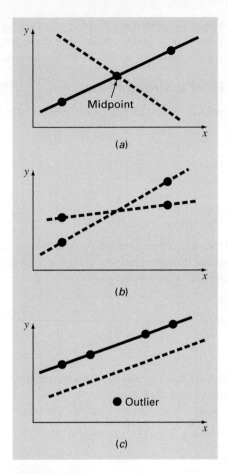

FIGURE 13.5
Examples of some criteria for "best fit" that are inadequate for regression: (a) minimizes the sum
of the residuals, (b) minimizes the sum of the absolute values of the residuals, and (c) minimizes
the maximum error of any individual point.

A third strategy for fitting a best line is the *minimax* criterion. In this technique, the line
is chosen that minimizes the maximum distance that an individual point falls from the line.
As depicted in Fig. 13.5c, this strategy is ill-suited for regression because it gives undue
influence to an outlier—that is, a single point with a large error. It should be noted that
the minimax principle is sometimes well-suited for fitting a simple function to a complicated function (Carnahan, Luther, and Wilkes, 1969).

A strategy that overcomes the shortcomings of the aforementioned approaches is to
minimize the sum of the squares of the residuals:

$$S_r = \sum_{i=1}^{n} e_i^2 = \sum_{i=1}^{n} (y_i - a_0 - a_1 x_i)^2 \qquad (13.12)$$

This criterion, which is called *least squares*, has a number of advantages, including that it yields a unique line for a given set of data. Before discussing these properties, we will present a technique for determining the values of a_0 and a_1 that minimize Eq. (13.12).

13.2.2 Least-Squares Fit of a Straight Line

To determine values for a_0 and a_1, Eq. (13.12) is differentiated with respect to each unknown coefficient:

$$\frac{\partial S_r}{\partial a_0} = -2 \sum (y_i - a_0 - a_1 x_i)$$

$$\frac{\partial S_r}{\partial a_1} = -2 \sum [(y_i - a_0 - a_1 x_i)x_i]$$

Note that we have simplified the summation symbols; unless otherwise indicated, all summations are from $i = 1$ to n. Setting these derivatives equal to zero will result in a minimum S_r. If this is done, the equations can be expressed as

$$0 = \sum y_i - \sum a_0 - \sum a_1 x_i$$

$$0 = \sum x_i y_i - \sum a_0 x_i - \sum a_1 x_i^2$$

Now, realizing that $\sum a_0 = n a_0$, we can express the equations as a set of two simultaneous linear equations with two unknowns (a_0 and a_1):

$$n \quad a_0 + \left(\sum x_i\right) a_1 = \sum y_i \tag{13.13}$$

$$\left(\sum x_i\right) a_0 + \left(\sum x_i^2\right) a_1 = \sum x_i y_i \tag{13.14}$$

These are called the *normal equations*. They can be solved simultaneously for

$$a_1 = \frac{n \sum x_i y_i - \sum x_i \sum y_i}{n \sum x_i^2 - \left(\sum x_i\right)^2} \tag{13.15}$$

This result can then be used in conjunction with Eq. (13.13) to solve for

$$a_0 = \bar{y} - a_1 \bar{x} \tag{13.16}$$

where $\bar{y}$ and $\bar{x}$ are the means of y and x, respectively.

EXAMPLE 13.2   Linear Regression

Problem Statement.   Fit a straight line to the values in Table 13.1.

Solution.   In this application, force is the dependent variable (y) and velocity is the independent variable (x). The data can be set up in tabular form and the necessary sums computed as in Table 13.4.

TABLE 13.4 Data and summations needed to compute the best-fit line for the data from Table 13.1.

i	x_i	y_i	x_i^2	$x_i y_i$
1	10	25	100	250
2	20	70	400	1,400
3	30	380	900	11,400
4	40	550	1,600	22,000
5	50	610	2,500	30,500
6	60	1,220	3,600	73,200
7	70	830	4,900	58,100
8	80	1,450	6,400	116,000
Σ	360	5,135	20,400	312,850

The means can be computed as

$$\bar{x} = \frac{360}{8} = 45 \qquad \bar{y} = \frac{5,135}{8} = 641.875$$

The slope and the intercept can then be calculated with Eqs. (13.15) and (13.16) as

$$a_1 = \frac{8(312,850) - 360(5,135)}{8(20,400) - (360)^2} = 19.47024$$

$$a_0 = 641.875 - 19.47024(45) = -234.2857$$

Using force and velocity in place of y and x, the least-squares fit is

$$F = -234.2857 + 19.47024v$$

The line, along with the data, is shown in Fig. 13.6.

FIGURE 13.6
Least-squares fit of a straight line to the data from Table 13.1

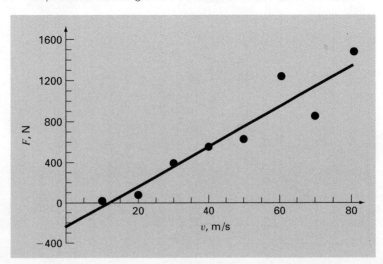

Notice that although the line fits the data well, the zero intercept means that the equation predicts physically unrealistic negative forces at low velocities. In Section 13.3, we will show how transformations can be employed to derive an alternative best-fit line that is more physically realistic.

13.2.3 Quantification of Error of Linear Regression

Any line other than the one computed in Example 13.2 results in a larger sum of the squares of the residuals. Thus, the line is unique and in terms of our chosen criterion is a "best" line through the points. A number of additional properties of this fit can be elucidated by examining more closely the way in which residuals were computed. Recall that the sum of the squares is defined as [Eq. (13.12)]

$$S_r = \sum_{i=1}^{n} (y_i - a_0 - a_1 x_i)^2 \tag{13.17}$$

Notice the similarity between this equation and Eq. (13.4)

$$S_t = \sum (y_i - \bar{y})^2 \tag{13.18}$$

In Eq. (13.18), the square of the residual represented the square of the discrepancy between the data and a single estimate of the measure of central tendency—the mean. In Eq. (13.17), the square of the residual represents the square of the vertical distance between the data and another measure of central tendency—the straight line (Fig. 13.7).

The analogy can be extended further for cases where (1) the spread of the points around the line is of similar magnitude along the entire range of the data and (2) the distribution of these points about the line is normal. It can be demonstrated that if these criteria

FIGURE 13.7
The residual in linear regression represents the vertical distance between a data point and the straight line.

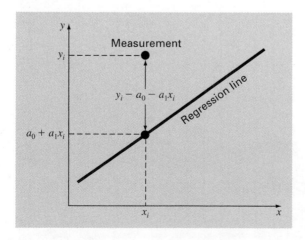

are met, least-squares regression will provide the best (i.e., the most likely) estimates of a_0 and a_1 (Draper and Smith, 1981). This is called the *maximum likelihood principle* in statistics. In addition, if these criteria are met, a "standard deviation" for the regression line can be determined as [compare with Eq. (13.3)]

$$s_{y/x} = \sqrt{\frac{S_r}{n - 2}} \tag{13.19}$$

where $s_{y/x}$ is called the *standard error of the estimate*. The subscript notation "y/x" designates that the error is for a predicted value of y corresponding to a particular value of x. Also, notice that we now divide by $n - 2$ because two data-derived estimates—a_0 and a_1— were used to compute S_r; thus, we have lost two degrees of freedom. As with our discussion of the standard deviation, another justification for dividing by $n - 2$ is that there is no such thing as the "spread of data" around a straight line connecting two points. Thus, for the case where $n = 2$, Eq. (13.19) yields a meaningless result of infinity.

Just as was the case with the standard deviation, the standard error of the estimate quantifies the spread of the data. However, $s_{y/x}$ quantifies the spread *around the regression line* as shown in Fig. 13.8b in contrast to the standard deviation s_y that quantified the spread *around the mean* (Fig. 13.8a).

These concepts can be used to quantify the "goodness" of our fit. This is particularly useful for comparison of several regressions (Fig. 13.9). To do this, we return to the original data and determine the total sum of the squares around the mean for the dependent variable (in our case, y). As was the case for Eq. (13.18), this quantity is designated S_t. This is the magnitude of the residual error associated with the dependent variable prior to regression. After performing the regression, we can compute S_r, the sum of the squares of the residuals around the regression line with Eq. (13.17). This characterizes the residual

FIGURE 13.8
Regression data showing (a) the spread of the data around the mean of the dependent variable and (b) the spread of the data around the best-fit line. The reduction in the spread in going from (a) to (b), as indicated by the bell-shaped curves at the right, represents the improvement due to linear regression.

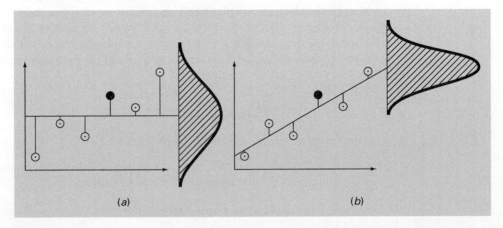

(a)   (b)

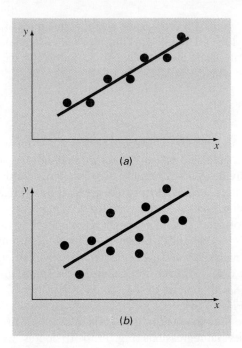

FIGURE 13.9
Examples of linear regression with (a) small and (b) large residual errors.

error that remains after the regression. It is, therefore, sometimes called the unexplained sum of the squares. The difference between the two quantities, $S_t - S_r$, quantifies the improvement or error reduction due to describing the data in terms of a straight line rather than as an average value. Because the magnitude of this quantity is scale-dependent, the difference is normalized to S_t to yield

$$r^2 = \frac{S_t - S_r}{S_t} \tag{13.20}$$

where r^2 is called the *coefficient of determination* and r is the *correlation coefficient* ($= \sqrt{r^2}$). For a perfect fit, $S_r = 0$ and $r^2 = 1$, signifying that the line explains 100% of the variability of the data. For $r^2 = 0$, $S_r = S_t$ and the fit represents no improvement. An alternative formulation for r that is more convenient for computer implementation is

$$r = \frac{n \sum (x_i y_i) - \left(\sum x_i \right) \left(\sum y_i \right)}{\sqrt{n \sum x_i^2 - \left(\sum x_i \right)^2} \sqrt{n \sum y_i^2 - \left(\sum y_i \right)^2}} \tag{13.21}$$

EXAMPLE 13.3   Estimation of Errors for the Linear Least-Squares Fit

Problem Statement.   Compute the total standard deviation, the standard error of the estimate, and the correlation coefficient for the fit in Example 13.2.

Solution.   The data can be set up in tabular form and the necessary sums computed as in Table 13.5.

TABLE 13.5 Data and summations needed to compute the goodness-of-fit statistics for the data from Table 13.1.

i	x_i	y_i	$a_0 + a_1x_i$	$(y_i - \bar{y})^2$	$(y_i - a_0 - a_1x_i)^2$
1	10	25	−39.58	380,535	4,171
2	20	70	155.12	327,041	7,245
3	30	380	349.82	68,579	911
4	40	550	544.52	8,441	30
5	50	610	739.23	1,016	16,699
6	60	1,220	933.93	334,229	81,837
7	70	830	1,128.63	35,391	89,180
8	80	1,450	1,323.33	653,066	16,044
Σ	360	5,135		1,808,297	216,118

The standard deviation is [Eq. (13.3)]

$$s_y = \sqrt{\frac{1,808,297}{8-1}} = 508.26$$

and the standard error of the estimate is [Eq. (13.19)]

$$s_{y/x} = \sqrt{\frac{216,118}{8-2}} = 189.79$$

Thus, because $s_{y/x} < s_y$, the linear regression model has merit. The extent of the improvement is quantified by [Eq. (13.20)]

$$r^2 = \frac{1,808,297 - 216,118}{1,808,297} = 0.8805$$

or $r = \sqrt{0.8805} = 0.9383$. These results indicate that 88.05% of the original uncertainty has been explained by the linear model.

Before proceeding, a word of caution is in order. Although the coefficient of determination provides a handy measure of goodness-of-fit, you should be careful not to ascribe more meaning to it than is warranted. Just because r^2 is "close" to 1 does not mean that the fit is necessarily "good." For example, it is possible to obtain a relatively high value of r^2 when the underlying relationship between y and x is not even linear. Draper and Smith (1981) provide guidance and additional material regarding assessment of results for linear regression. In addition, at the minimum, you should always inspect a plot of the data along with your regression curve.

A nice example was developed by Anscombe (1973). As in Fig. 13.10, he came up with four data sets consisting of 11 data points each. Although their graphs are very different, all have the same best-fit equation, $y = 3 + 0.5x$, and the same coefficient of determination, $r^2 = 0.67$! This example dramatically illustrates why developing plots is so valuable.

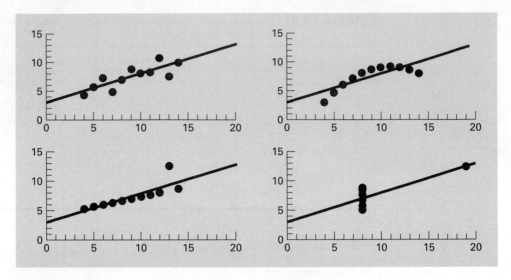

FIGURE 13.10
Anscombe's four data sets along with the best-fit line, $y = 3 + 0.5x$.

13.3 LINEARIZATION OF NONLINEAR RELATIONSHIPS

Linear regression provides a powerful technique for fitting a best line to data. However, it is predicated on the fact that the relationship between the dependent and independent variables is linear. This is not always the case, and the first step in any regression analysis should be to plot and visually inspect the data to ascertain whether a linear model applies. In some cases, techniques such as polynomial regression, which is described in Chap. 14, are appropriate. For others, transformations can be used to express the data in a form that is compatible with linear regression.

One example is the *exponential model:*

$$y = \alpha_1 e^{\beta_1 x} \tag{13.22}$$

where α_1 and β_1 are constants. This model is used in many fields of engineering and science to characterize quantities that increase (positive β_1) or decrease (negative β_1) at a rate that is directly proportional to their own magnitude. For example, population growth or radioactive decay can exhibit such behavior. As depicted in Fig. 13.11a, the equation represents a nonlinear relationship (for $\beta_1 \neq 0$) between y and x.

Another example of a nonlinear model is the simple *power equation:*

$$y = \alpha_2 x^{\beta_2} \tag{13.23}$$

where α_2 and β_2 are constant coefficients. This model has wide applicability in all fields of engineering and science. It is very frequently used to fit experimental data when the underlying model is not known. As depicted in Fig. 13.11b, the equation (for $\beta_2 \neq 0$) is nonlinear.

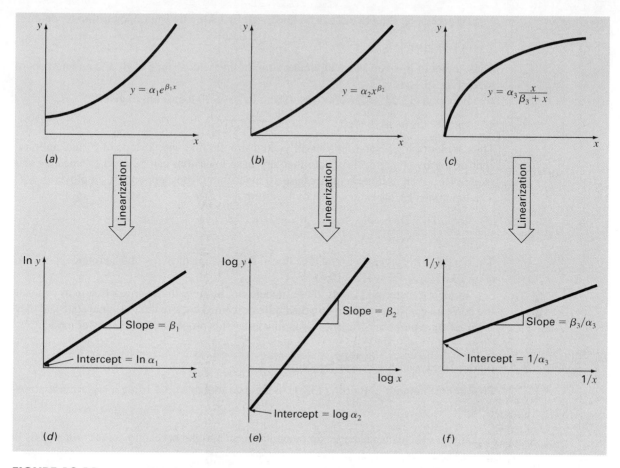

FIGURE 13.11
(a) The exponential equation, (b) the power equation, and (c) the saturation-growth-rate equation. Parts (d), (e), and (f) are linearized versions of these equations that result from simple transformations.

A third example of a nonlinear model is the *saturation-growth-rate equation:*

$$y = \alpha_3 \frac{x}{\beta_3 + x} \tag{13.24}$$

where α_3 and β_3 are constant coefficients. This model, which is particularly well-suited for characterizing population growth rate under limiting conditions, also represents a nonlinear relationship between y and x (Fig. 13.11c) that levels off, or "saturates," as x increases. It has many applications, particularly in biologically related areas of both engineering and science.

Nonlinear regression techniques are available to fit these equations to experimental data directly. However, a simpler alternative is to use mathematical manipulations to transform the equations into a linear form. Then linear regression can be employed to fit the equations to data.

For example, Eq. (13.22) can be linearized by taking its natural logarithm to yield

$$\ln y = \ln \alpha_1 + \beta_1 x \tag{13.25}$$

Thus, a plot of $\ln y$ versus x will yield a straight line with a slope of β_1 and an intercept of $\ln \alpha_1$ (Fig. 13.11d).

Equation (13.23) is linearized by taking its base-10 logarithm to give

$$\log y = \log \alpha_2 + \beta_2 \log x \tag{13.26}$$

Thus, a plot of $\log y$ versus $\log x$ will yield a straight line with a slope of β_2 and an intercept of $\log \alpha_2$ (Fig. 13.11e). Note that any base logarithm can be used to linearize this model. However, as done here, the base-10 logarithm is most commonly employed.

Equation (13.24) is linearized by inverting it to give

$$\frac{1}{y} = \frac{1}{\alpha_3} + \frac{\beta_3}{\alpha_3} \frac{1}{x} \tag{13.27}$$

Thus, a plot of $1/y$ versus $1/x$ will be linear, with a slope of β_3/α_3 and an intercept of $1/\alpha_3$ (Fig. 13.11f).

In their transformed forms, these models can be fit with linear regression to evaluate the constant coefficients. They can then be transformed back to their original state and used for predictive purposes. The following illustrates this procedure for the power model.

EXAMPLE 13.4   Fitting Data with the Power Equation

Problem Statement.   Fit Eq. (13.23) to the data in Table 13.1 using a logarithmic transformation.

Solution.   The data can be set up in tabular form and the necessary sums computed as in Table 13.6.

The means can be computed as

$$\bar{x} = \frac{12.606}{8} = 1.5757 \qquad \bar{y} = \frac{20.515}{8} = 2.5644$$

TABLE 13.6 Data and summations needed to fit the power model to the data from Table 13.1

i	x_i	y_i	$\log x_i$	$\log y_i$	$(\log x_i)^2$	$\log x_i \log y_i$
1	10	25	1.000	1.398	1.000	1.398
2	20	70	1.301	1.845	1.693	2.401
3	30	380	1.477	2.580	2.182	3.811
4	40	550	1.602	2.740	2.567	4.390
5	50	610	1.699	2.785	2.886	4.732
6	60	1220	1.778	3.086	3.162	5.488
7	70	830	1.845	2.919	3.404	5.386
8	80	1450	1.903	3.161	3.622	6.016
Σ			12.606	20.515	20.516	33.622

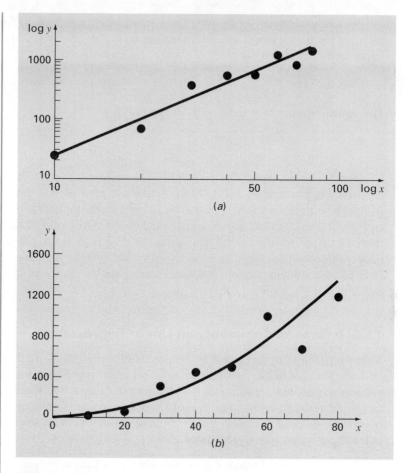

FIGURE 13.12
Least-squares fit of a power model to the data from Table 13.1. (a) The fit of the transformed data. (b) The power equation fit along with the data.

The slope and the intercept can then be calculated with Eqs. (13.15) and (13.16) as

$$a_1 = \frac{8(33.622) - 12.606(20.515)}{8(20.516) - (12.606)^2} = 1.9842$$

$$a_0 = 2.5644 - 1.9842(1.5757) = -0.5620$$

The least-squares fit is

$$\log y = -0.5620 + 1.9842 \log x$$

The fit along with the data is shown in Fig. 13.12a.

We can also display the fit using the untransformed coordinates. To do this, the coefficients of the power model are determined as $\alpha_2 = 10^{-0.5620} = 0.2741$ and $\beta_2 = 1.9842$. Using force and velocity in place of y and x, the least-squares fit is

$$F = 0.2741 v^{1.9842}$$

This equation, along with the data, is shown in Fig. 13.12b.

The fits in Example 13.4 (Fig. 13.12) should be compared with the one obtained previously in Example 13.2 (Fig. 13.6) using linear regression on the untransformed data. Although both results would appear to be acceptable, the transformed result has the advantage that it does not yield negative force predictions at low velocities. Further, it is known from the discipline of fluid mechanics that the drag force on an object moving through a fluid is often well described by a model with velocity squared. Thus, knowledge from the field you are studying often has a large bearing on the choice of the appropriate model equation you use for curve fitting.

13.3.1 General Comments on Linear Regression

Before proceeding to curvilinear and multiple linear regression, we must emphasize the introductory nature of the foregoing material on linear regression. We have focused on the simple derivation and practical use of equations to fit data. You should be cognizant of the fact that there are theoretical aspects of regression that are of practical importance but are beyond the scope of this book. For example, some statistical assumptions that are inherent in the linear least-squares procedures are

1. Each x has a fixed value; it is not random and is known without error.
2. The y values are independent random variables and all have the same variance.
3. The y values for a given x must be normally distributed.

Such assumptions are relevant to the proper derivation and use of regression. For example, the first assumption means that (1) the x values must be error-free and (2) the regression of y versus x is not the same as x versus y. You are urged to consult other references such as Draper and Smith (1981) to appreciate aspects and nuances of regression that are beyond the scope of this book.

13.4 COMPUTER APPLICATIONS

Linear regression is so commonplace that it can be implemented on most pocket calculators. In this section, we will show how a simple M-file can be developed to determine the slope and intercept as well as to create a plot of the data and the best-fit line. We will also show how linear regression can be implemented with the built-in `polyfit` function.

13.4.1 MATLAB M-file: `linregr`

An algorithm for linear regression can be easily developed (Fig. 13.13). The required summations are readily computed with MATLAB's sum function. These are then used to compute the slope and the intercept with Eqs. (13.15) and (13.16). The routine displays the intercept and slope, the coefficient of determination, and a plot of the best-fit line along with the measurements.

A simple example of the use of this M-file would be to fit the force-velocity data that was analyzed in Example 13.2:

```
>> x = [10 20 30 40 50 60 70 80];
>> y = [25 70 380 550 610 1220 830 1450];
>> linregr(x,y)

r2 =
    0.8805

ans =
    19.4702 -234.2857
```

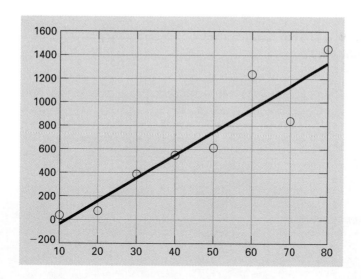

It can just as easily be used to fit the power model (Example 13.4) by applying the `log10` function to the data as in

```
>> linregr(log10(x),log10(y))

r2 =
    0.9481

ans =
    1.9842   -0.5620
```

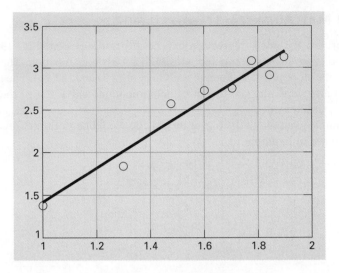

FIGURE 13.13
An M-file to implement linear regression.

```
function [a, r2] = linregr(x,y)
% linregr: linear regression curve fitting
%    [a, r2] = linregr(x,y): Least squares fit of straight
%                line to data by solving the normal equations
%
% input:
%    x = independent variable
%    y = dependent variable
% output:
%    a = vector of slope, a(1), and intercept, a(2)
%    r2 = coefficient of determination

n = length(x);
if length(y)~=n, error('x and y must be same length'); end
x = x(:); y = y(:);      % convert to column vectors
sx = sum(x); sy = sum(y);
sx2 = sum(x.*x); sxy = sum(x.*y); sy2 = sum(y.*y);
a(1) = (n*sxy-sx*sy)/(n*sx2-sx^2);
a(2) = sy/n-a(1)*sx/n;
r2 = ((n*sxy-sx*sy)/sqrt(n*sx2-sx^2)/sqrt(n*sy2-sy^2))^2;
% create plot of data and best fit line
xp = linspace(min(x),max(x),2);
yp = a(1)*xp+a(2);
plot(x,y,'o',xp,yp)
grid on
```

13.4.2 MATLAB Functions: `polyfit` and `polyval`

MATLAB has a built-in function `polyfit` that fits a least-squares nth-order polynomial to data. It can be applied as in

```
>> p = polyfit(x, y, n)
```

where x and y are the vectors of the independent and the dependent variables, respectively, and n = the order of the polynomial. The function returns a vector p containing the polynomial's coefficients. We should note that it represents the polynomial using decreasing powers of x as in the following representation:

$$f(x) = p_1 x^n + p_2 x^{n-1} + \cdots + p_n x + p_{n+1}$$

Because a straight line is a first-order polynomial, `polyfit(x,y,1)` will return the slope and the intercept of the best-fit straight line.

```
>> x = [10 20 30 40 50 60 70 80];
>> y = [25 70 380 550 610 1220 830 1450];
>> a = polyfit(x,y,1)

a =
   19.4702 -234.2857
```

Thus, the slope is 19.4702 and the intercept is -234.2857.

Another function, `polyval`, can then be used to compute a value using the coefficients. It has the general format:

```
>> y = polyval(p, x)
```

where p = the polynomial coefficients, and y = the best-fit value at x. For example,

```
>> y = polyval(a,45)

y =
   641.8750
```

13.5 CASE STUDY   ENZYME KINETICS

Background. *Enzymes* act as catalysts to speed up the rate of chemical reactions in living cells. In most cases, they convert one chemical, the *substrate,* into another, the *product.* The *Michaelis-Menten* equation is commonly used to describe such reactions:

$$v = \frac{v_m[S]}{k_s + [S]} \qquad (13.28)$$

where v = the initial reaction velocity, v_m = the maximum initial reaction velocity, $[S]$ = substrate concentration, and k_s = a half-saturation constant. As in Fig. 13.14, the equation describes a saturating relationship which levels off with increasing $[S]$. The graph also illustrates that the *half-saturation constant* corresponds to the substrate concentration at which the velocity is half the maximum.

13.5 CASE STUDY   continued

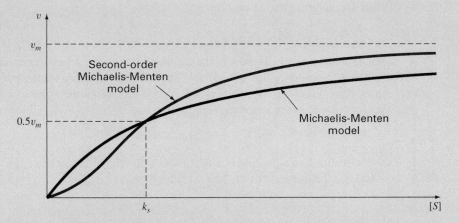

FIGURE 13.14
Two versions of the Michaelis-Menten model of enzyme kinetics.

Although the Michaelis-Menten model provides a nice starting point, it has been refined and extended to incorporate additional features of enzyme kinetics. One simple extension involves so-called *allosteric enzymes,* where the binding of a substrate molecule at one site leads to enhanced binding of subsequent molecules at other sites. For cases with two interacting bonding sites, the following second-order version often results in a better fit:

$$v = \frac{v_m[S]^2}{k_s^2 + [S]^2} \tag{13.29}$$

This model also describes a saturating curve but, as depicted in Fig. 13.14, the squared concentrations tend to make the shape more *sigmoid,* or S-shaped.

Suppose that you are provided with the following data:

$[S]$	1.3	1.8	3	4.5	6	8	9
v	0.07	0.13	0.22	0.275	0.335	0.35	0.36

Employ linear regression to fit this data with linearized versions of Eqs. (13.28) and (13.29). Aside from estimating the model parameters, assess the validity of the fits with both statistical measures and graphs.

Solution.   Equation (13.28), which is in the format of the saturation-growth-rate model (Eq. 13.24), can be linearized by inverting it to give (recall Eq. 13.27)

$$\frac{1}{v} = \frac{1}{v_m} + \frac{k_s}{v_m}\frac{1}{[S]}$$

13.5 CASE STUDY continued

The `linregr` function from Fig. 13.13 can then be used to determine the least-squares fit:

```
>> S=[1.3 1.8 3 4.5 6 8 9];
>> v=[0.07 0.13 0.22 0.275 0.335 0.35 0.36];
>> [a,r2]=linregr(1./S,1./v)

a =
   16.4022    0.1902
r2 =
    0.9344
```

The model coefficients can then be calculated as

```
>> vm=1/a(2)

vm =
    5.2570

>> ks=vm*a(1)

ks =
   86.2260
```

Thus, the best-fit model is

$$v = \frac{5.2570[S]}{86.2260 + [S]}$$

Although the high value of r^2 might lead you to believe that this result is acceptable, inspection of the coefficients might raise doubts. For example, the maximum velocity (5.2570) is much greater than the highest observed velocity (0.36). In addition, the half-saturation rate (86.2260) is much bigger than the maximum substrate concentration (9).

The problem is underscored when the fit is plotted along with the data. Figure 13.15a shows the transformed version. Although the straight line follows the upward trend, the data clearly appears to be curved. When the original equation is plotted along with the data in the untransformed version (Fig. 13.15b), the fit is obviously unacceptable. The data is clearly leveling off at about 0.36 or 0.37. If this is correct, an eyeball estimate would suggest that v_m should be about 0.36, and k_s should be in the range of 2 to 3.

Beyond the visual evidence, the poorness of the fit is also reflected by statistics like the coefficient of determination. For the untransformed case, a much less acceptable result of $r^2 = 0.6406$ is obtained.

The foregoing analysis can be repeated for the second-order model. Equation (13.28) can also be linearized by inverting it to give

$$\frac{1}{v} = \frac{1}{v_m} + \frac{k_s^2}{v_m}\frac{1}{[S]^2}$$

The `linregr` function from Fig. 13.13 can again be used to determine the least-squares fit:

```
>> [a,r2]=linregr(1./S.^2,1./v)

a =
   19.3760    2.4492
r2 =
    0.9929
```

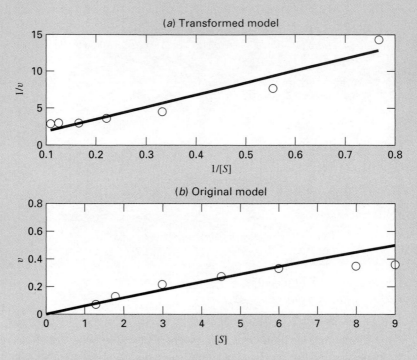

FIGURE 13.15
Plots of least-squares fit (line) of the Michaelis-Menten model along with data (points). The plot in (a) shows the transformed fit, and (b) shows how the fit looks when viewed in the untransformed, original form.

The model coefficients can then be calculated as

```
>> vm=1/a(2)

vm =
    0.4083
>> ks=sqrt(vm*a(1))

ks =
    2.8127
```

Substituting these values into Eq. (13.29) gives

$$v = \frac{0.4083[S]^2}{7.911 + [S]^2}$$

Although we know that a high r^2 does not guarantee of a good fit, the fact that it is very high (0.9929) is promising. In addition, the parameters values also seem consistent with the trends in the data; that is, the k_m is slightly greater than the highest observed velocity and the half-saturation rate is lower than the maximum substrate concentration (9).

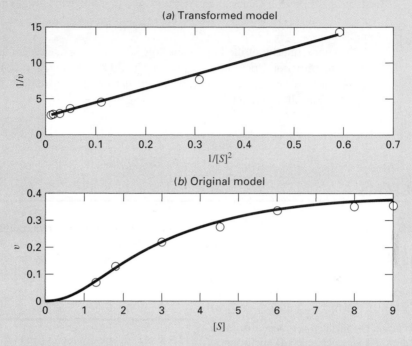

FIGURE 13.16
Plots of least-squares fit (line) of the second-order Michaelis-Menten model along with data (points). The plot in (a) shows the transformed fit, and (b) shows the untransformed, original form.

The adequacy of the fit can be assessed graphically. As in Fig. 13.16a, the transformed results appear linear. When the original equation is plotted along with the data in the untransformed version (Fig. 13.16b), the fit nicely follows the trend in the measurements. Beyond the graphs, the goodness of the fit is also reflected by the fact that the coefficient of determination for the untransformed case can be computed as $r^2 = 0.9896$.

Based on our analysis, we can conclude that the second-order model provides a good fit of this data set. This might suggest that we are dealing with an allosteric enzyme.

Beyond this specific result, there are a few other general conclusions that can be drawn from this case study. First, we should never solely rely on statistics such as r^2 as the sole basis of assessing goodness of fit. Second, regression equations should always be assessed graphically. And for cases where transformations are employed, a graph of the untransformed model and data should always be inspected.

Finally, although transformations may yield a decent fit of the transformed data, this does not always translate into an acceptable fit in the original format. The reason that this might occur is that minimizing squared residuals of transformed data is not the same as for the untransformed data. Linear regression assumes that the scatter of points around the

13.5 CASE STUDY continued

best-fit line follows a Gaussian distribution, and that the standard deviation is the same at every value of the dependent variable. These assumptions are rarely true after transforming data.

As a consequence of the last conclusion, some analysts suggest that rather than using linear transformations, nonlinear regression should be employed to fit curvilinear data. In this approach, a best-fit curve is developed that directly minimizes the untransformed residuals. We will describe how this is done in Chap. 14.

PROBLEMS

13.1 Given the data

8.8	9.5	9.8	9.4	10.0
9.4	10.1	9.2	11.3	9.4
10.0	10.4	7.9	10.4	9.8
9.8	9.5	8.9	8.8	10.6
10.1	9.5	9.6	10.2	8.9

Determine **(a)** the mean, **(b)** median, **(c)** mode, **(d)** range, **(e)** standard deviation, **(f)** variance, and **(g)** coefficient of variation.

13.2 Construct a histogram from the data from Prob. 13.1. Use a range from 7.5 to 11.5 with intervals of 0.5.

13.3 Given the data

28.65	26.55	26.65	27.65	27.35	28.35	26.85
28.65	29.65	27.85	27.05	28.25	28.85	26.75
27.65	28.45	28.65	28.45	31.65	26.35	27.75
29.25	27.65	28.65	27.65	28.55	27.65	27.25

Determine **(a)** the mean, **(b)** median, **(c)** mode, **(d)** range, **(e)** standard deviation, **(f)** variance, and **(g)** coefficient of variation.

(h) Construct a histogram. Use a range from 26 to 32 with increments of 0.5.

(i) Assuming that the distribution is normal, and that your estimate of the standard deviation is valid, compute the range (that is, the lower and the upper values) that encompasses 68% of the readings. Determine whether this is a valid estimate for the data in this problem.

13.4 Using the same approach as was employed to derive Eqs. (13.15) and (13.16), derive the least-squares fit of the following model:

$$y = a_1 x + e$$

That is, determine the slope that results in the least-squares fit for a straight line with a zero intercept. Fit the following data with this model and display the result graphically.

x	2	4	6	7	10	11	14	17	20
y	4	5	6	5	8	7	6	9	12

13.5 Use least-squares regression to fit a straight line to

x	0	2	4	6	9	11	12	15	17	19
y	5	6	7	6	9	8	7	10	12	12

Along with the slope and intercept, compute the standard error of the estimate and the correlation coefficient. Plot the data and the regression line. Then repeat the problem, but regress x versus y—that is, switch the variables. Interpret your results.

13.6 The following data was gathered to determine the relationship between pressure and temperature of a fixed volume of 1 kg of nitrogen. The volume is 10 m³.

T, °C	−40	0	40	80	120	160
p, N/m²	6900	8100	9300	10,500	11,700	12,900

Employ the ideal gas law $pV = nRT$ to determine R on the basis of this data. Note that for the law, T must be expressed in kelvins.

13.7 Beyond the examples in Fig. 13.11, there are other models that can be linearized using transformations. For example,

$$y = \alpha_4 x e^{\beta_4 x}$$

Linearize this model and use it to estimate α_4 and β_4 based on the following data. Develop a plot of your fit along with the data.

x	0.1	0.2	0.4	0.6	0.9	1.3	1.5	1.7	1.8
y	0.75	1.25	1.45	1.25	0.85	0.55	0.35	0.28	0.18

13.8 Fit a power model to the data from Table 13.1, but use natural logarithms to perform the transformations.

13.9 The concentration of *E. coli* bacteria in a swimming area is monitored after a storm:

t (hr)	4	8	12	16	20	24
c (CFU/100 mL)	1590	1320	1000	900	650	560

The time is measured in hours following the end of the storm and the unit CFU is a "colony forming unit." Use this data to estimate (**a**) the concentration at the end of the storm ($t = 0$) and (**b**) the time at which the concentration will reach 200 CFU/100 mL. Note that your choice of model should be consistent with the fact that negative concentrations are impossible and that the bacteria concentration always decreases with time.

13.10 Rather than using the base-*e* exponential model (Eq. 13.22), a common alternative is to employ a base-10 model:

$$y = \alpha_5 10^{\beta_5 x}$$

When used for curve fitting, this equation yields identical results to the base-*e* version, but the value of the exponent parameter (β_5) will differ from that estimated with Eq.13.22 (β_1). Use the base-10 version to solve Prob. 13.9. In addition, develop a formulation to relate β_1 to β_5.

13.11 On average, the surface area A of human beings is related to weight W and height H. Measurements on a number of individuals of height 180 cm and different weights (kg) give values of A (m^2) in the following table:

W (kg)	70	75	77	80	82	84	87	90
A (m^2)	2.10	2.12	2.15	2.20	2.22	2.23	2.26	2.30

Show that a power law $A = aW^b$ fits these data reasonably well. Evaluate the constants a and b, and predict what the surface area is for a 95-kg person.

13.12 Determine an equation to predict metabolism rate as a function of mass based on the following data:

Animal	Mass (kg)	Metabolism (watts)
Cow	400	270
Human	70	82
Sheep	45	50
Hen	2	4.8
Rat	0.3	1.45
Dove	0.16	0.97

13.13 Fit an exponential model to

x	0.4	0.8	1.2	1.6	2	2.3
y	800	975	1500	1950	2900	3600

Plot the data and the equation on both standard and semilogarithmic graph paper.

13.14 An investigator has reported the data tabulated below for an experiment to determine the growth rate of bacteria k (per d) as a function of oxygen concentration c (mg/L). It is known that such data can be modeled by the following equation:

$$k = \frac{k_{max} c^2}{c_s + c^2}$$

where c_s and k_{max} are parameters. Use a transformation to linearize this equation. Then use linear regression to estimate c_s and k_{max} and predict the growth rate at $c = 2$ mg/L.

c	0.5	0.8	1.5	2.5	4
k	1.1	2.4	5.3	7.6	8.9

13.15 Develop an M-file function to compute descriptive statistics for a vector of values. Have the function determine and display number of values, mean, median, mode, range, standard deviation, variance, and coefficient of variation. In addition, have it generate a histogram. Test it with the data from Prob. 13.3.

13.16 Modify the `linregr` function in Fig. 13.13 so that it (**a**) computes and returns the standard error of the estimate, and (**b**) uses the `subplot` function to also display a plot of the residuals (the predicted minus the measured y) versus x.

13.17 Develop an M-file function to fit a power model. Have the function return the best-fit coefficient α_2 and power β_2 along with the r^2 for the untransformed model. In addition, use the `subplot` function to display graphs of both the transformed and untransformed equations along with the data. Test it with the data from Prob. 13.12.

13.18 The following data shows the relationship between the viscosity of SAE 70 oil and temperature. After taking the log of the data, use linear regression to find the equation of the line that best fits the data and the r^2 value.

Temperature, °C	26.67	93.33	148.89	315.56
Viscosity, μ, N·s/m²	1.35	0.085	0.012	0.00075

13.19 You perform experiments and determine the following values of heat capacity c at various temperatures T for a gas:

T	−50	−30	0	60	90	110
c	1250	1280	1350	1480	1580	1700

Use regression to determine a model to predict c as a function of T.

13.20 It is known that the tensile strength of a plastic increases as a function of the time it is heat treated. The following data is collected:

Time	10	15	20	25	40	50	55	60	75
Tensile Strength	5	20	18	40	33	54	70	60	78

(a) Fit a straight line to this data and use the equation to determine the tensile strength at a time of 32 min.
(b) Repeat the analysis but for a straight line with a zero intercept.

13.21 The following data was taken from a stirred tank reactor for the reaction $A \to B$. Use the data to determine the best possible estimates for k_{01} and E_1 for the following kinetic model:

$$-\frac{dA}{dt} = k_{01}e^{-E_1/RT}A$$

where R is the gas constant and equals 0.00198 kcal/mol/K.

−dA/dt (moles/L/s)	460	960	2485	1600	1245
A (moles/L)	200	150	50	20	10
T (K)	280	320	450	500	550

13.22 Concentration data was taken at 15 time points for the polymerization reaction:

$$xA + yB \to A_xB_y$$

We assume the reaction occurs via a complex mechanism consisting of many steps. Several models have been hypothesized, and the sum of the squares of the residuals had been calculated for the fits of the models of the data. The results

are shown below. Which model best describes the data (statistically)? Explain your choice.

	Model A	Model B	Model C
S_r	135	105	100
Number of Model Parameters Fit	2	3	5

13.23 Below is data taken from a batch reactor of bacterial growth (after lag phase was over). The bacteria are allowed to grow as fast as possible for the first 2.5 hours, and then they are induced to produce a recombinant protein, the production of which slows the bacterial growth significantly. The theoretical growth of bacteria can be described by

$$\frac{dX}{dt} = \mu X$$

where X is the number of bacteria, and μ is the specific growth rate of the bacteria during exponential growth. Based on the data, estimate the specific growth rate of the bacteria during the first 2 hours of growth and during the next 4 hours of growth.

Time, h	0	1	2	3	4	5	6
[Cells], g/L	0.100	0.332	1.102	1.644	2.453	3.660	5.460

13.24 A transportation engineering study was conducted to determine the proper design of bike lanes. Data was gathered on bike-lane widths and average distance between bikes and passing cars. The data from 11 streets is

Distance, m	2.4	1.5	2.4	1.8	1.8	2.9	1.2	3	1.2
Lane Width, m	2.9	2.1	2.3	2.1	1.8	2.7	1.5	2.9	1.5

(a) Plot the data.
(b) Fit a straight line to the data with linear regression. Add this line to the plot.
(c) If the minimum safe average distance between bikes and passing cars is considered to be 2 m, determine the corresponding minimum lane width.

13.25 In water-resources engineering, the sizing of reservoirs depends on accurate estimates of water flow in the river that is being impounded. For some rivers, long-term historical records of such flow data are difficult to obtain. In contrast, meteorological data on precipitation is often available for many years past. Therefore, it is often useful to determine a relationship between flow and precipitation. This relationship can then be used to estimate flows for years when only precipitation measurements were made. The following data is available for a river that is to be dammed:

Precipitation, cm	88.9	108.5	104.1	139.7	127	94	116.8	99.1
Flow, m³/s		14.6	16.7	15.3	23.2	19.5	16.1	18.1	16.6

(a) Plot the data.
(b) Fit a straight line to the data with linear regression. Superimpose this line on your plot.
(c) Use the best-fit line to predict the annual water flow if the precipitation is 120 cm.
(d) If the drainage area is 1100 km², estimate what fraction of the precipitation is lost via processes such as evaporation, deep groundwater infiltration, and consumptive use.

13.26 The mast of a sailboat has a cross-sectional area of 10.65 cm² and is constructed of an experimental aluminum alloy. Tests were performed to define the relationship between stress and strain. The test results are

Strain, cm/cm	0.0032	0.0045	0.0055	0.0016	0.0085	0.0005
Stress, N/cm²	4970	5170	5500	3590	6900	1240

The stress caused by wind can be computed as F/A_c where F = force in the mast and A_c = mast's cross-sectional area. This value can then be substituted into Hooke's law to determine the mast's deflection, $\Delta L = $ strain $\times L$, where $L = $ the mast's length. If the wind force is 25,000 N, use the data to estimate the deflection of a 9-m mast.

13.27 The following data was taken from an experiment that measured the current in a wire for various imposed voltages:

V, V	2	3	4	5	7	10
i, A	5.2	7.8	10.7	13	19.3	27.5

(a) On the basis of a linear regression of this data, determine current for a voltage of 3.5 V. Plot the line and the data and evaluate the fit.
(b) Redo the regression and force the intercept to be zero.

13.28 An experiment is performed to determine the % elongation of electrical conducting material as a function of temperature. The resulting data is listed below. Predict the % elongation for a temperature of 400 °C.

Temperature, °C	200	250	300	375	425	475	600
% Elongation	7.5	8.6	8.7	10	11.3	12.7	15.3

13.29 The population p of a small community on the outskirts of a city grows rapidly over a 20-year period:

t	0	5	10	15	20
p	100	200	450	950	2000

As an engineer working for a utility company, you must forecast the population 5 years into the future in order to anticipate the demand for power. Employ an exponential model and linear regression to make this prediction.

13.30 The velocity u of air flowing past a flat surface is measured at several distances y away from the surface. Fit a curve to this data assuming that the velocity is zero at the surface ($y = 0$). Use your result to determine the shear stress (du/dy) at the surface.

y, m	0.002	0.006	0.012	0.018	0.024
u, m/s	0.287	0.899	1.915	3.048	4.299

13.31 *Andrade's equation* has been proposed as a model of the effect of temperature on viscosity:

$$\mu = De^{B/T_a}$$

where μ = dynamic viscosity of water (10^{-3} N·s/m²), T_a = absolute temperature (K), and D and B are parameters. Fit this model to the following data for water:

T	0	5	10	20	30	40
μ	1.787	1.519	1.307	1.002	0.7975	0.6529

14

General Linear Least-Squares and Nonlinear Regression

CHAPTER OBJECTIVES

This chapter takes the concept of fitting a straight line and extends it to (*a*) fitting a polynomial and (*b*) fitting a variable that is a linear function of two or more independent variables. We will then show how such applications can be generalized and applied to a broader group of problems. Finally, we will illustrate how optimization techniques can be used to implement nonlinear regression. Specific objectives and topics covered are

- Knowing how to implement polynomial regression.
- Knowing how to implement multiple linear regression.
- Understanding the formulation of the general linear least-squares model.
- Understanding how the general linear least-squares model can be solved with MATLAB using either the normal equations or left division.
- Understanding how to implement nonlinear regression with optimization techniques.

14.1 POLYNOMIAL REGRESSION

In Chap.13, a procedure was developed to derive the equation of a straight line using the least-squares criterion. Some data, although exhibiting a marked pattern such as seen in Fig. 14.1, is poorly represented by a straight line. For these cases, a curve would be better suited to fit the data. As discussed in Chap. 13, one method to accomplish this objective is to use transformations. Another alternative is to fit polynomials to the data using *polynomial regression.*

The least-squares procedure can be readily extended to fit the data to a higher-order polynomial. For example, suppose that we fit a second-order polynomial or quadratic:

$$y = a_0 + a_1x + a_2x^2 + e \tag{14.1}$$

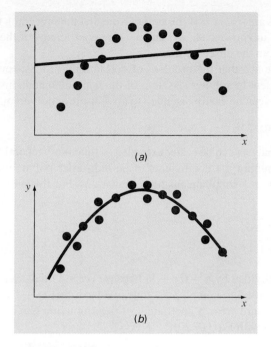

FIGURE 14.1

(a) Data that is ill-suited for linear least-squares regression. (b) Indication that a parabola is preferable.

For this case the sum of the squares of the residuals is

$$S_r = \sum_{i=1}^{n} \left(y_i - a_0 - a_1 x_i - a_2 x_i^2\right)^2 \tag{14.2}$$

To generate the least-squares fit, we take the derivative of Eq. (14.2) with respect to each of the unknown coefficients of the polynomial, as in

$$\frac{\partial S_r}{\partial a_0} = -2 \sum \left(y_i - a_0 - a_1 x_i - a_2 x_i^2\right)$$

$$\frac{\partial S_r}{\partial a_1} = -2 \sum x_i \left(y_i - a_0 - a_1 x_i - a_2 x_i^2\right)$$

$$\frac{\partial S_r}{\partial a_2} = -2 \sum x_i^2 \left(y_i - a_0 - a_1 x_i - a_2 x_i^2\right)$$

These equations can be set equal to zero and rearranged to develop the following set of normal equations:

$$(n)a_0 + \left(\sum x_i\right) a_1 + \left(\sum x_i^2\right) a_2 = \sum y_i$$

$$\left(\sum x_i\right) a_0 + \left(\sum x_i^2\right) a_1 + \left(\sum x_i^3\right) a_2 = \sum x_i y_i$$

$$\left(\sum x_i^2\right) a_0 + \left(\sum x_i^3\right) a_1 + \left(\sum x_i^4\right) a_2 = \sum x_i^2 y_i$$

where all summations are from $i = 1$ through n. Note that the preceding three equations are linear and have three unknowns: a_0, a_1, and a_2. The coefficients of the unknowns can be calculated directly from the observed data.

For this case, we see that the problem of determining a least-squares second-order polynomial is equivalent to solving a system of three simultaneous linear equations. The two-dimensional case can be easily extended to an mth-order polynomial as in

$$y = a_0 + a_1x + a_2x^2 + \cdots + a_mx^m + e$$

The foregoing analysis can be easily extended to this more general case. Thus, we can recognize that determining the coefficients of an mth-order polynomial is equivalent to solving a system of $m + 1$ simultaneous linear equations. For this case, the standard error is formulated as

$$s_{y/x} = \sqrt{\frac{S_r}{n - (m + 1)}} \tag{14.3}$$

This quantity is divided by $n - (m + 1)$ because $(m + 1)$ data-derived coefficients—$a_0, a_1, \ldots, a_m$—were used to compute S_r; thus, we have lost $m + 1$ degrees of freedom. In addition to the standard error, a coefficient of determination can also be computed for polynomial regression with Eq. (13.20).

EXAMPLE 14.1   Polynomial Regression

Problem Statement. Fit a second-order polynomial to the data in the first two columns of Table 14.1.

TABLE 14.1 Computations for an error analysis of the quadratic least-squares fit.

x_i	y_i	$(y_i - \bar{y})^2$	$(y_i - a_0 - a_1x_i - a_2x_i^2)^2$
0	2.1	544.44	0.14332
1	7.7	314.47	1.00286
2	13.6	140.03	1.08160
3	27.2	3.12	0.80487
4	40.9	239.22	0.61959
5	61.1	1272.11	0.09434
Σ	152.6	2513.39	3.74657

Solution. The following can be computed from the data:

$$m = 2 \qquad \sum x_i = 15 \qquad \sum x_i^4 = 979$$

$$n = 6 \qquad \sum y_i = 152.6 \qquad \sum x_i y_i = 585.6$$

$$\bar{x} = 2.5 \qquad \sum x_i^2 = 55 \qquad \sum x_i^2 y_i = 2488.8$$

$$\bar{y} = 25.433 \qquad \sum x_i^3 = 225$$

Therefore, the simultaneous linear equations are

$$\begin{bmatrix} 6 & 15 & 55 \\ 15 & 55 & 225 \\ 55 & 225 & 979 \end{bmatrix} \begin{Bmatrix} a_0 \\ a_1 \\ a_2 \end{Bmatrix} = \begin{Bmatrix} 152.6 \\ 585.6 \\ 2488.8 \end{Bmatrix}$$

These equations can be solved to evaluate the coefficients. For example, using MATLAB:

```
>> N = [6 15 55;15 55 225;55 225 979];
>> r = [152.6 585.6 2488.8];
>> a = N\r

a =
    2.4786
    2.3593
    1.8607
```

Therefore, the least-squares quadratic equation for this case is

$$y = 2.4786 + 2.3593x + 1.8607x^2$$

The standard error of the estimate based on the regression polynomial is [Eq. (14.3)]

$$s_{y/x} = \sqrt{\frac{3.74657}{6 - (2 + 1)}} = 1.1175$$

The coefficient of determination is

$$r^2 = \frac{2513.39 - 3.74657}{2513.39} = 0.99851$$

and the correlation coefficient is $r = 0.99925$.

FIGURE 14.2
Fit of a second-order polynomial.

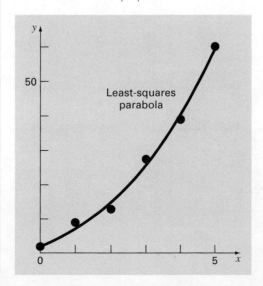

These results indicate that 99.851 percent of the original uncertainty has been explained by the model. This result supports the conclusion that the quadratic equation represents an excellent fit, as is also evident from Fig. 14.2.

14.2   MULTIPLE LINEAR REGRESSION

Another useful extension of linear regression is the case where y is a linear function of two or more independent variables. For example, y might be a linear function of x_1 and x_2, as in

$$y = a_0 + a_1 x_1 + a_2 x_2 + e$$

Such an equation is particularly useful when fitting experimental data where the variable being studied is often a function of two other variables. For this two-dimensional case, the regression "line" becomes a "plane" (Fig. 14.3).

As with the previous cases, the "best" values of the coefficients are determined by formulating the sum of the squares of the residuals:

$$S_r = \sum_{i=1}^{n} (y_i - a_0 - a_1 x_{1,i} - a_2 x_{2,i})^2 \tag{14.4}$$

and differentiating with respect to each of the unknown coefficients:

$$\frac{\partial S_r}{\partial a_0} = -2 \sum (y_i - a_0 - a_1 x_{1,i} - a_2 x_{2,i})$$

$$\frac{\partial S_r}{\partial a_1} = -2 \sum x_{1,i} (y_i - a_0 - a_1 x_{1,i} - a_2 x_{2,i})$$

$$\frac{\partial S_r}{\partial a_2} = -2 \sum x_{2,i} (y_i - a_0 - a_1 x_{1,i} - a_2 x_{2,i})$$

FIGURE 14.3
Graphical depiction of multiple linear regression where y is a linear function of x_1 and x_2.

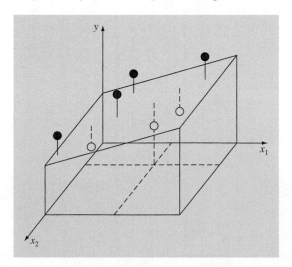

The coefficients yielding the minimum sum of the squares of the residuals are obtained by setting the partial derivatives equal to zero and expressing the result in matrix form as

$$
\begin{bmatrix}
n & \sum x_{1,i} & \sum x_{2,i} \\
\sum x_{1,i} & \sum x_{1,i}^2 & \sum x_{1,i}x_{2,i} \\
\sum x_{2,i} & \sum x_{1,i}x_{2,i} & \sum x_{2,i}^2
\end{bmatrix}
\begin{Bmatrix} a_0 \\ a_1 \\ a_2 \end{Bmatrix} =
\begin{Bmatrix} \sum y_i \\ \sum x_{1,i}y_i \\ \sum x_{2,i}y_i \end{Bmatrix}
\tag{14.5}
$$

EXAMPLE 14.2 Multiple Linear Regression

Problem Statement. The following data was calculated from the equation $y = 5 + 4x_1 - 3x_2$:

x_1	x_2	y
0	0	5
2	1	10
2.5	2	9
1	3	0
4	6	3
7	2	27

Use multiple linear regression to fit this data.

Solution. The summations required to develop Eq. (14.5) are computed in Table 14.2. Substituting them into Eq. (14.5) gives

$$
\begin{bmatrix}
6 & 16.5 & 14 \\
16.5 & 76.25 & 48 \\
14 & 48 & 54
\end{bmatrix}
\begin{Bmatrix} a_0 \\ a_1 \\ a_2 \end{Bmatrix} =
\begin{Bmatrix} 54 \\ 243.5 \\ 100 \end{Bmatrix}
\tag{14.6}
$$

which can be solved for

$$a_0 = 5 \qquad a_1 = 4 \qquad a_2 = -3$$

which is consistent with the original equation from which the data was derived.

The foregoing two-dimensional case can be easily extended to m dimensions, as in

$$y = a_0 + a_1 x_1 + a_2 x_2 + \cdots + a_m x_m + e$$

TABLE 14.2 Computations required to develop the normal equations for Example 14.2.

y	x_1	x_2	x_1^2	x_2^2	$x_1 x_2$	$x_1 y$	$x_2 y$
5	0	0	0	0	0	0	0
10	2	1	4	1	2	20	10
9	2.5	2	6.25	4	5	22.5	18
0	1	3	1	9	3	0	0
3	4	6	16	36	24	12	18
27	7	2	49	4	14	189	54
54	16.5	14	76.25	54	48	243.5	100

where the standard error is formulated as

$$s_{y/x} = \sqrt{\frac{S_r}{n - (m + 1)}}$$

and the coefficient of determination is computed as in Eq. (13.20).

Although there may be certain cases where a variable is linearly related to two or more other variables, multiple linear regression has additional utility in the derivation of power equations of the general form

$$y = a_0 x_1^{a_1} x_2^{a_2} \cdots x_m^{a_m}$$

Such equations are extremely useful when fitting experimental data. To use multiple linear regression, the equation is transformed by taking its logarithm to yield

$$\log y = \log a_0 + a_1 \log x_1 + a_2 \log x_2 + \cdots + a_m \log x_m$$

14.3 GENERAL LINEAR LEAST SQUARES

In the preceding pages, we have introduced three types of regression: simple linear, polynomial, and multiple linear. In fact, all three belong to the following general linear least-squares model:

$$y = a_0 z_0 + a_1 z_1 + a_2 z_2 + \cdots + a_m z_m + e \tag{14.7}$$

where $z_0, z_1, \ldots, z_m$ are $m + 1$ basis functions. It can easily be seen how simple linear and multiple linear regression fall within this model—that is, $z_0 = 1, z_1 = x_1, z_2 = x_2, \ldots, z_m = x_m$. Further, polynomial regression is also included if the basis functions are simple monomials as in $z_0 = 1, z_1 = x, z_2 = x^2, \ldots, z_m = x^m$.

Note that the terminology "linear" refers only to the model's dependence on its parameters—that is, the a's. As in the case of polynomial regression, the functions themselves can be highly nonlinear. For example, the z's can be sinusoids, as in

$$y = a_0 + a_1 \cos(\omega x) + a_2 \sin(\omega x)$$

Such a format is the basis of *Fourier analysis*.

On the other hand, a simple-looking model such as

$$y = a_0 (1 - e^{-a_1 x})$$

is truly nonlinear because it cannot be manipulated into the format of Eq. (14.7).

Equation (14.7) can be expressed in matrix notation as

$$\{y\} = [Z]\{a\} + \{e\} \tag{14.8}$$

where $[Z]$ is a matrix of the calculated values of the basis functions at the measured values of the independent variables:

$$[Z] = \begin{bmatrix} z_{01} & z_{11} & \cdots & z_{m1} \\ z_{02} & z_{12} & \cdots & z_{m2} \\ \vdots & \vdots & & \vdots \\ z_{0n} & z_{1n} & \cdots & z_{mn} \end{bmatrix}$$

where m is the number of variables in the model and n is the number of data points. Because $n \geq m + 1$, you should recognize that most of the time, $[Z]$ is not a square matrix.

The column vector $\{y\}$ contains the observed values of the dependent variable:

$$\{y\}^T = \lfloor y_1 \quad y_2 \quad \cdots \quad y_n \rfloor$$

The column vector $\{a\}$ contains the unknown coefficients:

$$\{a\}^T = \lfloor a_0 \quad a_1 \quad \cdots \quad a_m \rfloor$$

and the column vector $\{e\}$ contains the residuals:

$$\{e\}^T = \lfloor e_1 \quad e_2 \quad \cdots \quad e_n \rfloor$$

The sum of the squares of the residuals for this model can be defined as

$$S_r = \sum_{i=1}^{n} \left(y_i - \sum_{j=0}^{m} a_j z_{ji} \right)^2 \tag{14.9}$$

This quantity can be minimized by taking its partial derivative with respect to each of the coefficients and setting the resulting equation equal to zero. The outcome of this process is the normal equations that can be expressed concisely in matrix form as

$$[[Z]^T [Z]]\{a\} = \{[Z]^T \{y\}\} \tag{14.10}$$

It can be shown that Eq. (14.10) is, in fact, equivalent to the normal equations developed previously for simple linear, polynomial, and multiple linear regression.

The coefficient of determination and the standard error can also be formulated in terms of matrix algebra. Recall that r^2 is defined as

$$r^2 = \frac{S_t - S_r}{S_t} = 1 - \frac{S_r}{S_t}$$

Substituting the definitions of S_r and S_t gives

$$r^2 = 1 - \frac{\sum (y_i - \hat{y}_i)^2}{\sum (y_i - \bar{y}_i)^2}$$

where $\hat{y}$ = the prediction of the least-squares fit. The residuals between the best-fit curve and the data, $y_i - \hat{y}$, can be expressed in vector form as

$$\{y\} - [Z]\{a\}$$

Matrix algebra can then be used to manipulate this vector to compute both the coefficient of determination and the standard error of the estimate as illustrated in the following example.

EXAMPLE 14.3   Polynomial Regression with MATLAB

Problem Statement.   Repeat Example 14.1, but use matrix operations as described in this section.

Solution.   First, enter the data to be fit

```
>> x = [0 1 2 3 4 5]';
>> y = [2.1 7.7 13.6 27.2 40.9 61.1]';
```

Next, create the [Z] matrix:

```
>> Z = [ones(size(x)) x x.^2]
Z =
        1        0        0
        1        1        1
        1        2        4
        1        3        9
        1        4       16
        1        5       25
```

We can verify that $[Z]^T[Z]$ results in the coefficient matrix for the normal equations:

```
>> Z'*Z

ans =
        6       15       55
       15       55      225
       55      225      979
```

This is the same result we obtained with summations in Example 14.1. We can solve for the coefficients of the least-squares quadratic by implementing Eq. (14.10):

```
>> a = (Z'*Z)\(Z'*y)
ans =
    2.4786
    2.3593
    1.8607
```

In order to compute r^2 and $s_{y/x}$, first compute the sum of the squares of the residuals:

```
>> Sr = sum((y-Z*a).^2)

Sr =
    3.7466
```

Then r^2 can be computed as

```
>> r2 = 1-Sr/sum((y-mean(y)).^2)

r2 =
    0.9985
```

and $s_{y/x}$ can be computed as

```
>> syx = sqrt(Sr/(length(x)-length(a)))

syx =
    1.1175
```

Our primary motivation for the foregoing has been to illustrate the unity among the three approaches and to show how they can all be expressed simply in the same matrix notation. It also sets the stage for the next section where we will gain some insights into the preferred strategies for solving Eq. (14.10). The matrix notation will also have relevance when we turn to nonlinear regression in Section 14.5.

14.4   QR FACTORIZATION AND THE BACKSLASH OPERATOR

Generating a best fit by solving the normal equations is widely used and certainly adequate for many curve-fitting applications in engineering and science. It must be mentioned, however, that the normal equations can be ill-conditioned and hence sensitive to roundoff errors.

Two more advanced methods, *QR factorization* and *singular value decomposition,* are more robust in this regard. Although the description of these methods is beyond the scope of this text, we mention them here because they can be implemented with MATLAB.

Further, QR factorization is automatically used in two simple ways within MATLAB. First, for cases where you want to fit a polynomial, the built-in `polyfit` function automatically uses QR factorization to obtain its results.

Second, the general linear least-squares problem can be directly solved with the backslash operator. Recall that the general model is formulated as Eq. (14.8)

$$\{y\} = [Z]\{a\} \tag{14.11}$$

In Section 10.4, we used left division with the backslash operator to solve systems of linear algebraic equations where the number of equations equals the number of unknowns ($n = m$). For Eq. (14.8) as derived from general least squares, the number of equations is greater than the number of unknowns ($n > m$). Such systems are said to be *overdetermined.* When MATLAB senses that you want to solve such systems with left division, it automatically uses QR factorization to obtain the solution. The following example illustrates how this is done.

EXAMPLE 14.4   Implementing Polynomial Regression with `polyfit` and Left Division

Problem Statement.   Repeat Example 14.3, but use the built-in `polyfit` function and left division to calculate the coefficients.

Solution.   As in Example 14.3, the data can be entered and used to create the [Z] matrix as in

```
>> x = [0 1 2 3 4 5]';
>> y = [2.1 7.7 13.6 27.2 40.9 61.1]';
>> Z = [ones(size(x)) x x.^2];
```

The `polyfit` function can be used to compute the coefficients:

```
>> a = polyfit(x,y,2)

a =
    1.8607    2.3593    2.4786
```

The same result can also be calculated using the backslash:

```
>> a = Z\y

a =
    2.4786
    2.3593
    1.8607
```

As just stated, both these results are obtained automatically with QR factorization.

14.5 NONLINEAR REGRESSION

There are many cases in engineering and science where nonlinear models must be fit to data. In the present context, these models are defined as those that have a nonlinear dependence on their parameters. For example,

$$y = a_0(1 - e^{-a_1 x}) + e \tag{14.12}$$

This equation cannot be manipulated so that it conforms to the general form of Eq. (14.7).

As with linear least squares, nonlinear regression is based on determining the values of the parameters that minimize the sum of the squares of the residuals. However, for the nonlinear case, the solution must proceed in an iterative fashion.

There are techniques expressly designed for nonlinear regression. For example, the Gauss-Newton method uses a Taylor series expansion to express the original nonlinear equation in an approximate, linear form. Then least-squares theory can be used to obtain new estimates of the parameters that move in the direction of minimizing the residual. Details on this approach are provided elsewhere (Chapra and Canale, 2002).

An alternative is to use optimization techniques to directly determine the least-squares fit. For example, Eq. (14.12) can be expressed as an objective function to compute the sum of the squares:

$$f(a_0, a_1) = \sum_{i=1}^{n} [y_i - a_0(1 - e^{-a_1 x_i})]^2 \tag{14.13}$$

An optimization routine can then be used to determine the values of a_0 and a_1 that minimize the function.

As described previously in Sec. 7.3.1, MATLAB's `fminsearch` function can be used for this purpose. It has the general syntax

```
[x, fval] = fminsearch(fun,x0,options,p1,p2,...)
```

where x = a vector of the values of the parameters that minimize the function `fun`, `fval` = the value of the function at the minimum, $x0$ = a vector of the initial guesses for the parameters, `options` = a structure containing values of the optimization parameters as created with the `optimset` function (recall Section 6.4), and `p1`, `p2`, etc. = additional arguments that are passed to the objective function. Note that if `options` is omitted, MATLAB uses default values that are reasonable for most problems. If you would like to pass additional arguments (`p1`, `p2`, ...), but do not want to set the `options`, use empty brackets `[]` as a place holder.

EXAMPLE 14.5   Nonlinear Regression with MATLAB

Problem Statement. Recall that in Example 13.4, we fit the power model to data from Table 13.1 by linearization using logarithms. This yielded the model:

$$F = 0.2741 v^{1.9842}$$

Repeat this exercise, but use nonlinear regression. Employ initial guesses of 1 for the coefficients.

Solution. First, an M-file function must be created to compute the sum of the squares. The following file, called `fSSR.m`, is set up for the power equation:

```
function f = fSSR(a,xm,ym)
yp = a(1)*xm.^a(2);
f = sum((ym-yp).^2);
```

In command mode, the data can be entered as

```
>> x = [10 20 30 40 50 60 70 80];
>> y = [25 70 380 550 610 1220 830 1450];
```

The minimization of the function is then implemented by

```
>> fminsearch(@fSSR, [1, 1], [], x, y)

ans =
    2.5384    1.4359
```

The best-fit model is therefore

$$F = 2.5384v^{1.4359}$$

Both the original transformed fit and the present version are displayed in Fig. 14.4. Note that although the model coefficients are very different, it is difficult to judge which fit is superior based on inspection of the plot.

This example illustrates how different best-fit equations result when fitting the same model using nonlinear regression versus linear regression employing transformations. This is because the former minimizes the residuals of the original data whereas the latter minimizes the residuals of the transformed data.

FIGURE 14.4
Comparison of transformed and untransformed model fits for force versus velocity data from Table 13.1.

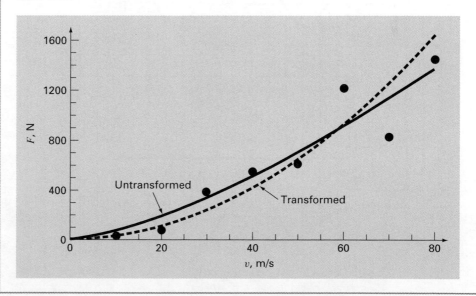

14.6 CASE STUDY   FITTING SINUSOIDS

Background.   Engineers and scientists often deal with systems that oscillate or vibrate. As might be expected, sinusoidal functions play a fundamental role in modeling such problem contexts.

In this discussion, we will use the term sinusoid to represent any waveform that can be described as a sine or cosine. There is no clear-cut convention for choosing either function, and in any case, the results will be identical. For this chapter, we will use the cosine, which is expressed generally as

$$f(t) = A_0 + C_1 \cos(\omega_0 t + \theta) \tag{14.14}$$

Thus, four parameters serve to characterize the sinusoid (Fig. 14.5a). The mean value A_0 sets the average height above the abscissa. The amplitude C_1 specifies the height of the

---

FIGURE 14.5

(a) A plot of the sinusoidal function $y(t) = A_0 + C_1\cos(\omega_0 t + \theta)$. For this case, $A_0 = 1.7$, $C_1 = 1$, $\omega_0 = 2\pi/(1.5 \text{ s})$, and $\theta = \pi/3$ radians. (b) An alternative expression of the same curve is $y(t) = A_0 + A_1 + A_1\cos(\omega_0 t) + B_1\sin(\omega_0 t)$. The three components of this function are depicted in (b), where $A_1 = 0.5$ and $B_1 = -0.866$. The summation of the three curves in (b) yields the single curve in (a).

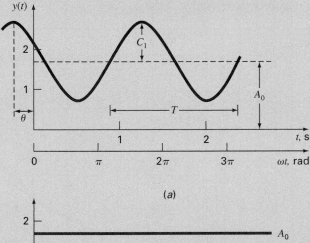

(a)

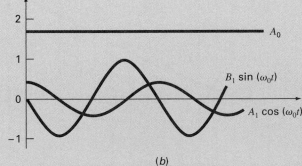

(b)

oscillation. The angular frequency ω_0 characterizes how often the cycles occur. Finally, the phase angle, or phase shift θ parameterizes the extent to which the sinusoid is shifted horizontally. It can be measured as the distance in radians from $t = 0$ to the point at which the cosine function begins a new cycle.

The angular frequency (in radians/time) is related to frequency f (in cycles/time) by

$$\omega_0 = 2\pi f \tag{14.15}$$

and frequency in turn is related to period T (in units of time) by

$$f = \frac{1}{T} \tag{14.16}$$

Although Eq. (14.14) is an adequate mathematical characterization of a sinusoid, it is awkward to work with from the standpoint of curve fitting because the phase shift is included in the argument of the cosine function. This means that it is not in the form of the general linear least-squares model (Eq. 14.7), and we must use nonlinear regression to estimate the coefficients.

This deficiency can be overcome by invoking the trigonometric identity:

$$C_1 \cos(\omega_0 t + \theta) = C_1[\cos(\omega_0 t)\cos(\theta) - \sin(\omega_0 t)\sin(\theta)] \tag{14.17}$$

Substituting Eq. (14.17) into Eq. (14.14) and collecting terms gives (Fig. 14.5b)

$$f(t) = A_0 + A_1 \cos(\omega_0 t) + B_1 \sin(\omega_0 t) \tag{14.18}$$

where

$$A_1 = C_1 \cos(\theta) \qquad B_1 = -C_1 \sin(\theta) \tag{14.19}$$

Dividing the two parts of Eq. (14.19) gives

$$\theta = \arctan\left(-\frac{B_1}{A_1}\right) \tag{14.20}$$

where if $A_1 < 0$, add π to θ. Squaring and summing the two parts of Eq. (14.19) leads to

$$C_1 = \sqrt{A_1^2 + B_1^2} \tag{14.21}$$

Thus, Eq. (14.18) represents an alternative formulation of Eq. (14.14) that still requires four parameters but that is cast in the format of a general linear model (Eq. 14.7). Thus, it can be simply applied as the basis for a least-squares fit.

The average monthly maximum air temperatures for Tucson, Arizona, have been tabulated as

Month	J	F	M	A	M	J	J	A	S	O	N	D
T, °C	18.9	21.1	23.3	27.8	32.2	37.2	36.1	34.4	29.4	23.3	18.9

14.6 CASE STUDY   continued

Observe that the July value is missing. Assuming each month is 30 days long, fit a sinusoid to this data. Use the resulting equation to predict the value in mid-July.

Solution.   In a similar fashion to Example 14.3, we can enter the data and create the [Z] matrix:

```
>> w0=2*pi/360;
>> t=[15 45 75 105 135 165 225 255 285 315 345]';
>> T=[18.9 21.1 23.3 27.8 32.2 37.2 36.1 34.4 29.4 23.3 18.9]';
>> Z=[ones(size(t)) cos(w0*t) sin(w0*t)];
```

The coefficients for the least-squares fit can then be computed as

```
>> a=(Z'*Z)\(Z'*T)

a =
   28.3878
   -9.2559
   -2.8002
```

The statistics can also be determined as

```
>> Sr=sum((T-Z*a).^2)

Sr =
    6.4398

>> r2=1-Sr/sum((T-mean(T)).^2)

r2 =
    0.9862

>> syx=sqrt(Sr/(length(t)-length(a)))

syx =
    0.8972
```

A plot can be developed as

```
>> tp=[0:360];
>> Tp=a(1)+a(2)*cos(w0*tp)+a(3)*sin(w0*tp);
>> plot(t,T,'o',tp,Tp)
```

The fit, which is displayed in Fig. 14.6, generally describes the data trends.

Equations (14.20) and (14.21) can be used to express the best-fit sinusoid in the more descriptive format of Eq. (14.14):

```
>> theta=atan2(-a(3),a(2))*360/(2*pi)

theta =
  163.1676
```

14.6 CASE STUDY continued

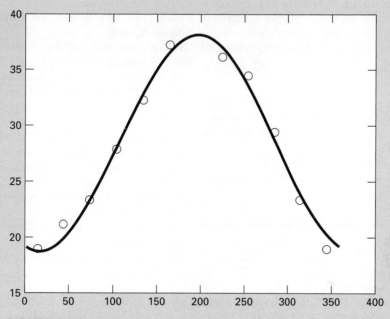

FIGURE 14.6
Least-squares fit of a sinusoid to temperature data for Tucson, Arizona.

```
>> C1=sqrt(a(2)^2+a(3)^2)

C1 =
   9.6702
```

Note that because it is more physically meaningful, the phase shift is expressed in days rather than in radians. Therefore, the final model is

$$R = 28.3878 + 9.6702 \cos\left(\frac{2\pi}{360}(t + 163.1676)\right)$$

We can see that the mean is 28.3878, and the amplitude is 9.6702. Hence, the maximum monthly temperature in Tucson ranges from about 18.7 to 38.1 °C. The phase shift tells us that the peak temperature occurs about 163 days prior to the start of the year, which translates to about day 197 (July 17). The value in mid-July ($t = 195$ d) can therefore be computed as

```
>> TpJul=a(1)+C1*cos(w0*(195+theta))

TpJul =
   38.0530
```

PROBLEMS

14.1 Fit a parabola to the data from Table 13.1. Determine the r^2 for the fit and comment on the efficacy of the result.
14.2 Using the same approach as was employed to derive Eqs. (13.15) and (13.16), derive the least-squares fit of the following model:

$$y = a_1 x + a_2 x^2 + e$$

That is, determine the coefficients that results in the least-squares fit for a second-order polynomial with a zero intercept. Test the approach by using it to fit the data from Table 13.1.
14.3 Fit a cubic polynomial to the following data:

x	3	4	5	7	8	9	11	12
y	1.6	3.6	4.4	3.4	2.2	2.8	3.8	4.6

Along with the coefficients, determine r^2 and $s_{y/x}$.
14.4 Develop an M-file to implement polynomial regression. Pass the M-file two vectors holding the x and y values along with the desired order m. Test it by solving Prob. 14.3.
14.5 For the data from Table P14.5, use polynomial regression to derive a predictive equation for dissolved oxygen concentration as a function of temperature for the case where the chloride concentration is equal to zero. Employ a polynomial that is of sufficiently high order that the predictions match the number of significant digits displayed in the table.
14.6 Use multiple linear regression to derive a predictive equation for dissolved oxygen concentration as a function of temperature and chloride based on the data from Table P14.5. Use the equation to estimate the concentration of dissolved oxygen for a chloride concentration of 15 g/L at $T = 12\,°C$.

TABLE P14.5 Dissolved oxygen concentration in water as a function of temperature (°C) and chloride concentration (g/L).

T, °C	c = 0 g/L	c = 10 g/L	c = 20 g/L
0	14.6	12.9	11.4
5	12.8	11.3	10.3
10	11.3	10.1	8.96
15	10.1	9.03	8.08
20	9.09	8.17	7.35
25	8.26	7.46	6.73
30	7.56	6.85	6.20

Note that the true value is 9.09 mg/L. Compute the percent relative error for your prediction. Explain possible causes for the discrepancy.
14.7 As compared with the models from Probs. 14.5 and 14.6, a somewhat more sophisticated model that accounts for the effect of both temperature and chloride on dissolved oxygen saturation can be hypothesized as being of the form

$$o = f_3(T) + f_1(c)$$

That is, a third-order polynomial in temperature and a linear relationship in chloride is assumed to yield superior results. Use the general linear least-squares approach to fit this model to the data in Table P14.5. Use the resulting equation to estimate the dissolved oxygen concentration for a chloride concentration of 15 g/L at $T = 12\,°C$. Note that the true value is 9.09 mg/L. Compute the percent relative error for your prediction.
14.8 Use multiple linear regression to fit

x_1	0	1	1	2	2	3	3	4	4
x_2	0	1	2	1	2	1	2	1	2
y	15.1	17.9	12.7	25.6	20.5	35.1	29.7	45.4	40.2

Compute the coefficients, the standard error of the estimate, and the correlation coefficient.
14.9 The following data was collected for the steady flow of water in a concrete circular pipe:

Experiment	Diameter, m	Slope, m/m	Flow, m³/s
1	0.3	0.001	0.04
2	0.6	0.001	0.24
3	0.9	0.001	0.69
4	0.3	0.01	0.13
5	0.6	0.01	0.82
6	0.9	0.01	2.38
7	0.3	0.05	0.31
8	0.6	0.05	1.95
9	0.9	0.05	5.66

Use multiple linear regression to fit the following model to this data:

$$Q = \alpha_0 D^{\alpha_1} S^{\alpha_2}$$

where Q = flow, D = diameter, and S = slope.
14.10 Three disease-carrying organisms decay exponentially in seawater according to the following model:

$$p(t) = Ae^{-1.5t} + Be^{-0.3t} + Ce^{-0.05t}$$

Estimate the initial concentration of each organism (A, B, and C) given the following measurements:

t	0.5	1	2	3	4	5	6	7	9
$p(t)$	6	4.4	3.2	2.7	2	1.9	1.7	1.4	1.1

14.11 The following model is used to represent the effect of solar radiation on the photosynthesis rate of aquatic plants:

$$P = P_m \frac{I}{I_{sat}} e^{-\frac{I}{I_{sat}}+1}$$

where P = the photosynthesis rate (mg m^{-3}d^{-1}), P_m = the maximum photosynthesis rate (mg m^{-3}d^{-1}), I = solar radiation (μE m^{-2}s^{-1}), and I_{sat} = optimal solar radiation (μE m^{-2}s^{-1}). Use nonlinear regression to evaluate P_m and I_{sat} based on the following data:

I	50	80	130	200	250	350	450	550	700
P	99	177	202	248	229	219	173	142	72

14.12 In Prob. 13.8 we used transformations to linearize and fit the following model:

$$y = \alpha_4 x e^{\beta_4 x}$$

Use nonlinear regression to estimate α_4 and β_4 based on the following data. Develop a plot of your fit along with the data.

x	0.1	0.2	0.4	0.6	0.9	1.3	1.5	1.7	1.8
y	0.75	1.25	1.45	1.25	0.85	0.55	0.35	0.28	0.18

14.13 Enzymatic reactions are used extensively to characterize biologically mediated reactions. The following is an example of a model that is used to fit such reactions:

$$v_0 = \frac{k_m[S]^3}{K + [S]^3}$$

where v_0 = the initial rate of the reaction (M/s), $[S]$ = the substrate concentration (M), and k_m and K are parameters. The following data can be fit with this model:

$[S]$, M	v_0, M/s
0.01	6.078×10^{-11}
0.05	7.595×10^{-9}
0.1	6.063×10^{-8}
0.5	5.788×10^{-6}
1	1.737×10^{-5}
5	2.423×10^{-5}
10	2.430×10^{-5}
50	2.431×10^{-5}
100	2.431×10^{-5}

(a) Use a transformation to linearize the model and evaluate the parameters. Display the data and the model fit on a graph.
(b) Perform the same evaluation as in **(a)** but use nonlinear regression.

14.14 Given the data

x	5	10	15	20	25	30	35	40	45	50
y	17	24	31	33	37	37	40	40	42	41

use least-squares regression to fit **(a)** a straight line, **(b)** a power equation, **(c)** a saturation-growth-rate equation, and **(d)** a parabola. For **(b)** and **(c)**, employ transformations to linearize the data. Plot the data along with all the curves. Is any one of the curves superior? If so, justify.

14.15 The following data represents the bacterial growth in a liquid culture over of number of days:

Day	0	4	8	12	16	20
Amount $\times$ 10^6		67.38	74.67	82.74	91.69	101.60	112.58

Find a best-fit equation to the data trend. Try several possibilities—polynomial, logarithmic, and exponential. Determine the best equation to predict the amount of bacteria after 30 days.

14.16 Derive the least-squares fit of the following model:

$$y = a_1 x + a_2 x^2 + e.$$

That is, determine the coefficients that results in the least-squares fit for a second-order polynomial with a zero intercept. Test the approach by using it to fit the data from Table 13.1.

14.17 Dynamic viscosity of water $\mu(10^{-3}$ N $\cdot$ s/m$^2)$ is related to temperature $T(^{\circ}$C) in the following manner:

T	0	5	10	20	30	40
μ	1.787	1.519	1.307	1.002	0.7975	0.6529

(a) Plot this data.
(b) Use linear interpolation to predict μ at $T = 7.5$ °C.
(c) Use polynomial regression to fit a parabola to the data in order to make the same prediction.

14.18 Use the following set of pressure-volume data to find the best possible virial constants (A_1 and A_2) for the following equation of state. $R = 82.05$ mL atm/gmol K, and $T = 303$ K.

$$\frac{PV}{RT} = 1 + \frac{A_1}{V} + \frac{A_2}{V^2}$$

P (atm)	0.985	1.108	1.363	1.631
V (mL)	25,000	22,200	18,000	15,000

14.19 Environmental scientists and engineers dealing with the impacts of acid rain must determine the value of the ion product of water K_w as a function of temperature. Scientists have suggested the following equation to model this relationship:

$$-\log_{10} K_w = \frac{a}{T_a} + b \log_{10} T_a + c T_a + d$$

where T_a = absolute temperature (K), and a, b, c, and d are parameters. Employ the following data and regression to estimate the parameters:

T (K)	K_w
0	1.164×10^{-15}
10	2.950×10^{-15}
20	6.846×10^{-15}
30	1.467×10^{-14}
40	2.929×10^{-14}

14.20 The distance required to stop an automobile consists of both thinking and braking components, each of which is a function of its speed. The following experimental data was

collected to quantify this relationship. Develop best-fit equations for both the thinking and braking components. Use these equations to estimate the total stopping distance for a car traveling at 110 km/h.

Speed, km/h	30	45	60	75	90	120
Thinking, m	5.6	8.5	11.1	14.5	16.7	22.4
Braking, m	5.0	12.3	21.0	32.9	47.6	84.7

14.21 The pH in a reactor varies over the course of a day. Use least-squares regression to fit a sinusoid to the following data. Use your fit to determine the mean, amplitude, and time of maximum pH.

Time, h	0	2	4	5	7	9	12	15	20	22	24
pH	7.6	7.2	7	6.5	7.5	7.2	8.9	9.1	8.9	7.9	7

14.22 The solar radiation for Tucson, Arizona, has been tabulated as

Time, mo	J	F	M	A	M	J	J	A	S	O	N	D
Radiation, W/m²	144	188	245	311	351	359	308	287	260	211	159	131

Assuming each month is 30 days long, fit a sinusoid to this data. Use the resulting equation to predict the radiation in mid-August.

15

Polynomial Interpolation

CHAPTER OBJECTIVES

The primary objective of this chapter is to introduce you to polynomial interpolation. Specific objectives and topics covered are

- Recognizing that evaluating polynomial coefficients with simultaneous equations is an ill-conditioned problem.
- Knowing how to evaluate polynomial coefficients and interpolate with MATLAB's `polyfit` and `polyval` functions.
- Knowing how to perform an interpolation with Newton's polynomial.
- Knowing how to perform an interpolation with a Lagrange polynomial.
- Knowing how to solve an inverse interpolation problem by recasting it as a roots problem.
- Appreciating the dangers of extrapolation.
- Recognizing that higher-order polynomials can manifest large oscillations.

YOU'VE GOT A PROBLEM

I f we want to improve the velocity prediction for the free-falling bungee jumper, we might expand our model to account for other factors beyond mass and the drag coefficient. As was previously mentioned in Section 2.7, the drag coefficient can itself be formulated as a function of other factors such as the area of the jumper and characteristics such as the air's density and viscosity.

Air density and viscosity are commonly presented in tabular form as a function of temperature. For example, Table 15.1 is reprinted from a popular fluid mechanics textbook (White, 1999).

Suppose that you desired the density at a temperature not included in the table. In such a case, you would have to interpolate. That is, you would have to estimate the value at the

TABLE 15.1 Density (ρ), dynamic viscosity (μ), and kinematic viscosity (v) as a function of temperature (T) at 1 atm as reported by White (1999).

T, °C	ρ, kg/m^3	μ, N · s/m^2	v, m^2/s
−40	1.52	1.51×10^{-5}	0.99×10^{-5}
0	1.29	1.71×10^{-5}	1.33×10^{-5}
20	1.20	1.80×10^{-5}	1.50×10^{-5}
50	1.09	1.95×10^{-5}	1.79×10^{-5}
100	0.946	2.17×10^{-5}	2.30×10^{-5}
150	0.835	2.38×10^{-5}	2.85×10^{-5}
200	0.746	2.57×10^{-5}	3.45×10^{-5}
250	0.675	2.75×10^{-5}	4.08×10^{-5}
300	0.616	2.93×10^{-5}	4.75×10^{-5}
400	0.525	3.25×10^{-5}	6.20×10^{-5}
500	0.457	3.55×10^{-5}	7.77×10^{-5}

desired temperature based on the densities that bracket it. The simplest approach is to determine the equation for the straight line connecting the two adjacent values and use this equation to estimate the density at the desired intermediate temperature. Although such *linear interpolation* is perfectly adequate in many cases, error can be introduced when the data exhibits significant curvature. In this chapter, we will explore a number of different approaches for obtaining adequate estimates for such situations.

15.1 INTRODUCTION TO INTERPOLATION

You will frequently have occasion to estimate intermediate values between precise data points. The most common method used for this purpose is polynomial interpolation. The general formula for an $(n - 1)$th-order polynomial can be written as

$$f(x) = a_1 + a_2x + a_3x^2 + \cdots + a_nx^{n-1} \tag{15.1}$$

For n data points, there is one and only one polynomial of order $(n - 1)$ that passes through all the points. For example, there is only one straight line (i.e., a first-order polynomial) that connects two points (Fig. 15.1a). Similarly, only one parabola connects a set of three points (Fig. 15.1b). *Polynomial interpolation* consists of determining the unique $(n - 1)$th-order polynomial that fits n data points. This polynomial then provides a formula to compute intermediate values.

Before proceeding, we should note that MATLAB represents polynomial coefficients in a different manner than Eq. (15.1). Rather than using increasing powers of x, it uses decreasing powers as in

$$f(x) = p_1x^{n-1} + p_2x^{n-2} + \cdots + p_{n-1}x + p_n \tag{15.2}$$

To be consistent with MATLAB, we will adopt this scheme in the following section.

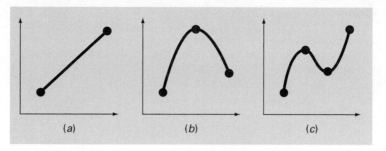

FIGURE 15.1
Examples of interpolating polynomials: (a) first-order (linear) connecting two points,
(b) second-order (quadratic or parabolic) connecting three points, and (c) third-order (cubic)
connecting four points.

15.1.1 Determining Polynomial Coefficients

A straightforward way for computing the coefficients of Eq. (15.2) is based on the fact that
n data points are required to determine the n coefficients. As in the following example, this
allows us to generate n linear algebraic equations that we can solve simultaneously for the
coefficients.

EXAMPLE 15.1   Determining Polynomial Coefficients with Simultaneous Equations

Problem Statement.   Suppose that we want to determine the coefficients of the parabola,
$f(x) = p_1 x^2 + p_2 x + p_3$, that passes through the last three density values from Table 15.1:

$$x_1 = 300 \quad f(x_1) = 0.616$$
$$x_2 = 400 \quad f(x_2) = 0.525$$
$$x_3 = 500 \quad f(x_3) = 0.457$$

Each of these pairs can be substituted into Eq. (15.2) to yield a system of three equations:

$$0.616 = p_1(300)^2 + p_2(300) + p_3$$
$$0.525 = p_1(400)^2 + p_2(400) + p_3$$
$$0.457 = p_1(500)^2 + p_2(500) + p_3$$

or in matrix form:

$$\begin{bmatrix} 90{,}000 & 300 & 1 \\ 160{,}000 & 400 & 1 \\ 250{,}000 & 500 & 1 \end{bmatrix} \begin{Bmatrix} p_1 \\ p_2 \\ p_3 \end{Bmatrix} = \begin{Bmatrix} 0.616 \\ 0.525 \\ 0.457 \end{Bmatrix}$$

Thus, the problem reduces to solving three simultaneous linear algebraic equations for
the three unknown coefficients. A simple MATLAB session can be used to obtain the

solution:

```
>> format long
>> A = [90000 300 1;160000 400 1;250000 500 1];
>> b = [0.616 0.525 0.457]';
>> p = A\b


p =
   0.00000115000000
  -0.00171500000000
   1.02700000000000
```

Thus, the parabola that passes exactly through the three points is

$$f(x) = 0.00000115x^2 - 0.001715x + 1.027$$

This polynomial then provides a means to determine intermediate points. For example, the value of density at a temperature of 350 °C can be calculated as

$$f(350) = 0.00000115(350)^2 - 0.001715(350) + 1.027 = 0.567625$$

Although the approach in Example 15.1 provides an easy way to perform interpolation, it has a serious deficiency. To understand this flaw, notice that the coefficient matrix in Example 15.1 has a decided structure. This can be seen clearly by expressing it in general terms:

$$\begin{bmatrix} x_1^2 & x_1 & 1 \\ x_2^2 & x_2 & 1 \\ x_3^2 & x_3 & 1 \end{bmatrix} \begin{Bmatrix} p_1 \\ p_2 \\ p_3 \end{Bmatrix} = \begin{Bmatrix} f(x_1) \\ f(x_2) \\ f(x_3) \end{Bmatrix} \tag{15.3}$$

Coefficient matrices of this form are referred to as *Vandermonde matrices*. Such matrices are very ill-conditioned. That is, their solutions are very sensitive to round-off errors. This can be illustrated by using MATLAB to compute the condition number for the coefficient matrix from Example 15.1 as

```
>> cond(A)

ans =

   5.8932e+006
```

This condition number, which is quite large for a 3×3 matrix, implies that about six digits of the solution would be questionable. The ill-conditioning becomes even worse as the number of simultaneous equations becomes larger.

As a consequence, there are alternative approaches that do not manifest this shortcoming. In this chapter, we will also describe two alternatives that are well-suited for computer implementation: the Newton and the Lagrange polynomials. Before doing this, however, we will first briefly review how the coefficients of the interpolating polynomial can be estimated directly with MATLAB's built-in functions.

15.1.2 MATLAB Functions: `polyfit` and `polyval`

Recall from Section 13.4.2, that the `polyfit` function can be used to perform polynomial regression. In such applications, the number of data points is greater than the number of coefficients being estimated. Consequently, the least-squares fit line does not necessarily pass through any of the points, but rather follows the general trend of the data.

For the case where the number of data points equals the number of coefficients, `polyfit` performs interpolation. That is, it returns the coefficients of the polynomial that pass directly through the data points. For example, it can be used to determine the coefficients of the parabola that passes through the last three density values from Table 15.1:

```
>> format long
>> T = [300 400 500];
>> density = [0.616 0.525 0.457];
>> p = polyfit(T,density,2)

p =
   0.00000115000000  -0.00171500000000   1.02700000000000
```

We can then use the `polyval` function to perform an interpolation as in

```
>> d = polyval(p,350)

d =
   0.56762500000000
```

These results agree with those obtained previously in Example 15.1 with simultaneous equations.

15.2 NEWTON INTERPOLATING POLYNOMIAL

There are a variety of alternative forms for expressing an interpolating polynomial beyond the familiar format of Eq. (15.2). Newton's interpolating polynomial is among the most popular and useful forms. Before presenting the general equation, we will introduce the first- and second-order versions because of their simple visual interpretation.

15.2.1 Linear Interpolation

The simplest form of interpolation is to connect two data points with a straight line. This technique, called *linear interpolation,* is depicted graphically in Fig. 15.2. Using similar triangles,

$$\frac{f_1(x) - f(x_1)}{x - x_1} = \frac{f(x_2) - f(x_1)}{x_2 - x_1} \tag{15.4}$$

which can be rearranged to yield

$$f_1(x) = f(x_1) + \frac{f(x_2) - f(x_1)}{x_2 - x_1}(x - x_1) \tag{15.5}$$

which is the *Newton linear-interpolation formula.* The notation $f_1(x)$ designates that this is a first-order interpolating polynomial. Notice that besides representing the slope of the line connecting the points, the term $[f(x_2) - f(x_1)]/(x_2 - x_1)$ is a finite-difference

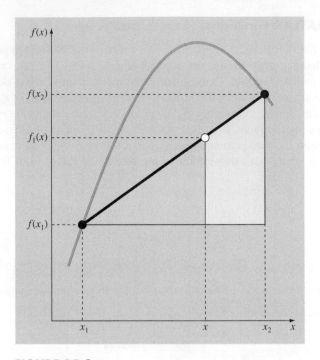

FIGURE 15.2
Graphical depiction of linear interpolation. The shaded areas indicate the similar triangles used
to derive the Newton linear-interpolation formula [Eq. (15.5)].

approximation of the first derivative [recall Eq. (4.20)]. In general, the smaller the interval
between the data points, the better the approximation. This is due to the fact that, as the
interval decreases, a continuous function will be better approximated by a straight line.
This characteristic is demonstrated in the following example.

EXAMPLE 15.2   Linear Interpolation

Problem Statement.   Estimate the natural logarithm of 2 using linear interpolation. First,
perform the computation by interpolating between $\ln 1 = 0$ and $\ln 6 = 1.791759$. Then,
repeat the procedure, but use a smaller interval from $\ln 1$ to $\ln 4$ (1.386294). Note that the
true value of $\ln 2$ is 0.6931472.

Solution.   We use Eq. (15.5) from $x_1 = 1$ to $x_2 = 6$ to give

$$f_1(2) = 0 + \frac{1.791759 - 0}{6 - 1}(2 - 1) = 0.3583519$$

which represents an error of $\varepsilon_t = 48.3\%$. Using the smaller interval from $x_1 = 1$ to $x_2 = 4$
yields

$$f_1(2) = 0 + \frac{1.386294 - 0}{4 - 1}(2 - 1) = 0.4620981$$

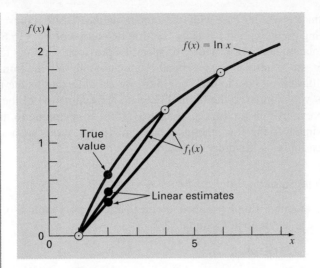

FIGURE 15.3
Two linear interpolations to estimate ln 2. Note how the smaller interval provides a better estimate.

Thus, using the shorter interval reduces the percent relative error to $\varepsilon_t = 33.3\%$. Both interpolations are shown in Fig. 15.3, along with the true function.

15.2.2 Quadratic Interpolation

The error in Example 15.2 resulted from approximating a curve with a straight line. Consequently, a strategy for improving the estimate is to introduce some curvature into the line connecting the points. If three data points are available, this can be accomplished with a second-order polynomial (also called a quadratic polynomial or a parabola). A particularly convenient form for this purpose is

$$f_2(x) = b_1 + b_2(x - x_1) + b_3(x - x_1)(x - x_2) \tag{15.6}$$

A simple procedure can be used to determine the values of the coefficients. For b_1, Eq. (15.6) with $x = x_1$ can be used to compute

$$b_1 = f(x_1) \tag{15.7}$$

Equation (15.7) can be substituted into Eq. (15.6), which can be evaluated at $x = x_2$ for

$$b_2 = \frac{f(x_2) - f(x_1)}{x_2 - x_1} \tag{15.8}$$

Finally, Eqs. (15.7) and (15.8) can be substituted into Eq. (15.6), which can be evaluated at $x = x_3$ and solved (after some algebraic manipulations) for

$$b_3 = \frac{\dfrac{f(x_3) - f(x_2)}{x_3 - x_2} - \dfrac{f(x_2) - f(x_1)}{x_2 - x_1}}{x_3 - x_1} \tag{15.9}$$

Notice that, as was the case with linear interpolation, b_2 still represents the slope of the line connecting points x_1 and x_2. Thus, the first two terms of Eq. (15.6) are equivalent to linear interpolation between x_1 and x_2, as specified previously in Eq. (15.5). The last term, $b_3(x - x_1)(x - x_2)$, introduces the second-order curvature into the formula.

Before illustrating how to use Eq. (15.6), we should examine the form of the coefficient b_3. It is very similar to the finite-difference approximation of the second derivative introduced previously in Eq. (4.27). Thus, Eq. (15.6) is beginning to manifest a structure that is very similar to the Taylor series expansion. That is, terms are added sequentially to capture increasingly higher-order curvature.

EXAMPLE 15.3   Quadratic Interpolation

Problem Statement.   Employ a second-order Newton polynomial to estimate ln 2 with the same three points used in Example 15.2:

$$x_1 = 1 \qquad f(x_1) = 0$$

$$x_2 = 4 \qquad f(x_2) = 1.386294$$

$$x_3 = 6 \qquad f(x_3) = 1.791759$$

Solution.   Applying Eq. (15.7) yields

$$b_1 = 0$$

Equation (15.8) gives

$$b_2 = \frac{1.386294 - 0}{4 - 1} = 0.4620981$$

FIGURE 15.4
The use of quadratic interpolation to estimate ln 2. The linear interpolation from $x = 1$ to 4 is also included for comparison.

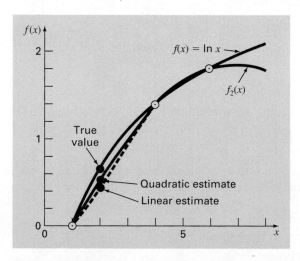

and Eq. (15.9) yields

$$b_3 = \frac{\dfrac{1.791759 - 1.386294}{6 - 4} - 0.4620981}{6 - 1} = -0.0518731$$

Substituting these values into Eq. (15.6) yields the quadratic formula

$$f_2(x) = 0 + 0.4620981(x - 1) - 0.0518731(x - 1)(x - 4)$$

which can be evaluated at $x = 2$ for $f_2(2) = 0.5658444$, which represents a relative error of $\varepsilon_t = 18.4\%$. Thus, the curvature introduced by the quadratic formula (Fig. 15.4) improves the interpolation compared with the result obtained using straight lines in Example 15.2 and Fig. 15.3.

15.2.3 General Form of Newton's Interpolating Polynomials

The preceding analysis can be generalized to fit an $(n - 1)$th-order polynomial to n data points. The $(n - 1)$th-order polynomial is

$$f_{n-1}(x) = b_1 + b_2(x - x_1) + \cdots + b_n(x - x_1)(x - x_2) \cdots (x - x_{n-1}) \qquad (15.10)$$

As was done previously with linear and quadratic interpolation, data points can be used to evaluate the coefficients $b_1, b_2, \ldots, b_n$. For an $(n - 1)$th-order polynomial, n data points are required: $[x_1, f(x_1)], [x_2, f(x_2)], \ldots, [x_n, f(x_n)]$. We use these data points and the following equations to evaluate the coefficients:

$$b_1 = f(x_1) \qquad (15.11)$$

$$b_2 = f[x_2, x_1] \qquad (15.12)$$

$$b_3 = f[x_3, x_2, x_1] \qquad (15.13)$$

$$\vdots$$

$$b_n = f[x_n, x_{n-1}, \ldots, x_2, x_1] \qquad (15.14)$$

where the bracketed function evaluations are finite divided differences. For example, the first finite divided difference is represented generally as

$$f[x_i, x_j] = \frac{f(x_i) - f(x_j)}{x_i - x_j} \qquad (15.15)$$

The second finite divided difference, which represents the difference of two first divided differences, is expressed generally as

$$f[x_i, x_j, x_k] = \frac{f[x_i, x_j] - f[x_j, x_k]}{x_i - x_k} \qquad (15.16)$$

Similarly, the nth finite divided difference is

$$f[x_n, x_{n-1}, \ldots, x_2, x_1] = \frac{f[x_n, x_{n-1}, \ldots, x_2] - f[x_{n-1}, x_{n-2}, \ldots, x_1]}{x_n - x_1} \qquad (15.17)$$

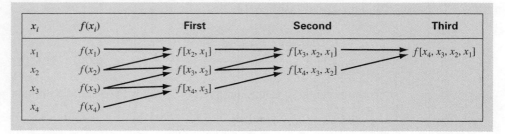

FIGURE 15.5
Graphical depiction of the recursive nature of finite divided differences. This representation is referred to as a divided difference table.

These differences can be used to evaluate the coefficients in Eqs. (15.11) through (15.14), which can then be substituted into Eq. (15.10) to yield the general form of Newton's interpolating polynomial:

$$f_{n-1}(x) = f(x_1) + (x - x_1)f[x_2, x_1] + (x - x_1)(x - x_2)f[x_3, x_2, x_1]$$
$$+ \cdots + (x - x_1)(x - x_2) \cdots (x - x_{n-1})f[x_n, x_{n-1}, \ldots, x_2, x_1] \quad (15.18)$$

We should note that it is not necessary that the data points used in Eq. (15.18) be equally spaced or that the abscissa values necessarily be in ascending order, as illustrated in the following example. Also, notice how Eqs. (15.15) through (15.17) are recursive— that is, higher-order differences are computed by taking differences of lower-order differences (Fig. 15.5). This property will be exploited when we develop an efficient M-file in Section 15.2.4 to implement the method.

EXAMPLE 15.4   Newton Interpolating Polynomial

Problem Statement.   In Example 15.3, data points at $x_1 = 1$, $x_2 = 4$, and $x_3 = 6$ were used to estimate ln 2 with a parabola. Now, adding a fourth point [$x_4 = 5$; $f(x_4) = 1.609438$], estimate ln 2 with a third-order Newton's interpolating polynomial.

Solution.   The third-order polynomial, Eq. (15.10) with $n = 4$, is

$$f_3(x) = b_1 + b_2(x - x_1) + b_3(x - x_1)(x - x_2) + b_4(x - x_1)(x - x_2)(x - x_3)$$

The first divided differences for the problem are [Eq. (15.15)]

$$f[x_2, x_1] = \frac{1.386294 - 0}{4 - 1} = 0.4620981$$

$$f[x_3, x_2] = \frac{1.791759 - 1.386294}{6 - 4} = 0.2027326$$

$$f[x_4, x_3] = \frac{1.609438 - 1.791759}{5 - 6} = 0.1823216$$

The second divided differences are [Eq. (15.16)]

$$f[x_3, x_2, x_1] = \frac{0.2027326 - 0.4620981}{6 - 1} = -0.05187311$$

$$f[x_4, x_3, x_2] = \frac{0.1823216 - 0.2027326}{5 - 4} = -0.02041100$$

The third divided difference is [Eq. (15.17) with $n = 4$]

$$f[x_4, x_3, x_2, x_1] = \frac{-0.02041100 - (-0.05187311)}{5 - 1} = 0.007865529$$

Thus, the divided difference table is

x_i	$f(x_i)$	First	Second	Third
1	0	0.4620981	−0.05187311	0.007865529
4	1.386294	0.2027326	−0.02041100	
6	1.791759	0.1823216		
5	1.609438			

The results for $f(x_1)$, $f[x_2, x_1]$, $f[x_3, x_2, x_1]$, and $f[x_4, x_3, x_2, x_1]$ represent the coefficients b_1, b_2, b_3, and b_4, respectively, of Eq. (15.10). Thus, the interpolating cubic is

$$f_3(x) = 0 + 0.4620981(x - 1) - 0.05187311(x - 1)(x - 4)$$
$$+ 0.007865529(x - 1)(x - 4)(x - 6)$$

which can be used to evaluate $f_3(2) = 0.6287686$, which represents a relative error of $\varepsilon_t = 9.3\%$. The complete cubic polynomial is shown in Fig. 15.6.

FIGURE 15.6
The use of cubic interpolation to estimate ln 2.

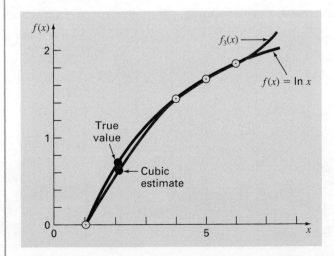

15.2.4 MATLAB M-file: `Newtint`

It is straightforward to develop an M-file to implement Newton interpolation. As in Fig. 15.7, the first step is to compute the finite divided differences and store them in an array. The differences are then used in conjunction with Eq. (15.18) to perform the interpolation.

An example of a session using the function would be to duplicate the calculation we just performed in Example 15.3:

```
>> format long
>> x = [1 4 6 5]';
```

FIGURE 15.7
An M-file to implement Newton interpolation.

```
function yint = Newtint(x,y,xx)
% Newtint: Newton interpolating polynomial
% yint = Newtint(x,y,xx): Uses an (n - 1)-order Newton
%   interpolating polynomial based on n data points (x, y)
%   to determine a value of the dependent variable (yint)
%   at a given value of the independent variable, xx.
% input:
%   x = independent variable
%   y = dependent variable
%   xx = value of independent variable at which
%        interpolation is calculated
% output:
%   yint = interpolated value of dependent variable

% compute the finite divided differences in the form of a
% difference table
n = length(x);
if length(y)~=n, error('x and y must be same length'); end
b = zeros(n,n);
% assign dependent variables to the first column of b.
b(:,1) = y(:);  % the (:) ensures that y is a column vector.
for j = 2:n
  for i = 1:n-j+1
    b(i,j) = (b(i+1,j-1)-b(i,j-1))/(x(i+j-1)-x(i));
  end
end
% use the finite divided differences to interpolate
xt = 1;
yint = b(1,1);
for j = 1:n-1
  xt = xt*(xx-x(j));
  yint = yint+b(1,j+1)*xt;
end
```

```
>> y = log(x);
>> Newtint(x,y,2)

ans =
     0.62876857890841
```

15.3 LAGRANGE INTERPOLATING POLYNOMIAL

Suppose we formulate a linear interpolating polynomial as the weighted average of the two values that we are connecting by a straight line:

$$f(x) = L_1 f(x_1) + L_2 f(x_2) \tag{15.19}$$

where the L's are the weighting coefficients. It is logical that the first weighting coefficient is the straight line that is equal to 1 at x_1 and 0 at x_2:

$$L_1 = \frac{x - x_2}{x_1 - x_2}$$

Similarly, the second coefficient is the straight line that is equal to 1 at x_2 and 0 at x_1:

$$L_2 = \frac{x - x_1}{x_2 - x_1}$$

Substituting these coefficients into Eq. 15.19 yields the straight line that connects the points (Fig. 15.8):

$$f_1(x) = \frac{x - x_2}{x_1 - x_2} f(x_1) + \frac{x - x_1}{x_2 - x_1} f(x_2) \tag{15.20}$$

where the nomenclature $f_1(x)$ designates that this is a first-order polynomial. Equation (15.20) is referred to as the *linear Lagrange interpolating polynomial.*

The same strategy can be employed to fit a parabola through three points. For this case three parabolas would be used with each one passing through one of the points and equaling zero at the other two. Their sum would then represent the unique parabola that connects the three points. Such a second-order Lagrange interpolating polynomial can be written as

$$f_2(x) = \frac{(x - x_2)(x - x_3)}{(x_1 - x_2)(x_1 - x_3)} f(x_1) + \frac{(x - x_1)(x - x_3)}{(x_2 - x_1)(x_2 - x_3)} f(x_2)$$
$$+ \frac{(x - x_1)(x - x_2)}{(x_3 - x_1)(x_3 - x_2)} f(x_3) \tag{15.21}$$

Notice how the first term is equal to $f(x_1)$ at x_1 and is equal to zero at x_2 and x_3. The other terms work in a similar fashion.

Both the first- and second-order versions as well as higher-order Lagrange polynomials can be represented concisely as

$$f_{n-1}(x) = \sum_{i=1}^{n} L_i(x) f(x_i) \tag{15.22}$$

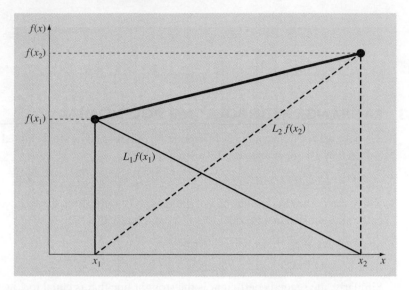

FIGURE 15.8
A visual depiction of the rationale behind Lagrange interpolating polynomials. The figure shows the first-order case. Each of the two terms of Eq. (15.20) passes through one of the points and is zero at the other. The summation of the two terms must, therefore, be the unique straight line that connects the two points.

where

$$L_i(x) = \prod_{\substack{j=1 \\ j \neq i}}^{n} \frac{x - x_j}{x_i - x_j} \tag{15.23}$$

where $n =$ the number of data points and $\prod$ designates the "product of."

EXAMPLE 15.5   **Lagrange Interpolating Polynomial**

Problem Statement. Use a Lagrange interpolating polynomial of the first and second order to evaluate the density of unused motor oil at $T = 15\ °C$ based on the following data:

$$x_1 = 0 \qquad f(x_1) = 3.85$$
$$x_2 = 20 \qquad f(x_2) = 0.800$$
$$x_3 = 40 \qquad f(x_3) = 0.212$$

Solution. The first-order polynomial [Eq. (15.20)] can be used to obtain the estimate at $x = 15$:

$$f_1(x) = \frac{15 - 20}{0 - 20}3.85 + \frac{15 - 0}{20 - 0}0.800 = 1.5625$$

In a similar fashion, the second-order polynomial is developed as [Eq. (15.21)]

$$f_2(x) = \frac{(15-20)(15-40)}{(0-20)(0-40)}3.85 + \frac{(15-0)(15-40)}{(20-0)(20-40)}0.800$$

$$+ \frac{(15-0)(15-20)}{(40-0)(40-20)}0.212 = 1.3316875$$

15.3.1 MATLAB M-file: Lagrange

It is straightforward to develop an M-file based on Eqs. (15.22) and (15.23). As in Fig. 15.9, the function is passed two vectors containing the independent (x) and the dependent (y) variables. It is also passed the value of the independent variable where you want to interpolate (xx). The order of the polynomial is based on the length of the x vector that is passed. If n values are passed, an $(n-1)$th order polynomial is fit.

FIGURE 15.9
An M-file to implement Lagrange interpolation.

```
function yint = Lagrange(x,y,xx)
% Lagrange: Lagrange interpolating polynomial
%   yint = Lagrange(x,y,xx): Uses an (n - 1)-order
%      Lagrange interpolating polynomial based on n data points
%      to determine a value of the dependent variable (yint) at
%      a given value of the independent variable, xx.
% input:
%   x = independent variable
%   y = dependent variable
%   xx = value of independent variable at which the
%        interpolation is calculated
% output:
%   yint = interpolated value of dependent variable

n = length(x);
if length(y)~=n, error('x and y must be same length'); end
s = 0;
for i = 1:n
  product = y(i);
  for j = 1:n
    if i ~= j
      product = product*(xx-x(j))/(x(i)-x(j));
    end
  end
  s = s+product;
end
yint = s;
```

An example of a session using the function would be to predict the density of air at 1 atm pressure at a temperature of 15 °C based on the first four values from Table 15.1. Because four values are passed to the function, a third-order polynomial would be implemented by the `Lagrange` function to give:

```
>> format long
>> T = [-40 0 20 50];
>> d = [1.52 1.29 1.2 1.09];
>> density = Lagrange(T,d,15)

density =
    1.22112847222222
```

15.4   INVERSE INTERPOLATION

As the nomenclature implies, the $f(x)$ and x values in most interpolation contexts are the dependent and independent variables, respectively. As a consequence, the values of the x's are typically uniformly spaced. A simple example is a table of values derived for the function $f(x) = 1/x$:

x	1	2	3	4	5	6	7
$f(x)$	1	0.5	0.3333	0.25	0.2	0.1667	0.1429

Now suppose that you must use the same data, but you are given a value for $f(x)$ and must determine the corresponding value of x. For instance, for the data above, suppose that you were asked to determine the value of x that corresponded to $f(x) = 0.3$. For this case, because the function is available and easy to manipulate, the correct answer can be determined directly as $x = 1/0.3 = 3.3333$.

Such a problem is called *inverse interpolation*. For a more complicated case, you might be tempted to switch the $f(x)$ and x values [i.e., merely plot x versus $f(x)$] and use an approach like Newton or Lagrange interpolation to determine the result. Unfortunately, when you reverse the variables, there is no guarantee that the values along the new abscissa [the $f(x)$'s] will be evenly spaced. In fact, in many cases, the values will be "telescoped." That is, they will have the appearance of a logarithmic scale with some adjacent points bunched together and others spread out widely. For example, for $f(x) = 1/x$ the result is

$f(x)$	0.1429	0.1667	0.2	0.25	0.3333	0.5	1
x	7	6	5	4	3	2	1

Such nonuniform spacing on the abscissa often leads to oscillations in the resulting interpolating polynomial. This can occur even for lower-order polynomials. An alternative strategy is to fit an nth-order interpolating polynomial, $f_n(x)$, to the original data [i.e., with $f(x)$ versus x]. In most cases, because the x's are evenly spaced, this polynomial will not be ill-conditioned. The answer to your problem then amounts to finding the value of x that makes this polynomial equal to the given $f(x)$. Thus, the interpolation problem reduces to a roots problem!

For example, for the problem just outlined, a simple approach would be to fit a quadratic polynomial to the three points: (2, 0.5), (3, 0.3333), and (4, 0.25). The result would be

$$f_2(x) = 0.041667x^2 - 0.375x + 1.08333$$

The answer to the inverse interpolation problem of finding the x corresponding to $f(x) = 0.3$ would therefore involve determining the root of

$$0.3 = 0.041667x^2 - 0.375x + 1.08333$$

For this simple case, the quadratic formula can be used to calculate

$$x = \frac{0.375 \pm \sqrt{(-0.375)^2 - 4(0.041667)0.78333}}{2(0.041667)} = \frac{5.704158}{3.295842}$$

Thus, the second root, 3.296, is a good approximation of the true value of 3.333. If additional accuracy were desired, a third- or fourth-order polynomial along with one of the root-location methods from Chaps. 5 or 6 could be employed.

15.5 EXTRAPOLATION AND OSCILLATIONS

Before leaving this chapter, there are two issues related to polynomial interpolation that must be addressed. These are extrapolation and oscillations.

15.5.1 Extrapolation

Extrapolation is the process of estimating a value of $f(x)$ that lies outside the range of the known base points, $x_1, x_2, \ldots, x_n$. As depicted in Fig. 15.10, the open-ended nature of

---

FIGURE 15.10
Illustration of the possible divergence of an extrapolated prediction. The extrapolation is based on fitting a parabola through the first three known points.

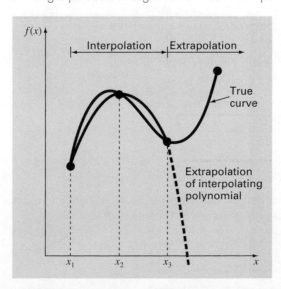

extrapolation represents a step into the unknown because the process extends the curve beyond the known region. As such, the true curve could easily diverge from the prediction. Extreme care should, therefore, be exercised whenever a case arises where one must extrapolate.

EXAMPLE 15.6   Dangers of Extrapolation

Problem Statement.   This example is patterned after one originally developed by Forsythe, Malcolm, and Moler.[1] The population in millions of the United States from 1920 to 2000 can be tabulated as

Date	1920	1930	1940	1950	1960	1970	1980	1990	2000
Population	106.46	123.08	132.12	152.27	180.67	205.05	227.23	249.46	281.42

Fit a seventh-order polynomial to the first 8 points (1920 to 1990). Use it to compute the population in 2000 by extrapolation and compare your prediction with the actual result.

Solution.   First, the data can be entered as

```
>> t = [1920:10:1990];
>> pop = [106.46 123.08 132.12 152.27 180.67 205.05 227.23
        249.46];
```

The `polyfit` function can be used to compute the coefficients

```
>> p = polyfit(t,pop,7)
```

However, when this is implemented, the following message is displayed:

```
Warning: Polynomial is badly conditioned. Remove repeated data
        points or try centering and scaling as described in HELP
        POLYFIT.
```

We can follow MATLAB's suggestion by scaling and centering the data values as in

```
>> ts = (t - 1955)/35;
```

Now `polyfit` works without an error message:

```
>> p = polyfit(ts,pop,7);
```

We can then use the polynomial coefficients along with the `polyval` function to predict the population in 2000 as

```
>> polyval(p,(2000-1955)/35)

ans =
  175.0800
```

which is much lower that the true value of 281.42. Insight into the problem can be gained by generating a plot of the data and the polynomial,

```
>> tt = linspace(1920,2000);
>> pp = polyval(p,(tt-1955)/35);
>> plot(t,pop,'o',tt,pp)
```

[1] Cleve Moler is one of the founders of The MathWorks, Inc., the makers of MATLAB.

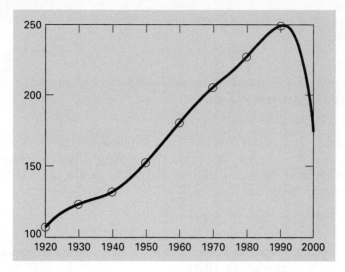

FIGURE 15.11
Use of a seventh-order polynomial to make a prediction of U.S. population in 2000 based on data from 1920 through 1990.

As in Fig. 15.11, the result indicates that the polynomial seems to fit the data nicely from 1920 to 1990. However, once we move beyond the range of the data into the realm of extrapolation, the seventh-order polynomial plunges to the erroneous prediction in 2000.

15.5.2 Oscillations

Although "more is better" in many contexts, it is absolutely not true for polynomial interpolation. Higher-order polynomials tend to be very ill-conditioned—that is, they tend to be highly sensitive to round-off error. The following example illustrates this point nicely.

EXAMPLE 15.7   Dangers of Higher-Order Polynomial Interpolation

Problem Statement.   In 1901, Carl Runge published a study on the dangers of higher-order polynomial interpolation. He looked at the following simple-looking function:

$$f(x) = \frac{1}{1 + 25x^2} \tag{15.24}$$

which is now called *Runge's function*. He took equidistantly spaced data points from this function over the interval [–1, 1]. He then used interpolating polynomials of increasing order and found that as he took more points, the polynomials and the original curve differed considerably. Further, the situation deteriorated greatly as the order was increased. Duplicate Runge's result by using the `polyfit` and `polyval` functions to fit fourth- and tenth-order polynomials to 5 and 11 equally spaced points generated with Eq. (15.24). Create plots of your results along with the sampled values and the complete Runge's function.

Solution. The five equally spaced data points can be generated as in

```
>> x = linspace(-1,1,5);
>> y = 1./(1+25*x.^2);
```

Next, a more finally spaced vector of xx values can be computed so that we can create a smooth plot of the results:

```
>> xx = linspace(-1,1);
```

Recall that linspace automatically creates 100 points if the desired number of points is not specified. The polyfit function can be used to generate the coefficients of the fourth-order polynomial, and the polval function can be used to generate the polynomial interpolation at the finely spaced values of xx:

```
>> p = polyfit(x,y,4);
>> y4 = polyval(p,xx);
```

Finally, we can generate values for Runge's function itself and plot them along with the polynomial fit and the sampled data:

```
>> yr = 1./(1+25*xx.^2);
>> plot(x,y,'o',xx,y4,xx,yr,'--')
```

As in Fig. 15.12, the polynomial does a poor job of following Runge's function.

Continuing with the analysis, the tenth-order polynomial can be generated and plotted with

```
>> x = linspace(-1,1,11);
>> y = 1./(1+25*x.^2);
```

FIGURE 15.12
Comparison of Runge's function (dashed line) with a fourth-order polynomial fit to 5 points sampled from the function.

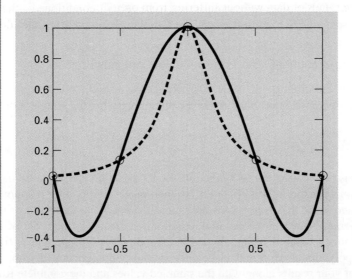

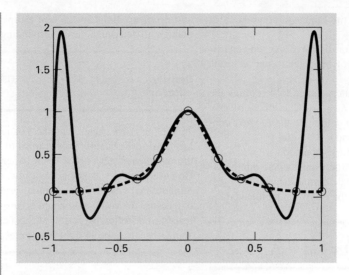

FIGURE 15.13
Comparison of Runge's function (dashed line) with a tenth-order polynomial fit to 11 points sampled from the function.

```
>> p = polyfit(x,y,10);
>> y10 = polyval(p,xx);
>> plot(x,y,'o',xx,y10,xx,yr,'--')
```

As in Fig. 15.13, the fit has gotten even worse, particularly at the ends of the interval!

Although there may be certain contexts where higher-order polynomials are necessary, they are usually to be avoided. In most engineering and scientific contexts, lower-order polynomials of the type described in this chapter can be used effectively to capture the curving trends of data without suffering from oscillations.

PROBLEMS

15.1 Given the data

x	1	2	2.5	3	4	5
f(x)	0	5	7	6.5	2	0

(a) Calculate $f(3.4)$ using Newton's interpolating polynomials of order 1 through 3. Choose the sequence of the points for your estimates to attain the best possible accuracy.
(b) Repeat (a) but use the Lagrange polynomial.

15.2 Given the data

x	1	2	3	5	6
f(x)	4.75	4	5.25	19.75	36

Calculate $f(4)$ using Newton's interpolating polynomials of order 1 through 4. Choose your base points to attain good accuracy. What do your results indicate regarding the order of the polynomial used to generate the data in the table?

15.3 Repeat Prob. 15.2 using the Lagrange polynomial of order 1 through 3.

15.4 Table P14.5 lists values for dissolved oxygen concentration in water as a function of temperature and chloride concentration.

(a) Use quadratic and cubic interpolation to determine the oxygen concentration for $T = 12\ °C$ and $c = 10$ g/L.

(b) Use linear interpolation to determine the oxygen concentration for $T = 12\ °C$ and $c = 15$ g/L.

(c) Repeat (b) but use quadratic interpolation.

15.5 Employ inverse interpolation using a cubic interpolating polynomial and bisection to determine the value of x that corresponds to $f(x) = 1.6$ for the following tabulated data:

x	1	2	3	4	5	6	7
$f(x)$	3.6	1.8	1.2	0.9	0.72	1.5	0.51429

15.6 Employ inverse interpolation to determine the value of x that corresponds to $f(x) = 0.93$ for the following tabulated data:

x	0	1	2	3	4	5
$f(x)$	0	0.5	0.8	0.9	0.941176	0.961538

Note that the values in the table were generated with the function $f(x) = x^2/(1 + x^2)$.

(a) Determine the correct value analytically.

(b) Use quadratic interpolation and the quadratic formula to determine the value numerically.

(c) Use cubic interpolation and bisection to determine the value numerically.

15.7 Use the portion of the given steam table for superheated water at 200 MPa to find (a) the corresponding entropy s for a specific volume v of 0.118 with linear interpolation, (b) the same corresponding entropy using quadratic interpolation, and (c) the volume corresponding to an entropy of 6.45 using inverse interpolation.

v, m^3/kg	0.10377	0.11144	0.12547
s, kJ/(kg K)	6.4147	6.5453	6.7664

15.8 The following data for the density of nitrogen gas versus temperature comes from a table that was measured with high precision. Use first- through fifth-order polynomials to estimate the density at a temperature of 330 K. What is your

best estimate? Employ this best estimate and inverse interpolation to determine the corresponding temperature.

T, K	200	250	300	350	400	450
Density, kg/m^3	1.708	1.367	1.139	0.967	0.854	0.759

15.9 Ohm's law states that the voltage drop V across an ideal resistor is linearly proportional to the current i flowing through the resister as in $V = iR$, where R is the resistance. However, real resistors may not always obey Ohm's law. Suppose that you performed some very precise experiments to measure the voltage drop and corresponding current for a resistor. The following results suggest a curvilinear relationship rather than the straight line represented by Ohm's law:

i	−2	−1	−0.5	0.5	1	2
V	−637	−96.5	−20.5	20.5	96.5	637

To quantify this relationship, a curve must be fit to the data. Because of measurement error, regression would typically be the preferred method of curve fitting for analyzing such experimental data. However, the smoothness of the relationship, as well as the precision of the experimental methods, suggests that interpolation might be appropriate. Use a fifth-order interpolating polynomial to fit the data and compute V for $i = 0.10$.

15.10 Bessel functions often arise in advanced engineering analyses such as the study of electric fields. Here are some selected values for the zero-order Bessel function of the first kind

x	1.8	2.0	2.2	2.4	2.6
$J_1(x)$	0.5815	0.5767	0.5560	0.5202	0.4708

Estimate $J_1(2.1)$ using third-, fourth-, and fifth-order interpolating polynomials. Determine the percent relative error for each case based on the true value, which can be determined with MATLAB's built-in function besselj.

15.11 Repeat Example 15.6 but using first-, second-, third-, and fourth-order interpolating polynomials to predict the population in 2000 based on the most recent data. That is, for the linear prediction use the data from 1980 and 1990, for the quadratic prediction use the data from 1970, 1980, and 1990, and so on. Which approach yields the best result?

15.12 The saturation concentration of dissolved oxygen in water as a function of temperature and chloride concentration is listed in Table P15.12. Use interpolation to estimate

TABLE P15.12 Dissolved oxygen concentration in water as a function of temperature (°C) and chloride concentration (g/L).

T, °C	Dissolved Oxygen (mg/L) for Temperature (°C) and Concentration of Chloride (g/L)		
	c = 0 g/L	c = 10 g/L	c = 20 g/L
0	14.6	12.9	11.4
5	12.8	11.3	10.3
10	11.3	10.1	8.96
15	10.1	9.03	8.08
20	9.09	8.17	7.35
25	8.26	7.46	6.73
30	7.56	6.85	6.20

the dissolved oxygen level for $T = 18$ °C with chloride $= 10$ g/L.

15.13 For the data in Table P15.12, use polynomial interpolation to derive a third-order predictive equation for dissolved oxygen concentration as a function of temperature for the case where chloride concentration is equal to 10 g/L. Use the equation to estimate the dissolved oxygen concentration for $T = 8$ °C.

15.14 The specific volume of a superheated steam is listed in steam tables for various temperatures. For example, at a pressure of 3000 lb/in², absolute:

T, °F	700	720	740	760	780
v, ft³/lbₘ	0.0977	0.12184	0.14060	0.15509	0.16643

Determine v at $T = 750$ °F.

15.15 The vertical stress σ_z under the corner of a rectangular area subjected to a uniform load of intensity q is given by the solution of Boussinesq's equation:

$$\sigma = \frac{q}{4\pi}\left[\frac{2mn\sqrt{m^2+n^2+1}}{m^2+n^2+1+m^2n^2}\frac{m^2+n^2+2}{m^2+n^2+1}\right.$$

$$\left. + \sin^{-1}\left(\frac{2mn\sqrt{m^2+n^2+1}}{m^2+n^2+1+m^2n^2}\right)\right]$$

Because this equation is inconvenient to solve manually, it has been reformulated as

$$\sigma_z = qf_z(m, n)$$

where $f_z(m, n)$ is called the influence value, and m and n are dimensionless ratios, with $m = a/z$ and $n = b/z$ and a and b are defined in Fig. P15.15. The influence value is then

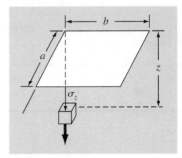

FIGURE P15.15

tabulated, a portion of which is given in Table P15.15. If $a = 4.6$ and $b = 14$, use a third-order interpolating polynomial to compute σ_z at a depth 10 m below the corner of a rectangular footing that is subject to a total load of 100 t (metric tons). Express your answer in tonnes per square meter. Note that q is equal to the load per area.

TABLE P15.15

m	n = 1.2	n = 1.4	n = 1.6
0.1	0.02926	0.03007	0.03058
0.2	0.05733	0.05894	0.05994
0.3	0.08323	0.08561	0.08709
0.4	0.10631	0.10941	0.11135
0.5	0.12626	0.13003	0.13241
0.6	0.14309	0.14749	0.15027
0.7	0.15703	0.16199	0.16515
0.8	0.16843	0.17389	0.17739

15.16 You measure the voltage drop V across a resistor for a number of different values of current i. The results are

i	0.25	0.75	1.25	1.5	2.0
V	−0.45	−0.6	0.70	1.88	6.0

Use first- through fourth-order polynomial interpolation to estimate the voltage drop for $i = 1.15$. Interpret your results.

15.17 The current in a wire is measured with great precision as a function of time:

t	0	0.1250	0.2500	0.3750	0.5000
i	0	6.24	7.75	4.85	0.0000

Determine i at $t = 0.23$.

15.18 The acceleration due to gravity at an altitude y above the surface of the earth is given by

y, m	0	30,000	60,000	90,000	120,000
g, m/s^2	9.8100	9.7487	9.6879	9.6278	9.5682

Compute g at $y = 55,000$ m.

TABLE P15.19 Temperatures (°C) at various points on a square heated plate.

	$x = 0$	$x = 2$	$x = 4$	$x = 6$	$x = 8$
$y = 0$	100.00	90.00	80.00	70.00	60.00
$y = 2$	85.00	64.49	53.50	48.15	50.00
$y = 4$	70.00	48.90	38.43	35.03	40.00
$y = 6$	55.00	38.78	30.39	27.07	30.00
$y = 8$	40.00	35.00	30.00	25.00	20.00

15.19 Temperatures are measured at various points on a heated plate (Table P15.19). Estimate the temperature at **(a)** $x = 4$, $y = 3.2$, and **(b)** $x = 4.3$, $y = 2.7$.

15.20 Use the portion of the given steam table for superheated H_2O at 200 MPa to **(a)** find the corresponding entropy s for a specific volume v of 0.108 m^3/kg with linear interpolation, **(b)** find the same corresponding entropy using quadratic interpolation, and **(c)** find the volume corresponding to an entropy of 6.6 using inverse interpolation.

v (m^3/kg)	0.10377	0.11144	0.12540
s (kJ/kg·K)	6.4147	6.5453	6.7664

16

Splines and Piecewise Interpolation

CHAPTER OBJECTIVES

The primary objective of this chapter is to introduce you to splines. Specific objectives and topics covered are

- Understanding that splines minimize oscillations by fitting lower-order polynomials to data in a piecewise fashion.
- Knowing how to develop code to perform a table lookup.
- Recognizing why cubic polynomials are preferable to quadratic and higher-order splines.
- Understanding the conditions that underlie a cubic spline fit.
- Understanding the differences between natural, clamped, and not-a-knot end conditions.
- Knowing how to fit a spline to data with MATLAB's built-in functions.
- Understanding how multidimensional interpolation is implemented with MATLAB.

16.1 INTRODUCTION TO SPLINES

In Chap. 15 $(n-1)$th-order polynomials were used to interpolate between n data points. For example, for eight points, we can derive a perfect seventh-order polynomial. This curve would capture all the meanderings (at least up to and including seventh derivatives) suggested by the points. However, there are cases where these functions can lead to erroneous results because of round-off error and oscillations. An alternative approach is to apply lower-order polynomials in a piecewise fashion to subsets of data points. Such connecting polynomials are called *spline functions*.

For example, third-order curves employed to connect each pair of data points are called *cubic splines*. These functions can be constructed so that the connections between

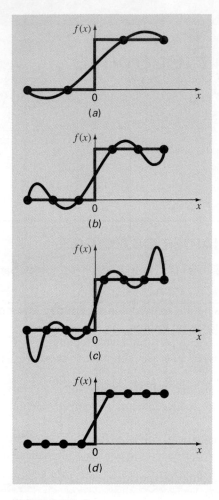

FIGURE 16.1
A visual representation of a situation where splines are superior to higher-order interpolating polynomials. The function to be fit undergoes an abrupt increase at $x = 0$. Parts (a) through (c) indicate that the abrupt change induces oscillations in interpolating polynomials. In contrast, because it is limited to straight-line connections, a linear spline (d) provides a much more acceptable approximation.

adjacent cubic equations are visually smooth. On the surface, it would seem that the third-order approximation of the splines would be inferior to the seventh-order expression. You might wonder why a spline would ever be preferable.

Figure 16.1 illustrates a situation where a spline performs better than a higher-order polynomial. This is the case where a function is generally smooth but undergoes an abrupt change somewhere along the region of interest. The step increase depicted in Fig. 16.1 is an extreme example of such a change and serves to illustrate the point.

Figure 16.1a through c illustrates how higher-order polynomials tend to swing through wild oscillations in the vicinity of an abrupt change. In contrast, the spline also connects the points, but because it is limited to lower-order changes, the oscillations are kept to a

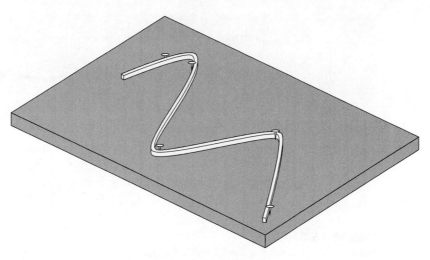

FIGURE 16.2
The drafting technique of using a spline to draw smooth curves through a series of points. Notice how, at the end points, the spline straightens out. This is called a "natural" spline.

minimum. As such, the spline usually provides a superior approximation of the behavior of functions that have local, abrupt changes.

The concept of the spline originated from the drafting technique of using a thin, flexible strip (called a *spline*) to draw smooth curves through a set of points. The process is depicted in Fig. 16.2 for a series of five pins (data points). In this technique, the drafter places paper over a wooden board and hammers nails or pins into the paper (and board) at the location of the data points. A smooth cubic curve results from interweaving the strip between the pins. Hence, the name "cubic spline" has been adopted for polynomials of this type.

In this chapter, simple linear functions will first be used to introduce some basic concepts and issues associated with spline interpolation. Then we derive an algorithm for fitting quadratic splines to data. This is followed by material on the cubic spline, which is the most common and useful version in engineering and science. Finally, we describe MATLAB's capabilities for piecewise interpolation including its ability to generate splines.

16.2 LINEAR SPLINES

The notation used for splines is displayed in Fig. 16.3. For n data points ($i = 1, 2, \ldots, n$), there are $n - 1$ intervals. Each interval i has its own spline function, $s_i(x)$. For linear splines, each function is merely the straight line connecting the two points at each end of the interval, which is formulated as

$$s_i(x) = a_i + b_i(x - x_i) \tag{16.1}$$

where a_i is the intercept, which is defined as

$$a_i = f_i \tag{16.2}$$

and b_i is the slope of the straight line connecting the points:

$$b_i = \frac{f_{i+1} - f_i}{x_{i+1} - x_i} \tag{16.3}$$

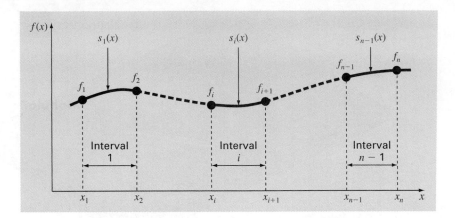

FIGURE 16.3
Notation used to derive splines. Notice that there are $n - 1$ intervals and n data points.

where f_i is shorthand for $f(x_i)$. Substituting Eqs. (16.1) and (16.2) into Eq. (16.3) gives

$$s_i(x) = f_i + \frac{f_{i+1} - f_i}{x_{i+1} - x_i}(x - x_i) \qquad (16.4)$$

These equations can be used to evaluate the function at any point between x_1 and x_n by first locating the interval within which the point lies. Then the appropriate equation is used to determine the function value within the interval. Inspection of Eq. (16.4) indicates that the linear spline amounts to using Newton's first-order polynomial [Eq. (15.5)] to interpolate within each interval.

EXAMPLE 16.1   First-Order Splines

Problem Statement.   Fit the data in Table 16.1 with first-order splines. Evaluate the function at $x = 5$.

TABLE 16.1 Data to be fit with spline functions.

i	x_i	f_i
1	3.0	2.5
2	4.5	1.0
3	7.0	2.5
4	9.0	0.5

Solution.   The data can be substituted into Eq. (16.4) to generate the linear spline functions. For example, for the second interval from $x = 4.5$ to $x = 7$, the function is

$$s_2(x) = 1.0 + \frac{2.5 - 1.0}{7.0 - 4.5}(x - 4.5)$$

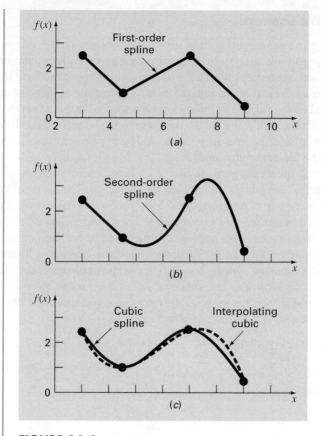

FIGURE 16.4
Spline fits of a set of four points. (a) Linear spline, (b) quadratic spline, and (c) cubic spline, with a cubic interpolating polynomial also plotted.

The equations for the other intervals can be computed, and the resulting first-order splines are plotted in Fig. 16.4a. The value at $x = 5$ is 1.3.

$$s_2(x) = 1.0 + \frac{2.5 - 1.0}{7.0 - 4.5}(5 - 4.5) = 1.3$$

Visual inspection of Fig. 16.4a indicates that the primary disadvantage of first-order splines is that they are not smooth. In essence, at the data points where two splines meet (called a *knot*), the slope changes abruptly. In formal terms, the first derivative of the function is discontinuous at these points. This deficiency is overcome by using higher-order polynomial splines that ensure smoothness at the knots by equating derivatives at these points, as will be discussed subsequently. Before doing that, the following section provides an application where linear splines are useful.

16.2.1 Table Lookup

A table lookup is a common task that is frequently encountered in engineering and science computer applications. It is useful for performing repeated interpolations from a table of independent and dependent variables. For example, suppose that you would like to set up an M-file that would use linear interpolation to determine air density at a particular temperature based on the data from Table 15.1. One way to do this would be to pass the M-file the temperature at which you want the interpolation to be performed along with the two adjoining values. A more general approach would be to pass in vectors containing all the data and have the M-file determine the bracket. This is called a *table lookup*.

Thus, the M-file would perform two tasks. First, it would search the independent variable vector to find the interval containing the unknown. Then it would perform the linear interpolation using one of the techniques described in this chapter or in Chap. 15.

For ordered data, there are two simple ways to find the interval. The first is called a *sequential search*. As the name implies, this method involves comparing the desired value with each element of the vector in sequence until the interval is located. For data in ascending order, this can be done by testing whether the unknown is less than the value being assessed. If so, we know that the unknown falls between this value and the previous one that we examined. If not, we move to the next value and repeat the comparison. Here is a simple M-file that accomplishes this objective:

```
function yi = TableLook(x, y, xx)

n = length(x);
if xx < x(1) | xx > x(n)
  error('Interpolation outside range')
end
% sequential search
i = 1;
while(1)
   if xx <= x(i + 1), break, end
   i = i + 1;
end
% linear interpolation
yi = y(i) + (y(i+1)-y(i))/(x(i+1)-x(i))*(xx-x(i));
```

The table's independent variables are stored in ascending order in the array x and the dependent variables stored in the array y. Before searching, an error trap is included to ensure that the desired value xx falls within the range of the x's. A while . . . break loop compares the value at which the interpolation is desired, xx, to determine whether it is less than the value at the top of the interval, $x(i+1)$. For cases where xx is in the second interval or higher, this will not test true at first. In this case the counter i is incremented by one so that on the next iteration, xx is compared with the value at the top of the second interval. The loop is repeated until the xx is less than or equal to the interval's upper bound, in which case the loop is exited. At this point, the interpolation can be performed simply as shown.

For situations for which there are lots of data, the sequential sort is inefficient because it must search through all the preceding points to find values. In these cases, a simple alternative is the *binary search*. Here is an M-file that performs a binary search followed by linear interpolation:

```
function yi = TableLookBin(x, y, xx)

n = length(x);
if xx < x(1) | xx > x(n)
  error('Interpolation outside range')
end
% binary search
iL = 1; iU = n;
while (1)
  if iU - iL <= 1, break, end
  iM = fix((iL + iU) / 2);
  if x(iM) < xx
    iL = iM;
  else
    iU = iM;
  end
end
% linear interpolation
yi = y(iL) + (y(iL+1)-y(iL))/(x(iL+1)-x(iL))*(xx - x(iL));
```

The approach is akin to the bisection method for root location. Just as in bisection, the index at the midpoint `iM` is computed as the average of the first or "lower" index `iL = 1` and the last or "upper" index `iU = n`. The unknown `xx` is then compared with the value of `x` at the midpoint `x(iM)` to assess whether it is in the lower half of the array or in the upper half. Depending on where it lies, either the lower or upper index is redefined as being the middle index. The process is repeated until the difference between the upper and the lower index is less than or equal to zero. At this point, the lower index lies at the lower bound of the interval containing `xx`, the loop terminates, and the linear interpolation is performed.

Here is a MATLAB session illustrating how the binary search function can be applied to calculate the air density at 350 °C based on the data from Table 15.1. The sequential search would be similar.

```
>> T = [-40 0 20 50 100 150 200 250 300 400 500];
>> density = [1.52 1.29 1.2 1.09 .946 .935 .746 .675 .616
              .525 .457];
>> TableLookBin(T,density,350)

ans =
   0.5705
```

This result can be verified by the hand calculation:

$$f(350) = 0.616 + \frac{0.525 - 0.616}{400 - 300}(350 - 300) = 0.5705$$

16.3 QUADRATIC SPLINES

To ensure that the nth derivatives are continuous at the knots, a spline of at least $n + 1$ order must be used. Third-order polynomials or cubic splines that ensure continuous first and second derivatives are most frequently used in practice. Although third and higher derivatives can be discontinuous when using cubic splines, they usually cannot be detected visually and consequently are ignored.

Because the derivation of cubic splines is somewhat involved, we have decided to first illustrate the concept of spline interpolation using second-order polynomials. These "quadratic splines" have continuous first derivatives at the knots. Although quadratic splines are not of practical importance, they serve nicely to demonstrate the general approach for developing higher-order splines.

The objective in quadratic splines is to derive a second-order polynomial for each interval between data points. The polynomial for each interval can be represented generally as

$$s_i(x) = a_i + b_i(x - x_i) + c_i(x - x_i)^2 \tag{16.5}$$

where the notation is as in Fig. 16.3. For n data points $(i = 1, 2, \ldots, n)$, there are $n - 1$ intervals and, consequently, $3(n - 1)$ unknown constants (the a's, b's, and c's) to evaluate. Therefore, $3n$ equations or conditions are required to evaluate the unknowns. These can be developed as follows:

1. The function must pass through all the points. This is called a *continuity condition*. It can be expressed mathematically as

$$f_i = a_i + b_i(x_i - x_i) + c_i(x_i - x_i)^2$$

which simplifies to

$$a_i = f_i \tag{16.6}$$

Therefore, the constant in each quadratic must be equal to the value of the dependent variable at the beginning of the interval. This result can be incorporated into Eq. (16.5):

$$s_i(x) = f_i + b_i(x - x_i) + c_i(x - x_i)^2$$

Note that because we have determined one of the coefficients, the number of conditions to be evaluated has now been reduced to $2(n - 1)$.

2. The function values of adjacent polynomials must be equal at the knots. This condition can be written for knot $i + 1$ as

$$f_i + b_i(x_{i+1} - x_i) + c_i(x_{i+1} - x_i)^2 = f_{i+1} + b_{i+1}(x_{i+1} - x_{i+1}) + c_{i+1}(x_{i+1} - x_{i+1})^2 \tag{16.7}$$

This equation can be simplified mathematically by defining the width of the ith interval as

$$h_i = x_{i+1} - x_i$$

Thus, Eq. (16.7) simplifies to

$$f_i + b_i h_i + c_i h_i^2 = f_{i+1} \tag{16.8}$$

This equation can be written for the nodes, $i = 1, \ldots, n - 1$. Since this amounts to $n - 1$ conditions, it means that there are $2(n - 1) - (n - 1) = n - 1$ remaining conditions.

3. The first derivatives at the interior nodes must be equal. This is an important condition, because it means that adjacent splines will be joined smoothly, rather than in the jagged fashion that we saw for the linear splines. Equation (16.5) can be differentiated to yield

$$s_i'(x) = b_i + 2c_i(x - x_i)$$

The equivalence of the derivatives at an interior node, $i + 1$ can therefore be written as

$$b_i + 2c_i h_i = b_{i+1} \tag{16.9}$$

Writing this equation for all the interior nodes amounts to $n - 2$ conditions. This means that there is $n - 1 - (n - 2) = 1$ remaining condition. Unless we have some additional information regarding the functions or their derivatives, we must make an arbitrary choice to successfully compute the constants. Although there are a number of different choices that can be made, we select the following condition.

4. Assume that the second derivative is zero at the first point. Because the second derivative of Eq. (16.5) is $2c_i$, this condition can be expressed mathematically as

$$c_1 = 0$$

The visual interpretation of this condition is that the first two points will be connected by a straight line.

EXAMPLE 16.2 Quadratic Splines

Problem Statement. Fit quadratic splines to the same data employed in Example 16.1 (Table 16.1). Use the results to estimate the value at $x = 5$.

Solution. For the present problem, we have four data points and $n = 3$ intervals. Therefore, after applying the continuity condition and the zero second-derivative condition, this means that $2(4 - 1) - 1 = 5$ conditions are required. Equation (16.8) is written for $i = 1$ through 3 (with $c_1 = 0$) to give

$$f_1 + b_1 h_1 = f_2$$
$$f_2 + b_2 h_2 + c_2 h_2^2 = f_3$$
$$f_3 + b_3 h_3 + c_3 h_3^2 = f_4$$

Continuity of derivatives, Eq. (16.9), creates an additional $3 - 1 = 2$ conditions (again, recall that $c_1 = 0$):

$$b_1 = b_2$$
$$b_2 + 2c_2 h_2 = b_3$$

The necessary function and interval width values are

$$f_1 = 2.5 \qquad h_1 = 4.5 - 3.0 = 1.5$$
$$f_2 = 1.0 \qquad h_2 = 7.0 - 4.5 = 2.5$$
$$f_3 = 2.5 \qquad h_3 = 9.0 - 7.0 = 2.0$$
$$f_4 = 0.5$$

These values can be substituted into the conditions which can be expressed in matrix form as

$$\begin{bmatrix} 1.5 & 0 & 0 & 0 & 0 \\ 0 & 2.5 & 6.25 & 0 & 0 \\ 0 & 0 & 0 & 2 & 4 \\ 1 & -1 & 0 & 0 & 0 \\ 0 & 1 & 5 & -1 & 0 \end{bmatrix} \begin{Bmatrix} b_1 \\ b_2 \\ c_2 \\ b_3 \\ c_3 \end{Bmatrix} = \begin{Bmatrix} -1.5 \\ 1.5 \\ -2 \\ 0 \\ 0 \end{Bmatrix}$$

These equations can be solved using MATLAB with the results:

$$b_1 = -1$$
$$b_2 = -1 \qquad\qquad c_2 = 0.64$$
$$b_3 = 2.2 \qquad\qquad c_3 = -1.6$$

These results, along with the values for the a's (Eq. 16.6), can be substituted into the original quadratic equations to develop the following quadratic splines for each interval:

$$s_1(x) = 2.5 - (x - 3)$$
$$s_2(x) = 1.0 - (x - 4.5) + 0.64(x - 4.5)^2$$
$$s_3(x) = 2.5 + 2.2(x - 7.0) - 1.6(x - 7.0)^2$$

Because $x = 5$ lies in the second interval, we use s_2 to make the prediction,

$$s_2(5) = 1.0 - (5 - 4.5) + 0.64(5 - 4.5)^2 = 0.66$$

The total quadratic spline fit is depicted in Fig. 16.4b. Notice that there are two shortcomings that detract from the fit: (1) the straight line connecting the first two points and (2) the spline for the last interval seems to swing too high. The cubic splines in the next section do not exhibit these shortcomings and, as a consequence, are better methods for spline interpolation.

16.4 CUBIC SPLINES

As stated at the beginning of the previous section, cubic splines are most frequently used in practice. The shortcomings of linear and quadratic splines have already been discussed. Quartic or higher-order splines are not used because they tend to exhibit the instabilities inherent in higher-order polynomials. Cubic splines are preferred because they provide the simplest representation that exhibits the desired appearance of smoothness.

The objective in cubic splines is to derive a third-order polynomial for each interval between knots as represented generally by

$$s_i(x) = a_i + b_i(x - x_i) + c_i(x - x_i)^2 + d_i(x - x_i)^3 \qquad (16.10)$$

Thus, for n data points ($i = 1, 2, \ldots, n$), there are $n - 1$ intervals and $4(n - 1)$ unknown coefficients to evaluate. Consequently, $4(n - 1)$ conditions are required for their evaluation.

The first conditions are identical to those used for the quadratic case. That is, they are set up so that the functions pass through the points and that the first derivatives at the knots are equal. In addition to these, conditions are developed to ensure that the second derivatives at the knots are also equal. This greatly enhances the fit's smoothness.

After these conditions are developed, two additional conditions are required to obtain the solution. This is a much nicer outcome than occurred for quadratic splines where we needed to specify a single condition. In that case, we had to arbitrarily specify a zero second derivative for the first interval, hence making the result asymmetric. For cubic splines, we are in the advantageous position of needing two additional conditions and can, therefore, apply them evenhandedly at both ends.

For cubic splines, these last two conditions can be formulated in several different ways. A very common approach is to assume that the second derivatives at the first and last knots are equal to zero. The visual interpretation of these conditions is that the function becomes a straight line at the end nodes. Specification of such an end condition leads to what is termed a "natural" spline. It is given this name because the drafting spline naturally behaves in this fashion (Fig. 16.2).

There are a variety of other end conditions that can be specified. Two of the more popular are the clamped condition and the not-a-knot conditions. We will describe these options in Section 16.4.2. For the following derivation, we will limit ourselves to natural splines.

Once the additional end conditions are specified, we would have the $4(n-1)$ conditions needed to evaluate the $4(n-1)$ unknown coefficients. Whereas it is certainly possible to develop cubic splines in this fashion, we will present an alternative approach that requires the solution of only $n-1$ equations. Further, the simultaneous equations will be tridiagonal and hence can be solved very efficiently. Although the derivation of this approach is less straightforward than for quadratic splines, the gain in efficiency is well worth the effort.

16.4.1 Derivation of Cubic Splines

As was the case with quadratic splines, the first condition is that the spline must pass through all the data points.

$$f_i = a_i + b_i(x_i - x_i) + c_i(x_i - x_i)^2 + d_i(x_i - x_i)^3$$

which simplifies to

$$a_i = f_i \tag{16.11}$$

Therefore, the constant in each cubic must be equal to the value of the dependent variable at the beginning of the interval. This result can be incorporated into Eq. (16.10):

$$s_i(x) = f_i + b_i(x - x_i) + c_i(x - x_i)^2 + d_i(x - x_i)^3 \tag{16.12}$$

Next, we will apply the condition that each of the cubics must join at the knots. For knot $i + 1$, this can be represented as

$$f_i + b_i h_i + c_i h_i^2 + d_i h_i^3 = f_{i+1} \tag{16.13}$$

where

$$h_i = x_{i+1} - x_i$$

The first derivatives at the interior nodes must be equal. Equation (16.12) is differentiated to yield

$$s_i'(x) = b_i + 2c_i(x - x_i) + 3d_i(x - x_i)^2 \qquad (16.14)$$

The equivalence of the derivatives at an interior node, $i + 1$ can therefore be written as

$$b_i + 2c_i h_i + 3d_i h_i^2 = b_{i+1} \qquad (16.15)$$

The second derivatives at the interior nodes must also be equal. Equation (16.14) can be differentiated to yield

$$s_i''(x) = 2c_i + 6d_i(x - x_i) \qquad (16.16)$$

The equivalence of the second derivatives at an interior node, $i + 1$ can therefore be written as

$$c_i + 3d_i h_i = c_{i+1} \qquad (16.17)$$

Next, we can solve Eq. (16.17) for d_i:

$$d_i = \frac{c_{i+1} - c_i}{3h_i} \qquad (16.18)$$

This can be substituted into Eq. (16.13) to give

$$f_i + b_i h_i + \frac{h_i^2}{3}(2c_i + c_{i+1}) = f_{i+1} \qquad (16.19)$$

Equation (16.18) can also be substituted into Eq. (16.15) to give

$$b_{i+1} = b_i + h_i(c_i + c_{i+1}) \qquad (16.20)$$

Equation (16.19) can be solved for

$$b_i = \frac{f_{i+1} - f_i}{h_i} - \frac{h_i}{3}(2c_i + c_{i+1}) \qquad (16.21)$$

The index of this equation can be reduced by 1:

$$b_{i-1} = \frac{f_i - f_{i-1}}{h_{i-1}} - \frac{h_{i-1}}{3}(2c_{i-1} + c_i) \qquad (16.22)$$

The index of Eq. (16.20) can also be reduced by 1:

$$b_i = b_{i-1} + h_{i-1}(c_{i-1} + c_i) \qquad (16.23)$$

Equations (16.21) and (16.22) can be substituted into Eq. (16.23) and the result simplified to yield

$$h_{i-1}c_{i-1} + 2(h_{i-1} - h_i)c_i + h_i c_{i+1} = 3\frac{f_{i+1} - f_i}{h_i} - 3\frac{f_i - f_{i-1}}{h_{i-1}} \qquad (16.24)$$

This equation can be made a little more concise by recognizing that the terms on the right-hand side are finite differences (recall Eq. 15.15):

$$f[x_i, x_j] = \frac{f_i - f_j}{x_i - x_j}$$

Therefore, Eq. (16.24) can be written as

$$h_{i-1}c_{i-1} + 2(h_{i-1} - h_i)c_i + h_i c_{i+1} = 3\left(f[x_{i+1}, x_i] - f[x_i, x_{i-1}]\right) \tag{16.25}$$

Equation (16.25) can be written for the interior knots, $i = 2, 3, \ldots, n - 2$, which results in $n - 3$ simultaneous tridiagonal equations with $n - 1$ unknown coefficients, $c_1, c_2, \ldots, c_{n-1}$. Therefore, if we have two additional conditions, we can solve for the c's. Once this is done, Eqs. (16.21) and (16.18) can be used to determine the remaining coefficients, b and d.

As stated previously, the two additional end conditions can be formulated in a number of ways. One common approach, the natural spline, assumes that the second derivatives at the end knots are equal to zero. To see how these can be integrated into the solution scheme, the second derivative at the first node (Eq. 16.16) can be set to zero as in

$$s_1''(x_1) = 0 = 2c_1 + 6d_1(x_1 - x_1)$$

Thus, this condition amounts to setting c_1 equal to zero.

The same evaluation can be made at the last node:

$$s_{n-1}''(x_n) = 0 = 2c_{n-1} + 6d_{n-1}h_{n-1} \tag{16.26}$$

Recalling Eq. (16.17), we can conveniently define an extraneous parameter c_n, in which case Eq. (16.26) becomes

$$c_{n-1} + 3d_{n-1}h_{n-1} = c_n = 0$$

Thus, to impose a zero second derivative at the last node, we set $c_n = 0$.

The final equations can now be written in matrix form as

$$\begin{bmatrix} 1 & & & & & \\ h_1 & 2(h_1 + h_2) & h_2 & & & \\ & & \ddots & & & \\ & & & h_{n-2} & 2(h_{n-2} + h_{n-1}) & h_{n-1} \\ & & & & & 1 \end{bmatrix} \begin{Bmatrix} c_1 \\ c_2 \\ \vdots \\ c_{n-1} \\ c_n \end{Bmatrix}$$

$$= \begin{Bmatrix} 0 \\ 3(f[x_3, x_2] - f[x_2, x_1]) \\ \vdots \\ 3(f[x_n, x_{n-1}] - f[x_{n-1}, x_{n-2}]) \\ 0 \end{Bmatrix} \tag{16.27}$$

As shown, the system is tridiagonal and hence efficient to solve.

EXAMPLE 16.3   Natural Cubic Splines

Problem Statement.   Fit cubic splines to the same data used in Examples 16.1 and 16.2 (Table 16.1). Utilize the results to estimate the value at $x = 5$.

Solution.   The first step is to employ Eq. (16.27) to generate the set of simultaneous equations that will be utilized to determine the c coefficients:

$$
\begin{bmatrix}
1 & & & \\
h_1 & 2(h_1 + h_2) & h_2 & \\
& h_2 & 2(h_2 + h_3) & h_3 \\
& & & 1
\end{bmatrix}
\begin{Bmatrix}
c_1 \\ c_2 \\ c_3 \\ c_4
\end{Bmatrix}
=
\begin{Bmatrix}
0 \\
3(f[x_3, x_2] - f[x_2, x_1]) \\
3(f[x_4, x_3] - f[x_3, x_2]) \\
0
\end{Bmatrix}
$$

The necessary function and interval width values are

$$
\begin{aligned}
f_1 &= 2.5 & h_1 &= 4.5 - 3.0 = 1.5 \\
f_2 &= 1.0 & h_2 &= 7.0 - 4.5 = 2.5 \\
f_3 &= 2.5 & h_3 &= 9.0 - 7.0 = 2.0 \\
f_4 &= 0.5
\end{aligned}
$$

These can be substituted to yield

$$
\begin{bmatrix}
1 & & & \\
1.5 & 8 & 2.5 & \\
& 2.5 & 9 & 2 \\
& & & 1
\end{bmatrix}
\begin{Bmatrix}
c_1 \\ c_2 \\ c_3 \\ c_4
\end{Bmatrix}
=
\begin{Bmatrix}
0 \\ 4.8 \\ -4.8 \\ 0
\end{Bmatrix}
$$

These equations can be solved using MATLAB with the results:

$$
\begin{aligned}
c_1 &= 0 & c_2 &= 0.839543726 \\
c_3 &= -0.766539924 & c_4 &= 0
\end{aligned}
$$

Equations (16.21) and (16.18) can be used to compute the b's and d's

$$
\begin{aligned}
b_1 &= -1.419771863 & d_1 &= 0.186565272 \\
b_2 &= -0.160456274 & d_2 &= -0.214144487 \\
b_3 &= 0.022053232 & d_3 &= 0.127756654
\end{aligned}
$$

These results, along with the values for the a's [Eq. (16.11)], can be substituted into Eq. (16.10) to develop the following cubic splines for each interval:

$$
\begin{aligned}
s_1(x) &= 2.5 - 1.419771863(x - 3) + 0.186565272(x - 3)^3 \\
s_2(x) &= 1.0 - 0.160456274(x - 4.5) + 0.839543726(x - 4.5)^2 \\
&\quad - 0.214144487(x - 4.5)^3 \\
s_3(x) &= 2.5 + 0.022053232(x - 7.0) - 0.766539924(x - 7.0)^2 \\
&\quad + 0.127756654(x - 7.0)^3
\end{aligned}
$$

The three equations can then be employed to compute values within each interval. For example, the value at $x = 5$, which falls within the second interval, is calculated as

$$
\begin{aligned}
s_2(5) &= 1.0 - 0.160456274(5 - 4.5) + 0.839543726(5 - 4.5)^2 - 0.214144487(5 - 4.5)^3 \\
&= 1.102889734.
\end{aligned}
$$

The total cubic spline fit is depicted in Fig. 16.4c.

The results of Examples 16.1 through 16.3 are summarized in Fig. 16.4. Notice the progressive improvement of the fit as we move from linear to quadratic to cubic splines. We have also superimposed a cubic interpolating polynomial on Fig. 16.4c. Although the cubic spline consists of a series of third-order curves, the resulting fit differs from that obtained using the third-order polynomial. This is due to the fact that the natural spline requires zero second derivatives at the end knots, whereas the cubic polynomial has no such constraint.

16.4.2 End Conditions

Although its graphical basis is appealing, the natural spline is only one of several end conditions that can be specified for splines. Two of the most popular are

* *Clamped End Condition.* This option involves specifying the first derivatives at the first and last nodes. This is sometimes called a "clamped" spline because it is what occurs when you clamp the end of a drafting spline so that it has a desired slope. For example, if zero first derivatives are specified, the spline will level off or become horizontal at the ends.
* *"Not-a-Knot" End Condition.* A third alternative is to force continuity of the third derivative at the second and the next-to-last knots. Since the spline already specifies that the function value and its first and second derivatives are equal at these knots, specifying continuous third derivatives means that the same cubic functions will apply to each of the first and last two adjacent segments. Since the first internal knots no longer represent the junction of two different cubic functions, they are no longer true knots. Hence, this case is referred to as the *"not-a-knot" condition.* It has the additional property that for four points, it yields the same result as is obtained using an ordinary cubic interpolating polynomial of the sort described in Chap. 15.

These conditions can be readily applied by using Eq. (16.25) for the interior knots, $i = 2, 3, \ldots, n-2$, and using first (1) and last equations ($n-1$) as written in Table 16.2.

Figure 16.5 shows a comparison of the three end conditions as applied to fit the data from Table 16.1. The clamped case is set up so that the derivatives at the ends are equal to zero.

As expected, the spline fit for the clamped case levels off at the ends. In contrast, the natural and not-a-knot cases follow the trend of the data points more closely. Notice how the natural spline tends to straighten out as would be expected because the second derivatives go to zero at the ends. Because it has nonzero second derivatives at the ends, the not-a-knot exhibits more curvature.

TABLE 16.2 The first and last equations needed to specify some commonly used end conditions for cubic splines.

Condition	First and Last Equations
Natural	$c_1 = 0, c_n = 0$
Clamped (where f_1' and f_n' are the specified first derivatives at the first and last nodes, respectively).	$2h_1c_1 + h_1c_2 = 3f[x_2, x_1] - 3f_1'$ $h_{n-1}c_{n-1} + 2h_{n-1}c_n = 3f_n' - 3f[x_n, x_{n-1}]$
Not-a-knot	$h_2c_1 - (h_1 + h_2)c_2 + h_1c_3 = 0$ $h_{n-1}c_{n-2} - (h_{n-2} + h_{n-1})c_{n-1} + h_{n-2}c_n = 0$

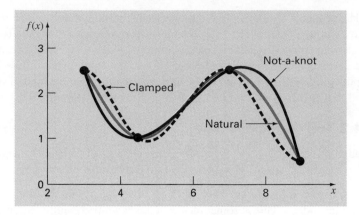

FIGURE 16.5
Comparison of the clamped (with zero first derivatives), not-a-knot, and natural splines for the data from Table 16.1.

16.5 PIECEWISE INTERPOLATION IN MATLAB

MATLAB has several built-in functions to implement piecewise interpolation. The `spline` function performs cubic spline interpolation as described in this chapter. The `pchip` function implements piecewise cubic Hermite interpolation. The `interp1` function can also implement spline and Hermite interpolation, but can also perform a number of other types of piecewise interpolation.

16.5.1 MATLAB Function: `spline`

Cubic splines can be easily computed with the built-in MATLAB function, `spline`. It has the general syntax,

$$yy = \text{spline}(x, y, xx) \tag{16.28}$$

where x and y = vectors containing the values that are to be interpolated, and yy = a vector containing the results of the spline interpolation as evaluated at the points in the vector xx.

By default, `spline` uses the not-a-knot condition. However, if y contains two more values than x has entries, then the first and last value in y are used as the derivatives at the end points. Consequently, this option provides the means to implement the clamped-end condition.

EXAMPLE 16.4   Splines in MATLAB

Problem Statement.   Runge's function is a notorious example of a function that cannot be fit well with polynomials (recall Example 15.7):

$$f(x) = \frac{1}{1 + 25x^2}$$

Use MATLAB to fit nine equally spaced data points sampled from this function in the interval $[-1, 1]$. Employ **(a)** a not-a-knot spline and **(b)** a clamped spline with end slopes of $f_1' = 1$ and $f_{n-1}' = -4$.

Solution.   **(a)** The nine equally spaced data points can be generated as in

```
>> x = linspace(-1,1,9);
>> y = 1./(1+25*x.^2);
```

Next, a more finely spaced vector of values can be generated so that we can create a smooth plot of the results as generated with the spline function:

```
>> xx = linspace(-1,1);
>> yy = spline(x,y,xx);
```

Recall that linspace automatically creates 100 points if the desired number of points are not specified. Finally, we can generate values for Runge's function itself and display them along with the spline fit and the original data:

```
>> yr = 1./(1+25*xx.^2);
>> plot(x,y,'o',xx,yy,xx,yr,'--')
```

As in Fig. 16.6, the not-a-knot spline does a nice job of following Runge's function without exhibiting wild oscillations between the points.

(b) The clamped condition can be implemented by creating a new vector yc that has the desired first derivatives as its first and last elements. The new vector can then be used to

---

FIGURE 16.6
Comparison of Runge's function (dashed line) with a 9-point not-a-knot spline fit generated with MATLAB (solid line).

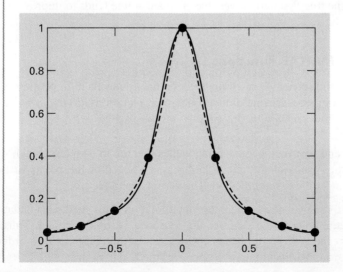

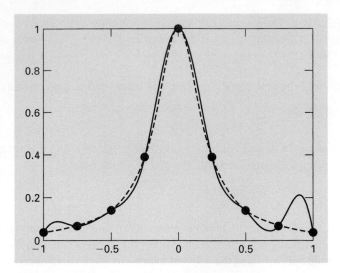

FIGURE 16.7
Comparison of Runge's function (dashed line) with a 9-point clamped end spline fit generated with MATLAB (solid line). Note that first derivatives of 1 and −4 are specified at the left and right boundaries, respectively.

generate and plot the spline fit:

```
>> yc = [1 y -4];
>> yyc = spline(x,yc,xx);
>> plot(x,y,'o',xx,yyc,xx,yr,'--')
```

As in Fig. 16.7, the clamped spline now exhibits some oscillations because of the artificial slopes that we have imposed at the boundaries. In other examples, where we have knowledge of the true first derivatives, the clamped spline tends to improve the fit.

16.5.2 MATLAB Function: `interp1`

The built-in function `interp1` provides a handy means to implement a number of different types of piecewise one-dimensional interpolation. It has the general syntax

$$yi = interp1(x, y, xi, 'method')$$

where x and y = vectors containing values that are to be interpolated, yi = a vector containing the results of the interpolation as evaluated at the points in the vector xi, and `'method'` = the desired method. The various methods are

- `'nearest'`—nearest neighbor interpolation. This method sets the value of an interpolated point to the value of the nearest existing data point. Thus, the interpolation looks like a series of plateaus, which can be thought of as zero-order polynomials.
- `'linear'`—linear interpolation. This method uses straight lines to connect the points.

- 'spline'—piecewise cubic spline interpolation. This is identical to the spline function.
- 'pchip' and 'cubic'—piecewise cubic Hermite interpolation.

If the 'method' argument is omitted, the default is linear interpolation.

The pchip option (short for "*p*iecewise *c*ubic *H*ermite *i*nterpolation") merits more discussion. As with cubic splines, pchip uses cubic polynomials to connect data points with continuous first derivatives. However, it differs from cubic splines in that the second derivatives are not necessarily continuous. Further, the first derivatives at the knots will not be the same as for cubic splines. Rather, they are expressly chosen so that the interpolation is "shape preserving." That is, the interpolated values do not tend to overshoot the data points as can sometimes happen with cubic splines.

Therefore, there are trade-offs between the spline and the pchip options. The results of using spline will generally appear smoother because the human eye can detect discontinuities in the second derivative. In addition, it will be more accurate if the data are values of a smooth function. On the other hand, pchip has no overshoots and less oscillation if the data are not smooth. These trade-offs, as well as those involving the other options, are explored in the following example.

EXAMPLE 16.5   Trade-Offs Using interp1

Problem Statement.   You perform a test drive on an automobile where you alternately accelerate the automobile and then hold it at a steady velocity. Note that you never decelerate during the experiment. The time series of spot measurements of velocity can be tabulated as

t	0	20	40	56	68	80	84	96	104	110
v	0	20	20	38	80	80	100	100	125	125

Use MATLAB's interp1 function to fit this data with (**a**) linear interpolation, (**b**) nearest neighbor, (**c**) cubic spline with not-a-knot end conditions, and (**d**) piecewise cubic Hermite interpolation.

Solution.   (**a**) The data can be entered, fit with linear interpolation, and plotted with the following commands:

```
>> t = [0 20 40 56 68 80 84 96 104 110];
>> v = [0 20 20 38 80 80 100 100 125 125];
>> tt = linspace(0,110);
>> vl = interp1(t,v,tt);
>> plot(t,v,'o',tt,vl)
```

The results (Fig. 16.8*a*) are not smooth, but do not exhibit any overshoot.

(**b**) The commands to implement and plot the nearest neighbor interpolation are

```
>> vn = interp1(t,v,tt,'nearest');
>> plot(t,v,'o',tt,vn)
```

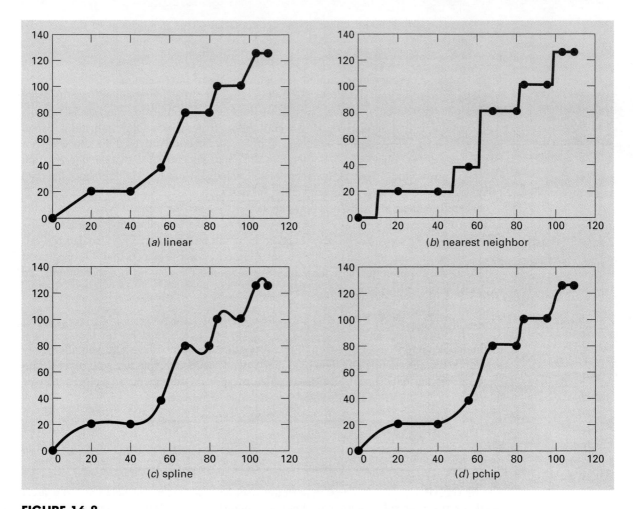

FIGURE 16.8
Use of several options of the `interp1` function to perform piecewise polynomial interpolation on a velocity time series for an automobile.

As in Fig. 16.8*b*, the results look like a series of plateaus. This option is neither a smooth nor an accurate depiction of the underlying process.

(c) The commands to implement the cubic spline are

```
>> vs = interp1(t,v,tt,'spline');
>> plot(t,v,'o',tt,vs)
```

These results (Fig. 16.8*c*) are quite smooth. However, severe overshoot occurs at several locations. This makes it appear that the automobile decelerated several times during the experiment.

(d) The commands to implement the piecewise cubic Hermite interpolation are

```
>> vh = interp1(t,v,tt,'pchip');
>> plot(t,v,'o',tt,vh)
```

For this case, the results (Fig. 16.8*d*) are physically realistic. Because of its shape-preserving nature, the velocities increase monotonically and never exhibit deceleration. Although the result is not as smooth as for the cubic splines, continuity of the first derivatives at the knots makes the transitions between points more gradual and hence more realistic.

16.6 MULTIDIMENSIONAL INTERPOLATION

The interpolation methods for one-dimensional problems can be extended to multidimensional interpolation. In this section, we will describe the simplest case of two-dimensional interpolation in Cartesian coordinates. In addition, we will describe MATLAB's capabilities for multidimensional interpolation.

16.6.1 Bilinear Interpolation

Two-dimensional interpolation deals with determining intermediate values for functions of two variables $z = f(x_i, y_i)$. As depicted in Fig. 16.9, we have values at four points: $f(x_1, y_1)$, $f(x_2, y_1)$, $f(x_1, y_2)$, and $f(x_2, y_2)$. We want to interpolate between these points

FIGURE 16.9
Graphical depiction of two-dimensional bilinear interpolation where an intermediate value (filled circle) is estimated based on four given values (open circles).

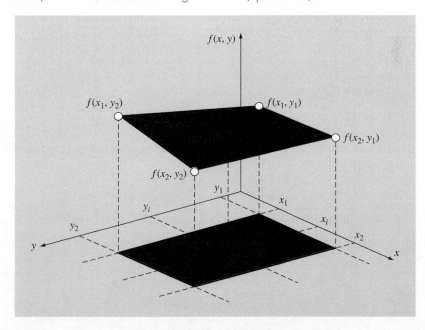

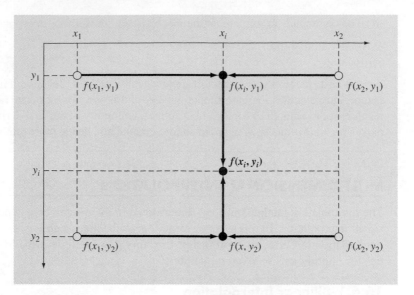

FIGURE 16.10
Two-dimensional bilinear interpolation can be implemented by first applying one-dimensional linear interpolation along the x dimension to determine values at x_i. These values can then be used to linearly interpolate along the y dimension to yield the final result at x_i, y_i.

to estimate the value at an intermediate point $f(x_i, y_i)$. If we use a linear function, the result is a plane connecting the points as in Fig. 16.9. Such functions are called *bilinear*.

A simple approach for developing the bilinear function is depicted in Fig. 16.10. First, we can hold the y value fixed and apply one-dimensional linear interpolation in the x direction. Using the Lagrange form, the result at (x_i, y_1) is

$$f(x_i, y_1) = \frac{x_i - x_2}{x_1 - x_2} f(x_1, y_1) + \frac{x_i - x_1}{x_2 - x_1} f(x_2, y_1) \tag{16.29}$$

and at (x_i, y_2) is

$$f(x_i, y_2) = \frac{x_i - x_2}{x_1 - x_2} f(x_1, y_2) + \frac{x_i - x_1}{x_2 - x_1} f(x_2, y_2) \tag{16.30}$$

These points can then be used to linearly interpolate along the y dimension to yield the final result:

$$f(x_i, y_i) = \frac{y_i - y_2}{y_1 - y_2} f(x_i, y_1) + \frac{y_i - y_1}{y_2 - y_1} f(x_i, y_2) \tag{16.31}$$

A single equation can be developed by substituting Eqs. (16.29) and (16.30) into Eq. (16.31) to give

$$f(x_i, y_i) = \frac{x_i - x_2}{x_1 - x_2} \frac{y_i - y_2}{y_1 - y_2} f(x_1, y_1) + \frac{x_i - x_1}{x_2 - x_1} \frac{y_i - y_2}{y_1 - y_2} f(x_2, y_1)$$

$$+ \frac{x_i - x_2}{x_1 - x_2} \frac{y_i - y_1}{y_2 - y_1} f(x_1, y_2) + \frac{x_i - x_1}{x_2 - x_1} \frac{y_i - y_1}{y_2 - y_1} f(x_2, y_2) \tag{16.32}$$

EXAMPLE 16.6   Bilinear Interpolation

Problem Statement. Suppose you have measured temperatures at a number of coordinates on the surface of a rectangular heated plate:

$$T(2, 1) = 60 \qquad\qquad T(9, 1) = 57.5$$
$$T(2, 6) = 55 \qquad\qquad T(9, 6) = 70$$

Use bilinear interpolation to estimate the temperature at $x_i = 5.25$ and $y_i = 4.8$.

Solution. Substituting these values into Eq. (16.32) gives

$$
\begin{aligned}
f(5.25, 4.8) = {} & \frac{5.25 - 9}{2 - 9}\frac{4.8 - 6}{1 - 6}60 + \frac{5.25 - 2}{9 - 2}\frac{4.8 - 6}{1 - 6}57.5 \\
& + \frac{5.25 - 9}{2 - 9}\frac{4.8 - 1}{6 - 1}55 + \frac{5.25 - 2}{9 - 2}\frac{4.8 - 1}{6 - 1}70 = 61.2143
\end{aligned}
$$

16.6.2 Multidimensional Interpolation in MATLAB

MATLAB has two built-in functions for two- and three-dimensional piecewise interpolation: `interp2` and `interp3`. As you might expect from their names, these functions operate in a similar fashion to `interp1` (Section 16.5.2). For example, a simple representation of the syntax of `interp2` is

```
zi = interp2(x, y, z, xi, yi, 'method')
```

where x and y = matrices containing the coordinates of the points at which the values in the matrix z are given, zi = a matrix containing the results of the interpolation as evaluated at the points in the matrices xi and yi, and `method` = the desired method. Note that the methods are identical to those used by `interp1`; that is, `linear`, `nearest`, `spline`, and `cubic`.

As with `interp1`, if the `method` argument is omitted, the default is linear interpolation. For example, `interp2` can be used to make the same evaluation as in Example 16.6 as

```
>> x=[2 9];
>> y=[1 6];
>> z=[60 57.5;55 70];
>> interp2(x,y,z,5.25,4.8)

ans =
    61.2143
```

16.7 CASE STUDY HEAT TRANSFER

Background. Lakes in the temperate zone can become thermally stratified during the summer. As depicted in Fig. 16.11, warm, buoyant water near the surface overlies colder, denser bottom water. Such stratification effectively divides the lake vertically into two layers: the *epilimnion* and the *hypolimnion*, separated by a plane called the *thermocline*.

Thermal stratification has great significance for environmental engineers and scientists studying such systems. In particular, the thermocline greatly diminishes mixing between the two layers. As a result, decomposition of organic matter can lead to severe depletion of oxygen in the isolated bottom waters.

The location of the thermocline can be defined as the inflection point of the temperature-depth curve—that is, the point at which $d^2T/dz^2 = 0$. It is also the point at which the absolute value of the first derivative or gradient is a maximum.

The temperature gradient is important in its own right because it can be used in conjunction with Fourier's law to determine the heat flux across the thermocline:

$$J = -D\rho C\frac{dT}{dz} \tag{16.33}$$

where J = heat flux [cal/(cm$^2 \cdot$ s)], α = an eddy diffusion coefficient (cm^2/s), ρ = density ($\cong 1$ g/cm^3), and C = specific heat [$\cong 1$ cal/(g $\cdot$ C)].

In this case study, natural cubic splines are employed to determine the thermocline depth and temperature gradient for Platte Lake, Michigan (Table 16.3). The latter is also used to determine the heat flux for the case where $\alpha = 0.01$ cm^2/s.

FIGURE 16.11
Temperature versus depth during summer for Platte Lake, Michigan.

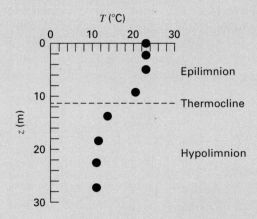

TABLE 16.3 Temperature versus depth during summer for Platte Lake, Michigan.

z, m	0	2.3	4.9	9.1	13.7	18.3	22.9	27.2
T, °C	22.8	22.8	22.8	20.6	13.9	11.7	11.1	11.1

16.7 CASE STUDY   continued

Solution.   As just described, we want to use natural spline end conditions to perform this analysis. Unfortunately, because it uses not-a-knot end conditions, the built-in MATLAB `spline` function does not meet our needs. Further, the `spline` function does not return the first and second derivatives we require for our analysis.

However, it is not difficult to develop our own M-file to implement a natural spline and return the derivatives. Such a code is shown in Fig. 16.12. After some preliminary error trapping, we set up and solve Eq. (16.27) for the second-order coefficients (c). Notice how

FIGURE 16.12
M-file to determine intermediate values and derivatives with a natural spline. Note that the `diff` function employed for error trapping is described in Section 19.7.1.

```
function [yy,dy,d2] = natspline(x,y,xx)
% natspline: natural spline with differentiation
%   [yy,dy,d2] = natspline(x,y,xx): uses a natural cubic spline
%   interpolation to find yy, the values of the underlying function
%   y at the points in the vector xx. The vector x specifies the
%   points at which the data y is given.
% input:
%   x = vector of independent variables
%   y = vector of dependent variables
%   xx = vector of desired values of dependent variables
% output:
%   yy = interpolated values at xx
%   dy = first derivatives at xx
%   d2 = second derivatives at xx

n = length(x);
if length(y)~=n, error('x and y must be same length'); end
if any(diff(x)<=0),error('x not strictly ascending'),end
m = length(xx);
b = zeros(n,n);
aa(1,1) = 1; aa(n,n) = 1;   %set up Eq. 16.27
bb(1)=0; bb(n)=0;
for i = 2:n-1
  aa(i,i-1) = h(x, i - 1);
  aa(i,i) = 2 * (h(x, i - 1) + h(x, i));
  aa(i,i+1) = h(x, i);
  bb(i) = 3 * (fd(i + 1, i, x, y) - fd(i, i - 1, x, y));
end
c=aa\bb';  %solve for c coefficients
for i = 1:n - 1   %solve for a, b and d coefficients
  a(i) = y(i);
  b(i) = fd(i + 1, i, x, y) - h(x, i) / 3 * (2 * c(i) + c(i + 1));
  d(i) = (c(i + 1) - c(i)) / 3 / h(x, i);
end
```

(continued)

```
for i = 1:m   %perform interpolations at desired values
  [yy(i),dy(i),d2(i)] = SplineInterp(x, n, a, b, c, d, xx(i));
end
end
function hh = h(x, i)
hh = x(i + 1) - x(i);
end
function fdd = fd(i, j, x, y)
fdd = (y(i) - y(j)) / (x(i) - x(j));
end
function [yyy,dyy,d2y]=SplineInterp(x, n, a, b, c, d, xi)
for ii = 1:n - 1
  if xi >= x(ii) - 0.000001 & xi <= x(ii + 1) + 0.000001
    yyy=a(ii)+b(ii)*(xi-x(ii))+c(ii)*(xi-x(ii))^2+d(ii)...
                                          *(xi-x(ii))^3;
    dyy=b(ii)+2*c(ii)*(xi-x(ii))+3*d(ii)*(xi-x(ii))^2;
    d2y=2*c(ii)+6*d(ii)*(xi-x(ii));
    break
  end
end
end
```

FIGURE 16.12 (*Continued*)

we use two subfunctions, h and fd, to compute the required finite differences. Once Eq. (16.27) is set up, we solve for the c's with back division. A loop is then employed to generate the other coefficients (a, b, and d).

At this point, we have all we need to generate intermediate values with the cubic equation:

$$f(x) = a_i + b_i(x - x_i) + c_i(x - x_i)^2 + d_i(x - x_i)^3$$

We can also determine the first and second derivatives by differentiating this equation twice to give

$$f'(x) = b_i + c_i(x - x_i) + 3d_i(x - x_i)^2$$
$$f''(x) = 2c_i + 6d_i(x - x_i)$$

As in Fig. 16.12, these equations can then be implemented in another subfunction, SplineInterp, to determine the values and the derivatives at the desired intermediate values.

Here is a script file that uses the natspline function to generate the spline and create plots of the results:

```
z = [0 2.3 4.9 9.1 13.7 18.3 22.9 27.2];
T=[22.8 22.8 22.8 20.6 13.9 11.7 11.1 11.1];
zz = linspace(z(1),z(length(z)));
```

16.7 CASE STUDY   continued

```
[TT,dT,dT2] = natspline(z,T,zz);
subplot(1,3,1),plot(T,z,'o',TT,zz)
title('(a) T'),legend('data','T')
set(gca,'YDir','reverse'),grid
subplot(1,3,2),plot(dT,zz)
title('(b) dT/dz')
set(gca,'YDir','reverse'),grid
subplot(1,3,3),plot(dT2,zz)
title('(c) d2T/dz2')
set(gca,'YDir','reverse'),grid
```

As in Fig. 16.13, the thermocline appears to be located at a depth of about 11.5 m. We can use root location (zero second derivative) or optimization methods (minimum first derivative) to refine this estimate. The result is that the thermocline is located at 11.35 m where the gradient is -1.61 °C/m.

FIGURE 16.13
Plots of (a) temperature, (b) gradient, and (c) second derivative versus depth (m) generated with the cubic spline program. The thermocline is located at the inflection point of the temperature-depth curve.

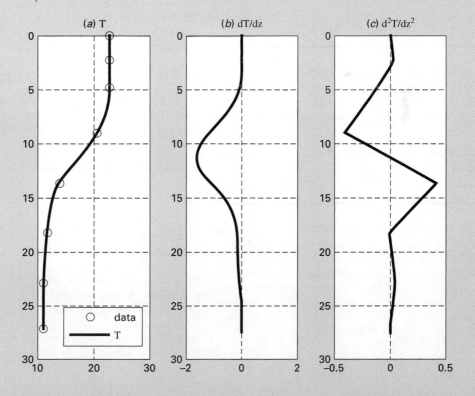

16.7 CASE STUDY continued

The gradient can be used to compute the heat flux across the thermocline with Eq. (16.33):

$$J = -0.01\frac{cm^2}{s} \times 1\frac{g}{cm^3} \times 1\frac{cal}{g \cdot °C} \times \left(-1.61\frac{°C}{m}\right) \times \frac{1\ m}{100\ cm} \times \frac{86{,}400\ s}{d} = 13.9\frac{cal}{cm^2 \cdot d}$$

The foregoing analysis demonstrates how spline interpolation can be used for engineering and scientific problem solving. However, it also is an example of numerical differentiation. As such, it illustrates how numerical approaches from different areas can be used in tandem for problem solving. We will be describing the topic of numerical differentiation in detail in Chap. 19.

PROBLEMS

16.1 Given the data

x	1	2	2.5	3	4	5
$f(x)$	1	5	7	8	2	1

Fit this data with **(a)** a cubic spline with natural end conditions, **(b)** a cubic spline with not-a-knot end conditions, and **(c)** piecewise cubic Hermite interpolation.

16.2 A reactor is thermally stratified as in the following table:

Depth, m	0	0.5	1	1.5	2	2.5	3
Temperature, °C	70	70	55	22	13	10	10

Based on these temperatures, the tank can be idealized as two zones separated by a strong temperature gradient or *thermocline*. The depth of the thermocline can be defined as the inflection point of the temperature-depth curve—that is, the point at which $d^2T/dz^2 = 0$. At this depth, the heat flux from the surface to the bottom layer can be computed with Fourier's law:

$$J = -k\frac{dT}{dz}$$

Use a clamped cubic spline fit with zero end derivatives to determine the thermocline depth. If $k = 0.01$ cal/ (s $\cdot$ cm $\cdot$ °C) compute the flux across this interface.

16.3 The following is the built-in `humps` function that MATLAB uses to demonstrate some of its numerical capabilities:

$$f(x) = \frac{1}{(x - 0.3)^2 + 0.01} + \frac{1}{(x - 0.9)^2 + 0.04} - 6$$

The `humps` function exhibits both flat and steep regions over a relatively short x range. Here are some values that have been generated at intervals of 0.1 over the range from $x = 0$ to 1:

x	0	0.1	0.2	0.3	0.4	0.5
$f(x)$	5.176	15.471	45.887	96.500	47.448	19.000

x	0.6	0.7	0.8	0.9	1
$f(x)$	11.692	12.382	17.846	21.703	16.000

Fit this data with a **(a)** cubic spline with not-a-knot end conditions and **(b)** piecewise cubic Hermite interpolation. In both cases, create a plot comparing the fit with the exact `humps` function.

16.4 Develop a plot of a cubic spline fit of the following data with **(a)** natural end conditions and **(b)** not-a-knot end conditions. In addition, develop a plot using **(c)** piecewise cubic Hermite interpolation.

x	0	100	200	400
$f(x)$	0	0.82436	1.00000	0.73576

x	600	800	1000
$f(x)$	0.40601	0.19915	0.09158

In each case, compare your plot with the following equation which was used to generate the data:

$$f(x) = \frac{x}{200} e^{-\frac{x}{200}+1}$$

16.5 The following data is sampled from the step function depicted in Fig. 16.1:

x	−1	−0.6	−0.2	0.2	0.6	1
$f(x)$	0	0	0	1	1	1

Fit this data with a **(a)** cubic spline with not-a-knot end conditions, **(b)** cubic spline with zero-slope clamped end conditions, and **(c)** piecewise cubic Hermite interpolation. In each case, create a plot comparing the fit with the step function.
16.6 Develop an M-file to compute a cubic spline fit with natural end conditions. Test your code by using it to duplicate Example 16.3.
16.7 The following data was generated with the fifth-order polynomial: $f(x) = 0.0185x^5 - 0.444x^4 + 3.9125x^3 - 15.456x^2 + 27.069x - 14.1$:

x	1	3	5	6	7	9
$f(x)$	1.000	2.172	4.220	5.430	4.912	9.120

(a) Fit this data with a cubic spline with not-a-knot end conditions. Create a plot comparing the fit with the function.
(b) Repeat **(a)** but use clamped end conditions where the end slopes are set at the exact values as determined by differentiating the function.
16.8 Bessel functions often arise in advanced engineering and scientific analyses such as the study of electric fields. These functions are usually not amenable to straightforward evaluation and, therefore, are often compiled in standard mathematical tables. For example,

x	1.8	2	2.2	2.4	2.6
$J_1(x)$	0.5815	0.5767	0.556	0.5202	0.4708

Estimate $J_1(2.1)$, **(a)** using an interpolating polynomial and **(b)** using cubic splines. Note that the true value is 0.5683.
16.9 The following data defines the sea-level concentration of dissolved oxygen for fresh water as a function of temperature:

T, °C	0	8	16	24	32	40
o, mg/L	14.621	11.843	9.870	8.418	7.305	6.413

Use MATLAB to fit the data with **(a)** piecewise linear interpolation, **(b)** a fifth-order polynomial, and **(c)** a spline. Display the results graphically and use each approach to estimate $o(27)$. Note that the exact result is 7.986 mg/L.
16.10 **(a)** Use MATLAB to fit a cubic spline to the following data:

x	0	2	4	7	10	12
y	20	20	12	7	6	6

Determine the value of y at $x = 1.5$. **(b)** Repeat **(a)**, but with zero first derivatives at the end knots.
16.12 Runge's function is written as

$$f(x) = \frac{1}{1 + 25x^2}$$

Generate five equidistantly spaced values of this function over the interval: [1, 1]. Fit this data with **(a)** a fourth-order polynomial, **(b)** a linear spline, and **(c)** a cubic spline. Present your results graphically.
16.13 Use MATLAB to generate eight points from the function

$$f(t) = \sin^2 t$$

from $t = 0$ to 2π. Fit this data using **(a)** cubic spline with not-a-knot end conditions, **(b)** cubic spline with derivative end conditions equal to the exact values calculated with differentiation, and **(c)** piecewise cubic hermite interpolation. Develop plots of each fit as well as plots of the absolute error (E_t = approximation − true) for each.

PART FIVE

Integration and Differentiation

5.1 OVERVIEW

In high school or during your first year of college, you were introduced to differential and integral calculus. There you learned techniques to obtain analytical or exact derivatives and integrals.

Mathematically, the *derivative* represents the rate of change of a dependent variable with respect to an independent variable. For example, if we are given a function $y(t)$ that specifies an object's position as a function of time, differentiation provides a means to determine its velocity, as in:

$$v(t) = \frac{d}{dt} y(t)$$

As in Fig. PT5.1*a*, the derivative can be visualized as the slope of a function.

Integration is the inverse of differentiation. Just as differentiation uses differences to quantify an instantaneous process, integration involves summing instantaneous information to give a total result over an interval. Thus, if we are provided with velocity as a function of time, integration can be used to determine the distance traveled:

$$y(t) = \int_0^t v(t)\, dt$$

As in Fig. PT5.1*b*, for functions lying above the abscissa, the integral can be visualized as the area under the curve of $v(t)$ from 0 to t. Consequently, just as a derivative can be thought of as a slope, an integral can be envisaged as a summation.

Because of the close relationship between differentiation and integration, we have opted to devote this part of the book to both processes. Among other

389

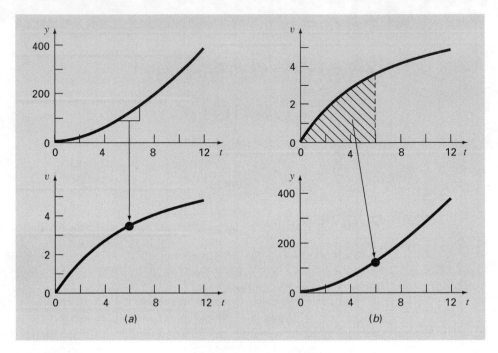

FIGURE PT5.1
The contrast between (a) differentiation and (b) integration.

things, this will provide the opportunity to highlight their similarities and differences from a numerical perspective. In addition, the material will have relevance to the next part of the book where we will cover differential equations.

Although differentiation is taught before integration in calculus, we reverse their order in the following chapters. We do this for several reasons. First, we have already introduced you to the basics of numerical differentiation in Chap. 4. Second, in part because it is much less sensitive to roundoff errors, integration represents a more highly developed area of numerical methods. Finally, although numerical differentiation is not as widely employed, it does have great significance for the solution of differential equations. Hence, it makes sense to cover it as the last topic prior to describing differential equations in Part Six.

5.2 PART ORGANIZATION

Chapter 17 is devoted to the most common approaches for numerical integration—the *Newton-Cotes formulas.* These relationships are based on replacing a complicated function or tabulated data with a simple polynomial that is easy to integrate. Three of the most widely used Newton-Cotes formulas are discussed in detail: the *trapezoidal rule, Simpson's 1/3 rule,* and *Simpson's 3/8 rule.* All these formulas are designed for cases where the data to be integrated is evenly spaced. In addition, we also include a discussion of numerical integration of unequally spaced data. This is a very important topic because many real-world applications deal with data that is in this form.

All the above material relates to *closed integration,* where the function values at the ends of the limits of integration are known. At the end of Chap. 17, we present *open integration formulas,* where the integration limits extend beyond the range of the known data. Although they are not commonly used for definite integration, open integration formulas are presented here because they are utilized in the solution of ordinary differential equations in Part Six.

The formulations covered in Chap. 17 can be employed to analyze both tabulated data and equations. *Chapter 18* deals with two techniques that are expressly designed to integrate equations and functions: *Romberg integration* and *Gauss quadrature*. Computer algorithms are provided for both of these methods. In addition, *adaptive integration* is discussed.

In *Chap. 19,* we present additional information on *numerical differentiation* to supplement the introductory material from Chap. 4. Topics include *high-accuracy finite-difference formulas, Richardson extrapolation,* and the differentiation of unequally spaced data. The effect of errors on both numerical differentiation and integration is also discussed.

17

Numerical Integration Formulas

CHAPTER OBJECTIVES

The primary objective of this chapter is to introduce you to numerical integration. Specific objectives and topics covered are

- Recognizing that Newton-Cotes integration formulas are based on the strategy of replacing a complicated function or tabulated data with a polynomial that is easy to integrate.
- Knowing how to implement the following single application Newton-Cotes formulas:
    Trapezoidal rule
    Simpson's 1/3 rule
    Simpson's 3/8 rule
- Knowing how to implement the following composite Newton-Cotes formulas:
    Trapezoidal rule
    Simpson's 1/3 rule
- Recognizing that even-segment–odd-point formulas like Simpson's 1/3 rule achieve higher than expected accuracy.
- Knowing how to use the trapezoidal rule to integrate unequally spaced data.
- Understanding the difference between open and closed integration formulas.

YOU'VE GOT A PROBLEM

Recall that the velocity of a free-falling bungee jumper as a function of time can be computed as

$$v(t) = \sqrt{\frac{gm}{c_d}} \tanh\left(\sqrt{\frac{gc_d}{m}}t\right) \tag{17.1}$$

Suppose that we would like to know the vertical distance z the jumper has fallen after a certain time t. This distance can be evaluated by integration:

$$z(t) = \int_0^t v(t)\, dt \qquad (17.2)$$

Substituting Eq. (17.1) into Eq. (17.2) gives

$$z(t) = \int_0^t \sqrt{\frac{gm}{c_d}} \tanh\left(\sqrt{\frac{gc_d}{m}}t\right) dt \qquad (17.3)$$

Thus, integration provides the means to determine the distance from the velocity. Calculus can be used to solve Eq. (17.3) for

$$z(t) = \frac{m}{c_d} \ln\left[\cosh\left(\sqrt{\frac{gc_d}{m}}t\right)\right] \qquad (17.4)$$

Although a closed form solution can be developed for this case, there are other functions that cannot be integrated analytically. Further, suppose that there was some way to measure the jumper's velocity at various times during the fall. These velocities along with their associated times could be assembled as a table of discrete values. In this situation, it would also be possible to integrate the discrete data to determine the distance. In both these instances, numerical integration methods are available to obtain solutions. Chapters 17 and 18 will introduce you to some of these methods.

17.1 INTRODUCTION AND BACKGROUND

17.1.1 What Is Integration?

According to the dictionary definition, to integrate means "to bring together, as parts, into a whole; to unite; to indicate the total amount. . . ." Mathematically, definite integration is represented by

$$I = \int_a^b f(x)\, dx \qquad (17.5)$$

which stands for the integral of the function $f(x)$ with respect to the independent variable x, evaluated between the limits $x = a$ to $x = b$.

As suggested by the dictionary definition, the "meaning" of Eq. (17.5) is the total value, or summation, of $f(x)\, dx$ over the range $x = a$ to b. In fact, the symbol $\int$ is actually a stylized capital S that is intended to signify the close connection between integration and summation.

Figure 17.1 represents a graphical manifestation of the concept. For functions lying above the x axis, the integral expressed by Eq. (17.5) corresponds to the area under the curve of $f(x)$ between $x = a$ and b.

Numerical integration is sometimes referred to as quadrature. This is an archaic term that originally meant the construction of a square having the same area as some curvilinear figure. Today, the term *quadrature* is generally taken to be synonymous with numerical definite integration.

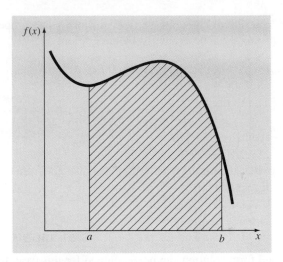

FIGURE 17.1
Graphical representation of the integral of $f(x)$ between the limits $x = a$ to b. The integral is equivalent to the area under the curve.

17.1.2 Integration in Engineering and Science

Integration has so many engineering and scientific applications that you were required to take integral calculus in your first year at college. Many specific examples of such applications could be given in all fields of engineering and science. A number of examples relate directly to the idea of the integral as the area under a curve. Figure 17.2 depicts a few cases where integration is used for this purpose.

Other common applications relate to the analogy between integration and summation. For example, a common application is to determine the mean of a continuous function. Recall that the mean of discrete of n discrete data points can be calculated by [Eq. (13.2)].

$$\text{Mean} = \frac{\sum_{i=1}^{n} y_i}{n} \tag{17.6}$$

where y_i are individual measurements. The determination of the mean of discrete points is depicted in Fig. 17.3a.

In contrast, suppose that y is a continuous function of an independent variable x, as depicted in Fig. 17.3b. For this case, there are an infinite number of values between a and b. Just as Eq. (17.6) can be applied to determine the mean of the discrete readings, you might also be interested in computing the mean or average of the continuous function $y = f(x)$ for the interval from a to b. Integration is used for this purpose, as specified by

$$\text{Mean} = \frac{\int_a^b f(x)\,dx}{b - a} \tag{17.7}$$

This formula has hundreds of engineering and scientific applications. For example, it is used to calculate the center of gravity of irregular objects in mechanical and civil engineering and to determine the root-mean-square current in electrical engineering.

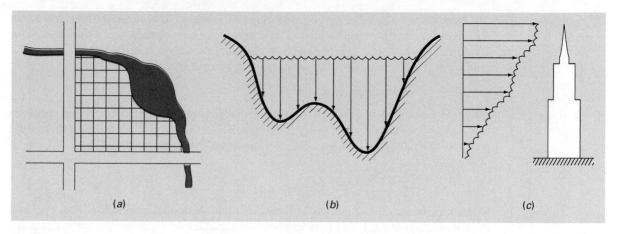

FIGURE 17.2
Examples of how integration is used to evaluate areas in engineering and scientific applications. (a) A surveyor might need to know the area of a field bounded by a meandering stream and two roads. (b) A hydrologist might need to know the cross-sectional area of a river. (c) A structural engineer might need to determine the net force due to a nonuniform wind blowing against the side of a skyscraper.

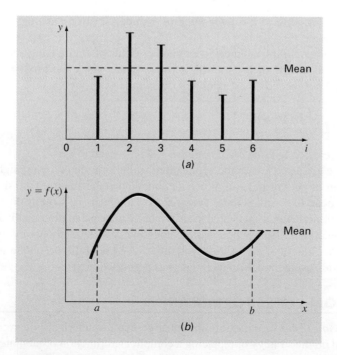

FIGURE 17.3
An illustration of the mean for (a) discrete and (b) continuous data.

Integrals are also employed by engineers and scientists to evaluate the total amount or quantity of a given physical variable. The integral may be evaluated over a line, an area, or a volume. For example, the total mass of chemical contained in a reactor is given as the product of the concentration of chemical and the reactor volume, or

Mass = concentration × volume

where concentration has units of mass per volume. However, suppose that concentration varies from location to location within the reactor. In this case, it is necessary to sum the products of local concentrations c_i and corresponding elemental volumes ΔV_i:

$$\text{Mass} = \sum_{i=1}^{n} c_i \Delta V_i$$

where n is the number of discrete volumes. For the continuous case, where $c(x, y, z)$ is a known function and x, y, and z are independent variables designating position in Cartesian coordinates, integration can be used for the same purpose:

$$\text{Mass} = \iiint c(x, y, z)\, dx\, dy\, dz$$

or

$$\text{Mass} = \iiint_V c(V)\, dV$$

which is referred to as a *volume integral*. Notice the strong analogy between summation and integration.

Similar examples could be given in other fields of engineering and science. For example, the total rate of energy transfer across a plane where the flux (in calories per square centimeter per second) is a function of position is given by

$$\text{Flux} = \iint_A \text{flux}\, dA$$

which is referred to as an *areal integral,* where A = area.

These are just a few of the applications of integration that you might face regularly in the pursuit of your profession. When the functions to be analyzed are simple, you will normally choose to evaluate them analytically. However, it is often difficult or impossible when the function is complicated, as is typically the case in more realistic examples. In addition, the underlying function is often unknown and defined only by measurement at discrete points. For both these cases, you must have the ability to obtain approximate values for integrals using numerical techniques as described next.

17.2 NEWTON-COTES FORMULAS

The *Newton-Cotes formulas* are the most common numerical integration schemes. They are based on the strategy of replacing a complicated function or tabulated data with a polynomial that is easy to integrate:

$$I = \int_a^b f(x)\, dx \cong \int_a^b f_n(x)\, dx \tag{17.8}$$

where $f_n(x) = $ a polynomial of the form

$$f_n(x) = a_0 + a_1 x + \cdots + a_{n-1} x^{n-1} + a_n x^n \qquad (17.9)$$

where n is the order of the polynomial. For example, in Fig. 17.4a, a first-order polynomial (a straight line) is used as an approximation. In Fig. 17.4b, a parabola is employed for the same purpose.

The integral can also be approximated using a series of polynomials applied piecewise to the function or data over segments of constant length. For example, in Fig. 17.5, three

FIGURE 17.4
The approximation of an integral by the area under (a) a straight line and (b) a parabola.

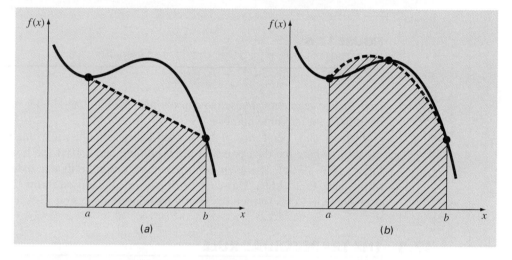

FIGURE 17.5
The approximation of an integral by the area under three straight-line segments.

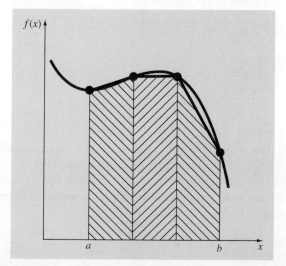

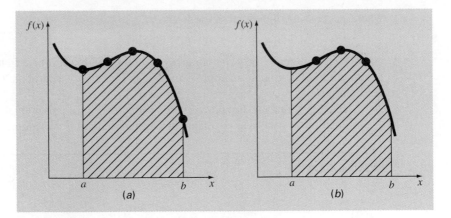

FIGURE 17.6
The difference between (a) closed and (b) open integration formulas.

straight-line segments are used to approximate the integral. Higher-order polynomials can be utilized for the same purpose.

Closed and open forms of the Newton-Cotes formulas are available. The *closed forms* are those where the data points at the beginning and end of the limits of integration are known (Fig. 17.6*a*). The *open forms* have integration limits that extend beyond the range of the data (Fig. 17.6*b*). This chapter emphasizes the closed forms. However, material on open Newton-Cotes formulas is briefly introduced in Section 17.7.

17.3 THE TRAPEZOIDAL RULE

The *trapezoidal rule* is the first of the Newton-Cotes closed integration formulas. It corresponds to the case where the polynomial in Eq. (17.8) is first-order:

$$I = \int_a^b \left[f(a) + \frac{f(b) - f(a)}{b - a}(x - a) \right] dx \qquad (17.10)$$

The result of the integration is

$$I = (b - a)\frac{f(a) + f(b)}{2} \qquad (17.11)$$

which is called the *trapezoidal rule*.

Geometrically, the trapezoidal rule is equivalent to approximating the area of the trapezoid under the straight line connecting $f(a)$ and $f(b)$ in Fig. 17.7. Recall from geometry that the formula for computing the area of a trapezoid is the height times the average of the bases. In our case, the concept is the same but the trapezoid is on its side. Therefore, the integral estimate can be represented as

$$I = \text{width} \times \text{average height} \qquad (17.12)$$

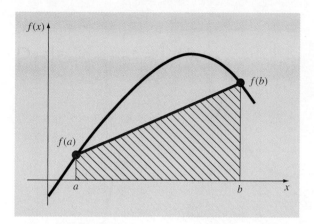

FIGURE 17.7
Graphical depiction of the trapezoidal rule.

or

$$I = (b - a) \times \text{average height} \qquad (17.13)$$

where, for the trapezoidal rule, the average height is the average of the function values at the end points, or $[f(a) + f(b)]/2$.

All the Newton-Cotes closed formulas can be expressed in the general format of Eq. (17.13). That is, they differ only with respect to the formulation of the average height.

17.3.1 Error of the Trapezoidal Rule

When we employ the integral under a straight-line segment to approximate the integral under a curve, we obviously can incur an error that may be substantial (Fig. 17.8). An estimate for the local truncation error of a single application of the trapezoidal rule is

$$E_t = -\frac{1}{12} f''(\xi)(b - a)^3 \qquad (17.14)$$

where ξ lies somewhere in the interval from a to b. Equation (17.14) indicates that if the function being integrated is linear, the trapezoidal rule will be exact because the second derivative of a straight line is zero. Otherwise, for functions with second- and higher-order derivatives (i.e., with curvature), some error can occur.

EXAMPLE 17.1   Single Application of the Trapezoidal Rule

Problem Statement.   Use Eq. (17.11) to numerically integrate

$$f(x) = 0.2 + 25x - 200x^2 + 675x^3 - 900x^4 + 400x^5$$

from $a = 0$ to $b = 0.8$. Note that the exact value of the integral can be determined analytically to be 1.640533.

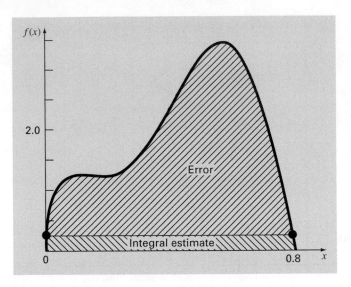

FIGURE 17.8
Graphical depiction of the use of a single application of the trapezoidal rule to approximate the integral of $f(x) = 0.2 + 25x - 200x^2 + 675x^3 - 900x^4 + 400x^5$ from $x = 0$ to 0.8.

Solution. The function values $f(0) = 0.2$ and $f(0.8) = 0.232$ can be substituted into Eq. (17.11) to yield

$$I = (0.8 - 0)\frac{0.2 + 0.232}{2} = 0.1728$$

which represents an error of $E_t = 1.640533 - 0.1728 = 1.467733$, which corresponds to a percent relative error of $\varepsilon_t = 89.5\%$. The reason for this large error is evident from the graphical depiction in Fig. 17.8. Notice that the area under the straight line neglects a significant portion of the integral lying above the line.

In actual situations, we would have no foreknowledge of the true value. Therefore, an approximate error estimate is required. To obtain this estimate, the function's second derivative over the interval can be computed by differentiating the original function twice to give

$$f''(x) = -400 + 4{,}050x - 10{,}800x^2 + 8{,}000x^3$$

The average value of the second derivative can be computed as [Eq. (17.7)]

$$\bar{f}''(x) = \frac{\int_0^{0.8} (-400 + 4{,}050x - 10{,}800x^2 + 8{,}000x^3)\,dx}{0.8 - 0} = -60$$

which can be substituted into Eq. (17.14) to yield

$$E_a = -\frac{1}{12}(-60)(0.8)^3 = 2.56$$

which is of the same order of magnitude and sign as the true error. A discrepancy does exist, however, because of the fact that for an interval of this size, the average second derivative is not necessarily an accurate approximation of $f''(\xi)$. Thus, we denote that the error is approximate by using the notation E_a, rather than exact by using E_t.

17.3.2 The Composite Trapezoidal Rule

One way to improve the accuracy of the trapezoidal rule is to divide the integration interval from a to b into a number of segments and apply the method to each segment (Fig. 17.9). The areas of individual segments can then be added to yield the integral for the entire interval. The resulting equations are called *composite,* or *multiple-application, integration formulas.*

Figure 17.9 shows the general format and nomenclature we will use to characterize composite integrals. There are $n + 1$ equally spaced base points $(x_0, x_1, x_2, \ldots, x_n)$. Consequently, there are n segments of equal width:

$$h = \frac{b - a}{n} \tag{17.15}$$

If a and b are designated as x_0 and x_n, respectively, the total integral can be represented as

$$I = \int_{x_0}^{x_1} f(x)\, dx + \int_{x_1}^{x_2} f(x)\, dx + \cdots + \int_{x_{n-1}}^{x_n} f(x)\, dx$$

FIGURE 17.9
Composite trapezoidal rule.

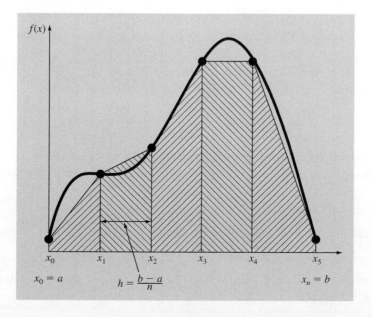

Substituting the trapezoidal rule for each integral yields

$$I = h\frac{f(x_0) + f(x_1)}{2} + h\frac{f(x_1) + f(x_2)}{2} + \cdots + h\frac{f(x_{n-1}) + f(x_n)}{2} \tag{17.16}$$

or, grouping terms:

$$I = \frac{h}{2}\left[f(x_0) + 2\sum_{i=1}^{n-1} f(x_i) + f(x_n) \right] \tag{17.17}$$

or, using Eq. (17.15) to express Eq. (17.17) in the general form of Eq. (17.13):

$$I = \underbrace{(b - a)}_{\text{Width}} \underbrace{\frac{f(x_0) + 2\sum_{i=1}^{n-1} f(x_i) + f(x_n)}{2n}}_{\text{Average height}} \tag{17.18}$$

Because the summation of the coefficients of $f(x)$ in the numerator divided by $2n$ is equal to 1, the average height represents a weighted average of the function values. According to Eq. (17.18), the interior points are given twice the weight of the two end points $f(x_0)$ and $f(x_n)$.

An error for the composite trapezoidal rule can be obtained by summing the individual errors for each segment to give

$$E_t = -\frac{(b - a)^3}{12n^3} \sum_{i=1}^{n} f''(\xi_i) \tag{17.19}$$

where $f''(\xi_i)$ is the second derivative at a point ξ_i located in segment i. This result can be simplified by estimating the mean or average value of the second derivative for the entire interval as

$$\bar{f}'' \cong \frac{\sum_{i=1}^{n} f''(\xi_i)}{n} \tag{17.20}$$

Therefore $\sum f''(\xi_i) \cong n\bar{f}''$ and Eq. (17.19) can be rewritten as

$$E_a = -\frac{(b - a)^3}{12n^2} \bar{f}'' \tag{17.21}$$

Thus, if the number of segments is doubled, the truncation error will be quartered. Note that Eq. (17.21) is an approximate error because of the approximate nature of Eq. (17.20).

EXAMPLE 17.2   Composite Application of the Trapezoidal Rule

Problem Statement.   Use the two-segment trapezoidal rule to estimate the integral of

$$f(x) = 0.2 + 25x - 200x^2 + 675x^3 - 900x^4 + 400x^5$$

from $a = 0$ to $b = 0.8$. Employ Eq. (17.21) to estimate the error. Recall that the exact value of the integral is 1.640533.

Solution. For $n = 2$ $(h = 0.4)$:

$$f(0) = 0.2 \qquad f(0.4) = 2.456 \qquad f(0.8) = 0.232$$

$$I = 0.8 \frac{0.2 + 2(2.456) + 0.232}{4} = 1.0688$$

$$E_t = 1.640533 - 1.0688 = 0.57173 \qquad \varepsilon_t = 34.9\%$$

$$E_a = -\frac{0.8^3}{12(2)^2}(-60) = 0.64$$

where -60 is the average second derivative determined previously in Example 17.1.

The results of the previous example, along with three- through ten-segment applications of the trapezoidal rule, are summarized in Table 17.1. Notice how the error decreases as the number of segments increases. However, also notice that the rate of decrease is gradual. This is because the error is inversely related to the square of n [Eq. (17.21)]. Therefore, doubling the number of segments quarters the error. In subsequent sections we develop higher-order formulas that are more accurate and that converge more quickly on the true integral as the segments are increased. However, before investigating these formulas, we will first discuss how MATLAB can be used to implement the trapezoidal rule.

17.3.3 MATLAB M-file: `trap`

A simple algorithm to implement the composite trapezoidal rule can be written as in Fig. 17.10. The function to be integrated is passed into the M-file along with the limits of integration and the number of segments. A loop is then employed to generate the integral following Eq. (17.18).

TABLE 17.1 Results for the composite trapezoidal rule to estimate the integral of $f(x) = 0.2 + 25x - 200x^2 + 675x^3 - 900x^4 + 400x^5$ from $x = 0$ to 0.8. The exact value is 1.640533.

n	h	I	ε_t (%)
2	0.4	1.0688	34.9
3	0.2667	1.3695	16.5
4	0.2	1.4848	9.5
5	0.16	1.5399	6.1
6	0.1333	1.5703	4.3
7	0.1143	1.5887	3.2
8	0.1	1.6008	2.4
9	0.0889	1.6091	1.9
10	0.08	1.6150	1.6

```
function I = trap(func,a,b,n,varargin)
% trap: composite trapezoidal rule quadrature
%   I = trap(func,a,b,n,p1,p2,...):
%                   composite trapezoidal rule
% input:
%   func = name of function to be integrated
%   a, b = integration limits
%   n = number of segments (default = 100)
%   p1,p2,... = additional parameters used by func
% output:
%   I = integral estimate

if nargin<3,error('at least 3 input arguments required'),end
if ~(b>a),error('upper bound must be greater than lower'),end
if nargin<4|isempty(n),n=100;end
x = a; h = (b - a)/n;
s=func(a,varargin{:});
for i = 1 : n-1
  x = x + h;
  s = s + 2*func(x,varargin{:});
end
s = s + func(b,varargin{:});
I = (b - a) * s/(2*n);
```

FIGURE 17.10
M-file to implement the composite trapezoidal rule.

An application of the M-file can be developed to determine the distance fallen by the free-falling bungee jumper in the first 3 s by evaluating the integral of Eq. (17.3). For this example, assume the following parameter values: $g = 9.81$ m/s^2, $m = 68.1$ kg, and $c_d = 0.25$ kg/m. Note that the exact value of the integral can be computed with Eq. (17.4) as 41.94805.

The function to be integrated can be developed as an M-file or with an anonymous function,

```
>> v=@(t) sqrt(9.81*68.1/0.25)*tanh(sqrt(9.81*0.25/68.1)*t)

v =
    @(t) sqrt(9.81*68.1/0.25)*tanh(sqrt(9.81*0.25/68.1)*t)
```

First, let's evaluate the integral with a crude five-segment approximation:

```
format long
>> trap(v,0,3,5)

ans =
  41.86992959072735
```

As would be expected, this result has a relatively high true error of 18.6%. To obtain a more accurate result, we can use a very fine approximation based on 10,000 segments:

```
>> trap(v,0,3,10000)

x =
   41.94804999917528
```

which is very close to the true value.

17.4 SIMPSON'S RULES

Aside from applying the trapezoidal rule with finer segmentation, another way to obtain a more accurate estimate of an integral is to use higher-order polynomials to connect the points. For example, if there is an extra point midway between $f(a)$ and $f(b)$, the three points can be connected with a parabola (Fig. 17.11a). If there are two points equally spaced between $f(a)$ and $f(b)$, the four points can be connected with a third-order polynomial (Fig. 17.11b). The formulas that result from taking the integrals under these polynomials are called *Simpson's rules.*

17.4.1 Simpson's 1/3 Rule

Simpson's 1/3 rule corresponds to the case where the polynomial in Eq. (17.8) is second-order:

$$I = \int_{x_0}^{x_2} \left[\frac{(x - x_1)(x - x_2)}{(x_0 - x_1)(x_0 - x_2)} f(x_0) + \frac{(x - x_0)(x - x_2)}{(x_1 - x_0)(x_1 - x_2)} f(x_1) \right.$$
$$\left. + \frac{(x - x_0)(x - x_1)}{(x_2 - x_0)(x_2 - x_1)} f(x_2) \right] dx$$

FIGURE 17.11
(a) Graphical depiction of Simpson's 1/3 rule: It consists of taking the area under a parabola connecting three points. (b) Graphical depiction of Simpson's 3/8 rule: It consists of taking the area under a cubic equation connecting four points.

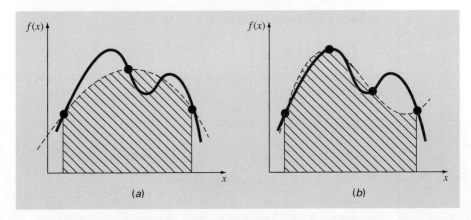

where a and b are designated as x_0 and x_2, respectively. The result of the integration is

$$I = \frac{h}{3}[f(x_0) + 4f(x_1) + f(x_2)] \tag{17.22}$$

where, for this case, $h = (b - a)/2$. This equation is known as *Simpson's 1/3 rule*. The label "1/3" stems from the fact that h is divided by 3 in Eq. (17.22). Simpson's 1/3 rule can also be expressed using the format of Eq. (17.13):

$$I = (b - a)\frac{f(x_0) + 4f(x_1) + f(x_2)}{6} \tag{17.23}$$

where $a = x_0$, $b = x_2$, and $x_1 = $ the point midway between a and b, which is given by $(a + b)/2$. Notice that, according to Eq. (17.23), the middle point is weighted by two-thirds and the two end points by one-sixth.

It can be shown that a single-segment application of Simpson's 1/3 rule has a truncation error of

$$E_t = -\frac{1}{90}h^5 f^{(4)}(\xi)$$

or, because $h = (b - a)/2$:

$$E_t = -\frac{(b - a)^5}{2880}f^{(4)}(\xi) \tag{17.24}$$

where ξ lies somewhere in the interval from a to b. Thus, Simpson's 1/3 rule is more accurate than the trapezoidal rule. However, comparison with Eq. (17.14) indicates that it is more accurate than expected. Rather than being proportional to the third derivative, the error is proportional to the fourth derivative. Consequently, Simpson's 1/3 rule is third-order accurate even though it is based on only three points. In other words, it yields exact results for cubic polynomials even though it is derived from a parabola!

EXAMPLE 17.3   Single Application of Simpson's 1/3 Rule

Problem Statement.   Use Eq. (17.23) to integrate

$$f(x) = 0.2 + 25x - 200x^2 + 675x^3 - 900x^4 + 400x^5$$

from $a = 0$ to $b = 0.8$. Employ Eq. (17.24) to estimate the error. Recall that the exact integral is 1.640533.

Solution. $n = 2 (h = 0.4)$:

$$f(0) = 0.2 \qquad f(0.4) = 2.456 \qquad f(0.8) = 0.232$$

$$I = 0.8\frac{0.2 + 4(2.456) + 0.232}{6} = 1.367467$$

$$E_t = 1.640533 - 1.367467 = 0.2730667 \qquad \varepsilon_t = 16.6\%$$

which is approximately five times more accurate than for a single application of the trapezoidal rule (Example 17.1). The approximate error can be estimated as

$$E_a = -\frac{0.8^5}{2880}(-2400) = 0.2730667$$

where -2400 is the average fourth derivative for the interval. As was the case in Example 17.1, the error is approximate (E_a) because the average fourth derivative is generally not an exact estimate of $f^{(4)}(\xi)$. However, because this case deals with a fifth-order polynomial, the result matches exactly.

17.4.2 The Composite Simpson's 1/3 Rule

Just as with the trapezoidal rule, Simpson's rule can be improved by dividing the integration interval into a number of segments of equal width (Fig. 17.12). The total integral can be represented as

$$I = \int_{x_0}^{x_2} f(x)\,dx + \int_{x_2}^{x_4} f(x)\,dx + \cdots + \int_{x_{n-2}}^{x_n} f(x)\,dx \tag{17.25}$$

FIGURE 17.12
Composite Simpson's 1/3 rule. The relative weights are depicted above the function values. Note that the method can be employed only if the number of segments is even.

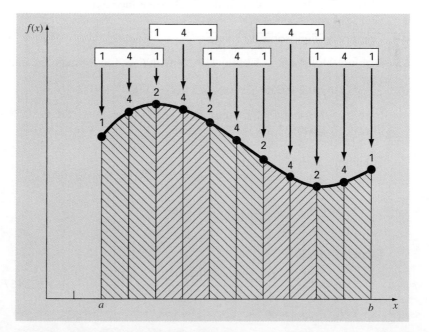

Substituting Simpson's 1/3 rule for each integral yields

$$I = 2h\frac{f(x_0) + 4f(x_1) + f(x_2)}{6} + 2h\frac{f(x_2) + 4f(x_3) + f(x_4)}{6}$$

$$+ \cdots + 2h\frac{f(x_{n-2}) + 4f(x_{n-1}) + f(x_n)}{6}$$

or, grouping terms and using Eq. (17.15):

$$I = (b - a)\frac{f(x_0) + 4\sum\limits_{i=1,3,5}^{n-1} f(x_i) + 2\sum\limits_{j=2,4,6}^{n-2} f(x_j) + f(x_n)}{3n} \tag{17.26}$$

Notice that, as illustrated in Fig. 17.12, an even number of segments must be utilized to implement the method. In addition, the coefficients "4" and "2" in Eq. (17.26) might seem peculiar at first glance. However, they follow naturally from Simpson's 1/3 rule. As illustrated in Fig. 17.12, the odd points represent the middle term for each application and hence carry the weight of four from Eq. (17.23). The even points are common to adjacent applications and hence are counted twice.

An error estimate for the composite Simpson's rule is obtained in the same fashion as for the trapezoidal rule by summing the individual errors for the segments and averaging the derivative to yield

$$E_a = -\frac{(b - a)^5}{180n^4}\bar{f}^{(4)} \tag{17.27}$$

where $f^{(4)}$ is the average fourth derivative for the interval.

EXAMPLE 17.4   Composite Simpson's 1/3 Rule

Problem Statement.   Use Eq. (17.26) with $n = 4$ to estimate the integral of

$$f(x) = 0.2 + 25x - 200x^2 + 675x^3 - 900x^4 + 400x^5$$

from $a = 0$ to $b = 0.8$. Employ Eq. (17.27) to estimate the error. Recall that the exact integral is 1.640533.

Solution. $n = 4(h = 0.2)$:

$$f(0) = 0.2 \qquad f(0.2) = 1.288$$
$$f(0.4) = 2.456 \qquad f(0.6) = 3.464$$
$$f(0.8) = 0.232$$

From Eq. (17.26):

$$I = 0.8\frac{0.2 + 4(1.288 + 3.464) + 2(2.456) + 0.232}{12} = 1.623467$$

$$E_t = 1.640533 - 1.623467 = 0.017067 \qquad \varepsilon_t = 1.04\%$$

The estimated error (Eq. 17.27) is

$$E_a = -\frac{(0.8)^5}{180(4)^4}(-2400) = 0.017067$$

which is exact (as was also the case for Example 17.3).

As in Example 17.4, the composite version of Simpson's 1/3 rule is considered superior to the trapezoidal rule for most applications. However, as mentioned previously, it is limited to cases where the values are equispaced. Further, it is limited to situations where there are an even number of segments and an odd number of points. Consequently, as discussed in Section 17.4.3, an odd-segment–even-point formula known as Simpson's 3/8 rule can be used in conjunction with the 1/3 rule to permit evaluation of both even and odd numbers of equispaced segments.

17.4.3 Simpson's 3/8 Rule

In a similar manner to the derivation of the trapezoidal and Simpson's 1/3 rule, a third-order Lagrange polynomial can be fit to four points and integrated to yield

$$I = \frac{3h}{8}\left[f(x_0) + 3f(x_1) + 3f(x_2) + f(x_3)\right]$$

where $h = (b-a)/3$. This equation is known as *Simpsons 3/8 rule* because h is multiplied by 3/8. It is the third Newton-Cotes closed integration formula. The 3/8 rule can also be expressed in the form of Eq. (17.13):

$$I = (b-a)\frac{f(x_0) + 3f(x_1) + 3f(x_2) + f(x_3)}{8} \tag{17.28}$$

Thus, the two interior points are given weights of three-eighths, whereas the end points are weighted with one-eighth. Simpson's 3/8 rule has an error of

$$E_t = -\frac{3}{80}h^5 f^{(4)}(\xi)$$

or, because $h = (b-a)/3$:

$$E_t = -\frac{(b-a)^5}{6480}f^{(4)}(\xi) \tag{17.29}$$

Because the denominator of Eq. (17.29) is larger than for Eq. (17.24), the 3/8 rule is somewhat more accurate than the 1/3 rule.

Simpson's 1/3 rule is usually the method of preference because it attains third-order accuracy with three points rather than the four points required for the 3/8 version. However, the 3/8 rule has utility when the number of segments is odd. For instance, in Example 17.4 we used Simpson's rule to integrate the function for four segments. Suppose that you desired an estimate for five segments. One option would be to use a composite version of the trapezoidal rule as was done in Example 17.2. This may not be advisable, however, because of the large truncation error associated with this method. An alternative would be to apply Simpson's 1/3 rule to the first two segments and Simpson's 3/8 rule to the last

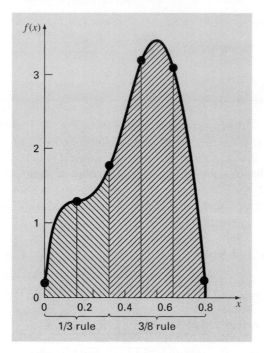

FIGURE 17.13
Illustration of how Simpson's 1/3 and 3/8 rules can be applied in tandem to handle multiple applications with odd numbers of intervals.

three (Fig. 17.13). In this way, we could obtain an estimate with third-order accuracy across the entire interval.

EXAMPLE 17.5   Simpson's 3/8 Rule

Problem Statement.   (a) Use Simpson's 3/8 rule to integrate

$$f(x) = 0.2 + 25x - 200x^2 + 675x^3 - 900x^4 + 400x^5$$

from $a = 0$ to $b = 0.8$. (b) Use it in conjunction with Simpson's 1/3 rule to integrate the same function for five segments.

Solution.   (a) A single application of Simpson's 3/8 rule requires four equally spaced points:

$f(0) = 0.2$ $f(0.2667) = 1.432724$
$f(0.5333) = 3.487177$ $f(0.8) = 0.232$

Using Eq. (17.28):

$$I = 0.8 \frac{0.2 + 3(1.432724 + 3.487177) + 0.232}{8} = 1.51970$$

(b) The data needed for a five-segment application ($h = 0.16$) is

$$f(0) = 0.2 \qquad f(0.16) = 1.296919$$
$$f(0.32) = 1.743393 \qquad f(0.48) = 3.186015$$
$$f(0.64) = 3.181929 \qquad f(0.80) = 0.232$$

The integral for the first two segments is obtained using Simpson's 1/3 rule:

$$I = 0.32 \frac{0.2 + 4(1.296919) + 1.743393}{6} = 0.3803237$$

For the last three segments, the 3/8 rule can be used to obtain

$$I = 0.48 \frac{1.743393 + 3(3.186015 + 3.181929) + 0.232}{8} = 1.264754$$

The total integral is computed by summing the two results:

$$I = 0.3803237 + 1.264754 = 1.645077$$

17.5 HIGHER-ORDER NEWTON-COTES FORMULAS

As noted previously, the trapezoidal rule and both of Simpson's rules are members of a family of integrating equations known as the Newton-Cotes closed integration formulas. Some of the formulas are summarized in Table 17.2 along with their truncation-error estimates.

Notice that, as was the case with Simpson's 1/3 and 3/8 rules, the five- and six-point formulas have the same order error. This general characteristic holds for the higher-point formulas and leads to the result that the even-segment–odd-point formulas (e.g., 1/3 rule and Boole's rule) are usually the methods of preference.

TABLE 17.2 Newton-Cotes closed integration formulas. The formulas are presented in the format of Eq. (17.13) so that the weighting of the data points to estimate the average height is apparent. The step size is given by $h = (b - a)/n$.

Segments (n)	Points	Name	Formula	Truncation Error
1	2	Trapezoidal rule	$(b - a) \dfrac{f(x_0) + f(x_1)}{2}$	$-(1/12)h^3 f''(\xi)$
2	3	Simpson's 1/3 rule	$(b - a) \dfrac{f(x_0) + 4f(x_1) + f(x_2)}{6}$	$-(1/90)h^5 f^{(4)}(\xi)$
3	4	Simpson's 3/8 rule	$(b - a) \dfrac{f(x_0) + 3f(x_1) + 3f(x_2) + f(x_3)}{8}$	$-(3/80)h^5 f^{(4)}(\xi)$
4	5	Boole's rule	$(b - a) \dfrac{7f(x_0) + 32f(x_1) + 12f(x_2) + 32f(x_3) + 7f(x_4)}{90}$	$-(8/945)h^7 f^{(6)}(\xi)$
5	6		$(b - a) \dfrac{19f(x_0) + 75f(x_1) + 50f(x_2) + 50f(x_3) + 75f(x_4) + 19f(x_5)}{288}$	$-(275/12,096)h^7 f^{(6)}(\xi)$

However, it must also be stressed that, in engineering and science practice, the higher-order (i.e., greater than four-point) formulas are not commonly used. Simpson's rules are sufficient for most applications. Accuracy can be improved by using the composite version. Furthermore, when the function is known and high accuracy is required, methods such as Romberg integration or Gauss quadrature, described in Chap. 18, offer viable and attractive alternatives.

17.6 INTEGRATION WITH UNEQUAL SEGMENTS

To this point, all formulas for numerical integration have been based on equispaced data points. In practice, there are many situations where this assumption does not hold and we must deal with unequal-sized segments. For example, experimentally derived data is often of this type. For these cases, one method is to apply the trapezoidal rule to each segment and sum the results:

$$I = h_1 \frac{f(x_0) + f(x_1)}{2} + h_2 \frac{f(x_1) + f(x_2)}{2} + \cdots + h_n \frac{f(x_{n-1}) + f(x_n)}{2} \qquad (17.30)$$

where h_i = the width of segment i. Note that this was the same approach used for the composite trapezoidal rule. The only difference between Eqs. (17.16) and (17.30) is that the h's in the former are constant.

EXAMPLE 17.6   Trapezoidal Rule with Unequal Segments

Problem Statement.   The information in Table 17.3 was generated using the same polynomial employed in Example 17.1. Use Eq. (17.30) to determine the integral for this data. Recall that the correct answer is 1.640533.

TABLE 17.3 Data for $f(x) = 0.2 + 25x - 200x^2 + 675x^3 - 900x^4 + 400x^5$, with unequally spaced values of x.

x	$f(x)$	x	$f(x)$
0.00	0.200000	0.44	2.842985
0.12	1.309729	0.54	3.507297
0.22	1.305241	0.64	3.181929
0.32	1.743393	0.70	2.363000
0.36	2.074903	0.80	0.232000
0.40	2.456000		

Solution.   Applying Eq. (17.30) yields

$$I = 0.12 \frac{0.2 + 1.309729}{2} + 0.10 \frac{1.309729 + 1.305241}{2}$$

$$+ \cdots + 0.10 \frac{2.363 + 0.232}{2} = 1.594801$$

which represents an absolute percent relative error of $\varepsilon_t = 2.8\%$.

17.6.1 MATLAB M-file: `trapuneq`

A simple algorithm to implement the trapezoidal rule for unequally spaced data can be written as in Fig. 17.14. Two vectors, x and y, holding the independent and dependent variables are passed into the M-file. Two error traps are included to ensure that (a) the two vectors are of the same length and (b) the x's are in ascending order.[1] A loop is employed to generate the integral. Notice that we have modified the subscripts from those of Eq. (17.30) to account for the fact that MATLAB does not allow zero subscripts in arrays.

An application of the M-file can be developed for the same problem that was solved in Example 17.6:

```
>> x = [0 .12 .22 .32 .36 .4 .44 .54 .64 .7 .8];
>> y = 0.2+25*x-200*x.^2+675*x.^3-900*x.^4+400*x.^5;
>> trapuneq(x,y)

ans =

    1.5948
```

which is identical to the result obtained in Example 17.6.

FIGURE 17.14
M-file to implement the trapezoidal rule for unequally spaced data.

```
function I = trapuneq(x,y)
% trapuneq: unequal spaced trapezoidal rule quadrature
%    I = trapuneq(x,y):
%    Applies the trapezoidal rule to determine the integral
%    for n data points (x, y) where x and y must be of the
%    same length and x must be monotonically ascending
% input:
%    x = vector of independent variables
%    y = vector of dependent variables
% output:
%    I = integral estimate

if nargin<2,error('at least 2 input arguments required'),end
if any(diff(x)<0),error('x not monotonically ascending'),end
n = length(x);
if length(y)~=n,error('x and y must be same length'); end
s = 0;
for k = 1:n-1
  s = s + (x(k+1)-x(k))*(y(k)+y(k+1))/2;
end
I = s;
```

---

[1] The `diff` function is described in Section 19.7.1.

17.6.2 MATLAB Functions: `trapz` and `cumtrapz`

MATLAB has a built-in function that evaluates integrals for data in the same fashion as the M-file we just presented in Fig. 17.14. It has the general syntax

```
z = trapz(x, y)
```

where the two vectors, x and y, hold the independent and dependent variables, respectively. Here is a simple MATLAB session that uses this function to integrate the data from Table 17.3:

```
>> x = [0 .12 .22 .32 .36 .4 .44 .54 .64 .7 .8];
>> y = 0.2+25*x-200*x.^2+675*x.^3-900*x.^4+400*x.^5;
>> trapz(x,y)

ans =

    1.5948
```

In addition, MATLAB has another function, `cumtrapz`, that computes the cumulative integral. A simple representation of its syntax is

```
z = cumtrapz(x, y)
```

where the two vectors, x and y, hold the independent and dependent variables, respectively, and $z = $ a vector whose elements $z(k)$ hold the integral from $x(1)$ to $x(k)$.

EXAMPLE 17.7   Using Numerical Integration to Compute Distance from Velocity

Problem Statement.   As described at the beginning of this chapter, a nice application of integration is to compute the distance $z(t)$ of an object based on its velocity $v(t)$ as in (recall Eq. 17.2):

$$z(t) = \int_0^t v(t)\, dt$$

Suppose that we had measurements of velocity at a series of discrete unequally spaced times during free fall. Use Eq. (17.2) to synthetically generate such information for a 70-kg jumper with a drag coefficient of 0.275 kg/m. Incorporate some random error by rounding the velocities to the nearest integer. Then use `cumtrapz` to determine the distance fallen and compare the results to the analytical solution (Eq. 17.4). In addition, develop a plot of the analytical and computed distances along with velocity on the same graph.

Solution.   Some unequally spaced times and rounded velocities can be generated as

```
>> format short g
>> t=[0 1 1.4 2 3 4.3 6 6.7 8];
>> g=9.81;m=70;cd=0.275;
>> v=round(sqrt(g*m/cd)*tanh(sqrt(g*cd/m)*t));
```

The distances can then be computed as

```
>> z=cumtrapz(t,v)

z =

    0    5    9.6    19.2    41.7    80.7    144.45    173.85    231.7
```

Thus, after 8 seconds, the jumper has fallen 231.7 m. This result is reasonably close to the analytical solution (Eq. 17.4):

$$z(t) = \frac{70}{0.275} \ln \left[\cosh \left(\sqrt{\frac{9.81(0.275)}{70}} 8 \right) \right] = 234.1$$

A graph of the numerical and analytical solutions along with both the exact and rounded velocities can be generated with the following commands:

```
>> ta=linspace(t(1),t(length(t)));
>> za=m/cd*log(cosh(sqrt(g*cd/m)*ta));
>> plot(ta,za,t,z,'o')
>> title('Distance versus time')
>> xlabel('t (s)'),ylabel('x (m)')
>> legend('analytical','numerical')
```

As in Fig. 17.15, the numerical and analytical results match fairly well.

FIGURE 17.15

Plot of distance versus time. The line was computed with the analytical solution, whereas the points were determined numerically with the `cumtrapz` function.

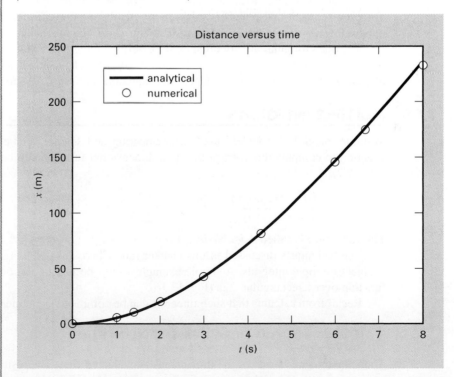

TABLE 17.4 Newton-Cotes open integration formulas. The formulas are presented in the format of Eq. (17.13) so that the weighting of the data points to estimate the average height is apparent. The step size is given by $h = (b - a)/n$.

Segments (n)	Points	Name	Formula	Truncation Error
2	1	Midpoint method	$(b-a)f(x_1)$	$(1/3)h^3 f''(\xi)$
3	2		$(b-a)\dfrac{f(x_1) + f(x_2)}{2}$	$(3/4)h^3 f''(\xi)$
4	3		$(b-a)\dfrac{2f(x_1) - f(x_2) + 2f(x_3)}{3}$	$(14/45)h^5 f^{(4)}(\xi)$
5	4		$(b-a)\dfrac{11f(x_1) + f(x_2) + f(x_3) + 11f(x_4)}{24}$	$(95/144)h^5 f^{(4)}(\xi)$
6	5		$(b-a)\dfrac{11f(x_1) - 14f(x_2) + 26f(x_3) - 14f(x_4) + 11f(x_5)}{20}$	$(41/140)h^7 f^{(6)}(\xi)$

17.7 OPEN METHODS

Recall from Fig. 17.6b that open integration formulas have limits that extend beyond the range of the data. Table 17.4 summarizes the *Newton-Cotes open integration formulas.* The formulas are expressed in the form of Eq. (17.13) so that the weighting factors are evident. As with the closed versions, successive pairs of the formulas have the same-order error. The even-segment-odd-point formulas are usually the methods of preference because they require fewer points to attain the same accuracy as the odd-segment–even-point formulas.

The open formulas are not often used for definite integration. However, they have utility for analyzing improper integrals. In addition, they will have relevance to our discussion of methods for solving ordinary differential equations in Chaps. 20 and 21.

17.8 MULTIPLE INTEGRALS

Multiple integrals are widely used in engineering and science. For example, a general equation to compute the average of a two-dimensional function can be written as [recall Eq. (17.7)]

$$\bar{f} = \frac{\int_c^d \left(\int_a^b f(x, y)\, dx \right)}{(d - c)(b - a)} \tag{17.31}$$

The numerator is called a *double integral.*

The techniques discussed in this chapter (and Chap. 18) can be readily employed to evaluate multiple integrals. A simple example would be to take the double integral of a function over a rectangular area (Fig. 17.16).

Recall from calculus that such integrals can be computed as iterated integrals:

$$\int_c^d \left(\int_a^b f(x, y)\, dx \right) dy = \int_a^b \left(\int_c^d f(x, y)\, dy \right) dx \tag{17.32}$$

Thus, the integral in one of the dimensions is evaluated first. The result of this first integration is integrated in the second dimension. Equation (17.32) states that the order of integration is not important.

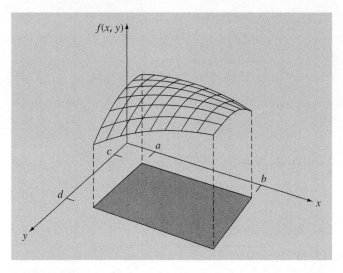

FIGURE 17.16
Double integral as the area under the function surface.

A numerical double integral would be based on the same idea. First, methods such as the composite trapezoidal or Simpson's rule would be applied in the first dimension with each value of the second dimension held constant. Then the method would be applied to integrate the second dimension. The approach is illustrated in the following example.

EXAMPLE 17.8   Using Double Integral to Determine Average Temperature

Problem Statement.   Suppose that the temperature of a rectangular heated plate is described by the following function:

$$T(x, y) = 2xy + 2x - x^2 - 2y^2 + 72$$

If the plate is 8 m long (x dimension) and 6 m wide (y dimension), compute the average temperature.

Solution.   First, let us merely use two-segment applications of the trapezoidal rule in each dimension. The temperatures at the necessary x and y values are depicted in Fig. 17.17. Note that a simple average of these values is 47.33. The function can also be evaluated analytically to yield a result of 58.66667.

To make the same evaluation numerically, the trapezoidal rule is first implemented along the x dimension for each y value. These values are then integrated along the y dimension to give the final result of 2688. Dividing this by the area yields the average temperature as $2688/(6 \times 8) = 56$.

Now we can apply a single-segment Simpson's 1/3 rule in the same fashion. This results in an integral of 2816 and an average of 58.66667, which is exact. Why does this occur? Recall that Simpson's 1/3 rule yielded perfect results for cubic polynomials. Since the highest-order term in the function is second order, the same exact result occurs for the present case.

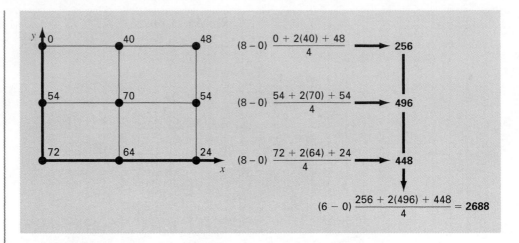

FIGURE 17.17
Numerical evaluation of a double integral using the two-segment trapezoidal rule.

For higher-order algebraic functions as well as transcendental functions, it would be necessary to use composite applications to attain accurate integral estimates. In addition, Chap. 18 introduces techniques that are more efficient than the Newton-Cotes formulas for evaluating integrals of given functions. These often provide a superior means to implement the numerical integrations for multiple integrals.

17.8.1 MATLAB Functions: `dblquad` and `triplequad`

MATLAB has functions to implement both double (`dblquad`) and triple (`triplequad`) integration. A simple representation of the syntax for `dblquad` is

```
q = dblquad(fun, xmin, xmax, ymin, ymax, tol)
```

where q is the double integral of the function *fun* over the ranges from *xmin* to *xmax* and *ymin* to *ymax*. If *tol* is not specified, a default tolerance of 1×10^{-6} is used.

Here is an example of how this function can be used to compute the double integral evaluated in Example 17.7:

```
>> q = dblquad(@(x,y) 2*x*y+2*x-x.^2-2*y.^2+72,0,8,0,6)

q =
        2816
```

17.9 CASE STUDY COMPUTING WORK WITH NUMERICAL INTEGRATION

Background. The calculation of work is an important component of many areas of engineering and science. The general formula is

Work = force × distance

When you were introduced to this concept in high school physics, simple applications were presented using forces that remained constant throughout the displacement. For example, if a force of 10 N was used to pull a block a distance of 5 m, the work would be calculated as 50 J (1 joule = 1 N · m).

Although such a simple computation is useful for introducing the concept, realistic problem settings are usually more complex. For example, suppose that the force varies during the course of the calculation. In such cases, the work equation is reexpressed as

$$W = \int_{x_0}^{x_n} F(x)\,dx \tag{17.33}$$

where W = work (J), x_0 and x_n = the initial and final positions (m), respectively, and $F(x)$ = a force that varies as a function of position (N). If $F(x)$ is easy to integrate, Eq. (17.33) can be evaluated analytically. However, in a realistic problem setting, the force might not be expressed in such a manner. In fact, when analyzing measured data, the force might be available only in tabular form. For such cases, numerical integration is the only viable option for the evaluation.

Further complexity is introduced if the angle between the force and the direction of movement also varies as a function of position (Fig. 17.18). The work equation can be modified further to account for this effect, as in

$$W = \int_{x_0}^{x_n} F(x)\cos[\theta(x)]\,dx \tag{17.34}$$

Again, if $F(x)$ and $\theta(x)$ are simple functions, Eq. (17.34) might be solved analytically. However, as in Fig. 17.18, it is more likely that the functional relationship is complicated. For this situation, numerical methods provide the only alternative for determining the integral.

Suppose that you have to perform the computation for the situation depicted in Fig. 17.18. Although the figure shows the continuous values for $F(x)$ and $\theta(x)$, assume that, because of experimental constraints, you are provided with only discrete measurements at x = 5-m intervals (Table 17.5). Use single- and multiple-application versions of the trapezoidal rule and Simpson's 1/3 and 3/8 rules to compute work for this data.

Solution. The results of the analysis are summarized in Table 17.6. A percent relative error ε_t was computed in reference to a true value of the integral of 129.52 that was estimated on the basis of values taken from Fig. 17.18 at 1-m intervals.

The results are interesting because the most accurate outcome occurs for the simple two-segment trapezoidal rule. More refined estimates using more segments, as well as Simpson's rules, yield less accurate results.

The reason for this apparently counterintuitive result is that the coarse spacing of the points is not adequate to capture the variations of the forces and angles. This is particularly

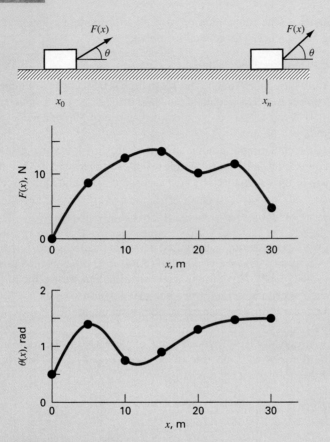

FIGURE 17.18
The case of a variable force acting on a block. For this case the angle, as well as the magnitude, of the force varies.

TABLE 17.5 Data for force $F(x)$ and angle $\theta(x)$ as a function of position x.

x, m	$F(x)$, N	θ, rad	$F(x)\cos\theta$
0	0.0	0.50	0.0000
5	9.0	1.40	1.5297
10	13.0	0.75	9.5120
15	14.0	0.90	8.7025
20	10.5	1.30	2.8087
25	12.0	1.48	1.0881
30	5.0	1.50	0.3537

17.9 CASE STUDY   continued

TABLE 17.6 Estimates of work calculated using the trapezoidal rule and Simpson's rules. The percent relative error ε_t as computed in reference to a true value of the integral (129.52 Pa) that was estimated on the basis of values at 1-m intervals.

Technique	Segments	Work	ε_t, %
Trapezoidal rule	1	5.31	95.9
	2	133.19	2.84
	3	124.98	3.51
	6	119.09	8.05
Simpson's 1/3 rule	2	175.82	35.75
	6	117.13	9.57
Simpson's 3/8 rule	3	139.93	8.04

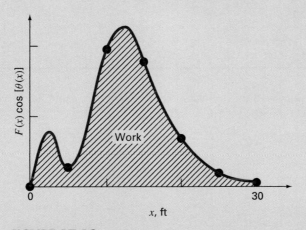

FIGURE 17.19
A continuous plot of $F(x) \cos [\theta(x)]$ versus position with the seven discrete points used to develop the numerical integration estimates in Table 17.6. Notice how the use of seven points to characterize this continuously varying function misses two peaks at $x = 2.5$ and 12.5 m.

evident in Fig. 17.19, where we have plotted the continuous curve for the product of $F(x)$ and $\cos [\theta(x)]$. Notice how the use of seven points to characterize the continuously varying function misses the two peaks at $x = 2.5$ and 12.5 m. The omission of these two points effectively limits the accuracy of the numerical integration estimates in Table 17.6. The fact that the two-segment trapezoidal rule yields the most accurate result is due to the chance positioning of the points for this particular problem (Fig. 17.20).

The conclusion to be drawn from Fig. 17.20 is that an adequate number of measurements must be made to accurately compute integrals. For the present case, if data were available at $F(2.5) \cos [\theta(2.5)] = 3.9007$ and $F(12.5) \cos [\theta(12.5)] = 11.3940$, we could

17.9 CASE STUDY continued

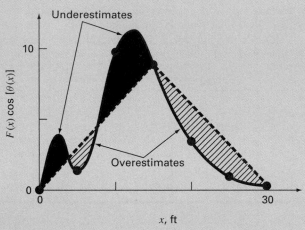

FIGURE 17.20
Graphical depiction of why the two-segment trapezoidal rule yields a good estimate of the integral for this particular case. By chance, the use of two trapezoids happens to lead to an even balance between positive and negative errors.

determine an improved integral estimate. For example, using the MATLAB `trapz` function, we could compute

```
>> x=[0 2.5 5 10 12.5 15 20 25 30];
>> y=[0 3.9007 1.5297 9.5120 11.3940 8.7025 2.8087 ...
                                    1.0881 0.3537];
>> trapz(x,y)

ans =
  132.6458
```

Including the two additional points yields an improved integral estimate of 132.6458 ($\varepsilon_t = 2.16\%$). Thus, the inclusion of the additional data incorporates the peaks that were missed previously and, as a consequence, lead to better results.

PROBLEMS

17.1 Derive Eq. (17.4) by integrating Eq. (17.3).
17.2 Evaluate the following integral:

$$\int_0^4 (1 - e^{-2x})\,dx$$

(a) analytically, **(b)** single application of the trapezoidal rule, **(c)** composite trapezoidal rule with $n = 2$ and 4, **(d)** single

application of Simpson's 1/3 rule, **(e)** composite Simpson's 1/3 rule with $n = 4$, and **(f)** Simpson's 3/8 rule. For each of the numerical estimates **(b)** through **(f)**, determine the percent relative error based on **(a)**.
17.3 Evaluate the following integral:

$$\int_0^{\pi/2} (6 + 3\cos x)\,dx$$

(a) analytically, (b) single application of the trapezoidal rule, (c) composite trapezoidal rule with $n = 2$ and 4, (d) single application of Simpson's 1/3 rule, (e) composite Simpson's 1/3 rule with $n = 4$, and (f) Simpson's 3/8 rule. For each of the numerical estimates (b) through (f), determine the percent relative error based on (a).

17.4 Evaluate the following integral:

$$\int_{-2}^{4} (1 - x - 4x^3 + 2x^5)\, dx$$

(a) analytically, (b) single application of the trapezoidal rule, (c) composite trapezoidal rule with $n = 2$ and 4, (d) single application of Simpson's 1/3 rule, (e) Simpson's 3/8 rule, and (f) Boole's rule. For each of the numerical estimates (b) through (f), determine the percent relative error based on (a).

17.5 The function

$$f(x) = 2e^{-1.5x}$$

can be used to generate the following table of unequally spaced data:

x	0	0.05	0.15	0.25	0.35	0.475	0.6
f(x)	2	1.8555	1.5970	1.3746	1.1831	0.9808	0.8131

Evaluate the integral from $a = 0$ to $b = 0.6$ using (a) analytical means, (b) the trapezoidal rule, and (c) a combination of the trapezoidal and Simpson's rules wherever possible to attain the highest accuracy. For (b) and (c), compute the percent relative error.

17.6 Evaluate the double integral

$$\int_{-1}^{1} \int_{0}^{2} (x^2 - 2y^2 + xy^3)\, dx\, dy$$

(a) analytically, (b) using the composite trapezoidal rule with $n = 2$, and (c) using single applications of Simpson's 1/3 rule. For (b) and (c), compute the percent relative error.

17.7 Evaluate the triple integral

$$\int_{-2}^{2} \int_{0}^{2} \int_{-3}^{1} (x^3 - 3yz)\, dx\, dy\, dz$$

(a) analytically, and (b) using single applications of Simpson's 1/3 rule. For (b), compute the percent relative error.

17.8 Determine the distance traveled from the following velocity data:

t	1	2	3.25	4.5	6	7	8	8.5	9.3	10
v	5	6	5.5	7	8.5	8	6	7	7	5

(a) Use the trapezoidal rule.
(b) Fit the data with a cubic equation using polynomial regression. Integrate the cubic equation to determine the distance.

17.9 Water exerts pressure on the upstream face of a dam as shown in Fig. P17.9. The pressure can be characterized by

$$p(z) = \rho g(D - z)$$

where $p(z)$ = pressure in pascals (or N/m^2) exerted at an elevation z meters above the reservoir bottom; ρ = density of water, which for this problem is assumed to be a constant 10^3 kg/m^3; g = acceleration due to gravity (9.81 m/s^2); and D = elevation (in m) of the water surface above the reservoir bottom. According to Eq. (P17.9), pressure increases linearly with depth, as depicted in Fig. P17.9a. Omitting atmospheric pressure (because it works against both sides of the dam face and essentially cancels out), the total force f_t can be determined by multiplying pressure times the area of the dam face (as shown in Fig. P17.9b). Because both pressure and area vary with elevation, the total force is obtained by evaluating

$$f_t = \int_{0}^{D} \rho g w(z)(D - z)\, dz$$

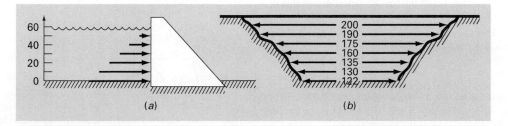

FIGURE P17.9
Water exerting pressure on the upstream face of a dam: (a) side view showing force increasing linearly with depth; (b) front view showing width of dam in meters.

where $w(z)$ = width of the dam face (m) at elevation z (Fig. P17.9b). The line of action can also be obtained by evaluating

$$d = \frac{\int_0^D \rho g z w(z)(D-z)\,dz}{\int_0^D \rho g w(z)(D-z)\,dz}$$

Use Simpson's rule to compute f_t and d.

17.10 The force on a sailboat mast can be represented by the following function:

$$f(z) = 200 \left(\frac{z}{7+z}\right) e^{-2.5z/H}$$

where z = the elevation above the deck and H = the height of the mast. The total force F exerted on the mast can be determined by integrating this function over the height of the mast:

$$F = \int_0^H f(z)\,dz$$

The line of action can also be determined by integration:

$$d = \frac{\int_0^H z f(z)\,dz}{\int_0^H f(z)\,dz}$$

(a) Use the composite trapezoidal rule to compute F and d for the case where $H = 30$ ($n = 6$).
(b) Repeat (a), but use the composite Simpson's 1/3 rule.

17.11 A wind force distributed against the side of a skyscraper is measured as

Height l, m	0	30	60	90	120
Force, $F(l)$, N/m	0	340	1200	1600	2700

Height l, m	150	180	210	240
Force, $F(l)$, N/m	3100	3200	3500	3800

Compute the net force and the line of action due to this distributed wind.

17.12 An 11-m beam is subjected to a load, and the shear force follows the equation

$$V(x) = 5 + 0.25x^2$$

where V is the shear force, and x is length in distance along the beam. We know that $V = dM/dx$, and M is the bending moment. Integration yields the relationship

$$M = M_o + \int_0^x V\,dx$$

If M_o is zero and $x = 11$, calculate M using **(a)** analytical integration, **(b)** multiple-application trapezoidal rule, and **(c)** multiple-application Simpson's rules. For (b) and (c) use 1-m increments.

17.13 The total mass of a variable density rod is given by

$$m = \int_0^L \rho(x) A_c(x)\,dx$$

where m = mass, $\rho(x)$ = density, $A_c(x)$ = cross-sectional area, x = distance along the rod and L = the total length of the rod. The following data has been measured for a 10-m length rod. Determine the mass in grams to the best possible accuracy.

x, m	0	2	3	4	6	8	10
ρ, g/cm³	4.00	3.95	3.89	3.80	3.60	3.41	3.30
A_c, cm²	100	103	106	110	120	133	150

17.14 A transportation engineering study requires that you determine the number of cars that pass through an intersection traveling during morning rush hour. You stand at the side of the road and count the number of cars that pass every 4 minutes at several times as tabulated below. Use the best numerical method to determine **(a)** the total number of cars that pass between 7:30 and 9:15, and **(b)** the rate of cars going through the intersection per minute. (Hint: Be careful with units.)

Time (hr)	7:30	7:45	8:00	8:15	8:45	9:15
Rate (cars per 4 min)	18	24	14	24	21	9

17.15 Determine the average value for the data in Fig. P17.15. Perform the integral needed for the average in the order shown by the following equation:

$$I = \int_{x_0}^{x_n} \left[\int_{y_0}^{y_m} f(x, y)\,dy \right] dx$$

17.16 Integration provides a means to compute how much mass enters or leaves a reactor over a specified time period, as in

$$M = \int_{t_1}^{t_2} Qc\,dt$$

where t_1 and t_2 = the initial and final times, respectively. This formula makes intuitive sense if you recall the analogy between integration and summation. Thus, the integral represents the summation of the product of flow times concentration to give the total mass entering or leaving from t_1 to t_2.

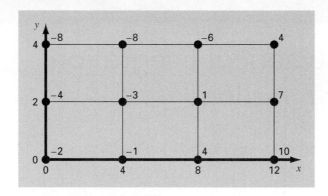

FIGURE P17.15

Use numerical integration to evaluate this equation for the data listed below:

t, min	0	10	20	30	35	40	45	50
Q, m³/min	4	4.8	5.2	5.0	4.6	4.3	4.3	5.0
c, mg/m³	10	35	55	52	40	37	32	34

17.17 The cross-sectional area of a channel can be computed as

$$A_c = \int_0^B H(y)\, dy$$

where B = the total channel width (m), H = the depth (m), and y = distance from the bank (m). In a similar fashion, the average flow Q (m³/s) can be computed as

$$Q = \int_0^B U(y) H(y)\, dy$$

where U = water velocity (m/s). Use these relationships and a numerical method to determine A_c and Q for the following data:

y, m	0	2	4	5	6	9
H, m	0.5	1.3	1.25	1.7	1	0.25
U, m/s	0.03	0.06	0.05	0.12	0.11	0.02

17.18 The average concentration of a substance $\bar{c}$(g/m³) in a lake where the area A_s(m²) varies with depth z(m) can be computed by integration:

$$\bar{c} = \frac{\int_0^Z c(z) A_s(z)\, dz}{\int_0^Z A_s(z)\, dz}$$

where Z = the total depth (m). Determine the average concentration based on the following data:

z, m	0	4	8	12	16
A, 10⁶ m²	9.8175	5.1051	1.9635	0.3927	0.0000
c, g/m³	10.2	8.5	7.4	5.2	4.1

17.19 As was done in Section 17.9, determine the work performed if a constant force of 1 N applied at an angle θ results in the following displacements. Use the cumtrapz function to determine the cumulative work and plot the result versus θ.

x, m	0	1	2.7	3.8	3.7	3	1.4
θ, rad	0	30	60	90	120	150	180

18

Numerical Integration of Functions

CHAPTER OBJECTIVES

The primary objective of this chapter is to introduce you to numerical methods for integrating given functions. Specific objectives and topics covered are

- Understanding how Richardson extrapolation provides a means to create a more accurate integral estimate by combining two less accurate estimates.
- Understanding how Gauss quadrature provides superior integral estimates by picking optimal abscissas at which to evaluate the function.
- Knowing how to use MATLAB's built-in functions `quad` and `quadl` to integrate functions.

18.1 INTRODUCTION

In Chap. 17, we noted that functions to be integrated numerically will typically be of two forms: a table of values or a function. The form of the data has an important influence on the approaches that can be used to evaluate the integral. For tabulated information, you are limited by the number of points that are given. In contrast, if the function is available, you can generate as many values of $f(x)$ as are required to attain acceptable accuracy.

At face value, the composite Simpson's 1/3 rule might seem to be a reasonable tool for such problems. Although it is certainly adequate for many problems, there are more efficient methods that are available. This chapter is devoted to three such techniques. Both capitalize on the ability to generate function values to develop efficient schemes for numerical integration.

The first technique is based on *Richardson extrapolation*, which is a method for combining two numerical integral estimates to obtain a third, more accurate value. The computational algorithm for implementing Richardson extrapolation in a highly efficient manner is called *Romberg integration*. This technique can be used to generate an integral estimate within a prespecified error tolerance.

The second method is called *Gauss quadrature*. Recall that, in Chap. 17, values of $f(x)$ for the Newton-Cotes formulas were determined at specified values of x. For example, if we used the trapezoidal rule to determine an integral, we were constrained to take the weighted average of $f(x)$ at the ends of the interval. Gauss-quadrature formulas employ x values that are positioned between the integration limits in such a manner that a much more accurate integral estimate results.

The third approach is called *adaptive quadrature*. This techniques applies composite Simpson's 1/3 rule to subintervals of the integration range in a way that allows error estimates to be computed. These error estimates are then used to determine whether more refined estimates are required for a subinterval. In this way, more refined segmentation is only used where it is necessary. Two built-in MATLAB functions that use adaptive quadrature are illustrated.

18.2 ROMBERG INTEGRATION

Romberg integration is one technique that is designed to attain efficient numerical integrals of functions. It is quite similar to the techniques discussed in Chap. 17 in the sense that it is based on successive application of the trapezoidal rule. However, through mathematical manipulations, superior results are attained for less effort.

18.2.1 Richardson Extrapolation

Techniques are available to improve the results of numerical integration on the basis of the integral estimates themselves. Generally called *Richardson extrapolation*, these methods use two estimates of an integral to compute a third, more accurate approximation.

The estimate and the error associated with the composite trapezoidal rule can be represented generally as

$$I = I(h) + E(h)$$

where I = the exact value of the integral, $I(h)$ = the approximation from an n-segment application of the trapezoidal rule with step size $h = (b - a)/n$, and $E(h)$ = the truncation error. If we make two separate estimates using step sizes of h_1 and h_2 and have exact values for the error:

$$I(h_1) + E(h_1) = I(h_2) + E(h_2) \tag{18.1}$$

Now recall that the error of the composite trapezoidal rule can be represented approximately by Eq. (17.21) [with $n = (b - a)/h$]:

$$E \cong -\frac{b - a}{12} h^2 \bar{f}'' \tag{18.2}$$

If it is assumed that $\bar{f}''$ is constant regardless of step size, Eq. (18.2) can be used to determine that the ratio of the two errors will be

$$\frac{E(h_1)}{E(h_2)} \cong \frac{h_1^2}{h_2^2} \tag{18.3}$$

This calculation has the important effect of removing the term $\bar{f}''$ from the computation. In so doing, we have made it possible to utilize the information embodied by Eq. (18.2)

without prior knowledge of the function's second derivative. To do this, we rearrange
Eq. (18.3) to give

$$E(h_1) \cong E(h_2)\left(\frac{h_1}{h_2}\right)^2$$

which can be substituted into Eq. (18.1):

$$I(h_1) + E(h_2)\left(\frac{h_1}{h_2}\right)^2 = I(h_2) + E(h_2)$$

which can be solved for

$$E(h_2) = \frac{I(h_1) - I(h_2)}{1 - (h_1/h_2)^2}$$

Thus, we have developed an estimate of the truncation error in terms of the integral estimates and their step sizes. This estimate can then be substituted into

$$I = I(h_2) + E(h_2)$$

to yield an improved estimate of the integral:

$$I = I(h_2) + \frac{1}{(h_1/h_2)^2 - 1}[I(h_2) - I(h_1)] \tag{18.4}$$

It can be shown (Ralston and Rabinowitz, 1978) that the error of this estimate is
$O(h^4)$. Thus, we have combined two trapezoidal rule estimates of $O(h^2)$ to yield a new estimate of $O(h^4)$. For the special case where the interval is halved ($h_2 = h_1/2$), this equation becomes

$$I = \frac{4}{3}I(h_2) - \frac{1}{3}I(h_1) \tag{18.5}$$

EXAMPLE 18.1   Richardson Extrapolation

Problem Statement.   Use Richardson extrapolation to evaluate the integral of $f(x) = 0.2 + 25x - 200x^2 + 675x^3 - 900x^4 + 400x^5$ from $a = 0$ to $b = 0.8$.

Solution.   Single and composite applications of the trapezoidal rule can be used to evaluate the integral:

Segments	h	Integral	ε_t
1	0.8	0.1728	89.5%
2	0.4	1.0688	34.9%
4	0.2	1.4848	9.5%

Richardson extrapolation can be used to combine these results to obtain improved estimates
of the integral. For example, the estimates for one and two segments can be combined

to yield

$$I = \frac{4}{3}(1.0688) - \frac{1}{3}(0.1728) = 1.367467$$

The error of the improved integral is $E_t = 1.640533 - 1.367467 = 0.273067 (\varepsilon_t = 16.6\%)$, which is superior to the estimates upon which it was based.

In the same manner, the estimates for two and four segments can be combined to give

$$I = \frac{4}{3}(1.4848) - \frac{1}{3}(1.0688) = 1.623467$$

which represents an error of $E_t = 1.640533 - 1.623467 = 0.017067 (\varepsilon_t = 1.0\%)$.

Equation (18.4) provides a way to combine two applications of the trapezoidal rule with error $O(h^2)$ to compute a third estimate with error $O(h^4)$. This approach is a subset of a more general method for combining integrals to obtain improved estimates. For instance, in Example 18.1, we computed two improved integrals of $O(h^4)$ on the basis of three trapezoidal rule estimates. These two improved integrals can, in turn, be combined to yield an even better value with $O(h^6)$. For the special case where the original trapezoidal estimates are based on successive halving of the step size, the equation used for $O(h^6)$ accuracy is

$$I = \frac{16}{15}I_m - \frac{1}{15}I_l \tag{18.6}$$

where I_m and I_l are the more and less accurate estimates, respectively. Similarly, two $O(h^6)$ results can be combined to compute an integral that is $O(h^8)$ using

$$I = \frac{64}{63}I_m - \frac{1}{63}I_l \tag{18.7}$$

EXAMPLE 18.2   Higher-Order Corrections

Problem Statement.   In Example 18.1, we used Richardson extrapolation to compute two integral estimates of $O(h^4)$. Utilize Eq. (18.6) to combine these estimates to compute an integral with $O(h^6)$.

Solution.   The two integral estimates of $O(h^4)$ obtained in Example 18.1 were 1.367467 and 1.623467. These values can be substituted into Eq. (18.6) to yield

$$I = \frac{16}{15}(1.623467) - \frac{1}{15}(1.367467) = 1.640533$$

which is the exact value of the integral.

18.2.2 The Romberg Integration Algorithm

Notice that the coefficients in each of the extrapolation equations [Eqs. (18.5), (18.6), and (18.7)] add up to 1. Thus, they represent weighting factors that, as accuracy increases, place relatively greater weight on the superior integral estimate. These formulations can be expressed in a general form that is well suited for computer implementation:

$$I_{j,k} = \frac{4^{k-1}I_{j+1,k-1} - I_{j,k-1}}{4^{k-1} - 1} \tag{18.8}$$

where $I_{j+1,k-1}$ and $I_{j,k-1}$ = the more and less accurate integrals, respectively, and $I_{j,k}$ = the improved integral. The index k signifies the level of the integration, where $k = 1$ corresponds to the original trapezoidal rule estimates, $k = 2$ corresponds to the $O(h^4)$ estimates, $k = 3$ to the $O(h^6)$, and so forth. The index j is used to distinguish between the more $(j + 1)$ and the less (j) accurate estimates. For example, for $k = 2$ and $j = 1$, Eq. (18.8) becomes

$$I_{1,2} = \frac{4I_{2,1} - I_{1,1}}{3}$$

which is equivalent to Eq. (18.5).

The general form represented by Eq. (18.8) is attributed to Romberg, and its systematic application to evaluate integrals is known as *Romberg integration*. Figure 18.1 is a graphical depiction of the sequence of integral estimates generated using this approach. Each matrix corresponds to a single iteration. The first column contains the trapezoidal rule evaluations that are designated $I_{j,1}$, where $j = 1$ is for a single-segment application (step size is $b - a$), $j = 2$ is for a two-segment application [step size is $(b - a)/2$], $j = 3$ is for a four-segment application [step size is $(b - a)/4$], and so forth. The other columns of the matrix are generated by systematically applying Eq. (18.8) to obtain successively better estimates of the integral.

For example, the first iteration (Fig. 18.1a) involves computing the one- and two-segment trapezoidal rule estimates ($I_{1,1}$ and $I_{2,1}$). Equation (18.8) is then used to compute the element $I_{1,2} = 1.367467$, which has an error of $O(h^4)$.

FIGURE 18.1
Graphical depiction of the sequence of integral estimates generated using Romberg integration. (a) First iteration. (b) Second iteration. (c) Third iteration.

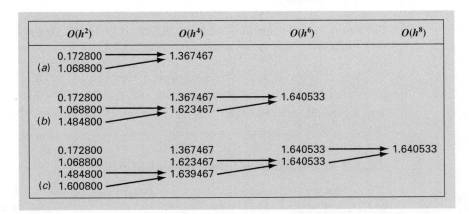

Now, we must check to determine whether this result is adequate for our needs. As in other approximate methods in this book, a termination, or stopping, criterion is required to assess the accuracy of the results. One method that can be employed for the present purposes is

$$|\varepsilon_a| = \left| \frac{I_{1,k} - I_{2,k-1}}{I_{1,k}} \right| \times 100\% \tag{18.9}$$

where ε_a = an estimate of the percent relative error. Thus, as was done previously in other iterative processes, we compare the new estimate with a previous value. For Eq. (18.9), the previous value is the most accurate estimate from the previous level of integration (i.e., the $k - 1$ level of integration with $j = 2$). When the change between the old and new values as represented by ε_a is below a prespecified error criterion ε_s, the computation is terminated. For Fig. 18.1a, this evaluation indicates the following percent change over the course of the first iteration:

$$|\varepsilon_a| = \left| \frac{1.367467 - 1.068800}{1.367467} \right| \times 100\% = 21.8\%$$

The object of the second iteration (Fig. 18.1b) is to obtain the $O(h^6)$ estimate—$I_{1,3}$. To do this, a four-segment trapezoidal rule estimate, $I_{3,1} = 1.4848$, is determined. Then it is combined with $I_{2,1}$ using Eq. (18.8) to generate $I_{2,2} = 1.623467$. The result is, in turn, combined with $I_{1,2}$ to yield $I_{1,3} = 1.640533$. Equation (18.9) can be applied to determine that this result represents a change of 1.0% when compared with the previous result $I_{2,2}$.

The third iteration (Fig. 18.1c) continues the process in the same fashion. In this case, an eight-segment trapezoidal estimate is added to the first column, and then Eq. (18.8) is applied to compute successively more accurate integrals along the lower diagonal. After only three iterations, because we are evaluating a fifth-order polynomial, the result ($I_{1,4} = 1.640533$) is exact.

Romberg integration is more efficient than the trapezoidal rule and Simpson's rules. For example, for determination of the integral as shown in Fig. 18.1, Simpson's 1/3 rule would require about a 48-segment application in double precision to yield an estimate of the integral to seven significant digits: 1.640533. In contrast, Romberg integration produces the same result based on combining one-, two-, four-, and eight-segment trapezoidal rules—that is, with only 15 function evaluations!

Figure 18.2 presents an M-file for Romberg integration. By using loops, this algorithm implements the method in an efficient manner. Note that the function uses another function `trap` to implement the composite trapezoidal rule evaluations (recall Fig. 17.10). Here is a MATLAB session showing how it can be used to determine the integral of the polynomial from Example 18.1:

```
>> f=@(x) 0.2+25*x-200*x^2+675*x^3-900*x^4+400*x^5;
>> romberg(f,0,0.8)

ans =
    1.6405
```

```
function [q,ea,iter]=romberg(func,a,b,es,maxit,varargin)
% romberg: Romberg integration quadrature
%   q = romberg(func,a,b,es,maxit,p1,p2,...):
%                    Romberg integration.
% input:
%   func = name of function to be integrated
%   a, b = integration limits
%   es = desired relative error (default = 0.000001%)
%   maxit = maximum allowable iterations (default = 30)
%   p1,p2,... = additional parameters used by func
% output:
%   q = integral estimate
%   ea = approximate relative error (%)
%   iter = number of iterations

if nargin<3,error('at least 3 input arguments required'),end
if nargin<4|isempty(es), es=0.000001;end
if nargin<5|isempty(maxit), maxit=50;end
n = 1;
I(1,1) = trap(func,a,b,n,varargin{:});
iter = 0;
while iter<maxit
  iter = iter+1;
  n = 2^iter;
  I(iter+1,1) = trap(func,a,b,n,varargin{:});
  for k = 2:iter+1
    j = 2+iter-k;
    I(j,k) = (4^(k-1)*I(j+1,k-1)-I(j,k-1))/(4^(k-1)-1);
  end
  ea = abs((I(1,iter+1)-I(2,iter))/I(1,iter+1))*100;
  if ea<=es, break; end
end
q = I(1,iter+1);
```

FIGURE 18.2
M-file to implement Romberg integration.

18.3 GAUSS QUADRATURE

In Chap. 17, we employed the Newton-Cotes equations. A characteristic of these formulas (with the exception of the special case of unequally spaced data) was that the integral estimate was based on evenly spaced function values. Consequently, the location of the base points used in these equations was predetermined or fixed.

For example, as depicted in Fig. 18.3a, the trapezoidal rule is based on taking the area under the straight line connecting the function values at the ends of the integration interval. The formula that is used to compute this area is

$$I \cong (b-a)\frac{f(a)+f(b)}{2}$$ (18.10)

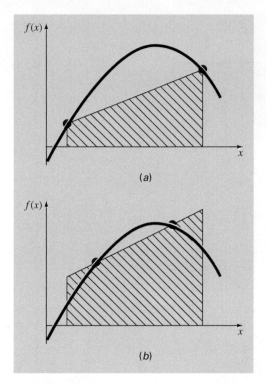

FIGURE 18.3
(a) Graphical depiction of the trapezoidal rule as the area under the straight line joining fixed end points. (b) An improved integral estimate obtained by taking the area under the straight line passing through two intermediate points. By positioning these points wisely, the positive and negative errors are better balanced, and an improved integral estimate results.

where a and b = the limits of integration and $b - a$ = the width of the integration interval. Because the trapezoidal rule must pass through the end points, there are cases such as Fig. 18.3a where the formula results in a large error.

Now, suppose that the constraint of fixed base points was removed and we were free to evaluate the area under a straight line joining any two points on the curve. By positioning these points wisely, we could define a straight line that would balance the positive and negative errors. Hence, as in Fig. 18.3b, we would arrive at an improved estimate of the integral.

Gauss quadrature is the name for a class of techniques to implement such a strategy. The particular Gauss quadrature formulas described in this section are called *Gauss-Legendre* formulas. Before describing the approach, we will show how numerical integration formulas such as the trapezoidal rule can be derived using the method of undetermined coefficients. This method will then be employed to develop the Gauss-Legendre formulas.

18.3.1 Method of Undetermined Coefficients

In Chap. 17, we derived the trapezoidal rule by integrating a linear interpolating polynomial and by geometrical reasoning. The method of undetermined coefficients offers a third approach that also has utility in deriving other integration techniques such as Gauss quadrature.

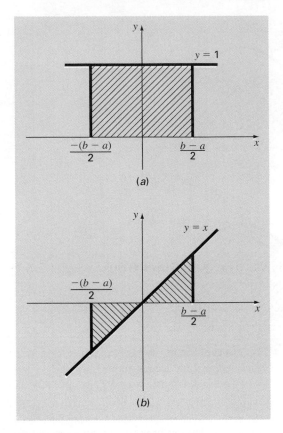

FIGURE 18.4
Two integrals that should be evaluated exactly by the trapezoidal rule: (a) a constant and (b) a straight line.

To illustrate the approach, Eq. (18.10) is expressed as

$$I \cong c_0 f(a) + c_1 f(b) \tag{18.11}$$

where the c's = constants. Now realize that the trapezoidal rule should yield exact results when the function being integrated is a constant or a straight line. Two simple equations that represent these cases are $y = 1$ and $y = x$ (Fig. 18.4). Thus, the following equalities should hold:

$$c_0 + c_1 = \int_{-(b-a)/2}^{(b-a)/2} 1 \, dx$$

and

$$-c_0 \frac{b-a}{2} + c_1 \frac{b-a}{2} = \int_{-(b-a)/2}^{(b-a)/2} x \, dx$$

or, evaluating the integrals,

$$c_0 + c_1 = b - a$$

and

$$-c_0 \frac{b-a}{2} + c_1 \frac{b-a}{2} = 0$$

These are two equations with two unknowns that can be solved for

$$c_0 = c_1 = \frac{b-a}{2}$$

which, when substituted back into Eq. (18.11), gives

$$I = \frac{b-a}{2} f(a) + \frac{b-a}{2} f(b)$$

which is equivalent to the trapezoidal rule.

18.3.2 Derivation of the Two-Point Gauss-Legendre Formula

Just as was the case for the previous derivation of the trapezoidal rule, the object of Gauss quadrature is to determine the coefficients of an equation of the form

$$I \cong c_0 f(x_0) + c_1 f(x_1) \tag{18.12}$$

where the c's = the unknown coefficients. However, in contrast to the trapezoidal rule that used fixed end points a and b, the function arguments x_0 and x_1 are not fixed at the end points, but are unknowns (Fig. 18.5). Thus, we now have a total of four unknowns that must be evaluated, and consequently, we require four conditions to determine them exactly.

Just as for the trapezoidal rule, we can obtain two of these conditions by assuming that Eq. (18.12) fits the integral of a constant and a linear function exactly. Then, to arrive at the other two conditions, we merely extend this reasoning by assuming that it also fits the integral of a parabolic ($y = x^2$) and a cubic ($y = x^3$) function. By doing this, we determine

FIGURE 18.5
Graphical depiction of the unknown variables x_0 and x_1 for integration by Gauss quadrature.

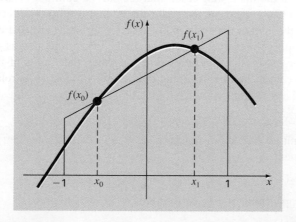

all four unknowns and in the bargain derive a linear two-point integration formula that is exact for cubics. The four equations to be solved are

$$c_0 + c_1 = \int_{-1}^{1} 1 \, dx = 2 \tag{18.13}$$

$$c_0 x_0 + c_1 x_1 = \int_{-1}^{1} x \, dx = 0 \tag{18.14}$$

$$c_0 x_0^2 + c_1 x_1^2 = \int_{-1}^{1} x^2 \, dx = \frac{2}{3} \tag{18.15}$$

$$c_0 x_0^3 + c_1 x_1^3 = \int_{-1}^{1} x^3 \, dx = 0 \tag{18.16}$$

Equations (18.13) through (18.16) can be solved simultaneously for the four unknowns. First, solve Eq. (18.14) for c_1 and substitute the result into Eq. (18.16), which can be solved for

$$x_0^2 = x_1^2$$

Since x_0 and x_1 cannot be equal, this means that $x_0 = -x_1$. Substituting this result into Eq. (18.14) yields $c_0 = c_1$. Consequently from Eq. (18.13) it follows that

$$c_0 = c_1 = 1$$

Substituting these results into Eq. (18.15) gives

$$x_0 = -\frac{1}{\sqrt{3}} = -0.5773503\ldots$$

$$x_1 = \frac{1}{\sqrt{3}} = 0.5773503\ldots$$

Therefore, the two-point Gauss-Legendre formula is

$$I = f\left(\frac{-1}{\sqrt{3}}\right) + f\left(\frac{1}{\sqrt{3}}\right) \tag{18.17}$$

Thus, we arrive at the interesting result that the simple addition of the function values at $x = -1/\sqrt{3}$ and $1/\sqrt{3}$ yields an integral estimate that is third-order accurate.

Notice that the integration limits in Eqs. (18.13) through (18.16) are from -1 to 1. This was done to simplify the mathematics and to make the formulation as general as possible. A simple change of variable can be used to translate other limits of integration into this form. This is accomplished by assuming that a new variable x_d is related to the original variable x in a linear fashion, as in

$$x = a_1 + a_2 x_d \tag{18.18}$$

If the lower limit, $x = a$, corresponds to $x_d = -1$, these values can be substituted into Eq. (18.18) to yield

$$a = a_1 + a_2(-1) \tag{18.19}$$

Similarly, the upper limit, $x = b$, corresponds to $x_d = 1$, to give

$$b = a_1 + a_2(1) \tag{18.20}$$

Equations (18.19) and (18.20) can be solved simultaneously for

$$a_1 = \frac{b+a}{2} \quad \text{and} \quad a_2 = \frac{b-a}{2} \tag{18.21}$$

which can be substituted into Eq. (18.18) to yield

$$x = \frac{(b+a)+(b-a)x_d}{2} \tag{18.22}$$

This equation can be differentiated to give

$$dx = \frac{b-a}{2} dx_d \tag{18.23}$$

Equations (18.22) and (18.23) can be substituted for x and dx, respectively, in the equation to be integrated. These substitutions effectively transform the integration interval without changing the value of the integral. The following example illustrates how this is done in practice.

EXAMPLE 18.3   Two-Point Gauss-Legendre Formula

Problem Statement.   Use Eq. (18.17) to evaluate the integral of

$$f(x) = 0.2 + 25x - 200x^2 + 675x^3 - 900x^4 + 400x^5$$

between the limits $x = 0$ to 0.8. The exact value of the integral is 1.640533.

Solution.   Before integrating the function, we must perform a change of variable so that the limits are from -1 to $+1$. To do this, we substitute $a = 0$ and $b = 0.8$ into Eqs. (18.22) and (18.23) to yield

$$x = 0.4 + 0.4x_d \quad \text{and} \quad dx = 0.4dx_d$$

Both of these can be substituted into the original equation to yield

$$\int_0^{0.8} (0.2 + 25x - 200x^2 + 675x^3 - 900x^4 + 400x^5)\, dx$$

$$= \int_{-1}^{1} [0.2 + 25(0.4 + 0.4x_d) - 200(0.4 + 0.4x_d)^2 + 675(0.4 + 0.4x_d)^3$$

$$- 900(0.4 + 0.4x_d)^4 + 400(0.4 + 0.4x_d)^5]0.4dx_d$$

Therefore, the right-hand side is in the form that is suitable for evaluation using Gauss quadrature. The transformed function can be evaluated at $x_d = -1/\sqrt{3}$ as 0.516741 and at $x_d = 1/\sqrt{3}$ as 1.305837. Therefore, the integral according to Eq. (18.17) is $0.516741 + 1.305837 = 1.822578$, which represents a percent relative error of -11.1%. This result is comparable in magnitude to a four-segment application of the trapezoidal rule or a single application of Simpson's 1/3 and 3/8 rules. This latter result is to be expected because Simpson's rules are also third-order accurate. However, because of the clever choice of base points, Gauss quadrature attains this accuracy on the basis of only two function evaluations.

TABLE 18.1 Weighting factors and function arguments used in Gauss-Legendre formulas.

Points	Weighting Factors	Function Arguments	Truncation Error
1	$c_0 = 2$	$x_0 = 0.0$	$\cong f^{(2)}(\xi)$
2	$c_0 = 1$ $c_1 = 1$	$x_0 = -1/\sqrt{3}$ $x_1 = 1/\sqrt{3}$	$\cong f^{(4)}(\xi)$
3	$c_0 = 5/9$ $c_1 = 8/9$ $c_2 = 5/9$	$x_0 = -\sqrt{3/5}$ $x_1 = 0.0$ $x_2 = \sqrt{3/5}$	$\cong f^{(6)}(\xi)$
4	$c_0 = (18 - \sqrt{30})/36$ $c_1 = (18 + \sqrt{30})/36$ $c_2 = (18 + \sqrt{30})/36$ $c_3 = (18 - \sqrt{30})/36$	$x_0 = -\sqrt{525 + 70\sqrt{30}}/35$ $x_1 = -\sqrt{525 - 70\sqrt{30}}/35$ $x_2 = \sqrt{525 - 70\sqrt{30}}/35$ $x_3 = \sqrt{525 + 70\sqrt{30}}/35$	$\cong f^{(8)}(\xi)$
5	$c_0 = (322 - 13\sqrt{70})/900$ $c_1 = (322 + 13\sqrt{70})/900$ $c_2 = 128/225$ $c_3 = (322 + 13\sqrt{70})/900$ $c_4 = (322 - 13\sqrt{70})/900$	$x_0 = -\sqrt{245 + 14\sqrt{70}}/21$ $x_1 = -\sqrt{245 - 14\sqrt{70}}/21$ $x_2 = 0.0$ $x_3 = \sqrt{245 - 14\sqrt{70}}/21$ $x_4 = \sqrt{245 + 14\sqrt{70}}/21$	$\cong f^{(10)}(\xi)$
6	$c_0 = 0.171324492379170$ $c_1 = 0.360761573048139$ $c_2 = 0.467913934572691$ $c_3 = 0.467913934572691$ $c_4 = 0.360761573048131$ $c_5 = 0.171324492379170$	$x_0 = -0.932469514203152$ $x_1 = -0.661209386466265$ $x_2 = -0.238619186083197$ $x_3 = 0.238619186083197$ $x_4 = 0.661209386466265$ $x_5 = 0.932469514203152$	$\cong f^{(12)}(\xi)$

18.3.3 Higher-Point Formulas

Beyond the two-point formula described in the previous section, higher-point versions can be developed in the general form

$$I \cong c_0 f(x_0) + c_1 f(x_1) + \cdots + c_{n-1} f(x_{n-1}) \tag{18.24}$$

where $n =$ the number of points. Values for c's and x's for up to and including the six-point formula are summarized in Table 18.1.

EXAMPLE 18.4   Three-Point Gauss-Legendre Formula

Problem Statement.   Use the three-point formula from Table 18.1 to estimate the integral for the same function as in Example 18.3.

Solution.   According to Table 18.1, the three-point formula is

$$I = 0.5555556 f(-0.7745967) + 0.8888889 f(0) + 0.5555556 f(0.7745967)$$

which is equal to

$$I = 0.2813013 + 0.8732444 + 0.4859876 = 1.640533$$

which is exact.

Because Gauss quadrature requires function evaluations at nonuniformly spaced points within the integration interval, it is not appropriate for cases where the function is unknown. Thus, it is not suited for engineering problems that deal with tabulated data. However, where the function is known, its efficiency can be a decided advantage. This is particularly true when numerous integral evaluations must be performed.

18.4 ADAPTIVE QUADRATURE

Although the composite Simpson's 1/3 rule can certainly be used to estimate the integral of given functions, it has the disadvantage that it uses equally spaced points. This constraint does not take into account that some functions have regions of relatively abrupt changes where more refined spacing might be required. Hence, to achieve a desired accuracy, the fine spacing must be applied everywhere even though it is only needed for the regions of sharp change. Adaptive quadrature methods remedy this situation by automatically adjusting the step size so that small steps are taken in regions of sharp variations and larger steps are taken where the function changes gradually.

Most of these techniques are based on applying the composite Simpson's 1/3 rule to subintervals in a manner similar to how the trapezoidal rule was used in Romberg integration. That is, the 1/3 rule is applied at two levels of refinement and the difference between these two levels is used to estimate the truncation error. If the truncation error is acceptable, no further refinement is required and the integral estimate for the subinterval is deemed acceptable. If the error estimate is too large, the step size is refined and the process repeated until the error falls to acceptable levels.

MATLAB includes two built-in functions to implement adaptive quadrature: `quad` and `quadl`. The following section describes how they can be applied.

18.4.1 MATLAB Functions: `quad` and `quadl`

MATLAB has two functions, both based on algorithms developed by Gander and Gautschi (2000), for implementing adaptive quadrature:

- **quad.** This function uses adaptive Simpson quadrature. It may be more efficient for low accuracies or nonsmooth functions.
- **quadl.** This function uses what is called *Lobatto quadrature*. It may be more efficient for high accuracies and smooth functions.

The following function syntax for the `quad` function is the same for the `quadl` function:

```
q = quad(fun, a, b, tol, trace, p1, p2, . . .)
```

where `fun` is the function to be integrated, a and b = the integration bounds, tol = the desired absolute error tolerance (default = 10^{-6}), `trace` is a variable that when set to a nonzero value causes additional computational detail to be displayed, and `p1, p2, . . .` are parameters that you want to pass to `fun`. It should be noted that array operators `.*`, `./` and `.^` should be used in the definition of `fun`. In addition, pass empty matrices for `tol` or `trace` to use the default values.

EXAMPLE 18.5   Adaptive Quadrature

Problem Statement.   Use quad to integrate the following function:

$$f(x) = \frac{1}{(x-q)^2 + 0.01} + \frac{1}{(x-r)^2 + 0.04} - s$$

between the limits $x = 0$ to 1. Note that for $q = 0.3$, $r = 0.9$, and $s = 6$, this is the built-in humps function that MATLAB uses to demonstrate some of its numerical capabilities. The humps function exhibits both flat and steep regions over a relatively short x range. Hence, it is useful for demonstrating and testing routines like quad and quadl. Note that the humps function can be integrated analytically between the given limits to yield an exact integral of 29.85832539549867.

Solution.   First, let's evaluate the integral in the simplest way possible, using the built-in version of humps along with the default tolerance:

```
>> format long
>> quad(@humps,0,1)

ans =
   29.85832612842764
```

Thus, the solution is correct to seven significant digits.

Next, we can solve the same problem, but using a looser tolerance and passing q, r, and s as parameters. First, we can develop an M-file for the function:

```
function y = myhumps(x,q,r,s)
y = 1./((x-q).^2 + 0.01) + 1./((x-r).^2+0.04) - s;
```

Then, we can integrate it with an error tolerance of 10^{-4} as in

```
>> quad(@myhumps,0,1,1e-4,[],0.3,0.9,6)

ans =
   29.85812133214492
```

Notice that because we used a larger tolerance, the result is now only accurate to five significant digits. However, although it would not be apparent from a single application, fewer function evaluations were made and, hence, the computation executes faster.

18.5 CASE STUDY   ROOT-MEAN-SQUARE CURRENT

Background.   Because it results in efficient energy transmission, the current in an AC circuit is often in the form of a sine wave:

$$i = i_{peak} \sin(\omega t)$$

where i = the current (A = C/s), i_{peak} = the peak current (A), ω = the angular frequency (radians/s) and t = time (s). The angular frequency is related to the period T(s) by $\omega = 2\pi/T$.

18.5 CASE STUDY continued

The power generated is related to the magnitude of the current. Integration can be used to determine the average current over one cycle:

$$\bar{i} = \frac{1}{T} \int_0^T i_{\text{peak}} \sin(\omega t)\, dt = \frac{i_{\text{peak}}}{T} (-\cos(2\pi) + \cos(0)) = 0$$

Despite the fact that the average is zero, such a current is capable of generating power. Therefore, an alternative to the average current must be derived.

To do this, electrical engineers and scientists determine the root mean square current i_{rms} (A), which is calculated as

$$i_{\text{rms}} = \sqrt{\frac{1}{T} \int_0^T i_{\text{peak}}^2 \sin^2(\omega t)\, dt} = \frac{i_{\text{peak}}}{\sqrt{2}} \tag{18.25}$$

Thus, as the name implies, the rms current is the square root of the mean of the squared current. Because $1/\sqrt{2} = 0.70707$, i_{rms} is equal to about 70% of the peak current for our assumed sinusoidal wave form.

This quantity has meaning because it is directly related to the average power absorbed by an element in an AC circuit. To understand this, recall that *Joule's law* states that the instantaneous power absorbed by a circuit element is equal to the product of the voltage across it and the current through it:

$$P = iV \tag{18.26}$$

where P = the power (W = J/s), and V = voltage (V = J/C). For a resistor, *Ohm's law* states that the voltage is directly proportional to the current:

$$V = iR \tag{18.27}$$

where R = the resistance (Ω = V/A = J $\cdot$ s/C^2). Substituting Eq. (18.27) into (18.26) gives

$$P = i^2 R \tag{18.28}$$

The average power can be determined by integrating Eq. (18.28) over a period with the result:

$$\bar{P} = i_{\text{rms}}^2 R$$

Thus, the AC circuit generates the equivalent power as a DC circuit with a constant current of i_{rms}.

Now, although the simple sinusoid is widely employed, it is by no means the only waveform that is used. For some of these forms, such as triangular or square waves, the i_{rms} can be evaluated analytically with closed-form integration. However, some waveforms must be analyzed with numerical integration methods.

In this case study, we will calculate the root-mean-square current of the waveform shown in Fig. 18.3 for $T = 1$ s. We will use both the Newton-Cotes formulas from Chap. 17 as well as the approaches described in this chapter.

Solution. The integral that must be evaluated is

$$i_{\text{rms}}^2 = \int_0^{1/2} (10e^{-t} \sin 2\pi t)^2 \, dt \tag{18.29}$$

For comparative purposes, the exact value of this integral to fifteen significant digits is 15.41260804810169.

Integral estimates for various applications of the trapezoidal rule and Simpson's 1/3 rule are listed in Table 18.2. Notice that Simpson's rule is more accurate than the trapezoidal rule. The value for the integral to seven significant digits is obtained using a 128-segment trapezoidal rule or a 32-segment Simpson's rule.

The M-file we developed in Fig. 18.2 can be used to evaluate the integral with Romberg integration:

```
>> format long
>> i2=@(t) (10*exp(-t).*sin(2*pi*t)).^2;
>> [q,ea,iter]=romberg(i2,0,.5)

q =
  15.41260804288977
ea =
    1.480058787326946e-008
iter =
    5
```

Thus, with the default stopping criterion of $\text{es} = 1 \times 10^{-6}$, we obtain a result that is correct to over nine significant figures in five iterations. We can obtain an even better result if we impose a more stringent stopping criterion:

```
>> [q,ea,iter]=romberg(i2,0,.5,1e-15)

q =
  15.41260804810169
ea =
    0
iter =
    7
```

Gauss quadrature can also be used to make the same estimate. First, a change in variable is performed by applying Eqs. (18.22) and (18.23) to yield

$$t = \frac{1}{4} + \frac{1}{4} t_d \qquad dt = \frac{1}{4} \, dt_d$$

These relationships can be substituted into Eq. (18.29) to yield

$$i_{\text{rms}}^2 = \int_{-1}^{1} \left[10e^{-(0.25+0.25t_d)} \sin 2\pi (0.25 + 0.25t_d) \right]^2 0.25 \, dt \tag{18.30}$$

TABLE 18.2 Values for the integral calculated using Newton-Cotes formulas.

Technique	Segments	Integral	ε_t (%)
Trapezoidal rule	1	0.0	100.0000
	2	15.163266493	1.6178
	4	15.401429095	0.0725
	8	15.411958360	4.22×10^{-3}
	16	15.412568151	2.59×10^{-4}
	32	15.412605565	1.61×10^{-5}
	64	15.412607893	1.01×10^{-6}
	128	15.412608038	6.28×10^{-8}
Simpson's 1/3 rule	2	20.217688657	31.1763
	4	15.480816629	0.4426
	8	15.415468115	0.0186
	16	15.412771415	1.06×10^{-3}
	32	15.412618037	6.48×10^{-5}

TABLE 18.3 Results of using various-point Gauss quadrature formulas to approximate the integral.

Points	Estimate	ε_t (%)
2	11.9978243	22.1
3	15.6575502	1.59
4	15.4058023	4.42×10^{-2}
5	15.4126391	2.01×10^{-4}
6	15.4126109	1.82×10^{-5}

For the two-point Gauss-Legendre formula, this function is evaluated at $t_d = -1/\sqrt{3}$ and $1/\sqrt{3}$, with the results being 7.684096 and 4.313728, respectively. These values can be substituted into Eq. (18.17) to yield an integral estimate of 11.99782, which represents an error of $\varepsilon_t = 22.1\%$.

The three-point formula is (Table 18.1)

$$I = 0.5555556(1.237449) + 0.8888889(15.16327) + 0.5555556(2.684915) = 15.65755$$

which has $\varepsilon_t = 1.6\%$. The results of using the higher-point formulas are summarized in Table 18.3.

Finally, the integral can be evaluated with the built-in MATLAB function quad and quadl:

```
>> irms2=quad(i2,0,.5)

irms2 =
  15.41260804934509
```

```
>> irms2=quadl(i2,0,.5)

irms2 =
   15.41260804809967
```

Both these results are very accurate, with quadl being a little better.

We can now compute the i_{rms} by merely taking the square root of the integral. For example, using the result computed with quadl, we get

```
>> irms=sqrt(irms2)

irms =
   3.92588945948554
```

This result could then be employed to guide other aspects of the design and operation of the circuit such as power dissipation computations.

As we did for the simple sinusoid in Eq. (18.25), an interesting calculation involves comparing this result with the peak current. Recognizing that this is an optimization problem, we can readily employ the fminbnd function to determine this value. Because we are looking for a maximum, we evaluate the negative of the function:

```
>> [tmax,imax]=fminbnd(@(t) -10*exp(-t).*sin(2*pi*t),0,.5)

tmax =
    0.22487940319321
imax =
   -7.88685387393258
```

A maximum current of 7.88685 A occurs at $t = 0.2249$ s. Hence, for this particular wave form, the root-mean-square value is about 49.8% of the maximum.

PROBLEMS

18.1 Use Romberg integration to evaluate

$$I = \int_1^2 \left(2x + \frac{3}{x}\right)^2 dx$$

to an accuracy of $\varepsilon_s = 0.5\%$. Your results should be presented in the format of Fig. 18.1. Use the analytical solution of the integral to determine the percent relative error of the result obtained with Romberg integration. Check that ε_t is less than ε_s.

18.2 Evaluate the following integral **(a)** analytically, **(b)** Romberg integration ($\varepsilon_s = 0.5\%$), **(c)** the three-point

Gauss quadrature formula, and **(d)** MATLAB quad function:

$$I = \int_0^8 -0.0547x^4 + 0.8646x^3 - 4.1562x^2 + 6.2917x + 2 \, dx$$

18.3 Evaluate the following integral with **(a)** Romberg integration ($\varepsilon_s = 0.5\%$), **(b)** the two-point Gauss quadrature formula, and **(c)** MATLAB quad and quadl functions:

$$I = \int_0^3 xe^x \, dx$$

18.4 There is no closed form solution for the error function

$$\text{erf}(a) = \frac{2}{\sqrt{\pi}} \int_0^a e^{-x^2} dx$$

Use the **(a)** two-point and **(b)** three-point Gauss-Legendre formulas to estimate erf(1.5). Determine the percent relative error for each case based on the true value, which can be determined with MATLAB's built-in function `erf`.

18.5 The force on a sailboat mast can be represented by the following function:

$$F = \int_0^H 200 \left(\frac{z}{7+z} \right) e^{-2.5z/H} dz$$

where z = the elevation above the deck and H = the height of the mast. Compute F for the case where $H = 30$ using **(a)** Romberg integration to a tolerance of $\varepsilon_s = 0.5\%$, **(b)** the two-point Gauss-Legendre formula, and **(c)** the MATLAB `quad` function.

18.6 The root-mean-square current can be computed as

$$I_{RMS} = \sqrt{\frac{1}{T} \int_0^T i^2(t) \, dt}$$

For $T = 1$, suppose that $i(t)$ is defined as

$$i(t) = 10e^{-t/T} \sin\left(2\pi \frac{t}{T} \right) \qquad \text{for } 0 \le t \le T/2$$

$$i(t) = 0 \qquad \text{for } T/2 \le t \le T$$

Evaluate the I_{RMS} using **(a)** Romberg integration to a tolerance of 0.1%, **(b)** the two- and three-point Gauss-Legendre formulas, and **(c)** the MATLAB `quad` function.

18.7 The velocity profile of a fluid in a circular pipe can be represented as

$$v = 10 \left(1 - \frac{r}{r_0} \right)^{1/n}$$

where v = velocity, r = radial distance measured out from the pipes centerline, r_0 = the pipe's radius, and n = a parameter. Determine the flow in the pipe if $r_0 = 0.75$ and $n = 7$ using **(a)** Romberg integration to a tolerance of 0.1%, **(b)** the two-point Gauss-Legendre formula, and **(c)** the MATLAB `quad` function. Note that flow is equal to velocity times area.

18.8 The amount of mass transported via a pipe over a period of time can be computed as

$$M = \int_{t_1}^{t_2} Q(t)c(t) \, dt$$

where M = mass (mg), t_1 = the initial time (min), t_2 = the final time (min), $Q(t)$ = flow rate (m³/min), and $c(t)$ = concentration (mg/m³). The following functional representations define the temporal variations in flow and concentration:

$$Q(t) = 9 + 4\cos^2(0.4t)$$

$$c(t) = 5e^{-0.5t} + 2e^{0.15t}$$

Determine the mass transported between $t_1 = 2$ and $t_2 = 8$ min with **(a)** Romberg integration to a tolerance of 0.1% and **(b)** the MATLAB `quad` function.

18.9 Evaluate the double integral

$$\int_{-2}^2 \int_0^4 (x^2 - 3y^2 + xy^3) \, dx \, dy$$

(a) analytically and **(b)** using the MATLAB `dblquad` function. Use `help` to understand how to implement the function.

18.10 Compute work as described in Sec. 17.9, but use the following equations for $F(x)$ and $\theta(x)$:

$$F(x) = 1.6x - 0.045x^2$$

$$\theta(x) = -0.00055x^3 + 0.0123x^2 + 0.13x$$

The force is in newtons and the angle is in radians. Perform the integration from $x = 0$ to 30 m.

18.11 Perform the same computation as in Sec. 18.5, but for the current as specified by

$$i(t) = 5e^{-1.25t} \sin 2\pi t \qquad \text{for } 0 \le t \le T/2$$

$$i(t) = 0 \qquad \text{for } T/2 < t \le T$$

where $T = 1$ s.

18.12 Compute the power absorbed by an element in a circuit as described in Sec. 18.5, but for a simple sinusoidal current $i = \sin(2\pi t/T)$ where $T = 1$ s.

(a) Assume that Ohm's law holds and $R = 5 \, \Omega$.

(b) Assume that Ohm's law does not hold and that voltage and current are related by the following nonlinear relationship: $V = (5i - 1.25i^3)$.

18.13 Suppose that the current through a resistor is described by the function

$$i(t) = (60 - t)^2 + (60 - t) \sin\left(\sqrt{t} \right)$$

and the resistance is a function of the current:

$$R = 10i + 2i^{2/3}$$

Compute the average voltage over $t = 0$ to 60 using the multiple-segment Simpson's 1/3 rule.

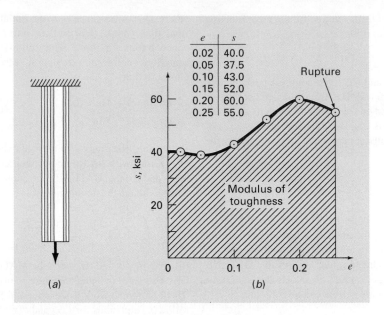

e	s
0.02	40.0
0.05	37.5
0.10	43.0
0.15	52.0
0.20	60.0
0.25	55.0

(a)   (b)

FIGURE P18.16
(a) A rod under axial loading and (b) the resulting stress-strain curve, where stress is in kips per square inch (10^3 lb/in^2), and strain is dimensionless.

18.14 If a capacitor initially holds no charge, the voltage across it as a function of time can be computed as

$$V(t) = \frac{1}{C} \int_0^t i(t)\, dt$$

If $C = 10^{-5}$ farad, use the following current data to develop a plot of voltage versus time:

t, s	0	0.2	0.4	0.6
i, 10^{-3} A	0.2	0.3683	0.3819	0.2282

t, s	0.8	1	1.2
i, 10^{-3} A	0.0486	0.0082	0.1441

18.15 The work done on an object is equal to the force times the distance moved in the direction of the force. The velocity of an object in the direction of a force is given by

$$v = 4t \qquad\qquad 0 \le t \le 4$$
$$v = 16 + (4 - t)^2 \qquad 4 \le t \le 14$$

where v is in m/s. Determine the work if a constant force of 200 N is applied for all t.

18.16 A rod subject to an axial load (Fig. P18.16a) will be deformed, as shown in the stress-strain curve in Fig. P18.16b.

The area under the curve from zero stress out to the point of rupture is called the *modulus of toughness* of the material. It provides a measure of the energy per unit volume required to cause the material to rupture. As such, it is representative of the material's ability to withstand an impact load. Use numerical integration to compute the modulus of toughness for the stress-strain curve seen in Fig. P18.16b.

18.17 If the velocity distribution of a fluid flowing through a pipe is known (Fig. P18.17), the flow rate Q (that is, the volume of water passing through the pipe per unit time) can be computed by $Q = \int v\, dA$, where v is the velocity, and A is the pipe's cross-sectional area. (To grasp the meaning of this relationship physically, recall the close connection between summation and integration.) For a circular pipe, $A = \pi r^2$ and $dA = 2\pi r\, dr$. Therefore,

$$Q = \int_0^r v(2\pi r)\, dr$$

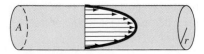

FIGURE P18.17

where r is the radial distance measured outward from the center of the pipe. If the velocity distribution is given by

$$v = 2 \left(1 - \frac{r}{r_0} \right)^{1/6}$$

where r_0 is the total radius (in this case, 3 cm), compute Q using the multiple-application trapezoidal rule. Discuss the results.

18.18 Using the following data, calculate the work done by stretching a spring that has a spring constant of $k = 300$ N/m to $x = 0.35$ m:

F, $10^3 \cdot$ N	0	0.01	0.028	0.046
x, m	0	0.05	0.10	0.15
F, $10^3 \cdot$ N	0.063	0.082	0.11	0.13
x, m	0.20	0.25	0.30	0.35

18.19 Evaluate the vertical distance traveled by a rocket if the vertical velocity is given by

$$v = 11t^2 - 5t \qquad\qquad 0 \le t \le 10$$
$$v = 1100 - 5t \qquad\qquad 10 \le t \le 20$$
$$v = 50t + 2(t - 20)^2 \qquad 20 \le t \le 30$$

18.20 The upward velocity of a rocket can be computed by the following formula:

$$v = u \ln \left(\frac{m_0}{m_0 - qt} \right) - gt$$

where $v =$ upward velocity, $u =$ velocity at which fuel is expelled relative to the rocket, $m_0 =$ initial mass of the rocket at time $t = 0$, $q =$ fuel consumption rate, and $g =$ downward acceleration of gravity (assumed constant $= 9.8$ m/s^2). If $u = 1800$ m/s, $m_0 = 160{,}000$ kg, and $q = 2500$ kg/s, determine how high the rocket will fly in 30 s.

18.21 The normal distribution is defined as

$$f(x) = \frac{1}{\sqrt{2\pi}} e^{-x^2/2}$$

(a) Use MATLAB to integrate this function from $x = -1$ to 1 and from -2 to 2.

(b) Use MATLAB to determine the inflection points of this function.

18.22 Use Romberg integration to evaluate

$$\int_0^2 \frac{e^x \sin x}{1 + x^2} \, dx$$

to an accuracy of $\varepsilon_s = 0.5\%$. Your results should be presented in the form of Fig. 18.1.

19

Numerical Differentiation

YOU'VE GOT A PROBLEM

Recall that the velocity of a free-falling bungee jumper as a function of time can be computed as

$$v(t) = \sqrt{\frac{gm}{c_d}} \tanh\left(\sqrt{\frac{gc_d}{m}}t\right) \tag{19.1}$$

At the beginning of Chap. 17, we used calculus to integrate this equation to determine the vertical distance z the jumper has fallen after a time t.

$$z(t) = \frac{m}{c_d} \ln\left[\cosh\left(\sqrt{\frac{gc_d}{m}}t\right)\right] \tag{19.2}$$

Now suppose that you were given the reverse problem. That is, you were asked to determine velocity based on the jumper's position as a function of time. Because it is the inverse of integration, differentiation could be used to make the determination:

$$v(t) = \frac{dz(t)}{dt} \tag{19.3}$$

Substituting Eq. (19.2) into Eq. (19.3) and differentiating would bring us back to Eq. (19.1).

Beyond velocity, you might also be asked to compute the jumper's acceleration. To do this, we could either take the first derivative of velocity, or the second derivative of displacement:

$$a(t) = \frac{dv(t)}{dt} = \frac{d^2 z(t)}{dt^2} \tag{19.4}$$

In either case, the result would be

$$a(t) = g \, \text{sech}^2 \left(\sqrt{\frac{g c_d}{m}} t \right) \tag{19.5}$$

Although a closed-form solution can be developed for this case, there are other functions that may be difficult or impossible to differentiate analytically. Further, suppose that there was some way to measure the jumper's position at various times during the fall. These distances along with their associated times could be assembled as a table of discrete values. In this situation, it would be useful to differentiate the discrete data to determine the velocity and the acceleration. In both these instances, numerical differentiation methods are available to obtain solutions. This chapter will introduce you to some of these methods.

19.1 INTRODUCTION AND BACKGROUND

19.1.1 What Is Differentiation?

Calculus is the mathematics of change. Because engineers and scientists must continuously deal with systems and processes that change, calculus is an essential tool of our profession. Standing at the heart of calculus is the mathematical concept of differentiation.

According to the dictionary definition, to *differentiate* means "to mark off by differences; distinguish; . . . to perceive the difference in or between." Mathematically, the *derivative,* which serves as the fundamental vehicle for differentiation, represents the rate of change of a dependent variable with respect to an independent variable. As depicted in Fig. 19.1, the mathematical definition of the derivative begins with a difference approximation:

$$\frac{\Delta y}{\Delta x} = \frac{f(x_i + \Delta x) - f(x_i)}{\Delta x} \tag{19.6}$$

where y and $f(x)$ are alternative representatives for the dependent variable and x is the independent variable. If Δx is allowed to approach zero, as occurs in moving from Fig. 19.1a to c, the difference becomes a derivative:

$$\frac{dy}{dx} = \lim_{\Delta x \to 0} \frac{f(x_i + \Delta x) - f(x_i)}{\Delta x} \tag{19.7}$$

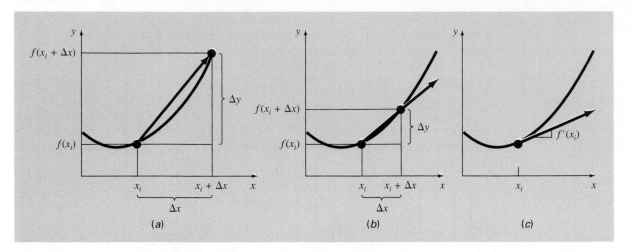

FIGURE 19.1

The graphical definition of a derivative: as Δx approaches zero in going from (a) to (c), the difference approximation becomes a derivative.

where dy/dx [which can also be designated as y' or $f'(x_i)$][1] is the first derivative of y with respect to x evaluated at x_i. As seen in the visual depiction of Fig. 19.1c, the derivative is the slope of the tangent to the curve at x_i.

The second derivative represents the derivative of the first derivative,

$$\frac{d^2y}{dx^2} = \frac{d}{dx}\left(\frac{dy}{dx}\right) \tag{19.8}$$

Thus, the second derivative tells us how fast the slope is changing. It is commonly referred to as the *curvature,* because a high value for the second derivative means high curvature.

Finally, partial derivatives are used for functions that depend on more than one variable. Partial derivatives can be thought of as taking the derivative of the function at a point with all but one variable held constant. For example, given a function f that depends on both x and y, the partial derivative of f with respect to x at an arbitrary point (x, y) is defined as

$$\frac{\partial f}{\partial x} = \lim_{\Delta x \to 0} \frac{f(x + \Delta x, y) - f(x, y)}{\Delta x} \tag{19.9}$$

Similarly, the partial derivative of f with respect to y is defined as

$$\frac{\partial f}{\partial y} = \lim_{\Delta y \to 0} \frac{f(x, y + \Delta y) - f(x, y)}{\Delta y} \tag{19.10}$$

To get an intuitive grasp of partial derivatives, recognize that a function that depends on two variables is a surface rather than a curve. Suppose you are mountain climbing and have access to a function f that yields elevation as a function of longitude (the east-west oriented

---

[1] The form dy/dx was devised by Leibnitz, whereas y' is attributed to Lagrange. Note that Newton used the so-called dot notation: $\dot{y}$. Today, the dot notation is usually used for time derivatives.

x axis) and latitude (the north-south oriented y axis). If you stop at a particular point (x_0, y_0), the slope to the east would be $\partial f(x_0, y_0)/\partial x$, and the slope to the north would be $\partial f(x_0, y_0)/\partial y$.

19.1.2 Differentiation in Engineering and Science

The differentiation of a function has so many engineering and scientific applications that you were required to take differential calculus in your first year at college. Many specific examples of such applications could be given in all fields of engineering and science. Differentiation is commonplace in engineering and science because so much of our work involves characterizing the changes of variables in both time and space. In fact, many of the laws and other generalizations that figure so prominently in our work are based on the predictable ways in which change manifests itself in the physical world. A prime example is Newton's second law, which is not couched in terms of the position of an object but rather in its change with respect to time.

Aside from such temporal examples, numerous laws involving the spatial behavior of variables are expressed in terms of derivatives. Among the most common of these are the *constitutive laws* that define how potentials or gradients influence physical processes. For example, *Fourier's law of heat conduction* quantifies the observation that heat flows from regions of high to low temperature. For the one-dimensional case, this can be expressed mathematically as

$$q = -k\frac{dT}{dx} \tag{19.11}$$

where $q(x)$ = heat flux (W/m²), k = coefficient of thermal conductivity [W/(m · K)], T = temperature (K), and x = distance (m). Thus, the derivative, or *gradient,* provides a measure of the intensity of the spatial temperature change, which drives the transfer of heat (Fig. 19.2).

FIGURE 19.2

Graphical depiction of a temperature gradient. Because heat moves "downhill" from high to low temperature, the flow in (a) is from left to right. However, due to the orientation of Cartesian coordinates, the slope is negative for this case. Thus, a negative gradient leads to a positive flow. This is the origin of the minus sign in Fourier's law of heat conduction. The reverse case is depicted in (b), where the positive gradient leads to a negative heat flow from right to left.

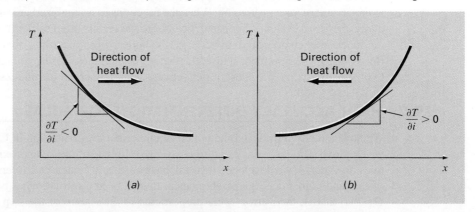

TABLE 19.1 The one-dimensional forms of some constitutive laws commonly used in engineering and science.

Law	Equation	Physical Area	Gradient	Flux	Proportionality
Fourier's law	$q = -k\dfrac{dT}{dx}$	Heat conduction	Temperature	Heat flux	Thermal Conductivity
Fick's law	$J = -D\dfrac{dc}{dx}$	Mass diffusion	Concentration	Mass flux	Diffusivity
D'Arcy's law	$q = -k\dfrac{dh}{dx}$	Flow through porous media	Head	Flow flux	Hydraulic Conductivity
Ohm's law	$J = -\sigma\dfrac{dV}{dx}$	Current flow	Voltage	Current flux	Electrical Conductivity
Newton's viscosity law	$\tau = \mu\dfrac{du}{dx}$	Fluids	Velocity	Shear Stress	Dynamic Viscosity
Hooke's law	$\sigma = E\dfrac{\Delta L}{L}$	Elasticity	Deformation	Stress	Young's Modulus

Similar laws provide workable models in many other areas of engineering and science, including the modeling of fluid dynamics, mass transfer, chemical reaction kinetics, electricity, and solid mechanics (Table 19.1). The ability to accurately estimate derivatives is an important facet of our capability to work effectively in these areas.

Beyond direct engineering and scientific applications, numerical differentiation is also important in a variety of general mathematical contexts including other areas of numerical methods. For example, recall that in Chap. 6 the secant method was based on a finite-difference approximation of the derivative. In addition, probably the most important application of numerical differentiation involves the solution of differential equations. We have already seen an example in the form of Euler's method in Chap. 1. In Chap. 22, we will investigate how numerical differentiation provides the basis for solving boundary-value problems of ordinary differential equations.

These are just a few of the applications of differentiation that you might face regularly in the pursuit of your profession. When the functions to be analyzed are simple, you will normally choose to evaluate them analytically. However, it is often difficult or impossible when the function is complicated. In addition, the underlying function is often unknown and defined only by measurement at discrete points. For both these cases, you must have the ability to obtain approximate values for derivatives, using numerical techniques as described next.

19.2 HIGH-ACCURACY DIFFERENTIATION FORMULAS

We have already introduced the notion of numerical differentiation in Chap. 4. Recall that we employed Taylor series expansions to derive finite-difference approximations of derivatives. In Chap. 4, we developed forward, backward, and centered difference approximations of first and higher derivatives. Remember that, at best, these estimates had errors that were $O(h^2)$—that is, their errors were proportional to the square of the step size. This level of accuracy is due to the number of terms of the Taylor series that were retained during the derivation of these formulas. We will now illustrate how high-accuracy finite-difference formulas can be generated by including additional terms from the Taylor series expansion.

For example, the forward Taylor series expansion can be written as [recall Eq. (4.13)]

$$f(x_{i+1}) = f(x_i) + f'(x_i)h + \frac{f''(x_i)}{2!}h^2 + \cdots \tag{19.12}$$

which can be solved for

$$f'(x_i) = \frac{f(x_{i+1}) - f(x_i)}{h} - \frac{f''(x_i)}{2!}h + O(h^2) \tag{19.13}$$

In Chap. 4, we truncated this result by excluding the second- and higher-derivative terms and were thus left with a forward-difference formula:

$$f'(x_i) = \frac{f(x_{i+1}) - f(x_i)}{h} + O(h) \tag{19.14}$$

In contrast to this approach, we now retain the second-derivative term by substituting the following forward-difference approximation of the second derivative [recall Eq. (4.27)]:

$$f''(x_i) = \frac{f(x_{i+2}) - 2f(x_{i+1}) + f(x_i)}{h^2} + O(h) \tag{19.15}$$

into Eq. (19.13) to yield

$$f'(x_i) = \frac{f(x_{i+1}) - f(x_i)}{h} - \frac{f(x_{i+2}) - 2f(x_{i+1}) + f(x_i)}{2h^2}h + O(h^2) \tag{19.16}$$

or, by collecting terms:

$$f'(x_i) = \frac{-f(x_{i+2}) + 4f(x_{i+1}) - 3f(x_i)}{2h} + O(h^2) \tag{19.17}$$

Notice that inclusion of the second-derivative term has improved the accuracy to $O(h^2)$. Similar improved versions can be developed for the backward and centered formulas as well as for the approximations of higher-order derivatives. The formulas are summarized in Fig. 19.3 through Fig. 19.5 along with the lower-order versions from Chap. 4. The following example illustrates the utility of these formulas for estimating derivatives.

EXAMPLE 19.1   High-Accuracy Differentiation Formulas

Problem Statement.   Recall that in Example 4.4 we estimated the derivative of

$$f(x) = -0.1x^4 - 0.15x^3 - 0.5x^2 - 0.25x + 1.2$$

at $x = 0.5$ using finite-differences and a step size of $h = 0.25$. The results are summarized in the following table. Note that the errors are based on the true value of $f'(0.5) = -0.9125$.

	Backward $O(h)$	Centered $O(h^2)$	Forward $O(h)$
Estimate	−0.714	−0.934	−1.155
ε_t	21.7%	−2.4%	−26.5%

Repeat this computation, but employ the high-accuracy formulas from Fig. 19.3 through Fig. 19.5.

First Derivative   Error

$$f'(x_i) = \frac{f(x_{i+1}) - f(x_i)}{h}$$ $O(h)$

$$f'(x_i) = \frac{-f(x_{i+2}) + 4f(x_{i+1}) - 3f(x_i)}{2h}$$ $O(h^2)$

Second Derivative

$$f''(x_i) = \frac{f(x_{i+2}) - 2f(x_{i+1}) + f(x_i)}{h^2}$$ $O(h)$

$$f''(x_i) = \frac{-f(x_{i+3}) + 4f(x_{i+2}) - 5f(x_{i+1}) + 2f(x_i)}{h^2}$$ $O(h^2)$

Third Derivative

$$f'''(x_i) = \frac{f(x_{i+3}) - 3f(x_{i+2}) + 3f(x_{i+1}) - f(x_i)}{h^3}$$ $O(h)$

$$f'''(x_i) = \frac{-3f(x_{i+4}) + 14f(x_{i+3}) - 24f(x_{i+2}) + 18f(x_{i+1}) - 5f(x_i)}{2h^3}$$ $O(h^2)$

Fourth Derivative

$$f''''(x_i) = \frac{f(x_{i+4}) - 4f(x_{i+3}) + 6f(x_{i+2}) - 4f(x_{i+1}) + f(x_i)}{h^4}$$ $O(h)$

$$f''''(x_i) = \frac{-2f(x_{i+5}) + 11f(x_{i+4}) - 24f(x_{i+3}) + 26f(x_{i+2}) - 14f(x_{i+1}) + 3f(x_i)}{h^4}$$ $O(h^2)$

FIGURE 19.3
Forward finite-difference formulas: two versions are presented for each derivative. The latter version incorporates more terms of the Taylor series expansion and is, consequently, more accurate.

Solution. The data needed for this example are

$x_{i-2} = 0$	$f(x_{i-2}) = 1.2$
$x_{i-1} = 0.25$	$f(x_{i-1}) = 1.1035156$
$x_i = 0.5$	$f(x_i) = 0.925$
$x_{i+1} = 0.75$	$f(x_{i+1}) = 0.6363281$
$x_{i+2} = 1$	$f(x_{i+2}) = 0.2$

The forward difference of accuracy $O(h^2)$ is computed as (Fig. 19.3)

$$f'(0.5) = \frac{-0.2 + 4(0.6363281) - 3(0.925)}{2(0.25)} = -0.859375 \qquad \varepsilon_t = 5.82\%$$

The backward difference of accuracy $O(h^2)$ is computed as (Fig. 19.4)

$$f'(0.5) = \frac{3(0.925) - 4(1.1035156) + 1.2}{2(0.25)} = -0.878125 \qquad \varepsilon_t = 3.77\%$$

The centered difference of accuracy $O(h^4)$ is computed as (Fig. 19.5)

$$f'(0.5) = \frac{-0.2 + 8(0.6363281) - 8(1.1035156) + 1.2}{12(0.25)} = -0.9125 \qquad \varepsilon_t = 0\%$$

First Derivative   Error

$$f'(x_i) = \frac{f(x_i) - f(x_{i-1})}{h}$$ $O(h)$

$$f'(x_i) = \frac{3f(x_i) - 4f(x_{i-1}) + f(x_{i-2})}{2h}$$ $O(h^2)$

Second Derivative

$$f''(x_i) = \frac{f(x_i) - 2f(x_{i-1}) + f(x_{i-2})}{h^2}$$ $O(h)$

$$f''(x_i) = \frac{2f(x_i) - 5f(x_{i-1}) + 4f(x_{i-2}) - f(x_{i-3})}{h^2}$$ $O(h^2)$

Third Derivative

$$f'''(x_i) = \frac{f(x_i) - 3f(x_{i-1}) + 3f(x_{i-2}) - f(x_{i-3})}{h^3}$$ $O(h)$

$$f'''(x_i) = \frac{5f(x_i) - 18f(x_{i-1}) + 24f(x_{i-2}) - 14f(x_{i-3}) + 3f(x_{i-4})}{2h^3}$$ $O(h^2)$

Fourth Derivative

$$f''''(x_i) = \frac{f(x_i) - 4f(x_{i-1}) + 6f(x_{i-2}) - 4f(x_{i-3}) + f(x_{i-4})}{h^4}$$ $O(h)$

$$f''''(x_i) = \frac{3f(x_i) - 14f(x_{i-1}) + 26f(x_{i-2}) - 24f(x_{i-3}) + 11f(x_{i-4}) - 2f(x_{i-5})}{h^4}$$ $O(h^2)$

FIGURE 19.4
Backward finite-difference formulas: two versions are presented for each derivative. The latter version incorporates more terms of the Taylor series expansion and is, consequently, more accurate.

As expected, the errors for the forward and backward differences are considerably more accurate than the results from Example 4.4. However, surprisingly, the centered difference yields the exact derivative at $x = 0.5$. This is because the formula based on the Taylor series is equivalent to passing a fourth-order polynomial through the data points.

19.3 RICHARDSON EXTRAPOLATION

To this point, we have seen that there are two ways to improve derivative estimates when employing finite differences: (1) decrease the step size or (2) use a higher-order formula that employs more points. A third approach, based on Richardson extrapolation, uses two derivative estimates to compute a third, more accurate, approximation.

Recall from Sec. 18.2.1 that Richardson extrapolation provided a means to obtain an improved integral estimate by the formula [Eq. (18.4)]

$$I = I(h_2) + \frac{1}{(h_1/h_2)^2 - 1}[I(h_2) - I(h_1)] \tag{19.18}$$

where $I(h_1)$ and $I(h_2)$ are integral estimates using two step sizes: h_1 and h_2. Because of its convenience when expressed as a computer algorithm, this formula is usually written for the case where $h_2 = h_1/2$, as in

$$I = \frac{4}{3}I(h_2) - \frac{1}{3}I(h_1) \tag{19.19}$$

First Derivative	Error
$f'(x_i) = \dfrac{f(x_{i+1}) - f(x_{i-1})}{2h}$	$O(h^2)$
$f'(x_i) = \dfrac{-f(x_{i+2}) + 8f(x_{i+1}) - 8f(x_{i-1}) + f(x_{i-2})}{12h}$	$O(h^4)$

Second Derivative

$$f''(x_i) = \frac{f(x_{i+1}) - 2f(x_i) + f(x_{i-1})}{h^2} \qquad O(h^2)$$

$$f''(x_i) = \frac{-f(x_{i+2}) + 16f(x_{i+1}) - 30f(x_i) + 16f(x_{i-1}) - f(x_{i-2})}{12h^2} \qquad O(h^4)$$

Third Derivative

$$f'''(x_i) = \frac{f(x_{i+2}) - 2f(x_{i+1}) + 2f(x_{i-1}) - f(x_{i-2})}{2h^3} \qquad O(h^2)$$

$$f'''(x_i) = \frac{-f(x_{i+3}) + 8f(x_{i+2}) - 13f(x_{i+1}) + 13f(x_{i-1}) - 8f(x_{i-2}) + f(x_{i-3})}{8h^3} \qquad O(h^4)$$

Fourth Derivative

$$f''''(x_i) = \frac{f(x_{i+2}) - 4f(x_{i+1}) + 6f(x_i) - 4f(x_{i-1}) + f(x_{i-2})}{h^4} \qquad O(h^2)$$

$$f''''(x_i) = \frac{-f(x_{i+3}) + 12f(x_{i+2}) + 39f(x_{i+1}) + 56f(x_i) - 39f(x_{i-1}) + 12f(x_{i-2}) + f(x_{i-3})}{6h^4} \qquad O(h^4)$$

FIGURE 19.5
Centered finite-difference formulas: two versions are presented for each derivative. The latter version incorporates more terms of the Taylor series expansion and is, consequently, more accurate.

In a similar fashion, Eq. (19.19) can be written for derivatives as

$$D = \frac{4}{3}D(h_2) - \frac{1}{3}D(h_1) \tag{19.20}$$

For centered difference approximations with $O(h^2)$, the application of this formula will yield a new derivative estimate of $O(h^4)$.

EXAMPLE 19.2   Richardson Extrapolation

Problem Statement.   Using the same function as in Example 19.1, estimate the first derivative at $x = 0.5$ employing step sizes of $h_1 = 0.5$ and $h_2 = 0.25$. Then use Eq. (19.20) to compute an improved estimate with Richardson extrapolation. Recall that the true value is -0.9125.

Solution.   The first-derivative estimates can be computed with centered differences as

$$D(0.5) = \frac{0.2 - 1.2}{1} = -1.0 \qquad \varepsilon_t = -9.6\%$$

and

$$D(0.25) = \frac{0.6363281 - 1.103516}{0.5} = -0.934375 \qquad \varepsilon_t = -2.4\%$$

The improved estimate can be determined by applying Eq. (19.20) to give

$$D = \frac{4}{3}(-0.934375) - \frac{1}{3}(-1) = -0.9125$$

which for the present case is exact.

The previous example yielded an exact result because the function being analyzed was a fourth-order polynomial. The exact outcome was due to the fact that Richardson extrapolation is actually equivalent to fitting a higher-order polynomial through the data and then evaluating the derivatives by centered divided differences. Thus, the present case matched the derivative of the fourth-order polynomial precisely. For most other functions, of course, this would not occur, and our derivative estimate would be improved but not exact. Consequently, as was the case for the application of Richardson extrapolation, the approach can be applied iteratively using a Romberg algorithm until the result falls below an acceptable error criterion.

19.4 DERIVATIVES OF UNEQUALLY SPACED DATA

The approaches discussed to this point are primarily designed to determine the derivative of a given function. For the finite-difference approximations of Sec. 19.2, the data had to be evenly spaced. For the Richardson extrapolation technique of Sec. 19.3, the data also had to be evenly spaced and generated for successively halved intervals. Such control of data spacing is usually available only in cases where we can use a function to generate a table of values.

In contrast, empirically derived information—that is, data from experiments or field studies—are often collected at unequal intervals. Such information cannot be analyzed with the techniques discussed to this point.

One way to handle nonequispaced data is to fit a Lagrange interpolating polynomial [recall Eq. (15.21)] to a set of adjacent points that bracket the location value at which you want to evaluate the derivative. Remember that this polynomial does not require that the points be equispaced. The polynomial can then be differentiated analytically to yield a formula that can be used to estimate the derivative.

For example, you can fit a second-order Lagrange polynomial to three adjacent points (x_0, y_0), (x_1, y_1), and (x_2, y_2). Differentiating the polynomial yields:

$$f'(x) = f(x_0)\frac{2x - x_1 - x_2}{(x_0 - x_1)(x_0 - x_2)} + f(x_1)\frac{2x - x_0 - x_2}{(x_1 - x_0)(x_1 - x_2)}$$
$$+ f(x_2)\frac{2x - x_0 - x_1}{(x_2 - x_0)(x_2 - x_1)} \tag{19.21}$$

where x is the value at which you want to estimate the derivative. Although this equation is certainly more complicated than the first-derivative approximation from Fig. 19.3 through Fig. 19.5, it has some important advantages. First, it can provide estimates anywhere within the range prescribed by the three points. Second, the points themselves do not have to be equally spaced. Third, the derivative estimate is of the same accuracy as the centered difference [Eq. (4.25)]. In fact, for equispaced points, Eq. (19.21) evaluated at $x = x_1$ reduces to Eq. (4.25).

EXAMPLE 19.3 Differentiating Unequally Spaced Data

Problem Statement. As in Fig. 19.6, a temperature gradient can be measured down into the soil. The heat flux at the soil-air interface can be computed with Fourier's law (Table 19.1):

$$q(z = 0) = -k \frac{dT}{dz}\bigg|_{z=0}$$

where $q(x)$ = heat flux (W/m²), k = coefficient of thermal conductivity for soil [= 0.5 W/(m · K)], T = temperature (K), and z = distance measured down from the surface into the soil (m). Note that a positive value for flux means that heat is transferred from the air to the soil. Use numerical differentiation to evaluate the gradient at the soil-air interface and employ this estimate to determine the heat flux into the ground.

Solution. Equation (19.21) can be used to calculate the derivative at the air-soil interface as

$$f'(0) = 13.5 \frac{2(0) - 0.0125 - 0.0375}{(0 - 0.0125)(0 - 0.0375)} + 12 \frac{2(0) - 0 - 0.0375}{(0.0125 - 0)(0.0125 - 0.0375)}$$

$$+ 10 \frac{2(0) - 0 - 0.0125}{(0.0375 - 0)(0.0375 - 0.0125)}$$

$$= -1440 + 1440 - 133.333 = -133.333 \text{ K/m}$$

which can be used to compute

$$q(z = 0) = -0.5 \frac{W}{m\,K}\left(-133.333 \frac{K}{m}\right) = 66.667 \frac{W}{m^2}$$

FIGURE 19.6
Temperature versus depth into the soil.

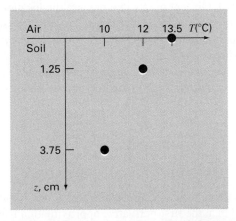

19.5 DERIVATIVES AND INTEGRALS FOR DATA WITH ERRORS

Aside from unequal spacing, another problem related to differentiating empirical data is that it usually includes measurement error. A shortcoming of numerical differentiation is that it tends to amplify errors in the data.

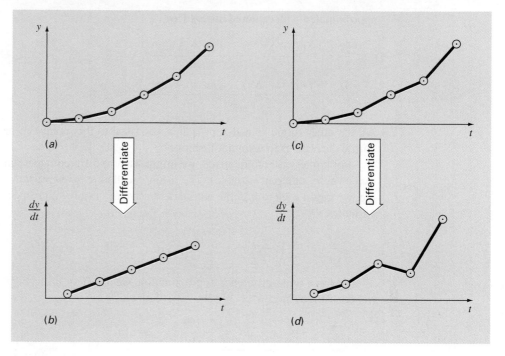

FIGURE 19.7
Illustration of how small data errors are amplified by numerical differentiation: (a) data with no error, (b) the resulting numerical differentiation of curve (a), (c) data modified slightly, and (d) the resulting differentiation of curve (c) manifesting increased variability. In contrast, the reverse operation of integration [moving from (d) to (c) by taking the area under (d)] tends to attenuate or smooth data errors.

Fig. 19.7a shows smooth, error-free data that when numerically differentiated yields a smooth result (Fig. 19.7b). In contrast, Fig. 19.7c uses the same data, but with alternating points raised and lowered slightly. This minor modification is barely apparent from Fig. 19.7c. However, the resulting effect in Fig. 19.7d is significant.

The error amplification occurs because differentiation is subtractive. Hence, random positive and negative errors tend to add. In contrast, the fact that integration is a summing process makes it very forgiving with regard to uncertain data. In essence, as points are summed to form an integral, random positive and negative errors cancel out.

As might be expected, the primary approach for determining derivatives for imprecise data is to use least-squares regression to fit a smooth, differentiable function to the data. In the absence of any other information, a lower-order polynomial regression might be a good first choice. Obviously, if the true functional relationship between the dependent and independent variable is known, this relationship should form the basis for the least-squares fit.

19.6 PARTIAL DERIVATIVES

Partial derivatives along a single dimension are computed in the same fashion as ordinary derivatives. For example, suppose that we want to determine to partial derivatives for a two-dimensional function $f(x, y)$. For equally spaced data, the partial first derivatives can be approximated with centered differences:

$$\frac{\partial f}{\partial x} = \frac{f(x + \Delta x, y) - f(x - \Delta x, y)}{2\Delta x} \tag{19.22}$$

$$\frac{\partial f}{\partial y} = \frac{f(x, y + \Delta y) - f(x, y - \Delta y)}{2\Delta y} \tag{19.23}$$

All the other formulas and approaches discussed to this point can be applied to evaluate partial derivatives in a similar fashion.

For higher-order derivatives, we might want to differentiate a function with respect to two or more different variables. The result is called a *mixed partial derivative*. For example, we might want to take the partial derivative of $f(x, y)$ with respect to both independent variables

$$\frac{\partial^2 f}{\partial x \partial y} = \frac{\partial}{\partial x}\left(\frac{\partial f}{\partial y}\right) \tag{19.24}$$

To develop a finite-difference approximation, we can first form a difference in x of the partial derivatives in y:

$$\frac{\partial^2 f}{\partial x \partial y} = \frac{\dfrac{\partial f}{\partial y}(x + \Delta x, y) - \dfrac{\partial f}{\partial y}(x - \Delta x, y)}{2\Delta x} \tag{19.25}$$

Then, we can use finite differences to evaluate each of the partials in y:

$$\frac{\partial^2 f}{\partial x \partial y} = \frac{\dfrac{f(x + \Delta x, y + \Delta y) - f(x + \Delta x, y - \Delta y)}{2\Delta y} - \dfrac{f(x - \Delta x, y + \Delta y) - f(x - \Delta x, y - \Delta y)}{2\Delta y}}{2\Delta x} \tag{19.26}$$

Collecting terms yields the final result

$$\frac{\partial^2 f}{\partial x \partial y} = \frac{f(x + \Delta x, y + \Delta y) - f(x + \Delta x, y - \Delta y) - f(x - \Delta x, y + \Delta y) + f(x - \Delta x, y - \Delta y)}{4\Delta x \Delta y} \tag{19.27}$$

19.7 NUMERICAL DIFFERENTIATION WITH MATLAB

MATLAB software has the ability to determine the derivatives of data based on two built-in functions: `diff` and `gradient`.

19.7.1 MATLAB Function: `diff`

When it is passed a one-dimensional vector of length n, the `diff` function returns a vector of length $n - 1$ containing the differences between adjacent elements. As described in the following example, these can then be employed to determine finite-difference approximations of first derivatives.

EXAMPLE 19.4   Using `diff` for Differentiation

Problem Statement.   Explore how the MATLAB `diff` function can be employed to differentiate the function

$$f(x) = 0.2 + 25x - 200x^2 + 675x^3 - 900x^4 + 400x^5$$

from $x = 0$ to 0.8. Compare your results with the exact solution:

$$f'(x) = 25 - 400x^2 + 2025x^2 - 3600x^3 + 2000x^4$$

Solution.   We can first express $f(x)$ as an anonymous function:

```
>> f=@(x) 0.2+25*x-200*x.^2+675*x.^3-900*x.^4+400*x.^5;
```

We can then generate a series of equally spaced values of the independent and dependent variables:

```
>> x=0:0.1:0.8;
>> y=f(x);
```

The `diff` function can be used to determine the differences between adjacent elements of each vector. For example,

```
>> diff(x)

ans =
  Columns 1 through 5
    0.1000    0.1000    0.1000    0.1000    0.1000
  Columns 6 through 8
    0.1000    0.1000    0.1000
```

As expected, the result represents the differences between each pair of elements of x. To compute divided-difference approximations of the derivative, we merely perform a vector division of the y differences by the x differences by entering

```
>> d=diff(y)./diff(x)

d =
  Columns 1 through 5
   10.8900   -0.0100    3.1900    8.4900    8.6900
  Columns 6 through 8
    1.3900  -11.0100  -21.3100
```

Note that because we are using equally spaced values, after generating the x values, we could have simply performed the above computation concisely as

```
>> d=diff(f(x))/0.1;
```

The vector d now contains derivative estimates corresponding to the midpoint between adjacent elements. Therefore, in order to develop a plot of our results, we must first generate a vector holding the x values for the midpoint of each interval:

```
>> n=length(x);
>> xm=(x(1:n-1)+x(2:n))./2;
```

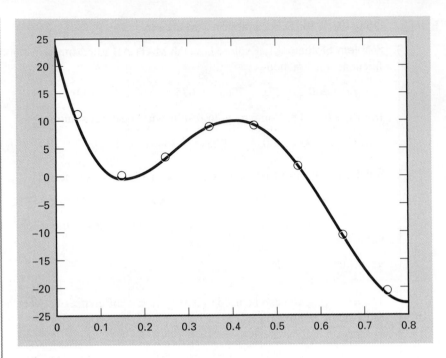

FIGURE 19.8
Comparison of the exact derivative (line) with numerical estimates (circles) computed with MATLAB's `diff` function.

As a final step, we can compute values for the analytical derivative at a finer level of resolution to include on the plot for comparison.

```
>> xa=0:.01:.8;
>> ya=25-400*xa+3*675*xa.^2-4*900*xa.^3+5*400*xa.^4;
```

A plot of the numerical and analytical estimates can be generated with

```
>> plot(xm,d,'o',xa,ya)
```

As displayed in Fig. 19.8, the results compare favorably for this case.

Note that aside from evaluating derivatives, the `diff` function comes in handy as a programming tool for testing certain characteristics of vectors. For example, the following statement displays an error message and terminates an M-file if it determines that a vector x has unequal spacing:

```
if any(diff(diff(x))~=0), error('unequal spacing'), end
```

Another common use is to detect whether a vector is in ascending or descending order. For example, the following code rejects a vector that is not in ascending order (monotonically increasing):

```
if any(diff(x)<=0), error('not in ascending order'), end
```

19.7.2 MATLAB Function: `gradient`

The `gradient` function also returns differences. However, it does so in a manner that is more compatible with evaluating derivatives at the values themselves rather than in the intervals between values. A simple representation of its syntax is

```
fx = gradient(f)
```

where f = a one-dimensional vector of length n, and fx is a vector of length n containing differences based on f. Just as with the `diff` function, the first value returned is the difference between the first and second value. However, for the intermediate values, a centered difference based on the adjacent values is returned

$$diff_i = \frac{f_{i+1} - f_{i-1}}{2} \tag{19.28}$$

The last value is then computed as the difference between the final two values. Hence, the results are akin to using centered differences for all the intermediate values, with forward and backward differences at the ends.

Note that the spacing between points is assumed to be one. If the vector represents equally spaced data, the following version divides all the results by the interval and hence returns the actual values of the derivatives,

```
fx = gradient(f, h)
```

where h = the spacing between points.

EXAMPLE 19.5   Using `gradient` for Differentiation

Problem Statement.   Use the `gradient` function to differentiate the same function that we analyzed in Example 19.4 with the `diff` function.

Solution.   In the same fashion as Example 19.4, we can generate a series of equally spaced values of the independent and dependent variables:

```
>> f=@(x) 0.2+25*x-200*x.^2+675*x.^3-900*x.^4+400*x.^5;
>> x=0:0.1:0.8;
>> y=f(x);
```

We can then use the `gradient` function to determine the derivatives as

```
>> dy=gradient(y,0.1)

dy =
  Columns 1 through 5
    10.8900     5.4400     1.5900     5.8400     8.5900
  Columns 6 through 9
     5.0400    -4.8100   -16.1600   -21.3100
```

As in Example 19.4, we can generate values for the analytical derivative and display both the numerical and analytical estimates on a plot:

```
>> xa=0:.01:.8;
>> ya=25-400*xa+3*675*xa.^2-4*900*xa.^3+5*400*xa.^4;
>> plot(x,dy,'o', xa,ya)
```

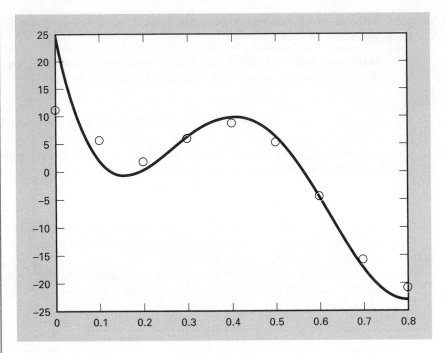

FIGURE 19.9
Comparison of the exact derivative (line) with numerical estimates (circles) computed with
MATLAB's `gradient` function.

As displayed in Fig. 19.9, the results are not as accurate as those obtained with the
`diff` function in Example 19.4. This is due to the fact that `gradient` employs intervals
that are two times (0.2) as wide as for those used for `diff` (0.1).

Beyond one-dimensional vectors, the `gradient` function is particularly well suited
for determining the partial derivatives of matrices. For example, for a two-dimensional matrix, `f`, the function can be invoked as

```
[fx,fy] = gradient(f, h)
```

where `fx` corresponds to the differences in the x (column) direction, and `fy` corresponds
to the differences in the y (row) direction, and `h` = the spacing between points. If `h` is
omitted, the spacing between points in both dimensions is assumed to be one. In the next
section, we will illustrate how `gradient` can be used to visualize vector fields.

19.8 CASE STUDY VISUALIZING FIELDS

Background. Beyond the determination of derivatives in one dimension, the `gradient` function is also quite useful for determining partial derivatives in two or more dimensions. In particular, it can be used in conjunction with other MATLAB functions to produce visualizations of vector fields.

To understand how this is done, we can return to our discussion of partial derivatives at the end of Section 19.1.1. Recall that we used mountain elevation as an example of a two-dimensional function. We can represent such a function mathematically as

$$z = f(x, y)$$

where z = elevation, x = distance measured along the east-west axis, and y = distance measured along the north-south axis.

For this example, the partial derivatives provide the slopes in the directions of the axes. However, if you were mountain climbing, you would probably be much more interested in determining the direction of the maximum slope. If we think of the two partial derivatives as component vectors, the answer is provided very neatly by

$$\nabla f = \frac{\partial f}{\partial x} i + \frac{\partial f}{\partial y} j$$

where ∇f is referred to as the *gradient* of f. This vector, which represents the steepest slope, has a magnitude

$$\sqrt{\left(\frac{\partial f}{\partial x}\right)^2 + \left(\frac{\partial f}{\partial y}\right)^2}$$

and a direction

$$\theta = \tan^{-1}\left(\frac{\partial f/\partial y}{\partial f/\partial x}\right)$$

where θ = the angle measured counterclockwise from the x axis.

Now suppose that we generate a grid of points in the x-y plane and used the foregoing equations to draw the gradient vector at each point. The result would be a field of arrows indicating the steepest route to the peak from any point. Conversely, if we plotted the negative of the gradient, it would indicate how a ball would travel as it rolled downhill from any point.

Such graphical representations are so useful that MATLAB has a special function, called `quiver`, to create such plots. A simple representation of its syntax is

```
quiver(x,y,u,v)
```

where `x` and `y` are matrices containing the position coordinates and `u` and `v` are matrices containing the partial derivatives. The following example demonstrates the use of `quiver` to visualize a field.

Employ the `gradient` function to determine to partial derivatives for the following two-dimensional function:

$$f(x, y) = y - x - 2x^2 - 2xy - y^2$$

from $x = -2$ to 2 and $y = 1$ to 3. Then use `quiver` to superimpose a vector field on a contour plot of the function.

Solution.   We can first express $f(x, y)$ as an anonymous function

```
>> f=@(x,y) y-x-2*x.^2-2.*x.*y-y.^2;
```

A series of equally spaced values of the independent and dependent variables can be generated as

```
>> [x,y]=meshgrid(-2:.25:0, 1:.25:3);
>> z=f(x,y);
```

The `gradient` function can be employed to determine the partial derivatives:

```
>> [fx,fy]=gradient(z,0.25);
```

We can then develop a contour plot of the results:

```
>> cs=contour(x,y,z);clabel(cs);hold on
```

As a final step, the resultant of the partial derivatives can be superimposed as vectors on the contour plot:

```
>> quiver(x,y,-fx,-fy);hold off
```

FIGURE 19.10
MATLAB generated contour plot of a two-dimensional function with the resultant of the partial derivatives displayed as arrows.

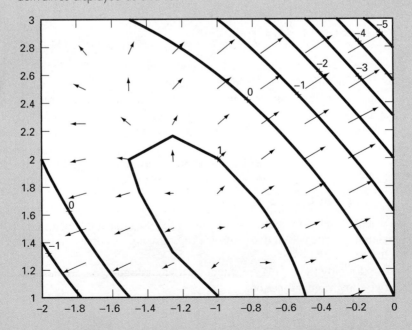

19.8 CASE STUDY continued

Note that we have displayed the negative of the resultants, in order that they point "downhill."

The result is shown in Fig. 19.10. The function's peak occurs at $x = -1$ and $y = 1.5$ and then drops away in all directions. As indicated by the lengthening arrows, the gradient drops off more steeply to the northeast and the southwest.

PROBLEMS

19.1 Compute forward and backward difference approximations of $O(h)$ and $O(h^2)$, and central difference approximations of $O(h^2)$ and $O(h^4)$ for the first derivative of $y = \cos x$ at $x = \pi/4$ using a value of $h = \pi/12$. Estimate the true percent relative error ε_t for each approximation.

19.2 Use centered difference approximations to estimate the first and second derivatives of $y = e^x$ at $x = 2$ for $h = 0.1$. Employ both $O(h^2)$ and $O(h^4)$ formulas for your estimates.

19.3 Use a Taylor series expansion to derive a centered finite-difference approximation to the third derivative that is second-order accurate. To do this, you will have to use four different expansions for the points x_{i-2}, x_{i-1}, x_{i+1}, and x_{i+2}. In each case, the expansion will be around the point x_i. The interval Δx will be used in each case of $i - 1$ and $i + 1$, and $2\Delta x$ will be used in each case of $i - 2$ and $i + 2$. The four equations must then be combined in a way to eliminate the first and second derivatives. Carry enough terms along in each expansion to evaluate the first term that will be truncated to determine the order of the approximation.

19.4 Use Richardson extrapolation to estimate the first derivative of $y = \cos x$ at $x = \pi/4$ using step sizes of $h_1 = \pi/3$ and $h_2 = \pi/6$. Employ centered differences of $O(h^2)$ for the initial estimates.

19.5 Repeat Prob. 19.4, but for the first derivative of $\ln x$ at $x = 5$ using $h_1 = 2$ and $h_2 = 1$.

19.6 Employ Eq. (19.21) to determine the first derivative of $y = 2x^4 - 6x^3 - 12x - 8$ at $x = 0$ based on values at $x_0 = -0.5$, $x_1 = 1$, and $x_2 = 2$. Compare this result with the true value and with an estimate obtained using a centered difference approximation based on $h = 1$.

19.7 Prove that for equispaced data points, Eq. (19.21) reduces to Eq. (4.25) at $x = x_1$.

19.8 Develop an M-file to apply a Romberg algorithm to estimate the derivative of a given function.

19.9 Develop an M-file to obtain first-derivative estimates for unequally spaced data. Test it with the following data:

x	0.6	1.5	1.6	2.5	3.5
$f(x)$	0.9036	0.3734	0.3261	0.08422	0.01596

where $f(x) = 5e^{-2x}x$. Compare your results with the true derivatives.

19.10 Develop an M-file function that computes first and second derivative estimates of order $O(h^2)$ based on the formulas in Figs. 19.3 through 19.5. The function's first line should be set up as

```
function  [dydx, d2ydx2] = diffeq(x,y)
```

where x and y are input vectors of length n containing the values of the independent and dependent variables, respectively, and dydx and dy2dx2 are output vectors of length n containing the first- and second-derivative estimates at each value of the independent variable. The function should generate a plot of dydx and dy2dx2 versus x. Have your M-file return an error message if **(a)** the input vectors are not the same length, or **(b)** the values for the independent variable are not equally spaced. Test your program with the data from Prob. 19.12.

19.11 The following data was collected for the distance traveled versus time for a rocket:

t, s	0	25	50	75	100	125
y, km	0	32	58	78	92	100

Use numerical differentiation to estimate the rocket's velocity and acceleration at each time.

19.12 A jet fighter's position on an aircraft carrier's runway was timed during landing:

t, s	0	0.52	1.04	1.75	2.37	3.25	3.83
x, m	153	185	208	249	261	271	273

where x is the distance from the end of the carrier. Estimate **(a)** velocity (dx/dt) and **(b)** acceleration (dv/dt) using numerical differentiation.

19.13 Use the following data to find the velocity and acceleration at $t = 10$ seconds:

Time, t, s	0	2	4	6	8	10	12	14	16
Position, x, m	0	0.7	1.8	3.4	5.1	6.3	7.3	8.0	8.4

Use second-order correct **(a)** centered finite-difference, **(b)** forward finite-difference, and **(c)** backward finite-difference methods.

19.14 A plane is being tracked by radar, and data is taken every second in polar coordinates θ and r.

t, s	200	202	204	206	208	210
θ, (rad)	0.75	0.72	0.70	0.68	0.67	0.66
r, m	5120	5370	5560	5800	6030	6240

At 206 seconds, use the centered finite-difference (second-order correct) to find the vector expressions for velocity $\vec{v}$ and acceleration $\vec{a}$. The velocity and acceleration given in polar coordinates are

$$\vec{v} = \dot{r}\vec{e}_r + r\dot{\theta}\vec{e}_\theta \quad \text{and} \quad \vec{a} = (\ddot{r} - r\dot{\theta}^2)\vec{e}_r + (r\ddot{\theta} + 2\dot{r}\dot{\theta})\vec{e}_\theta$$

19.15 Use regression to estimate the acceleration at each time for the following data with second-, third-, and fourth-order polynomials. Plot the results:

t	1	2	3.25	4.5	6	7	8	8.5	9.3	10
v	10	12	11	14	17	16	12	14	14	10

19.16 The normal distribution is defined as

$$f(x) = \frac{1}{\sqrt{2\pi}} e^{-x^2/2}$$

Use MATLAB to determine the inflection points of this function.

19.17 The following data was generated from the normal distribution:

x	-2	-1.5	-1	-0.5	0
$f(x)$	0.05399	0.12952	0.24197	0.35207	0.39894

x	0.5	1	1.5	2
$f(x)$	0.35207	0.24197	0.12952	0.05399

Use MATLAB to estimate the inflection points of this data.

19.18 Use the `diff(y)` command to develop a MATLAB M-file function to compute finite-difference approximations to the first and second derivative at each x value in the table below. Use finite-difference approximations that are second-order correct, $O(x^2)$:

x	0	1	2	3	4	5	6	7	8	9	10
y	1.4	2.1	3.3	4.8	6.8	6.6	8.6	7.5	8.9	10.9	10

19.19 The objective of this problem is to compare second-order accurate forward, backward, and centered finite-difference approximations of the first derivative of a function to the actual value of the derivative. This will be done for

$$f(x) = e^{-2x} - x$$

(a) Use calculus to determine the correct value of the derivative at $x = 2$.

(b) Develop an M-file function to evaluate the centered finite-difference approximations, starting with $x = 0.5$. Thus, for the first evaluation, the x values for the centered difference approximation will be $x = 2 \pm 0.5$ or $x = 1.5$ and 2.5. Then, decrease in increments of 0.1 down to a minimum value of $\Delta x = 0.01$.

(c) Repeat part **(b)** for the second-order forward and backward differences. (Note that these can be done at the same time that the centered difference is computed in the loop.)

(d) Plot the results of **(b)** and **(c)** versus x. Include the exact result on the plot for comparison.

19.20 You have to measure the flow rate of water through a small pipe. In order to do it, you place a bucket at the pipe's outlet and measure the volume in the bucket as a function of time as tabulated below. Estimate the flow rate at $t = 7$ s.

Time, s	0	1	5	8
Volume, cm³	0	1	8	16.4

19.21 The velocity v (m/s) of air flowing past a flat surface is measured at several distances y (m) away from the surface.

Use *Newton's viscosity law* to determine the shear stress τ (N/m^2) at the surface ($y = 0$),

$$\tau = \mu \frac{du}{dy}$$

Assume a value of dynamic viscosity $\mu = 1.8 \times 10^{-5}$ N $\cdot$ s/m^2.

y, m	0	0.002	0.006	0.012	0.018	0.024
u, m/s	0	0.287	0.899	1.915	3.048	4.299

19.22 *Fick's first diffusion law* states that

$$\text{Mass flux} = -D \frac{dc}{dx} \qquad \text{(P19.22)}$$

where mass flux = the quantity of mass that passes across a unit area per unit time (g/cm^2/s), D = a diffusion coefficient (cm^2/s), c = concentration (g/cm^3), and x = distance (cm). An environmental engineer measures the following concentration of a pollutant in the pore waters of sediments underlying a lake ($x = 0$ at the sediment-water interface and increases downward):

x, cm	0	1	3
c, 10^{-6} g/cm^3	0.06	0.32	0.6

Use the best numerical differentiation technique available to estimate the derivative at $x = 0$. Employ this estimate in conjunction with Eq. (P19.22) to compute the mass flux of pollutant out of the sediments and into the overlying waters ($D = 1.52 \times 10^{-6}$ cm^2/s). For a lake with 3.6×10^6 m^2 of sediments, how much pollutant would be transported into the lake over a year's time?

19.23 The following data was collected when a large oil tanker was loading:

t, min	0	10	20	30	45	60	75
V, 10^6 barrels	0.4	0.7	0.77	0.88	1.05	1.17	1.35

Calculate the flow rate Q (that is, dV/dt) for each time to the order of h^2.

19.24 *Fourier's law* is used routinely by architectural engineers to determine heat flow through walls. The following temperatures are measured from the surface ($x = 0$) into a stone wall:

x, m	0	0.08	0.16
T, °C	19	17	15

If the flux at $x = 0$ is 60 W/m^2, compute k.

19.25 The horizontal surface area A_s (m^2) of a lake at a particular depth can be computed from volume by differentiation:

$$A_s(z) = \frac{dV}{dz}(z)$$

where V = volume (m^3) and z = depth (m) as measured from the surface down to the bottom. The average concentration of a substance that varies with depth, $\bar{c}$ (g/m^3), can be computed by integration:

$$\bar{c} = \frac{\int_0^Z c(z) A_s(z)\, dz}{\int_0^Z A_s(z)\, dz}$$

where Z = the total depth (m). Determine the average concentration based on the following data:

z, m	0	4	8	12	16
V, 10^6 m^3	9.8175	5.1051	1.9635	0.3927	0.0000
c, g/m^3	10.2	8.5	7.4	5.2	4.1

19.26 *Faraday's law* characterizes the voltage drop across an inductor as

$$V_L = L \frac{di}{dt}$$

where V_L = voltage drop (V), L = inductance (in henrys; 1 H = 1 V $\cdot$ s/A), i = current (A), and t = time (s). Determine the voltage drop as a function of time from the following data for an inductance of 4 H.

t	0	0.1	0.2	0.3	0.5	0.7
i	0	0.16	0.32	0.56	0.84	2.0

19.27 Based on Faraday's law (Prob. 19.26), use the following voltage data to estimate the inductance if a current of 2 A is passed through the inductor over 400 milliseconds.

t, ms	0	10	20	40	60	80	120	180	280	400
V, volts	0	18	29	44	49	46	35	26	15	7

19.28 The rate of cooling of a body (Fig. P19.28) can be expressed as

$$\frac{dT}{dt} = -k(T - T_a)$$

where T = temperature of the body (°C), T_a = temperature of the surrounding medium (°C), and k = a proportionality constant (per minute). Thus, this equation (called *Newton's law of cooling*) specifies that the rate of cooling is proportional to

FIGURE P19.28

the difference in the temperatures of the body and of the surrounding medium. If a metal ball heated to 80 °C is dropped into water that is held constant at $T_a = 20$ °C, the temperature of the ball changes, as in

Time, min	0	5	10	15	20	25
T, °C	80	44.5	30.0	24.1	21.7	20.7

Utilize numerical differentiation to determine dT/dt at each value of time. Plot dT/dt versus $T - T_a$ and employ linear regression to evaluate k.

19.29 The enthalpy of a real gas is a function of pressure as described below. The data was taken for a real fluid. Estimate the enthalpy of the fluid at 400 K and 50 atm (evaluate the integral from 0.1 atm to 50 atm).

$$H \int_0^P \left(V - T \left(\frac{\partial V}{\partial T} \right)_P \right) dP$$

	V, L		
P, atm	**T = 350 K**	**T = 400 K**	**T = 450 K**
0.1	220	250	282.5
5	4.1	4.7	5.23
10	2.2	2.5	2.7
20	1.35	1.49	1.55
25	1.1	1.2	1.24
30	0.90	0.99	1.03
40	0.68	0.75	0.78
45	0.61	0.675	0.7
50	0.54	0.6	0.62

19.30 For fluid flow over a surface, the heat flux to the surface can be computed with Fourier's law: y = distance normal to the surface (m). The following measurements are made for air flowing over a flat plate where y = distance normal to the surface:

y, cm	0	1	3	5
T, K	900	480	270	210

If the plate's dimensions are 200 cm long and 50 cm wide, and $k = 0.028$ J/(s · m · K), **(a)** determine the flux at the surface, and **(b)** the heat transfer in watts. Note that $1\,J = 1\,W \cdot s$.

19.31 The pressure gradient for laminar flow through a constant radius tube is given by

$$\frac{dp}{dx} = -\frac{8\mu Q}{\pi r^4}$$

where p = pressure (N/m²), x = distance along the tube's centerline (m), μ = dynamic viscosity (N · s/m²), Q = flow (m³/s) and r = radius (m).

(a) Determine the pressure drop for a 10-cm length tube for a viscous liquid ($\mu = 0.005$ N · s/m², density $= \rho = 1 \times 10^3$ kg/m³) with a flow of 10×10^{-6} m³/s and the following varying radii along its length:

x, cm	0	2	4	5	6	7	10
r, mm	2	1.35	1.34	1.6	1.58	1.42	2

(b) Compare your result with the pressure drop that would have occurred if the tube had a constant radius equal to the average radius.

(c) Determine the average Reynolds number for the tube to verify that flow is truly laminar (Re $= \rho v D / \mu < 2100$ where v = velocity).

19.32 The following data for the specific heat of benzene was generated with an nth-order polynomial. Use numerical differentiation to determine n.

T, K	300	400	500	600
C_p, kJ/(kmol · K)	82.888	112.136	136.933	157.744

T, K	700	800	900	1000
C_p, kJ/(kmol · K)	175.036	189.273	200.923	210.450

19.33 The specific heat at constant pressure c_p [J/(kg · K)] of an ideal gas is related to enthalpy by

$$c_p = \frac{dh}{dT}$$

where h = enthalpy (kJ/kg), and T = absolute temperature (K). The following enthalpies are provided for carbon dioxide (CO_2) at several temperatures. Use these values to determine the specific heat in J/(kg · K) for each of the tabulated temperatures. Note that the atomic weights of carbon and oxygen are 12.011 and 15.9994 g/mol, respectively

T, K	750	800	900	1000
h, kJ/kmol	29,629	32,179	37,405	42,769

19.34 An nth-order rate law is often used to model chemical reactions that solely depend on the concentration of a single reactant:

$$\frac{dc}{dt} = -kc^n$$

where c = concentration (mole), t = time (min), n = reaction order (dimensionless), and k = reaction rate ($min^{-1} mole^{1-n}$). The *differential method* can be used to evaluate the parameters k and n. This involves applying a logarithmic transform to the rate law to yield,

$$\log\left(-\frac{dc}{dt}\right) = \log k + n \log c$$

Therefore, if the nth-order rate law holds, a plot of the $\log(-dc/dt)$ versus $\log c$ should yield a straight line with a slope of n and an intercept of $\log k$. Use the differential method and linear regression to determine k and n for the following data for the conversion of ammonium cyanate to urea:

t, min	0	5	15	30	45
c, mole	0.750	0.594	0.420	0.291	0.223

19.35 The sediment oxygen demand [SOD in units of g/(m^2 · d)] is an important parameter in determining the

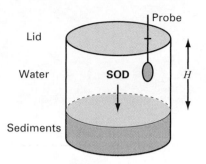

FIGURE P19.35

dissolved oxygen content of a natural water. It is measured by placing a sediment core in a cylindrical container (Fig. P19.35). After carefully introducing a layer of distilled, oxygenated water above the sediments, the container is covered to prevent gas transfer. A stirrer is used to mix the water gently, and an oxygen probe tracks how the water's oxygen concentration decreases over time. The SOD can then be computed as

$$SOD = -H\frac{do}{dt}$$

where H = the depth of water (m), o = oxygen concentration (g/m^3), and t = time (d).

Based on the following data and $H = 0.1$ m, use numerical differentiation to generate plots of **(a)** SOD versus time and **(b)** SOD versus oxygen concentration:

t, d	0	0.125	0.25	0.375	0.5	0.625	0.75
o, mg/L	10	7.11	4.59	2.57	1.15	0.33	0.03

PART SIX

Ordinary Differential Equations

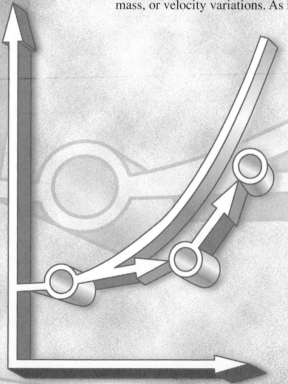

6.1 OVERVIEW

The fundamental laws of physics, mechanics, electricity, and thermodynamics are usually based on empirical observations that explain variations in physical properties and states of systems. Rather than describing the state of physical systems directly, the laws are usually couched in terms of spatial and temporal changes. These laws define mechanisms of change. When combined with continuity laws for energy, mass, or momentum, differential equations result. Subsequent integration of these differential equations results in mathematical functions that describe the spatial and temporal state of a system in terms of energy, mass, or velocity variations. As in Fig. PT6.1, the integration can be implemented analytically with calculus or numerically with the computer.

The free-falling bungee jumper problem introduced in Chap. 1 is an example of the derivation of a differential equation from a fundamental law. Recall that Newton's second law was used to develop an ODE describing the rate of change of velocity of a falling bungee jumper:

$$\frac{dv}{dt} = g - \frac{c_d}{m}v^2 \qquad \text{(PT6.1)}$$

where g is the gravitational constant, m is the mass, and c_d is a drag coefficient. Such equations, which are composed of an unknown function and its derivatives, are called *differential equations*. They are sometimes referred to as *rate equations* because they express the rate of change of a variable as a function of variables and parameters.

In Eq. (PT6.1), the quantity being differentiated v is called the *dependent variable*. The quantity with respect to which v is differentiated t is called the *independent variable*. When the function involves one independent variable, the equation is called an *ordinary*

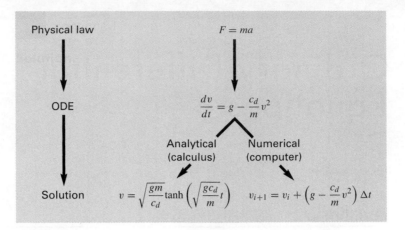

FIGURE PT6.1
The sequence of events in the development and solution of ODEs for engineering and science.
The example shown is for the velocity of the free-falling bungee jumper.

differential equation (or *ODE*). This is in contrast to a *partial differential equation* (or *PDE*) that involves two or more independent variables.

Differential equations are also classified as to their *order*. For example, Eq. (PT6.1) is called a *first-order equation* because the highest derivative is a first derivative. A *second-order equation* would include a second derivative. For example, the equation describing the position x of an unforced mass-spring system with damping is the second-order equation:

$$m\frac{d^2x}{dt^2} + c\frac{dx}{dt} + kx = 0 \tag{PT6.2}$$

where m is mass, c is a damping coefficient, and k is a spring constant. Similarly, an nth-order equation would include an nth derivative.

Higher-order differential equations can be reduced to a system of first-order equations. This is accomplished by defining the first derivative of the dependent variable as a new variable. For Eq. (PT6.2), this is done by creating a new variable v as the first derivative of displacement

$$v = \frac{dx}{dt} \tag{PT6.3}$$

where v is velocity. This equation can itself be differentiated to yield

$$\frac{dv}{dt} = \frac{d^2x}{dt^2} \tag{PT6.4}$$

Equations (PT6.3) and (PT6.4) can be substituted into Eq. (PT6.2) to convert it into a first-order equation:

$$m\frac{dv}{dt} + cv + kx = 0 \tag{PT6.5}$$

As a final step, we can express Eqs. (PT6.3) and (PT6.5) as rate equations:

$$\frac{dx}{dt} = v \tag{PT6.6}$$

$$\frac{dv}{dt} = -\frac{c}{m}v - \frac{k}{m}x \tag{PT6.7}$$

Thus, Eqs. (PT6.6) and (PT6.7) are a pair of first-order equations that are equivalent to the original second-order equation (Eq. PT6.2). Because other nth-order differential equations can be similarly reduced, this part of our book focuses on the solution of first-order equations.

A solution of an ordinary differential equation is a specific function of the independent variable and parameters that satisfies the original differential equation. To illustrate this concept, let us start with a simple fourth-order polynomial,

$$y = -0.5x^4 + 4x^3 - 10x^2 + 8.5x + 1 \tag{PT6.8}$$

Now, if we differentiate Eq. (PT6.8), we obtain an ODE:

$$\frac{dy}{dx} = -2x^3 + 12x^2 - 20x + 8.5 \tag{PT6.9}$$

This equation also describes the behavior of the polynomial, but in a manner different from Eq. (PT6.8). Rather than explicitly representing the values of y for each value of x, Eq. (PT6.9) gives the rate of change of y with respect to x (that is, the slope) at every value of x. Figure PT6.2 shows both the function and the derivative plotted versus x. Notice how the zero values of the derivatives correspond to the point at which the original function is flat—that is, where it has a zero slope. Also, the maximum absolute values of the derivatives are at the ends of the interval where the slopes of the function are greatest.

Although, as just demonstrated, we can determine a differential equation given the original function, the object here is to determine the original function given the differential equation. The original function then represents the solution.

Without computers, ODEs are usually solved analytically with calculus. For example, Eq. (PT6.9) could be multiplied by dx and integrated to yield

$$y = \int (-2x^3 + 12x^2 - 20x + 8.5)\,dx \tag{PT6.10}$$

The right-hand side of this equation is called an *indefinite integral* because the limits of integration are unspecified. This is in contrast to the *definite integrals* discussed previously in Part Five [compare Eq. (PT6.10) with Eq. (17.5)].

An analytical solution for Eq. (PT6.10) is obtained if the indefinite integral can be evaluated exactly in equation form. For this simple case, it is possible to do this with the result:

$$y = -0.5x^4 + 4x^3 - 10x^2 + 8.5x + C \tag{PT6.11}$$

which is identical to the original function with one notable exception. In the course of differentiating and then integrating, we lost the constant value of 1 in the original equation and gained the value C. This C is called a *constant of integration*. The fact that such an arbitrary constant appears indicates that the solution is not unique. In fact, it is but one of an

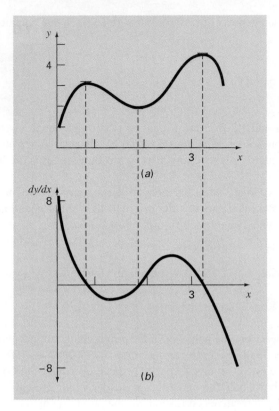

FIGURE PT6.2
Plots of (a) y versus x and (b) dy/dx versus x for the function $y = -0.5x^4 + 4x^3 - 10x^2 + 8.5x + 1$.

infinite number of possible functions (corresponding to an infinite number of possible values of C) that satisfy the differential equation. For example, Fig. PT6.3 shows six possible functions that satisfy Eq. (PT6.11).

Therefore, to specify the solution completely, a differential equation is usually accompanied by auxiliary conditions. For first-order ODEs, a type of auxiliary condition called an initial value is required to determine the constant and obtain a unique solution. For example, the original differential equation could be accompanied by the initial condition that at $x = 0$, $y = 1$. These values could be substituted into Eq. (PT6.11) to determine $C = 1$. Therefore, the unique solution that satisfies both the differential equation and the specified initial condition is

$$y = -0.5x^4 + 4x^3 - 10x^2 + 8.5x + 1$$

Thus, we have "pinned down" Eq. (PT6.11) by forcing it to pass through the initial condition, and in so doing, we have developed a unique solution to the ODE and have come full circle to the original function [Eq. (PT6.8)].

Initial conditions usually have very tangible interpretations for differential equations derived from physical problem settings. For example, in the bungee jumper problem, the

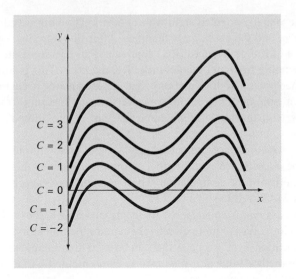

FIGURE PT6.3
Six possible solutions for the integral of $-2x^3 + 12x^2 - 20x + 8.5$. Each conforms to a different value of the constant of integration C.

initial condition was reflective of the physical fact that at time zero the vertical velocity was zero. If the bungee jumper had already been in vertical motion at time zero, the solution would have been modified to account for this initial velocity.

When dealing with an nth-order differential equation, n conditions are required to obtain a unique solution. If all conditions are specified at the same value of the independent variable (for example, at x or $t = 0$), then the problem is called an *initial-value problem*. This is in contrast to *boundary-value problems* where specification of conditions occurs at different values of the independent variable. Chapters 20 and 21 will focus on initial-value problems. Boundary-value problems are covered in Chap. 22.

6.2   PART ORGANIZATION

Chapter 20 is devoted to one-step methods for solving initial-value ODEs. As the name suggests, *one-step methods* compute a future prediction y_{i+1}, based only on information at a single point y_i and no other previous information. This is in contrast to *multistep approaches* that use information from several previous points as the basis for extrapolating to a new value.

With all but a minor exception, the one-step methods presented in Chap. 20 belong to what are called *Runge-Kutta techniques*. Although the chapter might have been organized around this theoretical notion, we have opted for a more graphical, intuitive approach to introduce the methods. Thus, we begin the chapter with *Euler's method*, which has a very straightforward graphical interpretation. In addition, because we have already introduced Euler's method in Chap. 1, our emphasis here is on quantifying its truncation error and describing its stability.

Next, we use visually oriented arguments to develop two improved versions of Euler's method—the *Heun* and the *midpoint* techniques. After this introduction, we formally develop the concept of Runge-Kutta (or RK) approaches and demonstrate how the foregoing techniques are actually first- and second-order RK methods. This is followed by a discussion of the higher-order RK formulations that are frequently used for engineering and scientific problem solving. In addition, we cover the application of one-step methods to *systems of ODEs*. Note that all the applications in Chap. 20 are limited to cases with a fixed step size.

In *Chap. 21,* we cover more advanced approaches for solving initial-value problems. First, we describe *adaptive RK methods* that automatically adjust the step size in response to the truncation error of the computation. These methods are especially pertinent as they are employed by MATLAB to solve ODEs.

Next, we discuss *multistep methods.* As mentioned above, these algorithms retain information of previous steps to more effectively capture the trajectory of the solution. They also yield the truncation error estimates that can be used to implement step-size control. We describe a simple method—the *non-self-starting Heun* method—to introduce the essential features of the multistep approaches.

Finally, the chapter ends with a description of *stiff ODEs*. These are both individual and systems of ODEs that have both fast and slow components to their solution. As a consequence, they require special solution approaches. We introduce the idea of an *implicit solution* technique as one commonly used remedy. We also describe MATLAB's built-in functions for solving stiff ODEs.

In *Chap. 22,* we focus on two approaches for obtaining solutions to *boundary-value problems:* the *shooting* and *finite-difference methods.* Aside from demonstrating how these techniques are implemented, we illustrate how they handle *derivative boundary conditions* and *nonlinear ODEs.*

20

Initial-Value Problems

CHAPTER OBJECTIVES

The primary objective of this chapter is to introduce you to solving initial-value problems for ODEs (ordinary differential equations). Specific objectives and topics covered are

- Understanding the meaning of local and global truncation errors and their relationship to step size for one-step methods for solving ODEs.
- Knowing how to implement the following Runge-Kutta (RK) methods for a single ODE:
  - Euler
  - Heun
  - Midpoint
  - Fourth-order RK
- Knowing how to iterate the corrector of Heun's method.
- Knowing how to implement the following Runge-Kutta methods for systems of ODEs:
  - Euler
  - Fourth-order RK

YOU'VE GOT A PROBLEM

We started this book with the problem of simulating the velocity of a free-falling bungee jumper. This problem amounted to formulating and solving an ordinary differential equation, the topic of this chapter. Now let's return to this problem and make it more interesting by computing what happens when the jumper reaches the end of the bungee cord.

To do this, we should recognize that the jumper will experience different forces depending on whether the cord is slack or stretched. If it is slack, the situation is that of free fall where the only forces are gravity and drag. However, because the jumper can now move up as well as down, the sign of the drag force must be modified so that it always tends to retard velocity,

$$\frac{dv}{dt} = g - \text{sign}(v)\frac{c_d}{m}v^2 \tag{20.1a}$$

where v is velocity (m/s), t is time (s), g is the acceleration due to gravity (9.81 m/s^2), c_d is the drag coefficient (kg/m), and m is mass (kg). The *signum function*,[1] sign, returns a -1 or a 1 depending on whether its argument is negative or positive, respectively. Thus, when the jumper is falling downward (positive velocity, sign = 1), the drag force will be negative and hence will act to reduce velocity. In contrast, when the jumper is moving upward (negative velocity, sign = -1), the drag force will be positive so that it again reduces the velocity.

Once the cord begins to stretch, it obviously exerts an upward force on the jumper. As done previously in Chap. 8, Hooke's law can be used as a first approximation of this force. In addition, a dampening force should also be included to account for frictional effects as the cord stretches and contracts. These factors can be incorporated along with gravity and drag into a second force balance that applies when the cord is stretched. The result is the following differential equation:

$$\frac{dv}{dt} = g - \text{sign}(v)\frac{c_d}{m}v^2 - \frac{k}{m}(x - L) - \frac{\gamma}{m}v \tag{20.1b}$$

where k is the cord's spring constant (N/m), x is vertical distance measured downward from the bungee jump platform (m), L is the length of the unstretched cord (m), and γ is a dampening coefficient (N · s/m).

Because Eq. (20.1b) only holds when the cord is stretched ($x > L$), the spring force will always be negative. That is, it will always act to pull the jumper back up. The dampening force increases in magnitude as the jumper's velocity increases and always acts to slow the jumper down.

If we want to simulate the jumper's velocity, we would initially solve Eq. (20.1a) until the cord was fully extended. Then, we could switch to Eq. (20.1b) for periods that the cord is stretched. Although this is fairly straightforward, it means that knowledge of the jumper's position is required. This can be done by formulating another differential equation for distance:

$$\frac{dx}{dt} = v \tag{20.2}$$

Thus, solving for the bungee jumper's velocity amounts to solving two ordinary differential equations where one of the equations takes different forms depending on the value

---

[1] Some computer languages represent the signum function as `sgn(x)`. As represented here, MATLAB uses the nomenclature `sign(x)`.

of one of the dependent variables. Chapters 20 and 21 explore methods for solving this and similar problems involving ODEs.

20.1 OVERVIEW

This chapter is devoted to solving ordinary differential equations of the form

$$\frac{dy}{dt} = f(t, y) \tag{20.3}$$

In Chap. 1, we developed a numerical method to solve such an equation for the velocity of the free-falling bungee jumper. Recall that the method was of the general form

New value = old value + slope × step size

or, in mathematical terms,

$$y_{i+1} = y_i + \phi h \tag{20.4}$$

where the slope ϕ is called an *increment function*. According to this equation, the slope estimate of ϕ is used to extrapolate from an old value y_i to a new value y_{i+1} over a distance h. This formula can be applied step by step to trace out the trajectory of the solution into the future. Such approaches are called *one-step methods* because the value of the increment function is based on information at a single point i. They are also referred to as *Runge-Kutta methods* after the two applied mathematicians who first discussed them in the early 1900s. Another class of methods called *multistep methods* use information from several previous points as the basis for extrapolating to a new value. We will describe multistep methods briefly in Chap. 21.

All one-step methods can be expressed in the general form of Eq. (20.4), with the only difference being the manner in which the slope is estimated. The simplest approach is to use the differential equation to estimate the slope in the form of the first derivative at t_i. In other words, the slope at the beginning of the interval is taken as an approximation of the average slope over the whole interval. This approach, called Euler's method, is discussed next. This is followed by other one-step methods that employ alternative slope estimates that result in more accurate predictions.

20.2 EULER'S METHOD

The first derivative provides a direct estimate of the slope at t_i (Fig. 20.1):

$$\phi = f(t_i, y_i)$$

where $f(t_i, y_i)$ is the differential equation evaluated at t_i and y_i. This estimate can be substituted into Eq. (20.1):

$$y_{i+1} = y_i + f(t_i, y_i)h \tag{20.5}$$

This formula is referred to as *Euler's method* (or the Euler-Cauchy or point-slope method). A new value of y is predicted using the slope (equal to the first derivative at the original value of t) to extrapolate linearly over the step size h (Fig. 20.1).

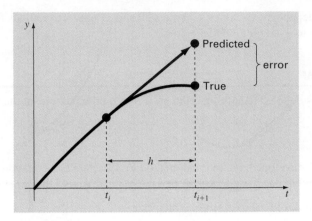

FIGURE 20.1
Euler's method.

EXAMPLE 20.1   Euler's Method

Problem Statement.   Use Euler's method to integrate $y' = 4e^{0.8t} - 0.5y$ from $t = 0$ to 4 with a step size of 1. The initial condition at $t = 0$ is $y = 2$. Note that the exact solution can be determined analytically as

$$y = \frac{4}{1.3}(e^{0.8t} - e^{-0.5t}) + 2e^{-0.5t}$$

Solution.   Equation (20.5) can be used to implement Euler's method:

$$y(1) = y(0) + f(0, 2)(1)$$

where $y(0) = 2$ and the slope estimate at $t = 0$ is

$$f(0, 2) = 4e^0 - 0.5(2) = 3$$

Therefore,

$$y(1) = 2 + 3(1) = 5$$

The true solution at $t = 1$ is

$$y = \frac{4}{1.3}\left(e^{0.8(1)} - e^{-0.5(1)}\right) + 2e^{-0.5(1)} = 6.19463$$

Thus, the percent relative error is

$$\varepsilon_t = \left| \frac{6.19463 - 5}{6.19463} \right| \times 100\% = 19.28\%$$

For the second step:

$$y(2) = y(1) + f(1, 5)(1)$$
$$= 5 + \left[4e^{0.8(1)} - 0.5(5)\right](1) = 11.40216$$

TABLE 20.1 Comparison of true and numerical values of the integral of $y' = 4e^{0.8t} - 0.5y$, with the initial condition that $y = 2$ at $t = 0$. The numerical values were computed using Euler's method with a step size of 1.

t	y_{true}	y_{Euler}	$|\varepsilon_t|$ (%)
0	2.00000	2.00000	
1	6.19463	5.00000	19.28
2	14.84392	11.40216	23.19
3	33.67717	25.51321	24.24
4	75.33896	56.84931	24.54

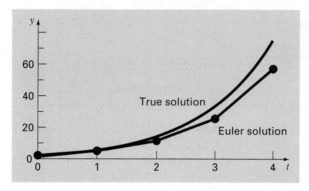

FIGURE 20.2
Comparison of the true solution with a numerical solution using Euler's method for the integral of $y' = 4e^{0.8t} - 0.5y$ from $t = 0$ to 4 with a step size of 1.0. The initial condition at $t = 0$ is $y = 2$.

The true solution at $t = 2.0$ is 14.84392 and, therefore, the true percent relative error is 23.19%. The computation is repeated, and the results compiled in Table 20.1 and Fig. 20.2. Note that although the computation captures the general trend of the true solution, the error is considerable. As discussed in the next section, this error can be reduced by using a smaller step size.

20.2.1 Error Analysis for Euler's Method

The numerical solution of ODEs involves two types of error (recall Chap. 4):

1. *Truncation,* or discretization, errors caused by the nature of the techniques employed to approximate values of y.
2. *Roundoff* errors caused by the limited numbers of significant digits that can be retained by a computer.

The truncation errors are composed of two parts. The first is a *local truncation error* that results from an application of the method in question over a single step. The second is a *propagated truncation error* that results from the approximations produced during the previous steps. The sum of the two is the total error. It is referred to as the *global trunca-tion error.*

Insight into the magnitude and properties of the truncation error can be gained by de-riving Euler's method directly from the Taylor series expansion. To do this, realize that the differential equation being integrated will be of the general form of Eq. (20.3), where $dy/dt = y'$, and t and y are the independent and the dependent variables, respectively. If the solution—that is, the function describing the behavior of y—has continuous derivatives, it can be represented by a Taylor series expansion about a starting value (t_i, y_i), as in [recall Eq. (4.13)]:

$$y_{i+1} = y_i + y_i'h + \frac{y_i''}{2!}h^2 + \cdots + \frac{y_i^{(n)}}{n!}h^n + R_n \tag{20.6}$$

where $h = t_{i+1} - t_i$ and $R_n =$ the remainder term, defined as

$$R_n = \frac{y^{(n+1)}(\xi)}{(n+1)!}h^{n+1} \tag{20.7}$$

where ξ lies somewhere in the interval from t_i to t_{i+1}. An alternative form can be devel-oped by substituting Eq. (20.3) into Eqs. (20.6) and (20.7) to yield

$$y_{i+1} = y_i + f(t_i, y_i)h + \frac{f'(t_i, y_i)}{2!}h^2 + \cdots + \frac{f^{(n-1)}(t_i, y_i)}{n!}h^n + O(h^{n+1}) \tag{20.8}$$

where $O(h^{n+1})$ specifies that the local truncation error is proportional to the step size raised to the $(n + 1)$th power.

By comparing Eqs. (20.5) and (20.8), it can be seen that Euler's method corresponds to the Taylor series up to and including the term $f(t_i, y_i)h$. Additionally, the comparison indicates that a truncation error occurs because we approximate the true solution using a fi-nite number of terms from the Taylor series. We thus truncate, or leave out, a part of the true solution. For example, the truncation error in Euler's method is attributable to the remain-ing terms in the Taylor series expansion that were not included in Eq. (20.5). Subtracting Eq. (20.5) from Eq. (20.8) yields

$$E_t = \frac{f'(t_i, y_i)}{2!}h^2 + \cdots + O(h^{n+1}) \tag{20.9}$$

where $E_t =$ the true local truncation error. For sufficiently small h, the higher-order terms in Eq. (20.9) are usually negligible, and the result is often represented as

$$E_a = \frac{f'(t_i, y_i)}{2!}h^2 \tag{20.10}$$

or

$$E_a = O(h^2) \tag{20.11}$$

where $E_a =$ the approximate local truncation error.

According to Eq. (20.11), we see that the local error is proportional to the square of the step size and the first derivative of the differential equation. It can also be demon-strated that the global truncation error is $O(h)$—that is, it is proportional to the step size

(Carnahan et al., 1969). These observations lead to some useful conclusions:

1. The global error can be reduced by decreasing the step size.
2. The method will provide error-free predictions if the underlying function (i.e., the solution of the differential equation) is linear, because for a straight line the second derivative would be zero.

This latter conclusion makes intuitive sense because Euler's method uses straight-line segments to approximate the solution. Hence, Euler's method is referred to as a *first-order method*.

It should also be noted that this general pattern holds for the higher-order one-step methods described in the following pages. That is, an nth-order method will yield perfect results if the underlying solution is an nth-order polynomial. Further, the local truncation error will be $O(h^{n+1})$ and the global error $O(h^n)$.

20.2.2 Stability of Euler's Method

In the preceding section, we learned that the truncation error of Euler's method depends on the step size in a predictable way based on the Taylor series. This is an accuracy issue.

The stability of a solution method is another important consideration that must be considered when solving ODEs. A numerical solution is said to be unstable if errors grow exponentially for a problem for which there is a bounded solution. The stability of a particular application can depend on three factors: the differential equation, the numerical method, and the step size.

Insight into the step size required for stability can be examined by studying a very simple ODE:

$$\frac{dy}{dt} = -ay \tag{20.12}$$

If $y(0) = y_0$, calculus can be used to determine the solution as

$$y = y_0 e^{-at}$$

Thus, the solution starts at y_0 and asymptotically approaches zero.

Now suppose that we use Euler's method to solve the same problem numerically:

$$y_{i+1} = y_i + \frac{dy_i}{dt} h$$

Substituting Eq. (20.12) gives

$$y_{i+1} = y_i - a y_i h$$

or

$$y_{i+1} = y_i (1 - ah) \tag{20.13}$$

The parenthetical quantity $1 - ah$ is called an *amplification factor*. If its absolute value is greater than unity, the solution will grow in an unbounded fashion. So clearly, the stability depends on the step size h. That is, if $h > 2/a$, $|y_i| \to \infty$ as $i \to \infty$. Based on this analysis, Euler's method is said to be *conditionally stable*.

Note that there are certain ODEs where errors always grow regardless of the method. Such ODEs are called *ill-conditioned*.

Inaccuracy and instability are often confused. This is probably because (a) both represent situations where the numerical solution breaks down and (b) both are affected by step size. However, they are distinct problems. For example, an inaccurate method can be very stable. We will return to the topic when we discuss stiff systems in Chap. 21.

20.2.3 MATLAB M-file Function: `eulode`

We have already developed a simple M-file to implement Euler's method for the falling bungee jumper problem in Chap. 3. Recall from Section 3.6, that this function used Euler's method to compute the velocity after a given time of free fall. Now, let's develop a more general, all-purpose algorithm.

Figure 20.3 shows an M-file that uses Euler's method to compute values of the dependent variable y over a range of values of the independent variable t. The name of the function holding the right-hand side of the differential equation is passed into the function as the

FIGURE 20.3
An M-file to implement Euler's method.

```
function [t,y] = eulode(dydt,tspan,y0,h,varargin)
% eulode: Euler ODE solver
%   [t,y] = eulode(dydt,tspan,y0,h,p1,p2,...):
%           uses Euler's method to integrate an ODE
% input:
%   dydt = name of the M-file that evaluates the ODE
%   tspan = [ti, tf] where ti and tf = initial and
%           final values of independent variable
%   y0 = initial value of dependent variable
%   h = step size
%   p1,p2,... = additional parameters used by dydt
% output:
%   t = vector of independent variable
%   y = vector of solution for dependent variable

if nargin<4,error('at least 4 input arguments required'),end
ti = tspan(1);tf = tspan(2);
if ~(tf>ti),error('upper limit must be greater than lower'),end
t = (ti:h:tf)'; n = length(t);
% if necessary, add an additional value of t
% so that range goes from t = ti to tf
if t(n)<tf
  t(n+1) = tf;
  n = n+1;
end
y = y0*ones(n,1); %preallocate y to improve efficiency
for i = 1:n-1 %implement Euler's method
  y(i+1) = y(i) + dydt(t(i),y(i),varargin{:})*(t(i+1)-t(i));
end
```

variable `dydt`. The initial and final values of the desired range of the independent variable is passed as a vector `tspan`. The initial value and the desired step size are passed as `y0` and `h`, respectively.

The function first generates a vector `t` over the desired range of the dependent variable using an increment of `h`. In the event that the step size is not evenly divisible into the range, the last value will fall short of the final value of the range. If this occurs, the final value is added to `t` so that the series spans the complete range. The length of the `t` vector is determined as `n`. In addition, a vector of the dependent variable `y` is preallocated with `n` values of the initial condition to improve efficiency.

At this point, Euler's method (Eq. 20.5) is implemented by a simple loop:

```
for i = 1:n-1
  y(i+1) = y(i) + dydt(t(i),y(i))*(t(i+1)-t(i),var);
end
```

Notice how a function is used to generate a value for the derivative at the appropriate values of the independent and dependent variables. Also notice how the time step is automatically calculated based on the difference between adjacent values in the vector `t`.

The ODE being solved can be set up in several ways. First, the differential equation can be defined as an anonymous function object. For example, for the ODE from Example 20.1:

```
>> dydt=@(t,y) 4*exp(0.8*t) - 0.5*y;
```

The solution can then be generated as

```
>> [t,y] = eulode(dydt,[0 4],2,1);
>> disp([t,y])
```

with the result (compare with Table 20.1):

```
     0    2.0000
1.0000    5.0000
2.0000   11.4022
3.0000   25.5132
4.0000   56.8493
```

Although using an anonymous function is feasible for the present case, there will be more complex problems where the definition of the ODE requires several lines of code. In such instances, creating a separate M-file is the only option.

20.3 IMPROVEMENTS OF EULER'S METHOD

A fundamental source of error in Euler's method is that the derivative at the beginning of the interval is assumed to apply across the entire interval. Two simple modifications are available to help circumvent this shortcoming. As will be demonstrated in Section 20.4, both modifications (as well as Euler's method itself) actually belong to a larger class of solution techniques called Runge-Kutta methods. However, because they have very straightforward graphical interpretations, we will present them prior to their formal derivation as Runge-Kutta methods.

20.3.1 Heun's Method

One method to improve the estimate of the slope involves the determination of two derivatives for the interval—one at the beginning and another at the end. The two derivatives are then averaged to obtain an improved estimate of the slope for the entire interval. This approach, called *Heun's method,* is depicted graphically in Fig. 20.4.

Recall that in Euler's method, the slope at the beginning of an interval

$$y_i' = f(t_i, y_i) \tag{20.14}$$

is used to extrapolate linearly to y_{i+1}:

$$y_{i+1}^0 = y_i + f(t_i, y_i)h \tag{20.15}$$

For the standard Euler method we would stop at this point. However, in Heun's method the y_{i+1}^0 calculated in Eq. (20.15) is not the final answer, but an intermediate prediction. This is why we have distinguished it with a superscript 0. Equation (20.15) is called a *predictor equation.* It provides an estimate that allows the calculation of a slope at the end of the interval:

$$y_{i+1}' = f\left(t_{i+1}, y_{i+1}^0\right) \tag{20.16}$$

Thus, the two slopes [Eqs. (20.14) and (20.16)] can be combined to obtain an average slope for the interval:

$$\bar{y}' = \frac{f(t_i, y_i) + f\left(t_{i+1}, y_{i+1}^0\right)}{2}$$

This average slope is then used to extrapolate linearly from y_i to y_{i+1} using Euler's method:

$$y_{i+1} = y_i + \frac{f(t_i, y_i) + f\left(t_{i+1}, y_{i+1}^0\right)}{2}h \tag{20.17}$$

which is called a *corrector equation.*

FIGURE 20.4
Graphical depiction of Heun's method. (a) Predictor and (b) corrector.

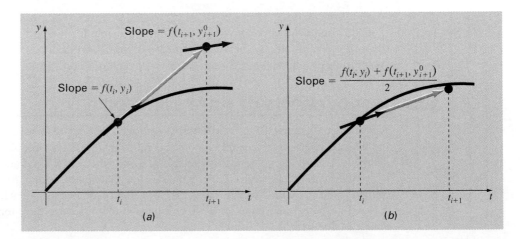

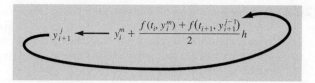

FIGURE 20.5
Graphical representation of iterating the corrector of Heun's method to obtain an improved estimate.

The Heun method is a *predictor-corrector approach*. As just derived, it can be expressed concisely as

Predictor (Fig. 20.4a): $y_{i+1}^0 = y_i^m + f(t_i, y_i)h$ (20.18)

Corrector (Fig. 20.4b): $y_{i+1}^j = y_i^m + \dfrac{f\left(t_i, y_i^m\right) + f\left(t_{i+1}, y_{i+1}^{j-1}\right)}{2} h$ (20.19)

$$(\text{for } j = 1, 2, \ldots, m)$$

Note that because Eq. (20.19) has y_{i+1} on both sides of the equal sign, it can be applied in an iterative fashion as indicated. That is, an old estimate can be used repeatedly to provide an improved estimate of y_{i+1}. The process is depicted in Fig. 20.5.

As with similar iterative methods discussed in previous sections of the book, a termination criterion for convergence of the corrector is provided by

$$|\varepsilon_a| = \left| \frac{y_{i+1}^j - y_{i+1}^{j-1}}{y_{i+1}^j} \right| \times 100\%$$

where y_{i+1}^{j-1} and y_{i+1}^j are the result from the prior and the present iteration of the corrector, respectively. It should be understood that the iterative process does not necessarily converge on the true answer but will converge on an estimate with a finite truncation error, as demonstrated in the following example.

EXAMPLE 20.2   Heun's Method

Problem Statement. Use Heun's method with iteration to integrate $y' = 4e^{0.8t} - 0.5y$ from $t = 0$ to 4 with a step size of 1. The initial condition at $t = 0$ is $y = 2$. Employ a stopping criterion of 0.00001% to terminate the corrector iterations.

Solution. First, the slope at (t_0, y_0) is calculated as

$$y_0' = 4e^0 - 0.5(2) = 3$$

Then, the predictor is used to compute a value at 1.0:

$$y_1^0 = 2 + 3(1) = 5$$

TABLE 20.2 Comparison of true and numerical values of the integral of $y' = 4e^{0.8t} - 0.5y$, with the initial condition that $y = 2$ at $t = 0$. The numerical values were computed using the Euler and Heun methods with a step size of 1. The Heun method was implemented both without and with iteration of the corrector.

				Without Iteration		With Iteration	
t	y_{true}	y_{Euler}	$\|\varepsilon_t\|$ (%)	y_{Heun}	$\|\varepsilon_t\|$ (%)	y_{Heun}	$\|\varepsilon_t\|$ (%)
0	2.00000	2.00000		2.00000		2.00000	
1	6.19463	5.00000	19.28	6.70108	8.18	6.36087	2.68
2	14.84392	11.40216	23.19	16.31978	9.94	15.30224	3.09
3	33.67717	25.51321	24.24	37.19925	10.46	34.74328	3.17
4	75.33896	56.84931	24.54	83.33777	10.62	77.73510	3.18

Note that this is the result that would be obtained by the standard Euler method. The true value in Table 20.2 shows that it corresponds to a percent relative error of 19.28%.

Now, to improve the estimate for y_{i+1}, we use the value y_1^0 to predict the slope at the end of the interval

$$y_1' = f(x_1, y_1^0) = 4e^{0.8(1)} - 0.5(5) = 6.402164$$

which can be combined with the initial slope to yield an average slope over the interval from $t = 0$ to 1:

$$\bar{y}' = \frac{3 + 6.402164}{2} = 4.701082$$

This result can then be substituted into the corrector [Eq. (20.19)] to give the prediction at $t = 1$:

$$y_1^1 = 2 + 4.701082(1) = 6.701082$$

which represents a true percent relative error of −8.18%. Thus, the Heun method without iteration of the corrector reduces the absolute value of the error by a factor of about 2.4 as compared with Euler's method. At this point, we can also compute an approximate error as

$$|\varepsilon_a| = \left| \frac{6.701082 - 5}{6.701082} \right| \times 100\% = 25.39\%$$

Now the estimate of y_1 can be refined by substituting the new result back into the right-hand side of Eq. (20.19) to give

$$y_1^2 = 2 + \frac{3 + 4e^{0.8(1)} - 0.5(6.701082)}{2}1 = 6.275811$$

which represents a true percent relative error of 1.31 percent and an approximate error of

$$|\varepsilon_a| = \left| \frac{6.275811 - 6.701082}{6.275811} \right| \times 100\% = 6.776\%$$

The next iteration gives

$$y_1^2 = 2 + \frac{3 + 4e^{0.8(1)} - 0.5(6.275811)}{2} 1 = 6.382129$$

which represents a true error of 3.03% and an approximate error of 1.666%.

The approximate error will keep dropping as the iterative process converges on a stable final result. In this example, after 12 iterations the approximate error falls below the stopping criterion. At this point, the result at $t = 1$ is 6.36087, which represents a true relative error of 2.68%. Table 20.2 shows results for the remainder of the computation along with results for Euler's method and for the Heun method without iteration of the corrector.

Insight into the local error of the Heun method can be gained by recognizing that it is related to the trapezoidal rule. In the previous example, the derivative is a function of both the dependent variable y and the independent variable t. For cases such as polynomials, where the ODE is solely a function of the independent variable, the predictor step [Eq. (20.18)] is not required and the corrector is applied only once for each iteration. For such cases, the technique is expressed concisely as

$$y_{i+1} = y_i + \frac{f(t_i) + f(t_{i+1})}{2} h \tag{20.20}$$

Notice the similarity between the second term on the right-hand side of Eq. (20.20) and the trapezoidal rule [Eq. (17.11)]. The connection between the two methods can be formally demonstrated by starting with the ordinary differential equation

$$\frac{dy}{dt} = f(t) \tag{20.21}$$

This equation can be solved for y by integration:

$$\int_{y_i}^{y_{i+1}} dy = \int_{t_i}^{t_{i+1}} f(t)\, dt \tag{20.22}$$

which yields

$$y_{i+1} - y_i = \int_{t_i}^{t_{i+1}} f(t)\, dt \tag{20.23}$$

or

$$y_{i+1} = y_i + \int_{t_i}^{t_{i+1}} f(t)\, dt \tag{20.24}$$

Now, recall that the trapezoidal rule [Eq. (17.11)] is defined as

$$\int_{t_i}^{t_{i+1}} f(t)\, dt = \frac{f(t_i) + f(t_{i+1})}{2} h \tag{20.25}$$

where $h = t_{i+1} - t_i$. Substituting Eq. (20.25) into Eq. (20.24) yields

$$y_{i+1} = y_i + \frac{f(t_i) + f(t_{i+1})}{2}h \tag{20.26}$$

which is equivalent to Eq. (20.20). For this reason, Heun's method is sometimes referred to as the trapezoidal rule.

Because Eq. (20.26) is a direct expression of the trapezoidal rule, the local truncation error is given by [recall Eq. (17.14)]

$$E_t = -\frac{f''(\xi)}{12}h^3 \tag{20.27}$$

where ξ is between t_i and t_{i+1}. Thus, the method is second order because the second derivative of the ODE is zero when the true solution is a quadratic. In addition, the local and global errors are $O(h^3)$ and $O(h^2)$, respectively. Therefore, decreasing the step size decreases the error at a faster rate than for Euler's method.

20.3.2 The Midpoint Method

Figure 20.6 illustrates another simple modification of Euler's method. Called the *midpoint method,* this technique uses Euler's method to predict a value of y at the midpoint of the interval (Fig. 20.6*a*):

$$y_{i+1/2} = y_i + f(t_i, y_i)\frac{h}{2} \tag{20.28}$$

Then, this predicted value is used to calculate a slope at the midpoint:

$$y'_{i+1/2} = f(t_{i+1/2}, y_{i+1/2}) \tag{20.29}$$

FIGURE 20.6
Graphical depiction of Heun's method. (a) Predictor and (b) corrector.

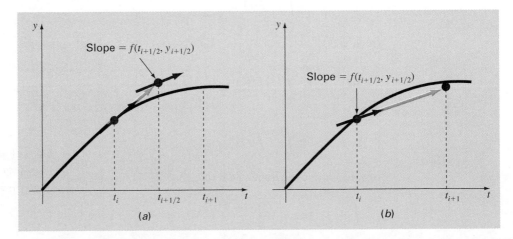

which is assumed to represent a valid approximation of the average slope for the entire interval. This slope is then used to extrapolate linearly from t_i to t_{i+1} (Fig. 20.6b):

$$y_{i+1} = y_i + f(t_{i+1/2}, y_{i+1/2})h \tag{20.30}$$

Observe that because y_{i+1} is not on both sides, the corrector [Eq. (20.30)] cannot be applied iteratively to improve the solution as was done with Heun's method.

As in our discussion of Heun's method, the midpoint method can also be linked to Newton-Cotes integration formulas. Recall from Table 17.4 that the simplest Newton-Cotes open integration formula, which is called the midpoint method, can be represented as

$$\int_a^b f(x)\,dx \cong (b-a)f(x_1) \tag{20.31}$$

where x_1 is the midpoint of the interval (a, b). Using the nomenclature for the present case, it can be expressed as

$$\int_{t_i}^{t_{i+1}} f(t)\,dt \cong hf(t_{i+1/2}) \tag{20.32}$$

Substitution of this formula into Eq. (20.24) yields Eq. (20.30). Thus, just as the Heun method can be called the trapezoidal rule, the midpoint method gets its name from the underlying integration formula on which it is based.

The midpoint method is superior to Euler's method because it utilizes a slope estimate at the midpoint of the prediction interval. Recall from our discussion of numerical differentiation in Section 4.3.4 that centered finite differences are better approximations of derivatives than either forward or backward versions. In the same sense, a centered approximation such as Eq. (20.29) has a local truncation error of $O(h^2)$ in comparison with the forward approximation of Euler's method, which has an error of $O(h)$. Consequently, the local and global errors of the midpoint method are $O(h^3)$ and $O(h^2)$, respectively.

20.4 RUNGE-KUTTA METHODS

Runge-Kutta (RK) methods achieve the accuracy of a Taylor series approach without requiring the calculation of higher derivatives. Many variations exist but all can be cast in the generalized form of Eq. (20.4):

$$y_{i+1} = y_i + \phi h \tag{20.33}$$

where ϕ is called an *increment function,* which can be interpreted as a representative slope over the interval. The increment function can be written in general form as

$$\phi = a_1 k_1 + a_2 k_2 + \cdots + a_n k_n \tag{20.34}$$

where the a's are constants and the k's are

$$k_1 = f(t_i, y_i) \tag{20.34a}$$

$$k_2 = f(t_i + p_1 h, y_i + q_{11}k_1 h) \tag{20.34b}$$

$$k_3 = f(t_i + p_2 h, y_i + q_{21}k_1 h + q_{22}k_2 h) \tag{20.34c}$$

$$\vdots$$

$$k_n = f(t_i + p_{n-1}h, y_i + q_{n-1,1}k_1 h + q_{n-1,2}k_2 h + \cdots + q_{n-1,n-1}k_{n-1}h) \tag{20.34d}$$

where the p's and q's are constants. Notice that the k's are recurrence relationships. That is, k_1 appears in the equation for k_2, which appears in the equation for k_3, and so forth. Because each k is a functional evaluation, this recurrence makes RK methods efficient for computer calculations.

Various types of Runge-Kutta methods can be devised by employing different numbers of terms in the increment function as specified by n. Note that the first-order RK method with $n = 1$ is, in fact, Euler's method. Once n is chosen, values for the a's, p's, and q's are evaluated by setting Eq. (20.33) equal to terms in a Taylor series expansion. Thus, at least for the lower-order versions, the number of terms n usually represents the order of the approach. For example, in Section 20.4.1, second-order RK methods use an increment function with two terms ($n = 2$). These second-order methods will be exact if the solution to the differential equation is quadratic. In addition, because terms with h^3 and higher are dropped during the derivation, the local truncation error is $O(h^3)$ and the global error is $O(h^2)$. In Section 20.4.2, the fourth-order RK method ($n = 4$) is presented for which global truncation error is $O(h^4)$.

20.4.1 Second-Order Runge-Kutta Methods

The second-order version of Eq. (20.33) is

$$y_{i+1} = y_i + (a_1 k_1 + a_2 k_2)h \tag{20.35}$$

where

$$k_1 = f(t_i, y_i) \tag{20.35a}$$
$$k_2 = f(t_i + p_1 h, y_i + q_{11} k_1 h) \tag{20.35b}$$

The values for a_1, a_2, p_1, and q_{11} are evaluated by setting Eq. (20.35) equal to a second-order Taylor series. By doing this, three equations can be derived to evaluate the four unknown constants (see Chapra and Canale, 2006, for details). The three equations are

$$a_1 + a_2 = 1 \tag{20.36}$$
$$a_2 p_1 = 1/2 \tag{20.37}$$
$$a_2 q_{11} = 1/2 \tag{20.38}$$

Because we have three equations with four unknowns, these equations are said to be underdetermined. We, therefore, must assume a value of one of the unknowns to determine the other three. Suppose that we specify a value for a_2. Then Eqs. (20.36) through (20.38) can be solved simultaneously for

$$a_1 = 1 - a_2 \tag{20.39}$$
$$p_1 = q_{11} = \frac{1}{2a_2} \tag{20.40}$$

Because we can choose an infinite number of values for a_2, there are an infinite number of second-order RK methods. Every version would yield exactly the same results if the solution to the ODE were quadratic, linear, or a constant. However, they yield different results when (as is typically the case) the solution is more complicated. Three of the most commonly used and preferred versions are presented next.

Heun Method without Iteration ($a_2 = 1/2$). If a_2 is assumed to be $1/2$, Eqs. (20.39) and (20.40) can be solved for $a_1 = 1/2$ and $p_1 = q_{11} = 1$. These parameters, when substituted into Eq. (20.35), yield

$$y_{i+1} = y_i + \left(\frac{1}{2}k_1 + \frac{1}{2}k_2\right)h \qquad (20.41)$$

where

$$k_1 = f(t_i, y_i) \qquad (20.41a)$$
$$k_2 = f(t_i + h, y_i + k_1h) \qquad (20.41b)$$

Note that k_1 is the slope at the beginning of the interval and k_2 is the slope at the end of the interval. Consequently, this second-order Runge-Kutta method is actually Heun's technique without iteration of the corrector.

The Midpoint Method ($a_2 = 1$). If a_2 is assumed to be 1, then $a_1 = 0$, $p_1 = q_{11} = 1/2$, and Eq. (20.35) becomes

$$y_{i+1} = y_i + k_2h \qquad (20.42)$$

where

$$k_1 = f(t_i, y_i) \qquad (20.42a)$$
$$k_2 = f(t_i + h/2, y_i + k_1h/2) \qquad (20.42b)$$

This is the midpoint method.

Ralston's Method ($a_2 = 2/3$). Ralston (1962) and Ralston and Rabinowitz (1978) determined that choosing $a_2 = 2/3$ provides a minimum bound on the truncation error for the second-order RK algorithms. For this version, $a_1 = 1/3$ and $p_1 = q_{11} = 3/4$, and Eq. (20.35) becomes

$$y_{i+1} = y_i + \left(\frac{1}{3}k_1 + \frac{2}{3}k_2\right)h \qquad (20.43)$$

where

$$k_1 = f(x_i, y_i) \qquad (20.43a)$$
$$k_2 = f\left(t_i + \frac{3}{4}h, y_i + \frac{3}{4}k_1h\right) \qquad (20.43b)$$

20.4.2 Classical Fourth-Order Runge-Kutta Method

The most popular RK methods are fourth order. As with the second-order approaches, there are an infinite number of versions. The following is the most commonly used form, and we therefore call it the *classical fourth-order RK method:*

$$y_{i+1} = y_i + \frac{1}{6}(k_1 + 2k_2 + 2k_3 + k_4)h \qquad (20.44)$$

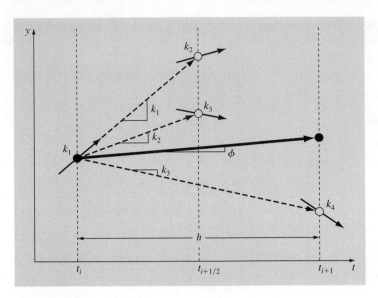

FIGURE 20.7
Graphical depiction of the slope estimates comprising the fourth-order RK method.

where

$$k_1 = f(t_i, y_i) \tag{20.44a}$$

$$k_2 = f\left(t_i + \frac{1}{2}h, y_i + \frac{1}{2}k_1h\right) \tag{20.44b}$$

$$k_3 = f\left(t_i + \frac{1}{2}h, y_i + \frac{1}{2}k_2h\right) \tag{20.44c}$$

$$k_4 = f(t_i + h, y_i + k_3h) \tag{20.44d}$$

Notice that for ODEs that are a function of t alone, the classical fourth-order RK method is similar to Simpson's 1/3 rule. In addition, the fourth-order RK method is similar to the Heun approach in that multiple estimates of the slope are developed to come up with an improved average slope for the interval. As depicted in Fig. 20.7, each of the k's represents a slope. Equation (20.44) then represents a weighted average of these to arrive at the improved slope.

EXAMPLE 20.3   Classical Fourth-Order RK Method

Problem Statement.   Employ the classical fourth-order RK method to integrate $y' = 4e^{0.8t} - 0.5y$ from $t = 0$ to 1 using a step size of 1 with $y(0) = 2$.

Solution.   For this case, the slope at the beginning of the interval is computed as

$$k_1 = f(0, 2) = 4e^{0.8(0)} - 0.5(2) = 3$$

This value is used to compute a value of y and a slope at the midpoint:

$$y(0.5) = 2 + 3(0.5) = 3.5$$
$$k_2 = f(0.5, 3.5) = 4e^{0.8(0.5)} - 0.5(3.5) = 4.217299$$

This slope in turn is used to compute another value of y and another slope at the midpoint:

$$y(0.5) = 2 + 4.217299(0.5) = 4.108649$$
$$k_3 = f(0.5, 4.108649) = 4e^{0.8(0.5)} - 0.5(4.108649) = 3.912974$$

Next, this slope is used to compute a value of y and a slope at the end of the interval:

$$y(1.0) = 2 + 3.912974(1.0) = 5.912974$$
$$k_4 = f(1.0, 5.912974) = 4e^{0.8(1.0)} - 0.5(5.912974) = 5.945677$$

Finally, the four slope estimates are combined to yield an average slope. This average slope is then used to make the final prediction at the end of the interval.

$$\phi = \frac{1}{6}[3 + 2(4.217299) + 2(3.912974) + 5.945677] = 4.201037$$

$$y(1.0) = 2 + 4.201037(1.0) = 6.201037$$

which compares favorably with the true solution of 6.194631 ($\varepsilon_t = 0.103\%$).

It is certainly possible to develop fifth- and higher-order RK methods. For example, Butcher's (1964) fifth-order RK method is written as

$$y_{i+1} = y_i + \frac{1}{90}(7k_1 + 32k_3 + 12k_4 + 32k_5 + 7k_6)h \tag{20.45}$$

where

$$k_1 = f(t_i, y_i) \tag{20.45a}$$

$$k_2 = f\left(t_i + \frac{1}{4}h, y_i + \frac{1}{4}k_1h\right) \tag{20.45b}$$

$$k_3 = f\left(t_i + \frac{1}{4}h, y_i + \frac{1}{8}k_1h + \frac{1}{8}k_2h\right) \tag{20.45c}$$

$$k_4 = f\left(t_i + \frac{1}{2}h, y_i - \frac{1}{2}k_2h + k_3h\right) \tag{20.45d}$$

$$k_5 = f\left(t_i + \frac{3}{4}h, y_i + \frac{3}{16}k_1h + \frac{9}{16}k_4h\right) \tag{20.45e}$$

$$k_6 = f\left(t_i + h, y_i - \frac{3}{7}k_1h + \frac{2}{7}k_2h + \frac{12}{7}k_3h - \frac{12}{7}k_4h + \frac{8}{7}k_5h\right) \tag{20.45f}$$

Note the similarity between Butcher's method and Boole's rule in Table 17.2. As expected, this method has a global truncation error of $O(h^5)$.

Although the fifth-order version provides more accuracy, notice that six function evaluations are required. Recall that up through the fourth-order versions, n function evaluations are required for an nth-order RK method. Interestingly, for orders higher than four, one or two additional function evaluations are necessary. Because the function evaluations account for the most computation time, methods of order five and higher are usually considered relatively less efficient than the fourth-order versions. This is one of the main reasons for the popularity of the fourth-order RK method.

20.5 SYSTEMS OF EQUATIONS

Many practical problems in engineering and science require the solution of a system of simultaneous ordinary differential equations rather than a single equation. Such systems may be represented generally as

$$\frac{dy_1}{dt} = f_1(t, y_1, y_2, \ldots, y_n)$$

$$\frac{dy_2}{dt} = f_2(t, y_1, y_2, \ldots, y_n)$$

$$\vdots \tag{20.46}$$

$$\frac{dy_n}{dt} = f_n(t, y_1, y_2, \ldots, y_n)$$

The solution of such a system requires that n initial conditions be known at the starting value of t.

An example is the calculation of the bungee jumper's velocity and position that we set up at the beginning of this chapter. For the free-fall portion of the jump, this problem amounts to solving the following system of ODEs:

$$\frac{dx}{dt} = v \tag{20.47}$$

$$\frac{dv}{dt} = g - \frac{c_d}{m}v^2 \tag{20.48}$$

If the stationary platform from which the jumper launches is defined as $x = 0$, the initial conditions would be $x(0) = v(0) = 0$.

20.5.1 Euler's Method

All the methods discussed in this chapter for single equations can be extended to systems of ODEs. Engineering applications can involve thousands of simultaneous equations. In each case, the procedure for solving a system of equations simply involves applying the one-step technique for every equation at each step before proceeding to the next step. This is best illustrated by the following example for Euler's method.

EXAMPLE 20.4   Solving Systems of ODEs with Euler's Method

Problem Statement.   Solve for the velocity and position of the free-falling bungee jumper using Euler's method. Assuming that at $t = 0$, $x = v = 0$, and integrate to $t = 10$ s with a step size of 2 s. As was done previously in Examples 1.1 and 1.2, the gravitational acceleration is 9.81 m/s^2, and the jumper has a mass of 68.1 kg with a drag coefficient of 0.25 kg/m.

Recall that the analytical solution for velocity is [Eq. (1.9)]:

$$v(t) = \sqrt{\frac{gm}{c_d}} \tanh\left(\sqrt{\frac{gc_d}{m}}\, t\right)$$

This result can be substituted into Eq. (20.47) which can be integrated to determine an analytical solution for distance as

$$x(t) = \frac{m}{c_d} \ln\left[\cosh\left(\sqrt{\frac{gc_d}{m}}\, t\right)\right]$$

Use these analytical solutions to compute the true relative errors of the results.

Solution.   The ODEs can be used to compute the slopes at $t = 0$ as

$$\frac{dx}{dt} = 0$$

$$\frac{dv}{dt} = 9.81 - \frac{0.25}{68.1}(0)^2 = 9.81$$

Euler's method is then used to compute the values at $t = 2$ s,

$$x = 0 + 0(2) = 0$$
$$v = 0 + 9.81(2) = 19.62$$

The analytical solutions can be computed as $x(2) = 19.16629$ and $v(2) = 18.72919$. Thus, the percent relative errors are 100% and 4.756%, respectively.

The process can be repeated to compute the results at $t = 4$ as

$$x = 0 + 19.62(2) = 39.24$$
$$v = 19.62 + \left(9.81 - \frac{0.25}{68.1}(19.62)^2\right)2 = 36.41368$$

Proceeding in a like manner gives the results displayed in Table 20.3.

TABLE 20.3 Distance and velocity of a free-falling bungee jumper as computed numerically with Euler's method.

t	x_{true}	v_{true}	x_{Euler}	v_{Euler}	$\varepsilon_t\,(x)$	$\varepsilon_t\,(v)$
0	0	0	0	0		
2	19.1663	18.7292	0	19.6200	100.00%	4.76%
4	71.9304	33.1118	39.2400	36.4137	45.45%	9.97%
6	147.9462	42.0762	112.0674	46.2983	24.25%	10.03%
8	237.5104	46.9575	204.6640	50.1802	13.83%	6.86%
10	334.1782	49.4214	305.0244	51.3123	8.72%	3.83%

Although the foregoing example illustrates how Euler's method can be implemented for systems of ODEs, the results are not very accurate because of the large step size. In addition, the results for distance are a bit unsatisfying because x does not change until the second iteration. Using a much smaller step greatly mitigates these deficiencies. As described next, using a higher-order solver provides decent results even with a relatively large step size.

20.5.2 Runge-Kutta Methods

Note that any of the higher-order RK methods in this chapter can be applied to systems of equations. However, care must be taken in determining the slopes. Figure 20.7 is helpful in visualizing the proper way to do this for the fourth-order method. That is, we first develop slopes for all variables at the initial value. These slopes (a set of k_1's) are then used to make predictions of the dependent variable at the midpoint of the interval. These midpoint values are in turn used to compute a set of slopes at the midpoint (the k_2's). These new slopes are then taken back to the starting point to make another set of midpoint predictions that lead to new slope predictions at the midpoint (the k_3's). These are then employed to make predictions at the end of the interval that are used to develop slopes at the end of the interval (the k_4's). Finally, the k's are combined into a set of increment functions [as in Eq. (20.44)] that are brought back to the beginning to make the final predictions. The following example illustrates the approach.

EXAMPLE 20.5   Solving Systems of ODEs with the Fourth-Order RK Method

Problem Statement.   Use the fourth-order RK method to solve for the same problem we addressed in Example 20.4.

Solution.   First, it is convenient to express the ODEs in the functional format of Eq. (20.46) as

$$\frac{dx}{dt} = f_1(t, x, v) = v$$

$$\frac{dv}{dt} = f_2(t, x, v) = g - \frac{c_d}{m}v^2$$

The first step in obtaining the solution is to solve for all the slopes at the beginning of the interval:

$$k_{1,1} = f_1(0,0,0) = 0$$

$$k_{1,2} = f_2(0,0,0) = 9.81 - \frac{0.25}{68.1}(0)^2 = 9.81$$

where $k_{i,j}$ is the ith value of k for the jth dependent variable. Next, we must calculate the first values of x and v at the midpoint of the first step:

$$x(1) = x(0) + k_{1,1}\frac{h}{2} = 0 + 0\frac{2}{2} = 0$$

$$v(1) = v(0) + k_{1,2}\frac{h}{2} = 0 + 9.81\frac{2}{2} = 9.81$$

which can be used to compute the first set of midpoint slopes:

$$k_{2,1} = f_1(1, 0, 9.81) = 9.8100$$
$$k_{2,2} = f_2(1, 0, 9.81) = 9.4567$$

These are used to determine the second set of midpoint predictions:

$$x(1) = x(0) + k_{2,1}\frac{h}{2} = 0 + 9.8100\frac{2}{2} = 9.8100$$

$$v(1) = v(0) + k_{2,2}\frac{h}{2} = 0 + 9.4567\frac{2}{2} = 9.4567$$

which can be used to compute the second set of midpoint slopes:

$$k_{3,1} = f_1(1, 9.8100, 9.4567) = 9.4567$$
$$k_{3,2} = f_2(1, 9.8100, 9.4567) = 9.4817$$

These are used to determine the predictions at the end of the interval:

$$x(2) = x(0) + k_{3,1}h = 0 + 9.4567(2) = 18.9134$$
$$v(2) = v(0) + k_{3,2}h = 0 + 9.4817(2) = 18.9634$$

which can be used to compute the endpoint slopes:

$$k_{4,1} = f_1(2, 18.9134, 18.9634) = 18.9634$$
$$k_{4,2} = f_2(2, 18.9134, 18.9634) = 8.4898$$

The values of k can then be used to compute [Eq. (20.44)]:

$$x(2) = 0 + \frac{1}{6}[0 + 2(9.8100 + 9.4567) + 18.9634]2 = 19.1656$$

$$v(2) = 0 + \frac{1}{6}[9.8100 + 2(9.4567 + 9.4817) + 8.4898]2 = 18.7256$$

Proceeding in a like manner for the remaining steps yields the values displayed in Table 20.4. In contrast to the results obtained with Euler's method, the fourth-order RK predictions are much closer to the true values. Further, a highly accurate, nonzero value is computed for distance on the first step.

TABLE 20.4 Distance and velocity of a free-falling bungee jumper as computed numerically with the fourth-order RK method.

t	x_{true}	v_{true}	x_{RK4}	v_{RK4}	$\varepsilon_t(x)$	$\varepsilon_t(v)$
0	0	0	0	0		
2	19.1663	18.7292	19.1656	18.7256	0.004%	0.019%
4	71.9304	33.1118	71.9311	33.0995	0.001%	0.037%
6	147.9462	42.0762	147.9521	42.0547	0.004%	0.051%
8	237.5104	46.9575	237.5104	46.9345	0.000%	0.049%
10	334.1782	49.4214	334.1626	49.4027	0.005%	0.038%

20.5.3 MATLAB M-file Function: `rk4sys`

Figure 20.8 shows an M-file called `rk4sys` that uses the fourth-order Runge-Kutta method to solve a system of ODEs. This code is similar in many ways to the function developed earlier (Fig. 20.3) to solve a single ODE with Euler's method. For example, it is passed the function name defining the ODEs through its argument.

FIGURE 20.8
An M-file to implement the RK4 method for a system of ODEs.

```
function [tp,yp] = rk4sys(dydt,tspan,y0,h,varargin)
% rk4sys: fourth-order Runge-Kutta for a system of ODEs
%   [t,y] = rk4sys(dydt,tspan,y0,h,p1,p2,...): integrates
%            a system of ODEs with fourth-order RK method
% input:
%   dydt = name of the M-file that evaluates the ODEs
%   tspan = [ti, tf]; initial and final times with output
%                     generated at interval of h, or
%         = [t0 t1 ... tf]; specific times where solution output
%   y0 = initial values of dependent variables
%   h = step size
%   p1,p2,... = additional parameters used by dydt
% output:
%   tp = vector of independent variable
%   yp = vector of solution for dependent variables

if nargin<4,error('at least 4 input arguments required'), end
if any(diff(tspan)<=0),error('tspan not ascending order'), end
n = length(tspan);
ti = tspan(1);tf = tspan(n);
if n == 2
  t = (ti:h:tf)'; n = length(t);
  if t(n)<tf
    t(n+1) = tf;
    n = n+1;
  end
else
  t = tspan;
end
tt = ti; y(1,:) = y0;
np = 1; tp(np) = tt; yp(np,:) = y(1,:);
i=1;
while(1)
  tend = t(np+1);
  hh = t(np+1) - t(np);
```

(Continued)

```
if hh>h,hh = h;end
while(1)
   if tt+hh>tend,hh = tend-tt;end
   k1 = dydt(tt,y(i,:),varargin{:})';
   ymid = y(i,:) + k1.*hh./2;
   k2 = dydt(tt+hh/2,ymid,varargin{:})';
   ymid = y(i,:) + k2*hh/2;
   k3 = dydt(tt+hh/2,ymid,varargin{:})';
   yend = y(i,:) + k3*hh;
   k4 = dydt(tt+hh,yend,varargin{:})';
   phi = (k1+2*(k2+k3)+k4)/6;
   y(i+1,:) = y(i,:) + phi*hh;
   tt = tt+hh;
   i=i+1;
   if tt>=tend,break,end
end
np = np+1; tp(np) = tt; yp(np,:) = y(i,:);
if tt>=tf,break,end
end
```

FIGURE 20.8 (*Continued*)

However, it has an additional feature that allows you to generate output in two ways, depending on how the input variable tspan is specified. As was the case for Fig. 20.3, you can set tspan = [ti tf], where ti and tf are the initial and final times, respectively. If done in this way, the routine automatically generates output values between these limits at equal spaced intervals h. Alternatively, if you want to obtain results at specific times, you can define tspan = [t0,t1,...,tf]. Note that in both cases, the tspan values must be in ascending order.

We can employ rk4sys to solve the same problem as in Example 20.5. First, we can develop an M-file to hold the ODEs:

```
function dy = dydtsys(t, y)
dy = [y(2);9.81-0.25/68.1*y(2)^2];
```

where $y(1)$ = distance (x) and $y(2)$ = velocity (v). The solution can then be generated as

```
>> [t y] = rk4sys(@dydtsys,[0 10],[0 0],2);
>> disp([t' y(:,1) y(:,2)])

        0          0          0
   2.0000    19.1656    18.7256
   4.0000    71.9311    33.0995
   6.0000   147.9521    42.0547
   8.0000   237.5104    46.9345
  10.0000   334.1626    49.4027
```

We can also use `tspan` to generate results at specific values of the independent variable. For example,

```
>> tspan=[0 6 10];
>> [t y] = rk4sys(@dydtsys,tspan,[0 0],2);
>> disp([t' y(:,1) y(:,2)])

          0          0          0
     6.0000   147.9521    42.0547
    10.0000   334.1626    49.4027
```

20.6 CASE STUDY   PREDATOR-PREY MODELS AND CHAOS

Background.   Engineers and scientists deal with a variety of problems involving systems of nonlinear ordinary differential equations. This case study focuses on two of these applications. The first relates to predator-prey models that are used to study species interactions. The second are equations derived from fluid dynamics that are used to simulate the atmosphere.

Predator-prey models were developed independently in the early part of the twentieth century by the Italian mathematician Vito Volterra and the American biologist Alfred Lotka. These equations are commonly called *Lotka-Volterra equations*. The simplest version is the following pairs of ODEs:

$$\frac{dx}{dt} = ax - bxy \tag{20.49}$$

$$\frac{dy}{dt} = -cy + dxy \tag{20.50}$$

where x and $y =$ the number of prey and predators, respectively, $a =$ the prey growth rate, $c =$ the predator death rate, and b and $d =$ the rates characterizing the effect of the predator-prey interactions on the prey death and the predator growth, respectively. The multiplicative terms (that is, those involving xy) are what make such equations nonlinear.

An example of a simple nonlinear model based on atmospheric fluid dynamics is the *Lorenz equations* created by the American meteorologist Edward Lorenz:

$$\frac{dx}{dt} = -\sigma x - \sigma y$$

$$\frac{dy}{dt} = rx - y - xz$$

$$\frac{dz}{dt} = -bz + xy$$

Lorenz developed these equations to relate the intensity of atmospheric fluid motion x to temperature variations y and z in the horizontal and vertical directions, respectively. As

20.6 CASE STUDY   continued

with the predator-prey model, the nonlinearities stem from the simple multiplicative terms: xz and xy.

Use numerical methods to obtain solutions for these equations. Plot the results to visualize how the dependent variables change temporally. In addition, graph the dependent variables versus each other to see whether any interesting patterns emerge.

Solution.   The following parameter values can be used for the predator-prey simulation: $a = 1.2$, $b = 0.6$, $c = 0.8$, and $d = 0.3$. Employ initial conditions of $x = 2$ and $y = 1$ and integrate from $t = 0$ to 30, using a step size of $h = 0.0625$.

First, we can develop a function to hold the differential equations:

```
function yp = predprey(t,y,a,b,c,d)
yp = [a*y(1)-b*y(1)*y(2);-c*y(2)+d*y(1)*y(2)];
```

The following script employs this function to generate solutions with both the Euler and the fourth-order RK methods. Note that the function `eulersys` was based on modifying the `rk4sys` function (Fig. 20.8). We will leave the development of such an M-file as a homework problem. In addition to displaying the solution as a time-series plot (x and y versus t), the script also generates a plot of y versus x. Such *phase-plane* plots are often useful in elucidating features of the model's underlying structure that may not be evident from the time series.

```
h=0.0625;tspan=[0 40];y0=[2 1];
a=1.2;b=0.6;c=0.8;d=0.3;
[t y] = eulersys(@predprey,tspan,y0,h,a,b,c,d);
subplot(2,2,1);plot(t,y(:,1),t,y(:,2),'--')
legend('prey','predator');title('(a) Euler time plot')
subplot(2,2,2);plot(y(:,1),y(:,2))
title('(b) Euler phase plane plot')
[t y] = rk4sys(@predprey,tspan,y0,h,a,b,c,d);
subplot(2,2,3);plot(t,y(:,1),t,y(:,2),'--')
title('(c) RK4 time plot')
subplot(2,2,4);plot(y(:,1),y(:,2))
title('(d) RK4 phase plane plot')
```

The solution obtained with Euler's method is shown at the top of Fig. 20.9. The time series (Fig. 20.9a) indicates that the amplitudes of the oscillations are expanding. This is reinforced by the phase-plane plot (Fig. 20.9b). Hence, these results indicate that the crude Euler method would require a much smaller time step to obtain accurate results.

In contrast, because of its much smaller truncation error, the RK4 method yields good results with the same time step. As in Fig. 20.9c, a cyclical pattern emerges in time. Because the predator population is initially small, the prey grows exponentially. At a certain point, the prey become so numerous that the predator population begins to grow. Eventually, the increased predators cause the prey to decline. This decrease, in turn, leads to a decrease of the predators. Eventually, the process repeats. Notice that, as expected, the

20.6 CASE STUDY continued

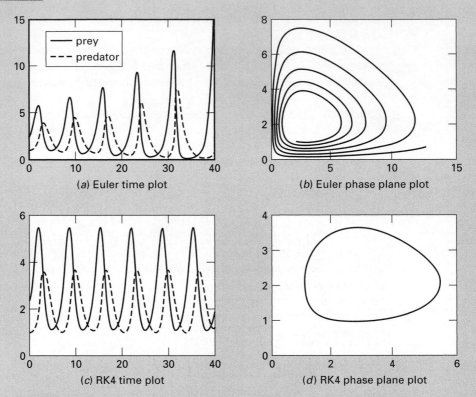

FIGURE 20.9
Solution for the Lotka-Volterra model. Euler's method (a) time-series and (b) phase-plane plots, and RK4 method (c) time-series and (d) phase-plane plots.

predator peak lags the prey. Also, observe that the process has a fixed period—that is, it repeats in a set time.

The phase-plane representation for the accurate RK4 solution (Fig. 20.9d) indicates that the interaction between the predator and the prey amounts to a closed counterclockwise orbit. Interestingly, there is a resting or *critical point* at the center of the orbit. The exact location of this point can be determined by setting Eqs. (20.49) and (20.50) to steady state $(dy/dt = dx/dt = 0)$ and solving for $(x, y) = (0, 0)$ and $(c/d, a/b)$. The former is the trivial result that if we start with neither predators nor prey, nothing will happen. The latter is the more interesting outcome that if the initial conditions are set at $x = c/d$ and $y = a/b$, the derivatives will be zero, and the populations will remain constant.

Now, let's use the same approach to investigate the trajectories of the Lorenz equations with the following parameter values: $a = 10, b = 8/3$, and $r = 28$. Employ initial conditions

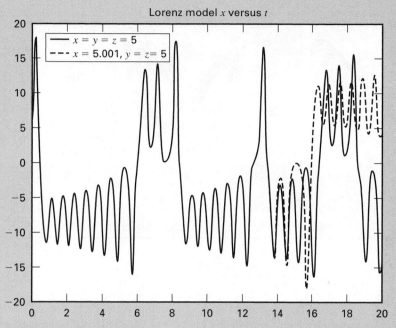

FIGURE 20.10
Time-domain representation of x versus t for the Lorenz equations. The solid time series is for the initial conditions (5, 5, 5). The dashed line is where the initial condition for x is perturbed slightly (5.001, 5, 5).

of $x = y = z = 5$ and integrate from $t = 0$ to 20. For this case, we will use the fourth-order RK method to obtain solutions with a constant time step of $h = 0.03125$.

The results are quite different from the behavior of the Lotka-Volterra equations. As in Fig. 20.10, the variable x seems to be undergoing an almost random pattern of oscillations, bouncing around from negative values to positive values. The other variables exhibit similar behavior. However, even though the patterns seem random, the frequency of the oscillation and the amplitudes seem fairly consistent.

An interesting feature of such solutions can be illustrated by changing the initial condition for x slightly (from 5 to 5.001). The results are superimposed as the dashed line in Fig. 20.10. Although the solutions track on each other for a time, after about $t = 15$ they diverge significantly. Thus, we can see that the Lorenz equations are quite sensitive to their initial conditions. The term *chaotic* is used to describe such solutions. In his original study, this led Lorenz to the conclusion that long-range weather forecasts might be impossible!

20.6 CASE STUDY   continued

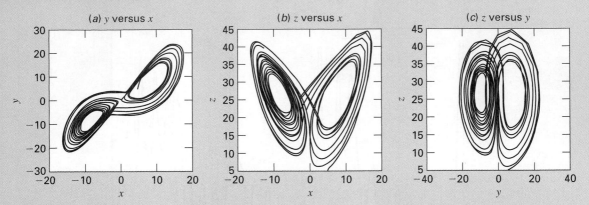

FIGURE 20.11
Phase-plane representation for the Lorenz equations. (a) xy, (b) xz, and (c) yz projections.

The sensitivity of a dynamical system to small perturbations of its initial conditions is sometimes called the *butterfly effect*. The idea is that the flapping of a butterfly's wings might induce tiny changes in the atmosphere that ultimately leads to a large-scale weather phenomenon like a tornado.

Although the time-series plots are chaotic, phase-plane plots reveal an underlying structure. Because we are dealing with three independent variables, we can generate projections. Figure 20.11 shows projections in the xy, xz, and the yz planes. Notice how a structure is manifest when perceived from the phase-plane perspective. The solution forms orbits around what appear to be critical points. These points are called *strange attractors* in the jargon of mathematicians who study such nonlinear systems.

Beyond the two-variable projections, MATLAB's plot3 function provides a vehicle to directly generate a three-dimensional phase-plane plot:

```
>> plot3(y(:,1),y(:,2),y(:,2))
>> xlabel('x');ylabel('y');zlabel('z');grid
```

As was the case for Fig. 20.11, the three-dimensional plot (Fig 20.12) depicts trajectories cycling in a definite pattern around a pair of critical points.

As a final note, the sensitivity of chaotic systems to initial conditions has implications for numerical computations. Beyond the initial conditions themselves, different step sizes or different algorithms (and in some cases, even different computers) can introduce small differences in the solutions. In a similar fashion to Fig. 20.10, these discrepancies will eventually lead to large deviations. Some of the problems in this chapter and in Chap. 21 are designed to demonstrate this issue.

20.6 CASE STUDY   continued

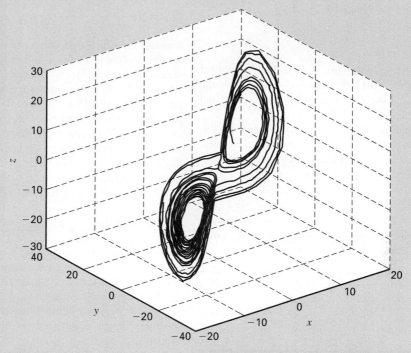

FIGURE 20.12
Three-dimensional phase-plane representation for the Lorenz equations generated with MATLAB's
`plot3` function.

PROBLEMS

20.1 Solve the following initial value problem over the interval from $t = 0$ to 2 where $y(0) = 1$. Display all your results on the same graph.

$$\frac{dy}{dt} = yt^2 - 1.1y$$

(a) Analytically.
(b) Using Euler's method with $h = 0.5$ and 0.25.
(c) Using the midpoint method with $h = 0.5$.
(d) Using the fourth-order RK method with $h = 0.5$.

20.2 Solve the following problem over the interval from $x = 0$ to 1 using a step size of 0.25 where $y(0) = 1$. Display all your results on the same graph.

$$\frac{dy}{dx} = (1 + 2x)\sqrt{y}$$

(a) Analytically.
(b) Using Euler's method.
(c) Using Heun's method without iteration.
(d) Using Ralston's method.
(e) Using the fourth-order RK method.

20.3 Solve the following problem over the interval from $t = 0$ to 3 using a step size of 0.5 where $y(0) = 1$. Display all your results on the same graph.

$$\frac{dy}{dt} = -y + t^2$$

Obtain your solutions with **(a)** Heun's method without iterating the corrector, **(b)** Heun's method with iterating the corrector until $\varepsilon_s < 0.1\%$, **(c)** the midpoint method, and **(d)** Ralston's method.

20.4 The growth of populations of organisms has many engineering and scientific applications. One of the simplest models assumes that the rate of change of the population p is proportional to the existing population at any time t:

$$\frac{dp}{dt} = k_g p \qquad \text{(P20.4.1)}$$

where k_g = the growth rate. The world population in millions from 1950 through 2000 was

t	1950	1955	1960	1965	1970	1975
p	2555	2780	3040	3346	3708	4087
t	1980	1985	1990	1995	2000	
p	4454	4850	5276	5686	6079	

(a) Assuming that Eq. (P20.4.1) holds, use the data from 1950 through 1970 to estimate k_g.
(b) Use the fourth-order RK method along with the results of **(a)** to stimulate the world population from 1950 to 2050 with a step size of 5 years. Display your simulation results along with the data on a plot.

20.5 Although the model in Prob. 20.4 works adequately when population growth is unlimited, it breaks down when factors such as food shortages, pollution, and lack of space inhibit growth. In such cases, the growth rate is not a constant, but can be formulated as

$$k_g = k_{gm}(1 - p/p_{max})$$

where k_{gm} = the maximum growth rate under unlimited conditions, p = population, and p_{max} = the maximum population. Note that p_{max} is sometimes called the *carrying capacity*. Thus, at low population density $p \ll p_{max}$, $k_g \rightarrow k_{gm}$. As p approaches p_{max}, the growth rate approaches zero. Using this growth rate formulation, the rate of change of population can be modeled as

$$\frac{dp}{dt} = k_{gm}(1 - p/p_{max})p$$

This is referred to as the *logistic model*. The analytical solution to this model is

$$p = p_0 \frac{p_{max}}{p_0 + (p_{max} - p_0)e^{-k_{gm}t}}$$

Simulate the world's population from 1950 to 2050 using **(a)** the analytical solution, and **(b)** the fourth-order RK method with a step size of 5 years. Employ the following initial conditions and parameter values for your simulation: p_0 (in 1950) = 2,555 million people, $k_{gm} = 0.026/\text{yr}$, and $p_{max} = 12,000$ million people. Display your results as a plot along with the data from Prob. 20.4.

20.6 Suppose that a projectile is launched upward from the earth's surface. Assume that the only force acting on the object is the downward force of gravity. Under these conditions, a force balance can be used to derive

$$\frac{dv}{dt} = -g(0)\frac{R^2}{(R + x)^2}$$

where v = upward velocity (m/s), t = time (s), x = altitude (m) measured upward from the earth's surface, $g(0)$ = the gravitational acceleration at the earth's surface ($\cong 9.8 \text{ m/s}^2$), and R = the earth's radius ($\cong 6.37 \times 10^6$ m). Recognizing that $dx/dt = v$, use Euler's method to determine the maximum height that would be obtained if $v(t = 0) = 1400$ m/s.

20.7 Solve the following pair of ODEs over the interval from $t = 0$ to 0.4 using a step size of 0.1. The initial conditions are $y(0) = 2$ and $z(0) = 4$. Obtain your solution with **(a)** Euler's method and **(b)** the fourth-order RK method. Display your results as a plot.

$$\frac{dy}{dt} = -2y + 4e^{-t}$$

$$\frac{dz}{dt} = -\frac{yz^2}{3}$$

20.8 The *van der Pol equation* is a model of an electronic circuit that arose back in the days of vacuum tubes:

$$\frac{d^2y}{dt^2} - (1 - y^2)\frac{dy}{dt} + y = 0$$

Given the initial conditions, $y(0) = y'(0) = 1$, solve this equation from $t = 0$ to 10 using Euler's method with a step size of **(a)** 0.2 and **(b)** 0.1. Plot both solutions on the same graph.

20.9 Given the initial conditions, $y(0) = 1$ and $y'(0) = 0$, solve the following initial-value problem from $t = 0$ to 4:

$$\frac{d^2y}{dt^2} + 9y = 0$$

Obtain your solutions with (a) Euler's method and (b) the fourth-order RK method. In both cases, use a step size of 0.1. Plot both solutions on the same graph along with the exact solution $y = \cos 3t$.

20.10 Develop an M-file to solve a single ODE with Heun's method with iteration. Design the M-file so that it creates a plot of the results. Test your program by using it to solve for population as described in Prob. 20.5. Employ a step size of 5 years and iterate the corrector until $\varepsilon_s < 0.1\%$.

20.11 Develop an M-file to solve a single ODE with the midpoint method. Design the M-file so that it creates a plot of the results. Test your program by using it to solve for population as described in Prob. 20.5. Employ a step size of 5 years.

20.12 Develop an M-file to solve a single ODE with the fourth-order RK method. Design the M-file so that it creates a plot of the results. Test your program by using it to solve Prob. 20.2. Employ a step size of 0.1.

20.13 Develop an M-file to solve a system of ODEs with Euler's method. Design the M-file so that it creates a plot of the results. Test your program by using it to solve Prob. 20.7 with a step size of 0.25.

20.14 Isle Royale National Park is a 210-square-mile archipelago composed of a single large island and many small islands in Lake Superior. Moose arrived around 1900, and by 1930, their population approached 3000, ravaging vegetation. In 1949, wolves crossed an ice bridge from Ontario. Since the late 1950s, the numbers of the moose and wolves have been tracked.

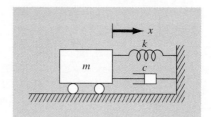

FIGURE P20.15

(a) Integrate the Lotka-Volterra equations (Section 20.6) from 1960 through 2020 using the following coefficient values: $a = 0.23$, $b = 0.0133$, $c = 0.4$, and $d = 0.0004$. Compare your simulation with the data using a time-series plot and determine the sum of the squares of the residuals between your model and the data for both the moose and the wolves.

(b) Develop a phase-plane plot of your solution.

20.15 The motion of a damped spring-mass system (Fig. P20.15) is described by the following ordinary differential equation:

$$m\frac{d^2x}{dt^2} + c\frac{dx}{dt} + kx = 0$$

where x = displacement from equilibrium position (m), t = time (s), m = 20-kg mass, and c = the damping coefficient (N · s/m). The damping coefficient c takes on three values

Year	Moose	Wolves	Year	Moose	Wolves	Year	Moose	Wolves
1959	563	20	1975	1355	41	1991	1313	12
1960	610	22	1976	1282	44	1992	1590	12
1961	628	22	1977	1143	34	1993	1879	13
1962	639	23	1978	1001	40	1994	1770	17
1963	663	20	1979	1028	43	1995	2422	16
1964	707	26	1980	910	50	1996	1163	22
1965	733	28	1981	863	30	1997	500	24
1966	765	26	1982	872	14	1998	699	14
1967	912	22	1983	932	23	1999	750	25
1968	1042	22	1984	1038	24	2000	850	29
1969	1268	17	1985	1115	22	2001	900	19
1970	1295	18	1986	1192	20	2002	1100	17
1971	1439	20	1987	1268	16	2003	900	19
1972	1493	23	1988	1335	12	2004	750	29
1973	1435	24	1989	1397	12	2005	540	30
1974	1467	31	1990	1216	15	2006	450	30

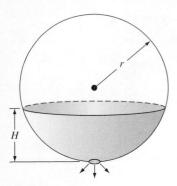

FIGURE P20.16
A spherical tank.

of 5 (underdamped), 40 (critically damped), and 200 (over-damped). The spring constant $k = 20$ N/m. The initial velocity is zero, and the initial displacement $x = 1$ m. Solve this equation using a numerical method over the time period $0 \leq t \leq 15$ s. Plot the displacement versus time for each of the three values of the damping coefficient on the same plot.

20.16 A spherical tank has a circular orifice in its bottom through which the liquid flows out (Fig. P20.16). The flow rate through the hole can be estimated as

$$Q_{out} = CA\sqrt{2gh}$$

where Q_{out} = outflow (m³/s), C = an empirically derived coefficient, A = the area of the orifice (m²), g = the gravitational constant (= 9.81 m/s²), and h = the depth of liquid in the tank. Use one of the numerical methods described in this chapter to determine how long it will take for the water to flow out of a 3-m diameter tank with an initial height of 2.75 m. Note that the orifice has a diameter of 3 cm and $C = 0.55$.

20.17 In the investigation of a homicide or accidental death, it is often important to estimate the time of death. From the experimental observations, it is known that the surface temperature of an object changes at a rate proportional to the difference between the temperature of the object and that of the surrounding environment or ambient temperature. This is known as Newton's law of cooling. Thus, if $T(t)$ is the temperature of the object at time t, and T_a is the constant ambient temperature:

$$\frac{dT}{dt} = -K(T - T_a)$$

where $K > 0$ is a constant of proportionality. Suppose that at time $t = 0$ a corpse is discovered and its temperature is measured to be T_o. We assume that at the time of death, the body temperature T_d was at the normal value of 37 °C.

Suppose that the temperature of the corpse when it was discovered was 29.5 °C, and that two hours later, it is 23.5 °C. The ambient temperature is 20 °C.
(a) Determine K and the time of death.
(b) Solve the ODE numerically and plot the results.

20.18 The reaction $A \rightarrow B$ takes place in two reactors in series. The reactors are well mixed but are not at steady state. The unsteady-state mass balance for each stirred tank reactor is shown below:

$$\frac{dCA_1}{dt} = \frac{1}{\tau}(CA_0 - CA_1) - kCA_1$$

$$\frac{dCB_1}{dt} = \frac{1}{\tau}CB_1 + kCA_1$$

$$\frac{dCA_2}{dt} = \frac{1}{\tau}(CA_1 - CA_2) - kCA_2$$

$$\frac{dCB_2}{dt} = \frac{1}{\tau}(CB_1 - CB_2) - kCB_2$$

where CA_0 = concentration of A at the inlet of the first reactor, CA_1 = concentration of A at the outlet of the first reactor (and inlet of the second), CA_2 = concentration of A at the outlet of the second reactor, CB_1 = concentration of B at the outlet of the first reactor (and inlet of the second), CB_2 = concentration of B in the second reactor, τ = residence time for each reactor, and k = the rate constant for reaction of A to produce B. If CA_0 is equal to 20, find the concentrations of A and B in both reactors during their first 10 minutes of operation. Use $k = 0.12$/min and $\tau = 5$ min and assume that the initial conditions of all the dependent variables are zero.

20.19 A nonisothermal batch reactor can be described by the following equations:

$$\frac{dC}{dt} = -e^{(-10/(T+273))}C$$

$$\frac{dT}{dt} = 1000e^{(-10/(T+273))}C - 10(T - 20)$$

where C is the concentration of the reactant and T is the temperature of the reactor. Initially, the reactor is at 15 °C and has a concentration of reactant C of 1.0 gmol/L. Find the concentration and temperature of the reactor as a function of time.

20.20 The following equation can be used to model the deflection of a sailboat mast subject to a wind force:

$$\frac{d^2y}{dz^2} = \frac{f(z)}{2EI}(L - z)^2$$

where $f(z)$ = wind force, E = modulus of elasticity, L = mast length, and I = moment of inertia. Note that the force

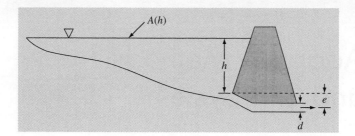

FIGURE P20.21

varies with height according to

$$f(z) = \frac{200z}{5+z}e^{-2z/30}$$

Calculate the deflection if $y = 0$ and $dy/dz = 0$ at $z = 0$. Use parameter values of $L = 30$, $E = 1.25 \times 10^8$, and $I = 0.05$ for your computation.

20.21 A pond drains through a pipe as shown in Fig. P20.21. Under a number of simplifying assumptions, the following differential equation describes how depth changes with time:

$$\frac{dh}{dt} = -\frac{\pi d^2}{4A(h)}\sqrt{2g(h+e)}$$

where h = depth (m), t = time (s), d = pipe diameter (m), $A(h)$ = pond surface area as a function of depth (m²), g = gravitational constant ($= 9.81$ m/s²), and e = depth of pipe outlet below the pond bottom (m). Based on the following area-depth table, solve this differential equation to determine how long it takes for the pond to empty, given that $h(0) = 6$ m, $d = 0.25$ m, $e = 1$ m.

h, m	6	5	4	3	2	1	0
A(h), m²	1.17	0.97	0.67	0.45	0.32	0.18	0

20.22 Engineers and scientists use mass-spring models to gain insight into the dynamics of structures under the influence of disturbances such as earthquakes. Figure P20.22 shows such a representation for a three-story building. For this case, the analysis is limited to horizontal motion of the structure. Using Newton's second law, force balances can be developed for this system as

$$\frac{d^2x_1}{dt^2} = -\frac{k_1}{m_1}x_1 + \frac{k_2}{m_1}(x_2 - x_1)$$

$$\frac{d^2x_2}{dt^2} = \frac{k_2}{m_2}(x_1 - x_2) + \frac{k_3}{m_2}(x_3 - x_2)$$

$$\frac{d^2x_3}{dt^2} = \frac{k_3}{m_3}(x_2 - x_3)$$

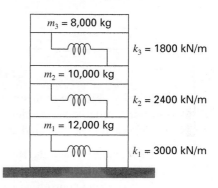

FIGURE P20.22

Simulate the dynamics of this structure from $t = 0$ to 20 s, given the initial condition that the velocity of the ground floor is $dx_1/dt = 1$ m/s, and all other initial values of displacements and velocities are zero. Present your results as two time-series plots of **(a)** displacements and **(b)** velocities. In addition, develop a three-dimensional phase-plane plot of the displacements.

20.23 Repeat the the same simulations as in Section 20.6 for the Lorenz equations but generate the solutions with the midpoint method.

20.24 Perform the same simulations as in Section 20.6 for the Lorenz equations but use a value of $r = 99.96$. Compare your results with those obtained in Section 20.6.

21

Adaptive Methods and Stiff Systems

CHAPTER OBJECTIVES

The primary objective of this chapter is to introduce you to more advanced methods for solving initial-value problems for ordinary differential equations. Specific objectives and topics covered are

- Understanding how the Runge-Kutta Fehlberg methods use RK methods of different orders to provide error estimates that are used to adjust the step size.
- Familiarizing yourself with the built-in MATLAB functions for solving ODEs.
- Learning how to adjust the options for MATLAB's ODE solvers.
- Learning how to pass parameters to MATLAB's ODE solvers.
- Understanding the difference between one-step and multistep methods for solving ODEs.
- Understanding what is meant by stiffness and its implications for solving ODEs.

21.1 ADAPTIVE RUNGE-KUTTA METHODS

To this point, we have presented methods for solving ODEs that employ a constant step size. For a significant number of problems, this can represent a serious limitation. For example, suppose that we are integrating an ODE with a solution of the type depicted in Fig. 21.1. For most of the range, the solution changes gradually. Such behavior suggests that a fairly large step size could be employed to obtain adequate results. However, for a localized region from $t = 1.75$ to 2.25, the solution undergoes an abrupt change. The practical consequence of dealing with such functions is that a very small step size would be required to accurately capture the impulsive behavior. If a constant step-size algorithm were employed, the smaller step size required for the region of abrupt change would have to be applied to the entire computation. As a consequence, a much smaller step size than necessary—and, therefore, many more calculations—would be wasted on the regions of gradual change.

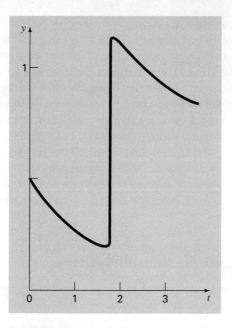

FIGURE 21.1
An example of a solution of an ODE that exhibits an abrupt change. Automatic step-size adjustment has great advantages for such cases.

Algorithms that automatically adjust the step size can avoid such overkill and hence be of great advantage. Because they "adapt" to the solution's trajectory, they are said to have *adaptive step-size control*. Implementation of such approaches requires that an estimate of the local truncation error be obtained at each step. This error estimate can then serve as a basis for either shortening or lengthening the step size.

Before proceeding, we should mention that aside from solving ODEs, the methods described in this chapter can also be used to evaluate definite integrals. The evaluation of the definite integral

$$I = \int_a^b f(x)\,dx$$

is equivalent to solving the differential equation

$$\frac{dy}{dx} = f(x)$$

for $y(b)$ given the initial condition $y(a) = 0$. Thus, the following techniques can be employed to efficiently evaluate definite integrals involving functions that are generally smooth but exhibit regions of abrupt change.

There are two primary approaches to incorporate adaptive step-size control into one-step methods. *Step halving* involves taking each step twice, once as a full step and then as two half steps. The difference in the two results represents an estimate of the local truncation error. The step size can then be adjusted based on this error estimate.

In the second approach, called *embedded RK methods,* the local truncation error is estimated as the difference between two predictions using different-order RK methods. These are currently the methods of choice because they are more efficient than step halving.

The embedded methods were first developed by Fehlberg. Hence, they are sometimes referred to as *RK-Fehlberg methods.* At face value, the idea of using two predictions of different order might seem too computationally expensive. For example, a fourth- and fifth-order prediction amounts to a total of 10 function evaluations per step [recall Eqs. (20.44) and (20.45)]. Fehlberg cleverly circumvented this problem by deriving a fifth-order RK method that employs most of the same function evaluations required for an accompanying fourth-order RK method. Thus, the approach yielded the error estimate on the basis of only six function evaluations!

21.1.1 MATLAB Functions for Nonstiff Systems

Since Fehlberg originally developed his approach, other even better approaches have been developed. Several of these are available as built-in functions in MATLAB.

ode23. The `ode23` function uses the BS23 algorithm (Bogacki and Shampine, 1989; Shampine, 1994), which simultaneously uses second- and third-order RK formulas to solve the ODE and make error estimates for step-size adjustment. The formulas to advance the solution are

$$y_{i+1} = y_i + \frac{1}{9}(2k_1 + 3k_2 + 4k_3)h \tag{21.1}$$

where

$$k_1 = f(t_i, y_i) \tag{21.1a}$$

$$k_2 = f\left(t_i + \frac{1}{2}h, y_i + \frac{1}{2}k_1h\right) \tag{21.1b}$$

$$k_3 = f\left(t_i + \frac{3}{4}h, y_i + \frac{3}{4}k_2h\right) \tag{21.1c}$$

The error is estimated as

$$E_{i+1} = \frac{1}{72}(-5k_1 + 6k_2 + 8k_3 - 9k_4)h \tag{21.2}$$

where

$$k_4 = f(t_{i+1}, y_{i+1}) \tag{21.2a}$$

Note that although there appear to be four function evaluations, there are really only three because after the first step, the k_1 for the present step will be the k_4 from the previous step. Thus, the approach yields a prediction and error estimate based on three evaluations rather than the five that would ordinarily result from using second- (two evaluations) and third-order (three evaluations) RK formulas in tandem.

After each step, the error is checked to determine whether it is within a desired tolerance. If it is, the value of y_{i+1} is accepted, and k_4 becomes k_1 for the next step. If the error is too large, the step is repeated with reduced step sizes until the estimated error satisfies

$$E \leq \max(\text{RelTol} \times |y|, \text{AbsTol}) \tag{21.3}$$

where RelTol is the relative tolerance (default $= 10^{-3}$) and AbsTol is the absolute tolerance (default $= 10^{-6}$). Observe that the criteria for the relative error uses a fraction rather than a percent relative error as we have done on many occasions prior to this point.

ode45. The ode45 function uses an algorithm developed by Dormand and Prince (1990), which simultaneously uses fourth- and fifth-order RK formulas to solve the ODE and make error estimates for step-size adjustment. MATLAB recommends that ode45 is the best function to apply as a "first try" for most problems.

ode113. The ode113 function uses a variable-order Adams-Bashforth-Moulton solver. It is useful for stringent error tolerances or computationally intensive ODE functions. Note that this is a multistep method as we will describe subsequently in Section 21.2.

These functions can be called in a number of different ways. The simplest approach is

```
[t, y] = ode45(odefun, tspan, y0)
```

where y is the solution array where each column is one of the dependent variables and each row corresponds to a time in the column vector t, $odefun$ is the name of the function returning a column vector of the right-hand-sides of the differential equations, $tspan$ specifies the integration interval, and $y0 = $ a vector containing the initial values.

Note that $tspan$ can be formulated in two ways. First, if it is entered as a vector of two numbers,

```
tspan = [ti tf];
```

the integration is performed from ti to tf. Second, to obtain solutions at specific times $t0, t1, \ldots, tn$ (all increasing or all decreasing), use

```
tspan = [t0 t1 ... tn];
```

Here is an example of how ode45 can be used to solve a single ODE, $y' = 4e^{0.8t} - 0.5y$ from $t = 0$ to 4 with an initial condition of $y(0) = 2$. Recall from Example 20.1 that the analytical solution at $t = 4$ is 75.33896. Representing the ODE as an anonymous function, ode45 can be used to generate the same result numerically as

```
>> dydt=@(t,y) 4*exp(0.8*t)-0.5*y;
>> [t,y]=ode45(dydt,[0 4],2);
>> y(length(t))

ans =
   75.3390
```

As described in the following example, the ODE is typically stored in its own M-file when dealing with systems of equations.

EXAMPLE 21.1   Using MATLAB to Solve a System of ODEs

Problem Statement.   Employ ode45 to solve the following set of nonlinear ODEs from $t = 0$ to 20:

$$\frac{dy_1}{dt} = 1.2y_1 - 0.6y_1 y_2 \qquad \frac{dy_2}{dt} = -0.8y_2 + 0.3y_1 y_2$$

where $y_1 = 2$ and $y_2 = 1$ at $t = 0$. Such equations are referred to as *predator-prey equations*.

Solution.   Before obtaining a solution with MATLAB, you must create a function to compute the right-hand side of the ODEs. One way to do this is to create an M-file as in

```
function yp = predprey(t,y)
yp = [1.2*y(1)-0.6*y(1)*y(2);-0.8*y(2)+0.3*y(1)*y(2)];
```

We stored this M-file under the name: predprey.m.

Next, enter the following commands to specify the integration range and the initial conditions:

```
>> tspan = [0 20];
>> y0 = [2, 1];
```

The solver can then be invoked by

```
>> [t,y] = ode45(@predprey, tspan, y0);
```

This command will then solve the differential equations in predprey.m over the range defined by tspan using the initial conditions found in y0. The results can be displayed by simply typing

```
>> plot(t,y)
```

which yields Fig. 21.2.

FIGURE 21.2
Solution of predator-prey model with MATLAB.

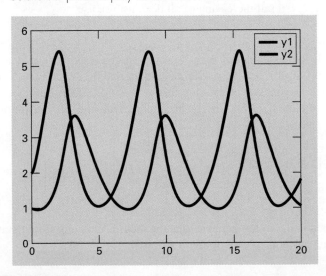

In addition to a time series plot, it is also instructive to generate a *phase-plane plot—*
that is, a plot of the dependent variables versus each other by

```
>> plot(y(:,1),y(:,2))
```

which yields Fig. 21.3.

FIGURE 21.3
State-space plot of predator-prey model with MATLAB.

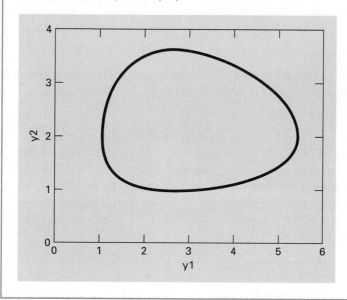

As in the previous example, the MATLAB solver uses default parameters to control various aspects of the integration. In addition, there is also no control over the differential equations' parameters. To have control over these features, additional arguments are included as in

```
[t, y] = ode45(odefun, tspan, y0, options, p1, p2,...)
```

where $options$ is a data structure that is created with the $odeset$ function to control features of the solution, and $p1, p2, \ldots$ are parameters that you want to pass into $odefun$.
The $odeset$ function has the general syntax

```
options = odeset('par₁',val₁,'par₂',val₂,...)
```

where the parameter par_i has the value val_i. A complete listing of all the possible parameters can be obtained by merely entering $odeset$ at the command prompt. Some commonly used parameters are

`'RelTol'`	Allows you to adjust the relative tolerance.
`'AbsTol'`	Allows you to adjust the absolute tolerance.
`'InitialStep'`	The solver automatically determines the initial step. This option allows you to set your own.
`'MaxStep'`	The maximum step defaults to one-tenth of the `tspan` interval. This option allows you to override this default.

EXAMPLE 21.2   Using `odeset` to Control Integration Options

Problem Statement.   Use `ode23` to solve the following ODE from $t = 0$ to 4:

$$\frac{dy}{dt} = 10e^{-(t-2)^2/[2(0.075)^2]} - 0.6y$$

where $y(0) = 0.5$. Obtain solutions for the default (10^{-3}) and for a more stringent (10^{-4}) relative error tolerance.

Solution.   First, we will create an M-file to compute the right-hand side of the ODE:

```
function yp = dydt(t, y)
yp = 10*exp(-(t-2)*(t-2)/(2*.075^2))-0.6*y;
```

Then, we can implement the solver without setting the options. Hence the default value for the relative error (10^{-3}) is automatically used:

```
>> ode23(@dydt, [0 4], 0.5);
```

Note that we have not set the function equal to output variables `[t, y]`. When we implement one of the ODE solvers in this way, MATLAB automatically creates a plot of the results displaying circles at the values it has computed. As in Fig. 21.4a, notice how `ode23` takes relatively large steps in the smooth regions of the solution whereas it takes smaller steps in the region of rapid change around $t = 2$.

We can obtain a more accurate solution by using the `odeset` function to set the relative error tolerance to 10^{-4}:

```
>> options=odeset('RelTol',1e-4);
>> ode23(@dydt, [0, 4], 0.5, options);
```

As in Fig. 21.4b, the solver takes more small steps to attain the increased accuracy.

FIGURE 21.4
Solution of ODE with MATLAB. For (b), a smaller relative error tolerance is used and hence many more steps are taken.

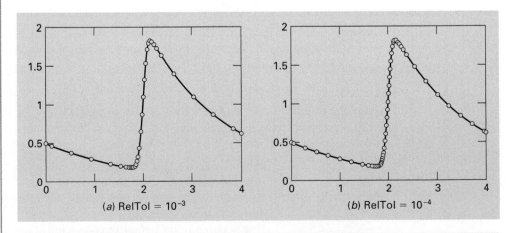

(a) RelTol = 10^{-3} (b) RelTol = 10^{-4}

21.2 MULTISTEP METHODS

The one-step methods described in the previous sections utilize information at a single point t_i to predict a value of the dependent variable y_{i+1} at a future point t_{i+1} (Fig. 21.5a). Alternative approaches, called *multistep methods* (Fig. 21.5b), are based on the insight that, once the computation has begun, valuable information from previous points is at our command. The curvature of the lines connecting these previous values provides information regarding the trajectory of the solution. Multistep methods exploit this information to solve ODEs. In this section, we will present a simple second-order method that serves to demonstrate the general characteristics of multistep approaches.

21.2.1 The Non-Self-Starting Heun Method

Recall that the Heun approach uses Euler's method as a predictor [Eq. (20.15)]:

$$y_{i+1}^0 = y_i + f(t_i, y_i)h \tag{21.4}$$

and the trapezoidal rule as a corrector [Eq. (20.17)]:

$$y_{i+1} = y_i + \frac{f(t_i, y_i) + f\left(t_{i+1}, y_{i+1}^0\right)}{2}h \tag{21.5}$$

Thus, the predictor and the corrector have local truncation errors of $O(h^2)$ and $O(h^3)$, respectively. This suggests that the predictor is the weak link in the method because it has the greatest error. This weakness is significant because the efficiency of the iterative corrector step depends on the accuracy of the initial prediction. Consequently, one way to improve Heun's method is to develop a predictor that has a local error of $O(h^3)$. This can be

FIGURE 21.5

Graphical depiction of the fundamental difference between (a) one-step and (b) multistep methods for solving ODEs.

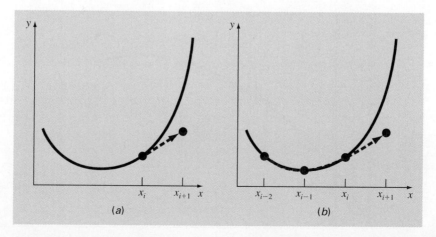

accomplished by using Euler's method and the slope at y_i, and extra information from a previous point y_{i-1}, as in

$$y_{i+1}^0 = y_{i-1} + f(t_i, y_i)2h \tag{21.6}$$

This formula attains $O(h^3)$ at the expense of employing a larger step size $2h$. In addition, note that the equation is not self-starting because it involves a previous value of the dependent variable y_{i-1}. Such a value would not be available in a typical initial-value problem. Because of this fact, Eqs. (21.5) and (21.6) are called the *non-self-starting Heun method*. As depicted in Fig. 21.6, the derivative estimate in Eq. (21.6) is now located at the midpoint rather than at the beginning of the interval over which the prediction is made. This centering improves the local error of the predictor to $O(h^3)$.

FIGURE 21.6
A graphical depiction of the non-self-starting Heun method. (*a*) The midpoint method that is used as a predictor. (*b*) The trapezoidal rule that is employed as a corrector.

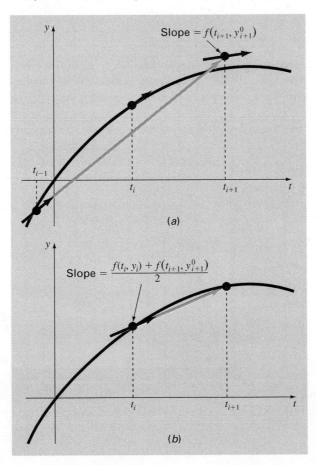

The non-self-starting Heun method can be summarized as

Predictor (Fig. 21.6a): $y_{i+1}^0 = y_{i-1}^m + f(t_i, y_i^m)2h$ (21.7)

Corrector (Fig. 21.6b): $y_{i+1}^j = y_i^m + \dfrac{f(t_i, y_i^m) + f\left(t_{i+1}, y_{i+1}^{j-1}\right)}{2}h$ (21.8)

(for $j = 1, 2, \ldots, m$)

where the superscripts denote that the corrector is applied iteratively from $j = 1$ to m to obtain refined solutions. Note that y_i^m and y_{i-1}^m are the final results of the corrector iterations at the previous time steps. The iterations are terminated based on an estimate of the approximate error,

$$|\varepsilon_a| = \left| \frac{y_{i+1}^j - y_{i+1}^{j-1}}{y_{i+1}^j} \right| \times 100\%$$ (21.9)

When $|\varepsilon_a|$ is less than a prespecified error tolerance ε_s, the iterations are terminated. At this point, $j = m$. The use of Eqs. (21.7) through (21.9) to solve an ODE is demonstrated in the following example.

EXAMPLE 21.3   Non-Self-Starting Heun's Method

Problem Statement. Use the non-self-starting Heun method to perform the same computations as were performed previously in Example 20.2 using Heun's method. That is, integrate $y' = 4e^{0.8t} - 0.5y$ from $t = 0$ to 4 with a step size of 1. As with Example 20.2, the initial condition at $t = 0$ is $y = 2$. However, because we are now dealing with a multistep method, we require the additional information that y is equal to -0.3929953 at $t = -1$.

Solution. The predictor [Eq. (21.7)] is used to extrapolate linearly from $t = -1$ to 1:

$$y_1^0 = -0.3929953 + \left[4e^{0.8(0)} - 0.5(2)\right]2 = 5.607005$$

The corrector [Eq. (21.8)] is then used to compute the value:

$$y_1^1 = 2 + \frac{4e^{0.8(0)} - 0.5(2) + 4e^{0.8(1)} - 0.5(5.607005)}{2}1 = 6.549331$$

which represents a true percent relative error of -5.73% (true value $= 6.194631$). This error is somewhat smaller than the value of -8.18% incurred in the self-starting Heun.

Now, Eq. (21.8) can be applied iteratively to improve the solution:

$$y_1^2 = 2 + \frac{3 + 4e^{0.8(1)} - 0.5(6.549331)}{2}1 = 6.313749$$

which represents an error of -1.92%. An approximate estimate of the error can be determined using Eq. (21.9):

$$|\varepsilon_a| = \left| \frac{6.313749 - 6.549331}{6.313749} \right| \times 100\% = 3.7\%$$

Equation (21.8) can be applied iteratively until ε_a falls below a prespecified value of ε_s. As was the case with the Heun method (recall Example 20.2), the iterations converge on a value of 6.36087 ($\varepsilon_t = -2.68\%$). However, because the initial predictor value is more accurate, the multistep method converges at a somewhat faster rate.

For the second step, the predictor is

$$y_1^0 = 2 + \left[4e^{0.8(1)} - 0.5(6.36087)\right]2 = 13.44346 \qquad \varepsilon_t = 9.43\%$$

which is superior to the prediction of 12.0826 ($\varepsilon_t = 18\%$) that was computed with the original Heun method. The first corrector yields 15.76693 ($\varepsilon_t = 6.8\%$), and subsequent iterations converge on the same result as was obtained with the self-starting Heun method: 15.30224 ($\varepsilon_t = -3.09\%$). As with the previous step, the rate of convergence of the corrector is somewhat improved because of the better initial prediction.

21.2.2 Error Estimates

Aside from providing increased efficiency, the non-self-starting Heun can also be used to estimate the local truncation error. As with the adaptive RK methods in Section 21.1, the error estimate then provides a criterion for changing the step size.

The error estimate can be derived by recognizing that the predictor is equivalent to the midpoint rule. Hence, its local truncation error is (Table 17.4)

$$E_p = \frac{1}{3}h^3 y^{(3)}(\xi_p) = \frac{1}{3}h^3 f''(\xi_p) \tag{21.10}$$

where the subscript p designates that this is the error of the predictor. This error estimate can be combined with the estimate of y_{i+1} from the predictor step to yield

$$\text{True value} = y_{i+1}^0 + \frac{1}{3}h^3 y^{(3)}(\xi_p) \tag{21.11}$$

By recognizing that the corrector is equivalent to the trapezoidal rule, a similar estimate of the local truncation error for the corrector is (Table 17.2)

$$E_c = -\frac{1}{12}h^3 y^{(3)}(\xi_c) = -\frac{1}{12}h^3 f''(\xi_c) \tag{21.12}$$

This error estimate can be combined with the corrector result y_{i+1} to give

$$\text{True value} = y_{i+1}^m - \frac{1}{12}h^3 y^{(3)}(\xi_c) \tag{21.13}$$

Equation (21.11) can be subtracted from Eq. (21.13) to yield

$$0 = y_{i+1}^m - y_{i+1}^0 - \frac{5}{12}h^3 y^{(3)}(\xi) \tag{21.14}$$

where ξ is now between t_{i-1} and t_i. Now, dividing Eq. (21.14) by 5 and rearranging the result gives

$$\frac{y_{i+1}^0 - y_{i+1}^m}{5} = -\frac{1}{12}h^3 y^{(3)}(\xi) \tag{21.15}$$

Notice that the right-hand sides of Eqs. (21.12) and (21.15) are identical, with the exception of the argument of the third derivative. If the third derivative does not vary appreciably over the interval in question, we can assume that the right-hand sides are equal, and therefore, the left-hand sides should also be equivalent, as in

$$E_c = -\frac{y_{i+1}^0 - y_{i+1}^m}{5} \tag{21.16}$$

Thus, we have arrived at a relationship that can be used to estimate the per-step truncation error on the basis of two quantities that are routine by-products of the computation: the predictor (y_{i+1}^0) and the corrector (y_{i+1}^m).

EXAMPLE 21.4   Estimate of Per-Step Truncation Error

Problem Statement.   Use Eq. (21.16) to estimate the per-step truncation error of Example 21.3. Note that the true values at $t = 1$ and 2 are 6.194631 and 14.84392, respectively.

Solution.   At $t_{i+1} = 1$, the predictor gives 5.607005 and the corrector yields 6.360865. These values can be substituted into Eq. (21.16) to give

$$E_c = -\frac{6.360865 - 5.607005}{5} = -0.150722$$

which compares well with the exact error,

$$E_t = 6.194631 - 6.360865 = -0.1662341$$

At $t_{i+1} = 2$, the predictor gives 13.44346 and the corrector yields 15.30224, which can be used to compute

$$E_c = -\frac{15.30224 - 13.44346}{5} = -0.37176$$

which also compares favorably with the exact error, $E_t = 14.84392 - 15.30224 = -0.45831$.

The foregoing has been a brief introduction to multistep methods. Additional information can be found elsewhere (e.g., Chapra and Canale, 2006). Although they still have their place for solving certain types of problems, multistep methods are usually not the method of choice for most problems routinely confronted in engineering and science. That said, they are still used. For example, the MATLAB function `ode113` is a multistep method. We have therefore included this section to introduce you to their basic principles.

21.3 STIFFNESS

Stiffness is a special problem that can arise in the solution of ordinary differential equations. A *stiff system* is one involving rapidly changing components together with slowly changing ones. In some cases, the rapidly varying components are ephemeral transients that die away quickly, after which the solution becomes dominated by the slowly varying

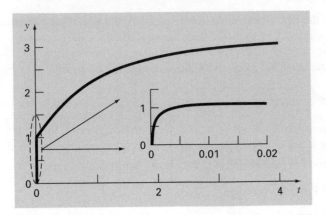

FIGURE 21.7
Plot of a stiff solution of a single ODE. Although the solution appears to start at 1, there is actually a fast transient from $y = 0$ to 1 that occurs in less than the 0.005 time unit. This transient is perceptible only when the response is viewed on the finer timescale in the inset.

components. Although the transient phenomena exist for only a short part of the integration interval, they can dictate the time step for the entire solution.

Both individual and systems of ODEs can be stiff. An example of a single stiff ODE is

$$\frac{dy}{dt} = -1000y + 3000 - 2000e^{-t} \tag{21.17}$$

If $y(0) = 0$, the analytical solution can be developed as

$$y = 3 - 0.998e^{-1000t} - 2.002e^{-t} \tag{21.18}$$

As in Fig. 21.7, the solution is initially dominated by the fast exponential term (e^{-1000t}). After a short period $(t < 0.005)$, this transient dies out and the solution becomes governed by the slow exponential (e^{-t}).

Insight into the step size required for stability of such a solution can be gained by examining the homogeneous part of Eq. (21.17):

$$\frac{dy}{dt} = -ay \tag{21.19}$$

If $y(0) = y_0$, calculus can be used to determine the solution as

$$y = y_0 e^{-at}$$

Thus, the solution starts at y_0 and asymptotically approaches zero.

Euler's method can be used to solve the same problem numerically:

$$y_{i+1} = y_i + \frac{dy_i}{dt}h$$

Substituting Eq. (21.19) gives

$$y_{i+1} = y_i - ay_i h$$

or

$$y_{i+1} = y_i(1 - ah) \tag{21.20}$$

The stability of this formula clearly depends on the step size h. That is, $|1 - ah|$ must be less than 1. Thus, if $h > 2/a$, $|y_i| \to \infty$ as $i \to \infty$.

For the fast transient part of Eq. (21.18), this criterion can be used to show that the step size to maintain stability must be $< 2/1000 = 0.002$. In addition, we should note that, whereas this criterion maintains stability (i.e., a bounded solution), an even smaller step size would be required to obtain an accurate solution. Thus, although the transient occurs for only a small fraction of the integration interval, it controls the maximum allowable step size.

Rather than using explicit approaches, implicit methods offer an alternative remedy. Such representations are called *implicit* because the unknown appears on both sides of the equation. An implicit form of Euler's method can be developed by evaluating the derivative at the future time:

$$y_{i+1} = y_i + \frac{dy_{i+1}}{dt}h$$

This is called the *backward, or implicit, Euler's method.* Substituting Eq. (21.19) yields

$$y_{i+1} = y_i - ay_{i+1}h$$

which can be solved for

$$y_{i+1} = \frac{y_i}{1 + ah} \tag{21.21}$$

For this case, regardless of the size of the step, $|y_i| \to 0$ as $i \to \infty$. Hence, the approach is called *unconditionally stable.*

EXAMPLE 21.5   Explicit and Implicit Euler

Problem Statement.   Use both the explicit and implicit Euler methods to solve Eq. (21.17), where $y(0) = 0$. **(a)** Use the explicit Euler with step sizes of 0.0005 and 0.0015 to solve for y between $t = 0$ and 0.006. **(b)** Use the implicit Euler with a step size of 0.05 to solve for y between 0 and 0.4.

Solution.   **(a)** For this problem, the explicit Euler's method is

$$y_{i+1} = y_i + (-1000y_i + 3000 - 2000e^{-t_i})h$$

The result for $h = 0.0005$ is displayed in Fig. 21.8a along with the analytical solution. Although it exhibits some truncation error, the result captures the general shape of the analytical solution. In contrast, when the step size is increased to a value just below the stability limit ($h = 0.0015$), the solution manifests oscillations. Using $h > 0.002$ would result in a totally unstable solution—that is, it would go infinite as the solution progressed.

(b) The implicit Euler's method is

$$y_{i+1} = y_i + (-1000y_{i+1} + 3000 - 2000e^{-t_{i+1}})h$$

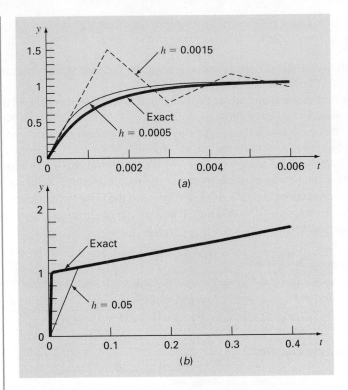

FIGURE 21.8
Solution of a stiff ODE with (a) the explicit and (b) implicit Euler methods.

Now because the ODE is linear, we can rearrange this equation so that y_{i+1} is isolated on the left-hand side:

$$y_{i+1} = \frac{y_i + 3000h - 2000he^{-t_{i+1}}}{1 + 1000h}$$

The result for $h = 0.05$ is displayed in Fig. 21.8b along with the analytical solution. Notice that even though we have used a much bigger step size than the one that induced instability for the explicit Euler, the numerical result tracks nicely on the analytical solution.

Systems of ODEs can also be stiff. An example is

$$\frac{dy_1}{dt} = -5y_1 + 3y_2 \tag{21.22a}$$

$$\frac{dy_2}{dt} = 100y_1 - 301y_2 \tag{21.22b}$$

For the initial conditions $y_1(0) = 52.29$ and $y_2(0) = 83.82$, the exact solution is

$$y_1 = 52.96e^{-3.9899t} - 0.67e^{-302.0101t} \tag{21.23a}$$

$$y_2 = 17.83e^{-3.9899t} + 65.99e^{-302.0101t} \tag{21.23b}$$

Note that the exponents are negative and differ by about two orders of magnitude. As with the single equation, it is the large exponents that respond rapidly and are at the heart of the system's stiffness.

An implicit Euler's method for systems can be formulated for the present example as

$$y_{1,i+1} = y_{1,i} + (-5y_{1,i+1} + 3y_{2,i+1})h \tag{21.24a}$$

$$y_{2,i+1} = y_{2,i} + (100y_{1,i+1} - 301y_{2,i+1})h \tag{21.24b}$$

Collecting terms gives

$$(1 + 5h)y_{1,i+1} - 3y_{2,i+1} = y_{1,i} \tag{21.25a}$$

$$-100y_{1,i+1} + (1 + 301h)y_{2,i+1} = y_{2,i} \tag{21.25b}$$

Thus, we can see that the problem consists of solving a set of simultaneous equations for each time step.

For nonlinear ODEs, the solution becomes even more difficult since it involves solving a system of nonlinear simultaneous equations (recall Section 12.2). Thus, although stability is gained through implicit approaches, a price is paid in the form of added solution complexity.

21.3.1 MATLAB Functions for Stiff Systems

MATLAB has a number of built-in functions for solving stiff systems of ODEs. These are

ode15s. This function is a variable-order solver based on numerical differentiation formulas. It is a multistep solver that optionally uses the Gear backward differentiation formulas. This is used for stiff problems of low to medium accuracy.

ode23s. This function is based on a modified Rosenbrock formula of order 2. Because it is a one-step solver, it may be more efficient than `ode15s` at crude tolerances. It can solve some kinds of stiff problems better than `ode15s`.

ode23t. This function is an implementation of the trapezoidal rule with a "free" inter-polant. This is used for moderately stiff problems with low accuracy where you need a solution without numerical damping.

ode23tb. This is an implementation of an implicit Runge-Kutta formula with a first stage that is a trapezoidal rule and a second stage that is a backward differentiation formula of order 2. This solver may also be more efficient than `ode15s` at crude tolerances.

EXAMPLE 21.6 MATLAB for Stiff ODEs

Problem Statement. The van der Pol equation is a model of an electronic circuit that arose back in the days of vacuum tubes,

$$\frac{d^2y_1}{dt^2} - \mu(1 - y_1^2)\frac{dy_1}{dt} + y_1 = 0 \tag{E21.6.1}$$

The solution to this equation becomes progressively stiffer as μ gets large. Given the initial conditions, $y_1(0) = dy_1/dt = 1$, use MATLAB to solve the following two cases: **(a)** for $\mu = 1$, use `ode45` to solve from $t = 0$ to 20; and **(b)** for $\mu = 1000$, use `ode23s` to solve from $t = 0$ to 6000.

Solution. **(a)** The first step is to convert the second-order ODE into a pair of first-order ODEs by defining

$$\frac{dy_1}{dt} = y_2$$

Using this equation, Eq. (E21.6.1) can be written as

$$\frac{dy_2}{dt} = \mu\left(1 - y_1^2\right)y_2 - y_1 = 0$$

An M-file can now be created to hold this pair of differential equations:

```
function yp = vanderpol(t,y,mu)
yp = [y(2);mu*(1-y(1)^2)*y(2)-y(1)];
```

Notice how the value of μ is passed as a parameter. As in Example 21.1, ode45 can be invoked and the results plotted:

```
>> [t,y] = ode45(@vanderpol,[0 20],[1 1],[],1);
>> plot(t,y(:,1),'-',t,y(:,2),'--')
>> legend('y1','y2');
```

Observe that because we are not specifying any options, we must use open brackets [] as a place holder. The smooth nature of the plot (Fig. 21.9a) suggests that the van der Pol equation with $\mu = 1$ is not a stiff system.

(b) If a standard solver like ode45 is used for the stiff case ($\mu = 1000$), it will fail miserably (try it, if you like). However, ode23s does an efficient job:

```
>> [t,y] = ode23s(@vanderpol,[0 6000],[1 1],[],1000);
>> plot(t,y(:,1))
```

FIGURE 21.9
Solutions for van der Pol's equation. (a) Nonstiff form solved with ode45 and (b) stiff form solved with ode23s.

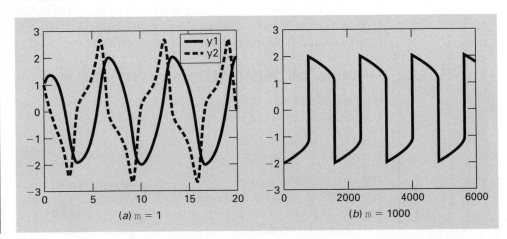

(a) m = 1   (b) m = 1000

We have only displayed the y_1 component because the result for y_2 has a much larger scale. Notice how this solution (Fig. 21.9b) has much sharper edges than is the case in Fig. 21.9a. This is a visual manifestation of the "stiffness" of the solution.

21.4 MATLAB APPLICATION: BUNGEE JUMPER WITH CORD

In this section, we will use MATLAB to solve for the vertical dynamics of a jumper connected to a stationary platform with a bungee cord. As developed at the beginning of Chap. 20, the problem consisted of solving two coupled ODEs for vertical position and velocity. The differential equation for position is

$$\frac{dx}{dt} = v \tag{21.26}$$

The differential equation for velocity is different depending on whether the jumper has fallen to a distance where the cord is fully extended and begins to stretch. Thus, if the distance fallen is less than the cord length, the jumper is only subject to gravitational and drag forces,

$$\frac{dv}{dt} = g - \text{sign}(v)\frac{c_d}{m}v^2 \tag{21.27a}$$

Once the cord begins to stretch, the spring and dampening forces of the cord must also be included:

$$\frac{dv}{dt} = g - \text{sign}(v)\frac{c_d}{m}v^2 - \frac{k}{m}(x - L) - \frac{\gamma}{m}v \tag{21.27b}$$

The following example shows how MATLAB can be used to solve this problem.

EXAMPLE 21.7   Bungee Jumper with Cord

Problem Statement.   Determine the position and velocity of a bungee jumper with the following parameters: $L = 30$ m, $g = 9.81$ m/s^2, $m = 68.1$ kg, $c_d = 0.25$ kg/m, $k = 40$ N/m, and $\gamma = 8$ N $\cdot$ s/m. Perform the computation from $t = 0$ to 50 s and assume that the initial conditions are $x(0) = v(0) = 0$.

Solution.   The following M-file can be set up to compute the right-hand sides of the ODEs:

```
function dydt = bungee(t,y,L,cd,m,k,gamma)
g = 9.81;
cord = 0;
if y(1) > L %determine if the cord exerts a force
  cord = k/m*(y(1)-L)+gamma/m*y(2);
end
dydt = [y(2); g - sign(y(2))*cd/m*y(2)^2 - cord];
```

Notice that the derivatives are returned as a column vector because this is the format required by the MATLAB solvers.

Because these equations are not stiff, we can use `ode45` to obtain the solutions and display them on a plot:

```
>> [t,y] = ode45(@bungee,[0 50],[0 0],[],30,0.25,68.1,40,8);
>> plot(t,-y(:,1),'-',t,y(:,2),':')
>> legend('x (m)','v (m/s)')
```

As in Fig. 21.10, we have reversed the sign of distance for the plot so that negative distance is in the downward direction. Notice how the simulation captures the jumper's bouncing motion.

FIGURE 21.10
Plot of distance and velocity of a bungee jumper.

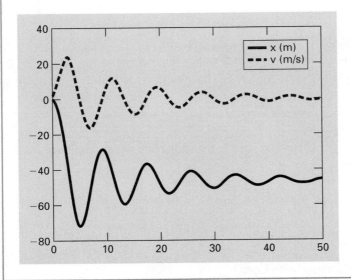

21.5 CASE STUDY   PLINY'S INTERMITTENT FOUNTAIN

Background. The Roman natural philosopher, Pliny the Elder, purportedly had an intermittent fountain in his garden. As in Fig. 21.11, water enters a cylindrical tank at a constant flow rate Q_{in} and fills until the water reaches y_{high}. At this point, water siphons out of the tank through a circular discharge pipe, producing a fountain at the pipe's exit. The fountain runs until the water level decreases to y_{low}, whereupon the siphon fills with air and the fountain stops. The cycle then repeats as the tank fills until the water reaches y_{high}, and the fountain flows again.

When the siphon is running, the outflow Q_{out} can be computed with the following formula based on *Torricelli's law*:

$$Q_{out} = C\sqrt{2gy}\pi r^2 \tag{21.28}$$

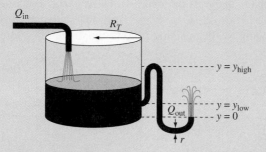

FIGURE 21.11
An intermittent fountain.

Neglecting the volume of water in the pipe, compute and plot the level of the water in the tank as a function of time over 100 seconds. Assume an initial condition of an empty tank $y(0) = 0$, and employ the following parameters for your computation:

$$R_T = 0.05 \text{ m} \qquad r = 0.007 \text{ m} \qquad y_{\text{low}} = 0.025 \text{ m}$$
$$y_{\text{high}} = 0.1 \text{ m} \qquad C = 0.6 \qquad g = 9.81 \text{ m/s}^2$$
$$Q_{\text{in}} = 50 \times 10^{-6} \text{ m}^3/\text{s}$$

Solution. When the fountain is running, the rate of change in the tank's volume $V (\text{m}^3)$ is determined by a simple balance of inflow minus the outflow:

$$\frac{dV}{dt} = Q_{\text{in}} - Q_{\text{out}} \tag{21.29}$$

where $V =$ volume (m^3). Because the tank is cylindrical, $V = \pi R_t^2 y$. Substituting this relationship along with Eq. (21.28) into Eq. (21.29) gives

$$\frac{dy}{dt} = \frac{Q_{\text{in}} - C\sqrt{2gy}\pi r^2}{\pi R_t^2} \tag{21.30}$$

When the fountain is not running, the second term in the numerator goes to zero. We can incorporate this mechanism in the model by introducing a new dimensionless variable *siphon* that equals zero when the fountain is off and equals one when it is flowing:

$$\frac{dy}{dt} = \frac{Q_{\text{in}} - siphon \times C\sqrt{2gy}\pi r^2}{\pi R_t^2} \tag{21.31}$$

In the present context, *siphon* can be thought of as a switch that turns the fountain off and on. Such two-state variables are called *Boolean* or *logical variables,* where zero is equivalent to false and one is equivalent to true.

Next we must relate *siphon* to the dependent variable y. First, *siphon* is set to zero whenever the level falls below y_{low}. Conversely, *siphon* is set to one whenever the level rises above y_{high}. The following M-file function follows this logic in computing the derivative:

```
function dy = Plinyode(t,y)
global siphon
Rt = 0.05; r = 0.007; yhi = 0.1; ylo = 0.025;
C = 0.6; g = 9.81; Qin = 0.00005;
if y(1) <= ylo
  siphon = 0;
elseif y(1) >= yhi
  siphon = 1;
end
Qout = siphon * C * sqrt(2 * g * y(1)) * pi * r ^ 2;
dy = (Qin - Qout) / (pi * Rt ^ 2);
```

Notice that because its value must be maintained between function calls, `siphon` is declared as a global variable. Although the use of global variables is not encouraged (particularly in larger programs), it is useful in the present context.

The following script employs the built-in `ode45` function to integrate `Plinyode` and generate a plot of the solution:

```
global siphon
siphon = 0;
tspan = [0 100]; y0 = 0;
[tp,yp]=ode45(@Plinyode,tspan,y0);
plot(tp,yp)
xlabel('time, (s)')
ylabel('water level in tank, (m)')
```

As shown in Fig. 21.12, the result is clearly incorrect. Except for the original filling period, the level seems to start emptying prior to reaching y_{high}. Similarly, when it is draining, the siphon shuts off well before the level drops to y_{low}.

At this point, suspecting that the problem demands more firepower than the trusty `ode45` routine, you might be tempted to use one of the other MATLAB ODE solvers such as `ode23s` or `ode23tb`. But if you did, you would discover that although these routines yield somewhat different results, they would still generate incorrect solutions.

The difficulty arises because the ODE is discontinuous at the point that the siphon switches on or off. For example, as the tank is filling, the derivative is dependent only on the constant inflow and for the present parameters has a constant value of 6.366×10^{-3} m/s. However, as soon as the level reaches y_{high}, the outflow kicks in and the derivative abruptly drops to -1.013×10^{-2} m/s. Although the adaptive step-size routines used by MATLAB work marvelously for many problems, they often get heartburn when dealing with such discontinuities. Because they infer the behavior of the solution by comparing the results of different steps, a discontinuity represents something akin to stepping into a deep pothole on a dark street.

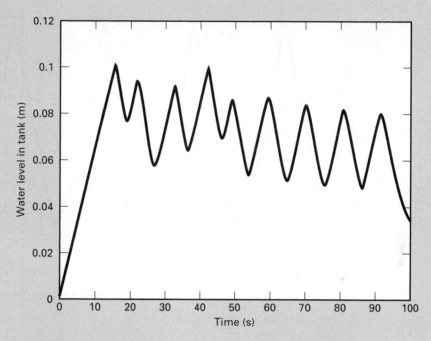

FIGURE 21.12
The level in Pliny's fountain versus time as simulated with ode45.

At this point, your first inclination might be to just give up. After all, if it's too hard for MATLAB, no reasonable person could expect you to come up with a solution. Because professional engineers and scientists rarely get away with such excuses, your only recourse is to develop a remedy based on your knowledge of numerical methods.

Because the problem results from adaptively stepping across a discontinuity, you might revert to a simpler approach and use a constant, small step size. If you think about it, that's precisely the approach you would take if you were traversing a dark, pothole-filled street. We can implement this solution strategy by merely replacing ode45 with the constant-step rk4sys function from Chap. 20 (Fig. 20.8). For the script outlined above, the fourth line would be formulated as

```
[tp,yp] = rk4sys(@Plinyode,tspan,y0,0.0625);
```

As in Fig. 21.13, the solution now evolves as expected. The tank fills to y_{high} and then empties until it reaches y_{low}, when the cycle repeats.

There are a two take-home messages that can be gleaned from this case study. First, although it's human nature to think the opposite, simpler is sometimes better. After all, to paraphrase Einstein, "Everything should be as simple as possible, but no simpler." Second, you should never blindly believe every result generated by the computer. You've probably heard the old chestnut, "garbage in, garbage out" in reference to the impact of data quality

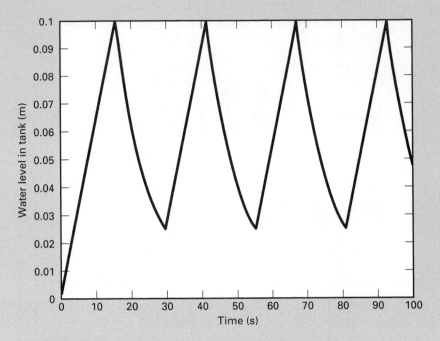

FIGURE 21.13
The level in Pliny's fountain versus time as simulated with a small, constant step size using the
rk4sys function (Fig. 20.8).

on the validity of computer output. Unfortunately, some individuals think that regardless of
what went in (the data) and what's going on inside (the algorithm), it's always "gospel out."
Situations like the one depicted in Fig. 21.12 are particularly dangerous—that is, although
the output is incorrect, it's not obviously wrong. That is, the simulation does not go unstable or yield negative levels. In fact, the solution moves up and down in the manner of an
intermittent fountain, albeit incorrectly.

Hopefully, this case study illustrates that even a great piece of software such as
MATLAB is not foolproof. Hence, sophisticated engineers and scientists always examine
numerical output with a healthy skepticism based on their considerable experience and
knowledge of the problems they are solving.

PROBLEMS

21.1 Repeat the the same simulations as in Section 21.5 for Pliny's fountain, but generate the solutions with `ode23`, `ode23s`, and `ode113`. Use `subplot` to develop a vertical three-pane plot of the time series.

21.2 The following ODEs have been proposed as a model of an epidemic:

$$\frac{dS}{dt} = -aSI$$

$$\frac{dI}{dt} = aSI - rI$$

$$\frac{dR}{dt} = rI$$

where S = the susceptible individuals, I = the infected, R = the recovered, a = the infection rate, and r = the recovery rate. A city has 10,000 people, all of whom are susceptible.

(a) If a single infectious individual enters the city at $t = 0$, compute the progression of the epidemic until the number of infected individuals falls below 10. Use the following parameters: $a = 0.002/$(person $\cdot$ week) and $r = 0.15/$d. Develop time-series plots of all the state variables. Also generate a phase-plane plot of S versus I versus R.

(b) Suppose that after recovery, there is a loss of immunity that causes recovered individuals to become susceptible. This reinfection mechanism can be computed as ρR, where ρ = the reinfection rate. Modify the model to include this mechanism and repeat the computations in (a) using $\rho = 0.03/$d.

21.3 Solve the following initial-value problem over the interval from $t = 2$ to 3:

$$\frac{dy}{dt} = -0.5y + e^{-t}$$

Use the non-self-starting Heun method with a step size of 0.5 and initial conditions of $y(1.5) = 5.222138$ and $y(2.0) = 4.143883$. Iterate the corrector to $\varepsilon_s = 0.1\%$. Compute the percent relative errors for your results based on the exact solutions obtained analytically: $y(2.5) = 3.273888$ and $y(3.0) = 2.577988$.

21.4 Solve the following initial-value problem over the interval from $t = 0$ to 0.5:

$$\frac{dy}{dt} = yt^2 - y$$

Use the fourth-order RK method to predict the first value at $t = 0.25$. Then use the non-self-starting Heun method to make the prediction at $t = 0.5$. Note: $y(0) = 1$.

21.5 Given

$$\frac{dy}{dt} = -100,000y + 99,999e^{-t}$$

(a) Estimate the step size required to maintain stability using the explicit Euler method.

(b) If $y(0) = 0$, use the implicit Euler to obtain a solution from $t = 0$ to 2 using a step size of 0.1.

21.6 Given

$$\frac{dy}{dt} = 30(\sin t - y) + 3\cos t$$

If $y(0) = 0$, use the implicit Euler to obtain a solution from $t = 0$ to 4 using a step size of 0.4.

21.7 Given

$$\frac{dx_1}{dt} = 999x_1 + 1999x_2$$

$$\frac{dx_2}{dt} = -1000x_1 - 2000x_2$$

If $x_1(0) = x_2(0) = 1$, obtain a solution from $t = 0$ to 0.2 using a step size of 0.05 with the (a) explicit and (b) implicit Euler methods.

21.8 The following nonlinear, parasitic ODE was suggested by Hornbeck (1975):

$$\frac{dy}{dt} = 5(y - t^2)$$

If the initial condition is $y(0) = 0.08$, obtain a solution from $t = 0$ to 5:

(a) Analytically.

(b) Using the fourth-order RK method with a constant step size of 0.03125.

(c) Using the MATLAB function `ode45`.

(d) Using the MATLAB function `ode23s`.

(e) Using the MATLAB function `ode23tb`.

Present your results in graphical form.

21.9 Recall from Example 17.5 that the following `humps` function exhibits both flat and steep regions over a relatively short x range,

$$f(x) = \frac{1}{(x - 0.3)^2 + 0.01} + \frac{1}{(x - 0.9)^2 + 0.04} - 6$$

Determine the value of the definite integral of this function between $x = 0$ and 1 using (a) the `quad` and (b) the `ode45` functions.

21.10 The oscillations of a swinging pendulum can be simulated with the following nonlinear model:

$$\frac{d^2\theta}{dt^2} + \frac{g}{l}\sin\theta = 0$$

where θ = the angle of displacement, g = the gravitational constant, and l = the pendulum length. For small angular displacements, the $\sin\theta$ is approximately equal to θ and the model can be linearized as

$$\frac{d^2\theta}{dt^2} + \frac{g}{l}\theta = 0$$

Use `ode45` to solve for θ as a function of time for both the linear and nonlinear models where $l = 0.6$ m and $g = 9.81$ m/s^2. First, solve for the case where the initial condition is for a small displacement ($\theta = \pi/8$ and $d\theta/dt = 0$). Then repeat the calculation for a large displacement ($\theta = \pi/2$ and $d\theta/dt = 0$). For each case, plot the linear and nonlinear simulations on the same plot.

21.11 The following system is a classic example of stiff ODEs that can occur in the solution of chemical reaction kinetics:

$$\frac{dc_1}{dt} = -0.013c_1 - 1000c_1c_3$$

$$\frac{dc_2}{dt} = -2500c_2c_3$$

$$\frac{dc_3}{dt} = -0.013c_1 - 1000c_1c_3 - 2500c_2c_3$$

Solve these equations from $t = 0$ to 50 with initial conditions $c_1(0) = c_2(0) = 1$ and $c_3(0) = 0$. If you have access to MATLAB software, use both standard (e.g., `ode45`) and stiff (e.g., `ode23s`) functions to obtain your solutions.

21.12 The following second-order ODE is considered to be stiff:

$$\frac{d^2y}{dx^2} = -1001\frac{dy}{dx} - 1000y$$

Solve this differential equation **(a)** analytically and **(b)** numerically for $x = 0$ to 5. For **(b)** use an implicit approach with $h = 0.5$. Note that the initial conditions are $y(0) = 1$ and $y'(0) = 0$. Display both results graphically.

21.13 Consider the thin rod of length l moving in the x-y plane as shown in Fig. P21.13. The rod is fixed with a pin on one end and a mass at the other. Note that $g = 9.81$ m/s^2 and $l = 0.5$ m. This system can be solved using

$$\ddot{\theta} - \frac{g}{l}\theta = 0$$

FIGURE P21.13

Let $\theta(0) = 0$ and $\dot{\theta}(0) = 0.25$ rad/s. Solve using any method studied in this chapter. Plot the angle versus time and the angular velocity versus time. (**Hint:** Decompose the second-order ODE.)

21.14 Given the first-order ODE:

$$\frac{dx}{dt} = -700x - 1000e^{-t}$$

$$x(t = 0) = 4$$

Solve this stiff differential equation using a numerical method over the time period $0 \le t \le 5$. Also solve analytically and plot the analytic and numerical solution for both the fast transient and slow transition phase of the time scale.

21.15 Solve the following differential equation from $t = 0$ to 2

$$\frac{dy}{dt} = -10y$$

with the initial condition $y(0) = 1$. Use the following techniques to obtain your solutions: **(a)** analytically, **(b)** the explicit Euler method, and **(c)** the implicit Euler method. For **(b)** and **(c)** use $h = 0.1$ and 0.2. Plot your results.

21.16 The Lotka-Volterra equations described in Section 20.6 have been refined to include additional factors that impact predator-prey dynamics. For example, over and above predation, prey population can be limited by other factors such as space. Space limitation can be incorporated into the model as a carrying capacity (recall the logistic model described in Prob. 20.5) as in

$$\frac{dx}{dt} = a\left(1 - \frac{x}{K}\right)x - bxy$$

$$\frac{dy}{dt} = -cy + dxy$$

where K = the carrying capacity. Use the same parameter values and initial conditions as in Section 20.5 to integrate these equations from $t = 0$ to 100 using `ode45`.

(a) Employ a very large value of $K = 10^8$ to validate that you obtain the same results as in Section 20.5.

(b) Compare (a) with the more realistic carrying capacity of $K = 20{,}000$. Discuss your results.

21.17 Two masses are attached to a wall by linear springs (Fig. P21.17). Force balances based on Newton's second law can be written as

$$\frac{d^2 x_1}{dt^2} = -\frac{k_1}{m_1}(x_1 - L_1) + \frac{k_2}{m_1}(x_2 - x_1 - w_1 - L_2)$$

$$\frac{d^2 x_2}{dt^2} = -\frac{k_2}{m_2}(x_2 - x_1 - w_1 - L_2)$$

where k = the spring constants, m = mass, L = the length of the unstretched spring, and w = the width of the mass. Compute the positions of the masses as a function of time using the following parameter values: $k_1 = k_2 = 5$, $m_1 = m_2 = 2$,

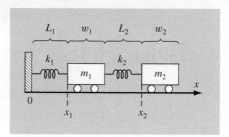

FIGURE P21.17

$w_1 = w_2 = 5$, and $L_1 = L_2 = 2$. Set the initial conditions as $x_1 = L_1$ and $x_2 = L_1 + w_1 + L_2 + 6$. Perform the simulation from $t = 0$ to 20. Construct time-series plots of both the displacements and the velocities. In addition, produce a phase-plane plot of x_1 versus x_2.

22

Boundary-Value Problems

CHAPTER OBJECTIVES

The primary objective of this chapter is to introduce you to solving boundary-value problems for ODEs. Specific objectives and topics covered are

- Understanding the difference between initial-value and boundary-value problems
- Knowing how to express an nth-order ODE as a system of n first-order ODEs.
- Knowing how to implement the shooting method for linear ODEs by using linear interpolation to generate accurate "shots."
- Understanding how derivative boundary conditions are incorporated into the shooting method.
- Knowing how to solve nonlinear ODEs with the shooting method by using root location to generate accurate "shots."
- Knowing how to implement the finite-difference method.
- Understanding how derivative boundary conditions are incorporated into the finite-difference method.
- Knowing how to solve nonlinear ODEs with the finite-difference method by using root location methods for systems of nonlinear algebraic equations.

YOU'VE GOT A PROBLEM

To this point, we have been computing the velocity of a free-falling bungee jumper by integrating a single ODE:

$$\frac{dv}{dt} = g - \frac{c_d}{m}v^2 \tag{22.1}$$

Suppose that rather than velocity, you are asked to determine the position of the jumper as a function of time. One way to do this is to recognize that velocity is the first derivative of distance:

$$\frac{dx}{dt} = v \tag{22.2}$$

Thus, by solving the system of two ODEs represented by Eqs. (22.1) and (22.2), we can simultaneously determine both the velocity and the position.

However, because we are now integrating two ODEs, we require two conditions to obtain the solution. We are already familiar with one way to do this for the case where we have values for both position and velocity at the initial time:

$$x(t = 0) = x_i$$
$$v(t = 0) = v_i$$

Given such conditions, we can easily integrate the ODEs using the numerical techniques described in Chaps. 20 and 21. This is referred to as an *initial-value problem*.

But what if we do not know values for both position and velocity at $t = 0$? Let's say that we know the initial position but rather than having the initial velocity, we want the jumper to be at a specified position at a later time. In other words:

$$x(t = 0) = x_i$$
$$x(t = t_f) = x_f$$

Because the two conditions are given at different values of the independent variable, this is called a *boundary-value problem*.

Such problems require special solution techniques. Some of these are related to the methods for initial value problems that were described in the previous two chapters. However, others employ entirely different strategies to obtain solutions. This chapter is designed to introduce you to the more common of these methods.

22.1 INTRODUCTION AND BACKGROUND

22.1.1 What Are Boundary-Value Problems?

An ordinary differential equation is accompanied by auxiliary conditions, which are used to evaluate the constants of integration that result during the solution of the equation. For an nth-order equation, n conditions are required. If all the conditions are specified at the same value of the independent variable, then we are dealing with an *initial-value problem* (Fig. 22.1a). To this point, the material in Part Six (Chaps. 20 and 21) has been devoted to this type of problem.

In contrast, there are often cases when the conditions are not known at a single point but rather are given at different values of the independent variable. Because these values are often specified at the extreme points or boundaries of a system, they are customarily referred to as *boundary-value problems* (Fig. 22.1b). A variety of significant engineering applications fall within this class. In this chapter, we discuss some of the basic approaches for solving such problems.

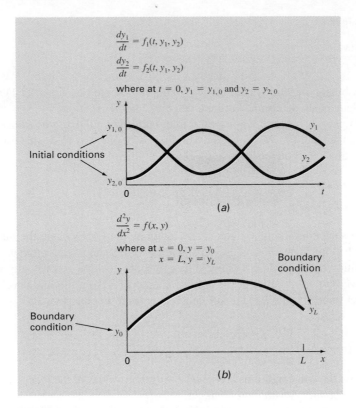

FIGURE 22.1
Initial-value versus boundary-value problems. (a) An initial-value problem where all the conditions are specified at the same value of the independent variable. (b) A boundary-value problem where the conditions are specified at different values of the independent variable.

22.1.2 Boundary-Value Problems in Engineering and Science

At the beginning of this chapter, we showed how the determination of the position and velocity of a falling object could be formulated as a boundary-value problem. For that example, a pair of ODEs was integrated in time. Although other time-variable examples can be developed, boundary-value problems arise more naturally when integrating in space. This occurs because auxiliary conditions are often specified at different positions in space.

A case in point is the simulation of the steady-state temperature distribution for a long, thin rod positioned between two constant-temperature walls (Fig. 22.2). The rod's cross-sectional dimensions are small enough so that radial temperature gradients are minimal and, consequently, temperature is a function exclusively of the axial coordinate x. Heat is transferred along the rod's longitudinal axis by conduction and between the rod and the surrounding gas by convection. For this example, radiation is assumed to be negligible.[1]

---

[1] We incorporate radiation into this problem later in this chapter in Example 22.4.

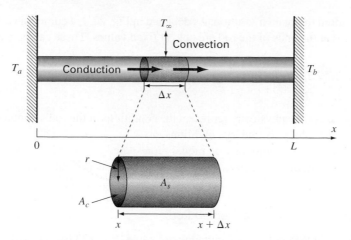

FIGURE 22.2
A heat balance for a differential element of a heated rod subject to conduction and convection.

As depicted in Fig. 22.2, a heat balance can be taken around a differential element of thickness Δx as

$$0 = q(x)A_c - q(x + \Delta x)A_c + hA_s(T_\infty - T) \qquad (22.3)$$

where $q(x)$ = flux into the element due to conduction [J/(m$^2 \cdot$ s)]; $q(x + \Delta x)$ = flux out of the element due to conduction [J/(m$^2 \cdot$ s)]; A_c = cross-sectional area [m^2] = πr^2, r = the radius [m]; h = the convection heat transfer coefficient [J/(m$^2 \cdot$ K $\cdot$ s)]; A_s = the element's surface area [m^2] = $2\pi r \Delta x$; T_∞ = the temperature of the surrounding gas [K]; and T = the rod's temperature [K].

Equation (22.3) can be divided by the element's volume ($\pi r^2 \Delta x$) to yield

$$0 = \frac{q(x) - q(x + \Delta x)}{\Delta x} + \frac{2h}{r}(T_\infty - T)$$

Taking the limit $\Delta x \to 0$ gives

$$0 = -\frac{dq}{dx} + \frac{2h}{r}(T_\infty - T) \qquad (22.4)$$

The flux can be related to the temperature gradient by *Fourier's law:*

$$q = -k\frac{dT}{dx} \qquad (22.5)$$

where k = the coefficient of thermal conductivity [J/(s $\cdot$ m $\cdot$ K)]. Equation (22.5) can be differentiated with respect to x, substituted into Eq. (22.4), and the result divided by k to yield,

$$0 = \frac{d^2T}{dx^2} + h'(T_\infty - T) \qquad (22.6)$$

where h' = a bulk heat-transfer parameter reflecting the relative impacts of convection and conduction [m^{-2}] = $2h/(rk)$.

Equation (22.6) represents a mathematical model that can be used to compute the temperature along the rod's axial dimension. Because it is a second-order ODE, two conditions are required to obtain a solution. As depicted in Fig. 22.2, a common case is where the temperatures at the ends of the rod are held at fixed values. These can be expressed mathematically as

$$T(0) = T_a$$
$$T(L) = T_b$$

The fact that they physically represent the conditions at the rod's "boundaries" is the origin of the terminology: boundary conditions.

Given these conditions, the model represented by Eq. (22.6) can be solved. Because this particular ODE is linear, an analytical solution is possible as illustrated in the following example.

EXAMPLE 22.1   Analytical Solution for a Heated Rod

Problem Statement.   Use calculus to solve Eq. (22.6) for a 10-m rod with $h' = 0.05 \, \text{m}^{-2} [h = 1 \, \text{J/(m}^2 \cdot \text{K} \cdot \text{s)}, \, r = 0.2 \, \text{m}, \, k = 200 \, \text{J/(s} \cdot \text{m} \cdot \text{K)}], \, T_\infty = 200 \, \text{K}$, and the boundary conditions:

$$T(0) = 300 \, \text{K} \qquad\qquad T(10) = 400 \, \text{K}$$

Solution.   This ODE can be solved in a number of ways. A straightforward approach is to first express the equation as

$$\frac{d^2 T}{dx^2} - h'T = -h'T_\infty$$

Because this is a linear ODE with constant coefficients, the general solution can be readily obtained by setting the right-hand side to zero and assuming a solution of the form $T = e^{\lambda x}$. Substituting this solution along with its second derivative into the homogeneous form of the ODE yields

$$\lambda^2 e^{\lambda x} - h'e^{\lambda x} = 0$$

which can be solved for $\lambda = \pm\sqrt{h'}$. Thus, the general solution is

$$T = Ae^{\lambda x} + Be^{-\lambda x}$$

where A and B are constants of integration. Using the method of undetermined coefficients we can derive the particular solution $T = T_\infty$. Therefore, the total solution is

$$T = T_\infty + Ae^{\lambda x} + Be^{-\lambda x}$$

The constants can be evaluated by applying the boundary conditions

$$T_a = T_\infty + A + B$$
$$T_b = T_\infty + Ae^{\lambda L} + Be^{-\lambda L}$$

These two equations can be solved simultaneously for

$$A = \frac{(T_a - T_\infty)e^{-\lambda L} - (T_b - T_\infty)}{e^{-\lambda L} - e^{\lambda L}}$$

$$B = \frac{(T_b - T_\infty) - (T_a - T_\infty)e^{\lambda L}}{e^{-\lambda L} - e^{\lambda L}}$$

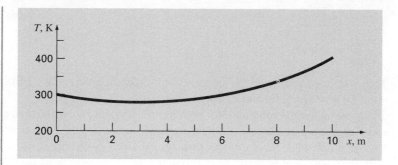

FIGURE 22.3
Analytical solution for the heated rod.

Substituting the parameter values from this problem gives $A = 20.4671$ and $B = 79.5329$. Therefore, the final solution is

$$T = 200 + 20.4671e^{\sqrt{0.05}x} + 79.5329e^{-\sqrt{0.05}x} \qquad (22.7)$$

As can be seen in Fig. 22.3, the solution is a smooth curve connecting the two boundary temperatures. The temperature in the middle is depressed due to the convective heat loss to the cooler surrounding gas.

In the following sections, we will illustrate numerical approaches for solving the same problem we just solved analytically in Example 22.1. The exact analytical solution will be useful in assessing the accuracy of the solutions obtained with the approximate, numerical methods.

22.2 THE SHOOTING METHOD

The shooting method is based on converting the boundary-value problem into an equivalent initial-value problem. A trial-and-error approach is then implemented to develop a solution for the initial-value version that satisfies the given boundary conditions.

Although the method can be employed for higher-order and nonlinear equations, it is nicely illustrated for a second-order, linear ODE such as the heated rod described in the previous section:

$$0 = \frac{d^2T}{dx^2} + h'(T_\infty - T) \qquad (22.8)$$

subject to the boundary conditions

$$T(0) = T_a$$
$$T(L) = T_b$$

We convert this boundary-value problem into an initial-value problem by defining the rate of change of temperature, or *gradient*, as

$$\frac{dT}{dx} = z \qquad (22.9)$$

and reexpressing Eq. (22.8) as

$$\frac{dz}{dx} = -h'(T_\infty - T) \tag{22.10}$$

Thus, we have converted the single second-order equation (Eq. 22.8) into a pair of first-order ODEs (Eqs. 22.9 and 22.10).

If we had initial conditions for both T and z, we could solve these equations as an initial-value problem with the methods described in Chaps. 20 and 21. However, because we only have an initial value for one of the variables $T(0) = T_a$ we simply make a guess for the other $z(0) = z_{a1}$ and then perform the integration.

After performing the integration, we will have generated a value of T at the end of the interval, which we will call T_{b1}. Unless we are incredibly lucky, this result will differ from the desired result T_b.

Now, let's say that the value of T_{b1} is too high ($T_{b1} > T_b$), it would make sense that a lower value of the initial slope $z(0) = z_{a2}$ might result in a better prediction. Using this new guess, we can integrate again to generate a second result at the end of the interval T_{b2}. We could then continue guessing in a trial-and-error fashion until we arrived at a guess for $z(0)$ that resulted in the correct value of $T(L) = T_b$.

At this point, the origin of the name *shooting method* should be pretty clear. Just as you would adjust the angle of a cannon in order to hit a target, we are adjusting the trajectory of our solution by guessing values of $z(0)$ until we hit our target $T(L) = T_b$.

Although we could certainly keep guessing, a more efficient strategy is possible for linear ODEs. In such cases, the trajectory of the perfect shot z_a is linearly related to the results of our two erroneous shots (z_{a1}, T_{b1}) and (z_{a2}, T_{b2}). Consequently, linear interpolation can be employed to arrive at the required trajectory:

$$z_a = z_{a1} + \frac{z_{a2} - z_{a1}}{T_{b2} - T_{b1}}(T_b - T_{b1}) \tag{22.11}$$

The approach can be illustrated by an example.

EXAMPLE 22.2   The Shooting Method for a Linear ODE

Problem Statement.   Use the shooting method to solve Eq. (22.6) for the same conditions as Example 22.1: $L = 10$ m, $h' = 0.05$ m^{-2}, $T_\infty = 200$ K, $T(0) = 300$ K, and $T(10) = 400$ K.

Solution.   Equation (22.6) is first expressed as a pair of first-order ODEs:

$$\frac{dT}{dx} = z$$

$$\frac{dz}{dx} = -0.05(200 - T)$$

Along with the initial value for temperature $T(0) = 300$ K, we arbitrarily guess a value of $z_{a1} = -5$ K/m for the initial value for $z(0)$. The solution is then obtained by integrating the pair of ODEs from $x = 0$ to 10. We can do this with MATLAB's `ode45` function by first setting up an M-file to hold the differential equations:

```
function dy=Ex2302(x,y)
dy=[y(2);-0.05*(200-y(1))];
```

We can then generate the solution as

```
>> [t,y]=ode45(@Ex2302,[0 10],[300,-5]);
>> Tb1=y(length(y))

Tb1 =
  569.7539
```

Thus, we obtain a value at the end of the interval of $T_{b1} = 569.7539$ (Fig. 22.4a), which differs from the desired boundary condition of $T_b = 400$. Therefore, we make another guess $z_{a2} = -20$ and perform the computation again. This time, the result of $T_{b2} = 259.5131$ is obtained (Fig. 22.4b).

FIGURE 22.4

Temperature (K) versus distance (m) computed with the shooting method: (a) the first "shot," (b) the second "shot," and (c) the final exact "hit."

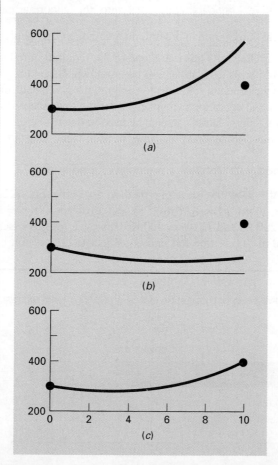

Now, because the original ODE is linear, we can use Eq. (22.11) to determine the correct trajectory to yield the perfect shot:

$$z_a = -5 + \frac{-20 - (-5)}{259.5131 - 569.7539}(400 - 569.7539) = -13.2075$$

This value can then be used in conjunction with `ode45` to generate the correct solution, as depicted in Fig. 22.4c.

Although it is not obvious from the graph, the analytical solution is also plotted on Fig. 22.4c. Thus, the shooting method yields a solution that is virtually indistinguishable from the exact result.

22.2.1 Derivative Boundary Conditions

The fixed or *Dirichlet boundary condition* discussed to this point is but one of several types that are used in engineering and science. A common alternative is the case where the derivative is given. This is commonly referred to as a *Neumann boundary condition*.

Because it is already set up to compute both the dependent variable and its derivative, incorporating derivative boundary conditions into the shooting method is relatively straightforward.

Just as with the fixed-boundary condition case, we first express the second-order ODE as a pair of first-order ODEs. At this point, one of the required initial conditions, whether the dependent variable or its derivative, will be unknown. Based on guesses for the missing initial condition, we generate solutions to compute the given end condition. As with the initial condition, this end condition can either be for the dependent variable or its derivative. For linear ODEs, interpolation can then be used to determine the value of the missing initial condition required to generate the final, perfect "shot" that hits the end condition.

EXAMPLE 22.3   The Shooting Method with Derivative Boundary Conditions

Problem Statement. Use the shooting method to solve Eq. (22.6) for the rod in Example 22.1: $L = 10$ m, $h' = 0.05$ m^{-2} [$h = 1$ J/(m$^2 \cdot$ K $\cdot$ s), $r = 0.2$ m, $k = 200$ J/ (s $\cdot$ m $\cdot$ K)], $T_\infty = 200$ K, and $T(10) = 400$ K. However, for this case, rather than having a fixed temperature of 300 K, the left end is subject to convection as in Fig. 22.5. For

FIGURE 22.5
A rod with a convective boundary condition at one end and a fixed temperature at the other.

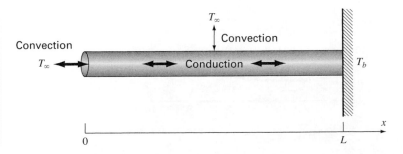

simplicity, we will assume that the convection heat transfer coefficient for the end area is the same as for the rod's surface.

Solution. As in Example 22.2, Eq. (22.6) is first expressed as

$$\frac{dT}{dx} = z$$

$$\frac{dz}{dx} = -0.05(200 - T)$$

Although it might not be obvious, convection through the end is equivalent to specifying a gradient boundary condition. In order to see this, we must recognize that because the system is at steady state, convection must equal conduction at the rod's left boundary ($x = 0$). Using Fourier's law (Eq. 22.5) to represent conduction, the heat balance at the end can be formulated as

$$hA_c(T_\infty - T(0)) = -kA_c\frac{dT}{dx}(0) \tag{22.12}$$

This equation can be solved for the gradient

$$\frac{dT}{dx}(0) = \frac{h}{k}(T(0) - T_\infty) \tag{22.13}$$

If we guess a value for temperature, we can see that this equation specifies the gradient.

The shooting method is implemented by arbitrarily guessing a value for $T(0)$. If we choose a value of $T(0) = T_{a1} = 300$ K, Eq. (22.13) then yields the initial value for the gradient

$$z_{a1} = \frac{dT}{dx}(0) = \frac{1}{200}(300 - 200) = 0.5$$

The solution is obtained by integrating the pair of ODEs from $x = 0$ to 10. We can do this with MATLAB's ode45 function by first setting up an M-file to hold the differential equations in the same fashion as in Example 22.2. We can then generate the solution as

```
>> [t,y]=ode45(@Ex2302,[0 10],[300,0.5]);
>> Tb1=y(length(y))

Tb1 =
  683.5088
```

As expected, the value at the end of the interval of $T_{b1} = 683.5088$ K differs from the desired boundary condition of $T_b = 400$. Therefore, we make another guess $T_{a2} = 150$ K, which corresponds to $z_{a2} = -0.25$, and perform the computation again.

```
>> [t,y]=ode45(@Ex2302,[0 10],[150,-0.25]);
>> Tb2=y(length(y))

Tb2 =
  -41.7544
```

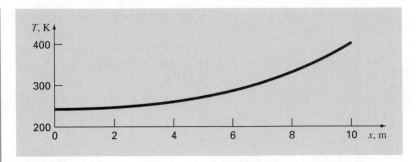

FIGURE 22.6
The solution of a second-order ODE with a convective boundary condition at one end and a fixed temperature at the other.

Linear interpolation can then be employed to compute the correct initial temperature:

$$T_a = 300 + \frac{150 - 300}{-41.7544 - 683.5088}(400 - 683.5088) = 241.3643 \text{ K}$$

which corresponds to a gradient of $z_a = 0.2068$. Using these initial conditions, ode45 can be employed to generate the correct solution, as depicted in Fig. 22.6.

Note that we can verify that our boundary condition has been satisfied by substituting the initial conditions into Eq. (22.12) to give

$$1 \frac{\text{J}}{\text{m}^2\text{K s}}\pi \times (0.2 \text{ m})^2 \times (200 \text{ K} - 241.3643 \text{ K}) = -200 \frac{\text{J}}{\text{m K s}}\pi \times (0.2 \text{ m})^2 \times 0.2068 \frac{\text{K}}{\text{m}}$$

which can be evaluated to yield -5.1980 J/s $= -5.1980$ J/s. Thus, conduction and convection are equal and transfer heat out of the left end of the rod at a rate of 5.1980 W.

22.2.2 The Shooting Method for Nonlinear ODEs

For nonlinear boundary-value problems, linear interpolation or extrapolation through two solution points will not necessarily result in an accurate estimate of the required boundary condition to attain an exact solution. An alternative is to perform three applications of the shooting method and use a quadratic interpolating polynomial to estimate the proper boundary condition. However, it is unlikely that such an approach would yield the exact answer, and additional iterations would be necessary to home in on the solution.

Another approach for a nonlinear problem involves recasting it as a roots problem. Recall that the general goal of a roots problem is to find the value of x that makes the function $f(x) = 0$. Now, let us use the heated rod problem to understand how the shooting method can be recast in this form.

First, recognize that the solution of the pair of differential equations is also a "function" in the sense that we guess a condition at the left-hand end of the rod z_a, and the integration yields a prediction of the temperature at the right-hand end T_b. Thus, we can think of the integration as

$$T_b = f(z_a)$$

That is, it represents a process whereby a guess of z_a yields a prediction of T_b. Viewed in this way, we can see that what we desire is the value of z_a that yields a specific value of T_b. If, as in the example, we desire $T_b = 400$, the problem can be posed as

$$400 = f(z_a)$$

By bringing the goal of 400 over to the right-hand side of the equation, we generate a new function $res(z_a)$ that represents the difference, or *residual,* between what we have, $f(z_a)$, and what we want, 400.

$$res(z_a) = f(z_a) - 400$$

If we drive this new function to zero, we will obtain the solution. The next example illustrates the approach.

EXAMPLE 22.4   The Shooting Method for Nonlinear ODEs

Problem Statement.   Although it served our purposes for illustrating the shooting method, Eq. (22.6) was not a completely realistic model for a heated rod. For one thing, such a rod would lose heat by mechanisms such as radiation that are nonlinear.

Suppose that the following nonlinear ODE is used to simulate the temperature of the heated rod:

$$0 = \frac{d^2 T}{dx^2} + h'(T_\infty - T) + \sigma'(T_\infty^4 - T^4)$$

where σ' = a bulk heat-transfer parameter reflecting the relative impacts of radiation and conduction = 2.7×10^{-9} K^{-3} m^{-2}. This equation can serve to illustrate how the shooting method is used to solve a two-point nonlinear boundary-value problem. The remaining problem conditions are as specified in Example 22.2: $L = 10$ m, $h' = 0.05$ m^{-2}, $T_\infty = 200$ K, $T(0) = 300$ K, and $T(10) = 400$ K.

Solution.   Just as with the linear ODE, the nonlinear second-order equation is first expressed as two first-order ODEs:

$$\frac{dT}{dx} = z$$

$$\frac{dz}{dx} = -0.05(200 - T) - 2.7 \times 10^{-9}(1.6 \times 10^9 - T^4)$$

An M-file can be developed to compute the right-hand sides of these equations:

```
function dy=dydxn(x,y)
dy=[y(2);-0.05*(200-y(1))-2.7e-9*(1.6e9-y(1)^4)];
```

Next, we can build a function to hold the residual that we will try to drive to zero as

```
function r=res(za)
[x,y]=ode45(@dydxn,[0 10],[300 za]);
r=y(length(x),1)-400;
```

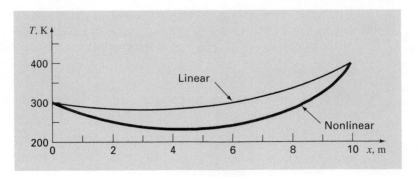

FIGURE 22.7
The result of using the shooting method to solve a nonlinear problem.

Notice how we use the `ode45` function to solve the two ODEs to generate the temperature at the rod's end: `y(length(x),1)`. We can then find the root with the `fzero` function:

```
>> fzero(@res,-50)

ans =
  -41.7434
```

Thus, we see that if we set the initial trajectory $z(0) = -41.7434$, the residual function will be driven to zero and the temperature boundary condition $T(10) = 400$ at the end of the rod should be satisfied. This can be verified by generating the entire solution and plotting the temperatures versus x:

```
>> [x,y]=ode45(@dydxn,[0 10],[300 fzero(@res,-50)]);
>> plot(x,y(:,1))
```

The result is shown in Fig. 22.7 along with the original linear case from Example 22.2. As expected, the nonlinear case is depressed lower than the linear model due to the additional heat lost to the surrounding gas by radiation.

22.3 FINITE-DIFFERENCE METHODS

The most common alternatives to the shooting method are finite-difference approaches. In these techniques, finite differences (Chap. 19) are substituted for the derivatives in the original equation. Thus, a linear differential equation is transformed into a set of simultaneous algebraic equations that can be solved using the methods from Part Three.

We can illustrate the approach for the heated rod model (Eq. 22.6):

$$0 = \frac{d^2T}{dx^2} + h'(T_\infty - T) \tag{22.14}$$

The solution domain is first divided into a series of nodes (Fig. 22.8). At each node, finite-difference approximations can be written for the derivatives in the equation. For example,

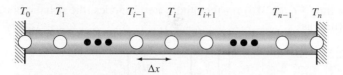

FIGURE 22.8
In order to implement the finite-difference approach, the heated rod is divided into a series of nodes.

at node i, the second derivative can be represented by (Eq. 19.15):

$$\frac{d^2 T}{dx^2} = \frac{T_{i-1} - 2T_i + T_{i+1}}{\Delta x^2} \tag{22.15}$$

This approximation can be substituted into Eq. (22.14) to give

$$\frac{T_{i-1} - 2T_i + T_{i+1}}{\Delta x^2} + h'(T_\infty - T_i) = 0$$

Thus, the differential equation has been converted into an algebraic equation. Collecting terms gives

$$-T_{i-1} + (2 + h'\Delta x^2)T_i - T_{i+1} = h'\Delta x^2 T_\infty \tag{22.16}$$

This equation can be written for each of the $n-1$ interior nodes of the rod. The first and last nodes T_0 and T_n, respectively, are specified by the boundary conditions. Therefore, the problem reduces to solving $n-1$ simultaneous linear algebraic equations for the $n-1$ unknowns.

Before providing an example, we should mention two nice features of Eq. (22.16). First, observe that since the nodes are numbered consecutively, and since each equation consists of a node (i) and its adjoining neighbors ($i-1$ and $i+1$), the resulting set of linear algebraic equations will be tridiagonal. As such, they can be solved with the efficient algorithms that are available for such systems (recall Sec. 9.4).

Further, inspection of the coefficients on the left-hand side of Eq. (22.16) indicates that the system of linear equations will also be diagonally dominant. Hence, convergent solutions can also be generated with iterative techniques like the Gauss-Seidel method (Sec. 12.1).

EXAMPLE 22.5 Finite-Difference Approximation of Boundary-Value Problems

Problem Statement. Use the finite-difference approach to solve the same problem as in Examples 22.1 and 22.2. Use four interior nodes with a segment length of $\Delta x = 2$ m.

Solution. Employing the parameters in Example 22.1 and $\Delta x = 2$ m, we can write Eq. (22.16) for each of the rod's interior nodes. For example, for node 1:

$$-T_0 + 2.2T_1 - T_2 = 40$$

Substituting the boundary condition $T_0 = 300$ gives

$$2.2T_1 - T_2 = 340$$

After writing Eq. (22.16) for the other interior nodes, the equations can be assembled in matrix form as

$$\begin{bmatrix} 2.2 & -1 & 0 & 0 \\ -1 & 2.2 & -1 & 0 \\ 0 & -1 & 2.2 & -1 \\ 0 & 0 & -1 & 2.2 \end{bmatrix} \begin{Bmatrix} T_1 \\ T_2 \\ T_3 \\ T_4 \end{Bmatrix} = \begin{Bmatrix} 340 \\ 40 \\ 40 \\ 440 \end{Bmatrix}$$

Notice that the matrix is both tridiagonal and diagonally dominant.

MATLAB can be used to generate the solution:

```
>> A=[2.2 -1 0 0;
-1 2.2 -1 0;
0 -1 2.2 -1;
0 0 -1 2.2];
>> b=[340 40 40 440]';
>> T=A\b

T =
  283.2660
  283.1853
  299.7416
  336.2462
```

Table 22.1 provides a comparison between the analytical solution (Eq. 22.7) and the numerical solutions obtained with the shooting method (Example 22.2) and the finite-difference method (Example 22.5). Note that although there are some discrepancies, the numerical approaches agree reasonably well with the analytical solution. Further, the biggest discrepancy occurs for the finite-difference method due to the coarse node spacing we used in Example 22.5. Better agreement would occur if a finer nodal spacing had been used.

TABLE 22.1 Comparison of the exact analytical solution for temperature with the results obtained with the shooting and finite-difference methods.

x	Analytical Solution	Shooting Method	Finite Difference
0	300	300	300
2	282.8634	282.8889	283.2660
4	282.5775	282.6158	283.1853
6	299.0843	299.1254	299.7416
8	335.7404	335.7718	336.2462
10	400	400	400

22.3.1 Derivative Boundary Conditions

As mentioned in our discussion of the shooting method, the fixed or *Dirichlet boundary condition* is but one of several types that are used in engineering and science. A common

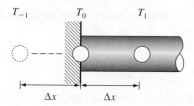

FIGURE 22.9
A boundary node at the left end of a heated rod. To approximate the derivative at the boundary, an imaginary node is located a distance Δx to the left of the rod's end.

alternative, called the *Neumann boundary condition,* is the case where the derivative is given.

We can use the heated rod introduced earlier in this chapter to demonstrate how a derivative boundary condition can be incorporated into the finite-difference approach:

$$0 = \frac{d^2T}{dx^2} + h'(T_\infty - T)$$

However, in contrast to our previous discussions, we will prescribe a derivative boundary condition at one end of the rod:

$$\frac{dT}{dx}(0) = T'_a$$

$$T(L) = T_b$$

Thus, we have a derivative boundary condition at one end of the solution domain and a fixed boundary condition at the other.

Just as in the previous section, the rod is divided into a series of nodes and a finite-difference version of the differential equation (Eq. 22.16) is applied to each interior node. However, because its temperature is not specified, the node at the left end must also be included. Fig. 22.9 depicts the node (0) at the left edge of a heated plate for which the derivative boundary condition applies. Writing Eq. (22.16) for this node gives

$$-T_{-1} + (2 + h'\Delta x^2)T_0 - T_1 = h'\Delta x^2 T_\infty \qquad (22.17)$$

Notice that an imaginary node (-1) lying to the left of the rod's end is required for this equation. Although this exterior point might seem to represent a difficulty, it actually serves as the vehicle for incorporating the derivative boundary condition into the problem. This is done by representing the first derivative in the x dimension at (0) by the centered difference (Eq. 4.25):

$$\frac{dT}{dx} = \frac{T_1 - T_{-1}}{2\Delta x}$$

which can be solved for

$$T_{-1} = T_1 - 2\Delta x \frac{dT}{dx}$$

Now we have a formula for T_{-1} that actually reflects the impact of the derivative. It can be substituted into Eq. (22.17) to give

$$(2 + h'\Delta x^2)T_0 - 2T_1 = h'\Delta x^2 T_\infty - 2\Delta x \frac{dT}{dx} \tag{22.18}$$

Consequently, we have incorporated the derivative into the balance.

A common example of a derivative boundary condition is the situation where the end of the rod is insulated. In this case, the derivative is set to zero. This conclusion follows directly from Fourier's law (Eq. 22.5), because insulating a boundary means that the heat flux (and consequently the gradient) must be zero. The following example illustrates how the solution is affected by such boundary conditions.

EXAMPLE 22.6   Incorporating Derivative Boundary Conditions

Problem Statement. Generate the finite-difference solution for a 10-m rod with $\Delta x = 2$ m, $h' = 0.05$ m^{-2}, $T_\infty = 200$ K, and the boundary conditions: $T_a' = 0$ and $T_b = 400$ K. Note that the first condition means that the slope of the solution should approach zero at the rod's left end. Aside from this case, also generate the solution for $dT/dx = -20$ at $x = 0$.

Solution. Equation (22.18) can be used to represent node 0 as

$$2.2T_0 - 2T_1 = 40$$

We can write Eq. (22.16) for the interior nodes. For example, for node 1,

$$-T_0 + 2.2T_1 - T_2 = 40$$

A similar approach can be used for the remaining interior nodes. The final system of equations can be assembled in matrix form as

$$\begin{bmatrix} 2.2 & -2 & & & \\ -1 & 2.2 & -1 & & \\ & -1 & 2.2 & -1 & \\ & & -1 & 2.2 & -1 \\ & & & -1 & 2.2 \end{bmatrix} \begin{Bmatrix} T_0 \\ T_1 \\ T_2 \\ T_3 \\ T_4 \end{Bmatrix} = \begin{Bmatrix} 40 \\ 40 \\ 40 \\ 40 \\ 440 \end{Bmatrix}$$

These equations can be solved for

$$T_0 = 243.0278$$
$$T_1 = 247.3306$$
$$T_2 = 261.0994$$
$$T_3 = 287.0882$$
$$T_4 = 330.4946$$

As displayed in Fig. 22.10, the solution is flat at $x = 0$ due to the zero derivative condition and then curves upward to the fixed condition of $T = 400$ at $x = 10$.

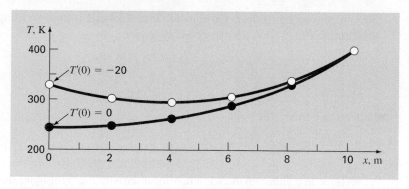

FIGURE 22.10
The solution of a second-order ODE with a derivative boundary condition at one end and a fixed boundary condition at the other. Two cases are shown reflecting different derivative values at $x = 0$.

For the case where the derivative at $x = 0$ is set to -20, the simultaneous equations are

$$
\begin{bmatrix}
2.2 & -2 & & & \\
-1 & 2.2 & -1 & & \\
& -1 & 2.2 & -1 & \\
& & -1 & 2.2 & -1 \\
& & & -1 & 2.2
\end{bmatrix}
\begin{Bmatrix}
T_0 \\ T_1 \\ T_2 \\ T_3 \\ T_4
\end{Bmatrix}
=
\begin{Bmatrix}
120 \\ 40 \\ 40 \\ 40 \\ 440
\end{Bmatrix}
$$

which can be solved for

$$T_0 = 328.2710$$
$$T_1 = 301.0981$$
$$T_2 = 294.1448$$
$$T_3 = 306.0204$$
$$T_4 = 339.1002$$

As in Fig. 22.10, the solution at $x = 0$ now curves downward due to the negative derivative we imposed at the boundary.

22.3.2 Finite-Difference Approaches for Nonlinear ODEs

For nonlinear ODEs, the substitution of finite differences yields a system of nonlinear simultaneous equations. Thus, the most general approach to solving such problems is to use root location methods for systems of equations such as the Newton-Raphson method described in Sec. 12.2.2. Although this approach is certainly feasible, an adaptation of successive substitution can sometimes provide a simpler alternative.

The heated rod with convection and radiation introduced in Example 22.4 provides a nice vehicle for demonstrating this approach,

$$0 = \frac{d^2T}{dx^2} + h'(T_\infty - T) + \sigma''(T_\infty^4 - T^4)$$

We can convert this differential equation into algebraic form by writing it for a node i and substituting Eq. (22.15) for the second derivative:

$$0 = \frac{T_{i-1} - 2T_i + T_{i+1}}{\Delta x^2} + h'(T_\infty - T_i) + \sigma''(T_\infty^4 - T_i^4)$$

Collecting terms gives

$$-T_{i-1} + (2 + h'\Delta x^2)T_i - T_{i+1} = h'\Delta x^2 T_\infty + \sigma''\Delta x^2(T_\infty^4 - T_i^4)$$

Notice that although there is a nonlinear term on the right-hand side, the left-hand side is expressed in the form of a linear algebraic system that is diagonally dominant. If we assume that the unknown nonlinear term on the right is equal to its value from the previous iteration, the equation can be solved for

$$T_i = \frac{h'\Delta x^2 T_\infty + \sigma''\Delta x^2(T_\infty^4 - T_i^4) + T_{i-1} + T_{i+1}}{2 + h'\Delta x^2} \qquad (22.19)$$

As in the Gauss-Seidel method, we can use Eq. (22.19) to successively calculate the temperature of each node and iterate until the process converges to an acceptable tolerance. Although this approach will not work for all cases, it converges for many ODEs derived from physically based systems. Hence, it can sometimes prove useful for solving problems routinely encountered in engineering and science.

EXAMPLE 22.7   The Finite-Difference Method for Nonlinear ODEs

Problem Statement.   Use the finite-difference approach to simulate the temperature of a heated rod subject to both convection and radiation:

$$0 = \frac{d^2T}{dx^2} + h'(T_\infty - T) + \sigma''(T_\infty^4 - T^4)$$

where $\sigma' = 2.7 \times 10^{-9}$ K^{-3}m^{-2}, $L = 10$ m, $h' = 0.05$ m^{-2}, $T_\infty = 200$ K, $T(0) = 300$ K, and $T(10) = 400$ K. Use four interior nodes with a segment length of $\Delta x = 2$ m. Recall that we solved the same problem with the shooting method in Example 22.4.

Solution.   Using Eq. (22.19) we can successively solve for the temperatures of the rod's interior nodes. As with the standard Gauss-Seidel technique, the initial values of the interior nodes are zero with the boundary nodes set at the fixed conditions of $T_0 = 300$ and $T_5 = 400$. The results for the first iteration are

$$T_1 = \frac{0.05(2)^2\,200 + 2.7 \times 10^{-9'}(2)^2(200^4 - 0^4) + 300 + 0}{2 + 0.05(2)^2} = 159.2432$$

$$T_2 = \frac{0.05(2)^2\,200 + 2.7 \times 10^{-9'}(2)^2(200^4 - 0^4) + 159.2432 + 0}{2 + 0.05(2)^2} = 97.9674$$

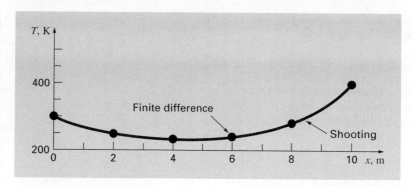

FIGURE 22.11
The filled circles are the result of using the finite-difference method to solve a nonlinear problem. The line generated with the shooting method in Example 22.4 is shown for comparison.

$$T_3 = \frac{0.05(2)^2\,200 + 2.7 \times 10^{-9'}(2)^2(200^4 - 0^4) + 97.9674 + 0}{2 + 0.05(2)^2} = 70.4461$$

$$T_4 = \frac{0.05(2)^2\,200 + 2.7 \times 10^{-9'}(2)^2(200^4 - 0^4) + 70.4461 + 400}{2 + 0.05(2)^2} = 226.8704$$

The process can be continued until we converge on the final result:

$T_0 = 300$
$T_1 = 250.4827$
$T_2 = 236.2962$
$T_3 = 245.7596$
$T_4 = 286.4921$
$T_5 = 400$

These results are displayed in Fig. 22.11 along with the result generated in Example 22.4 with the shooting method.

PROBLEMS

22.1 A steady-state heat balance for a rod can be represented as

$$\frac{d^2T}{dx^2} - 0.15T = 0$$

Obtain a solution for a 10-m rod with $T(0) = 240$ and $T(10) = 150$ **(a)** analytically, **(b)** with the shooting method, and **(c)** using the finite-difference approach with $\Delta x = 1$.

22.2 Repeat Prob. 22.1 but with the right end insulated and the left end temperature fixed at 240.
22.3 Use the shooting method to solve

$$7\frac{d^2y}{dx^2} - 2\frac{dy}{dx} - y + x = 0$$

with the boundary conditions $y(0) = 5$ and $y(20) = 8$.
22.4 Solve Prob. 22.3 with the finite-difference approach using $\Delta x = 2$.

22.5 The following nonlinear differential equation was solved in Examples 22.4 and 22.7.

$$0 = \frac{d^2 T}{dx^2} + h'(T_\infty - T) + \sigma''(T_\infty^4 - T^4) \qquad \text{(P22.5)}$$

Such equations are sometimes linearized to obtain an approximate solution. This is done by employing a first-order Taylor series expansion to linearize the quartic term in the equation as

$$\sigma' T^4 = \sigma' \overline{T}^4 + 4\sigma' \overline{T}^3 (T - \overline{T})$$

where $\overline{T}$ is a base temperature about which the term is linearized. Substitute this relationship into Eq. (P22.5), and then solve the resulting linear equation with the finite-difference approach. Employ $\overline{T} = 300$, $\Delta x = 1$ m, and the parameters from Example 22.4 to obtain your solution. Plot your results along with those obtained for the nonlinear version in Example 22.4.

22.6 Develop an M-file to implement the shooting method for a linear second-order ODE. Test the program by duplicating Example 22.2.

22.7 Develop an M-file to implement the finite-difference approach for solving a linear second-order ODE with Dirichlet boundary conditions. Test it by duplicating Example 22.5.

22.8 An insulated heated rod with a uniform heat source can be modeled with the *Poisson equation:*

$$\frac{d^2 T}{dx^2} = -f(x)$$

Given a heat source $f(x) = 25\,°\text{C/m}^2$ and the boundary conditions $T(x = 0) = 40\,°\text{C}$ and $T(x = 10) = 200\,°\text{C}$, solve for the temperature distribution with **(a)** the shooting method and **(b)** the finite-difference method ($\Delta x = 2$).

22.9 Repeat Prob. 22.8, but for the following spatially varying heat source: $f(x) = 0.12x^3 - 2.4x^2 + 12x$.

22.10 The temperature distribution in a tapered conical cooling fin (Fig. P22.10) is described by the following differential equation, which has been nondimensionalized:

$$\frac{d^2 u}{dx^2} + \left(\frac{2}{x}\right)\left(\frac{du}{dx} - pu\right) = 0$$

where $u = $ temperature ($0 \le u \le 1$), $x = $ axial distance ($0 \le x \le 1$), and p is a nondimensional parameter that describes the heat transfer and geometry:

$$p = \frac{hL}{k}\sqrt{1 + \frac{4}{2m^2}}$$

where $h = $ a heat transfer coefficient, $k = $ thermal conductivity, $L = $ the length or height of the cone, and $m = $ the slope

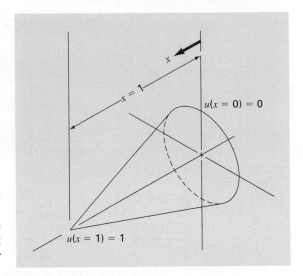

FIGURE P22.10

of the cone wall. The equation has the boundary conditions:

$$u(x = 0) = 0 \qquad u(x = 1) = 1$$

Solve this equation for the temperature distribution using finite-difference methods. Use second-order accurate finite-difference formulas for the derivatives. Write a computer program to obtain the solution and plot temperature versus axial distance for various values of $p = 10, 20, 50,$ and 100.

22.11 Compound A diffuses through a 4-cm-long tube and reacts as it diffuses. The equation governing diffusion with reaction is

$$D\frac{d^2 A}{dx^2} - kA = 0$$

At one end of the tube ($x = 0$), there is a large source of A that results in a fixed concentration of 0.1 M. At the other end of the tube there is a material that quickly absorbs any A, making the concentration 0 M. If $D = 1.5 \times 10^{-6}$ cm^2/s and $k = 5 \times 10^{-6}$ s^{-1}, what is the concentration of A as a function of distance in the tube?

22.12 The following differential equation describes the steady-state concentration of a substance that reacts with first-order kinetics in an axially dispersed plug-flow reactor (Fig. P22.12):

$$D\frac{d^2 c}{dx^2} - U\frac{dc}{dx} - kc = 0$$

where $D = $ the dispersion coefficient (m^2/hr), $c = $ concentration (mol/L), $x = $ distance (m), $U = $ the velocity (m/hr),

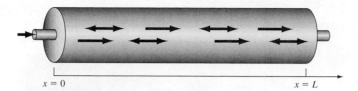

FIGURE P22.12
An axially dispersed plug-flow reactor.

and k = the reaction rate (/hr). The boundary conditions can be formulated as

$$Uc_{in} = Uc(x = 0) - D\frac{dc}{dx}(x = 0)$$

$$\frac{dc}{dx}(x = L) = 0$$

where c_{in} = the concentration in the inflow (mol/L), L = the length of the reactor (m). These are called *Danckwerts boundary conditions*.

Use the finite-difference approach to solve for concentration as a function of distance given the following parameters: $D = 5000$ m²/hr, $U = 100$ m/hr, $k = 2$/hr, $L = 100$ m, and $c_{in} = 100$ mol/L. Employ centered finite-difference approximations with $\Delta x = 10$ m to obtain your solutions. Compare your numerical results with the analytical solution:

$$c = \frac{Uc_{in}}{(U - D\lambda_1)\lambda_2 e^{\lambda_2 L} - (U - D\lambda_2)\lambda_1 e^{\lambda_1 L}}$$

$$\times (\lambda_2 e^{\lambda_2 L} e^{\lambda_1 x} - \lambda_1 e^{\lambda_1 L} e^{\lambda_2 x})$$

where

$$\frac{\lambda_1}{\lambda_2} = \frac{U}{2D}\left(1 \pm \sqrt{1 + \frac{4kD}{U^2}}\right)$$

22.13 A series of first-order, liquid-phase reactions create a desirable product (B) and an undesirable byproduct (C):

$$A \xrightarrow{k_1} B \xrightarrow{k_2} C$$

If the reactions take place in an axially dispersed plug-flow reactor (Fig. P22.12), steady-state mass balances can be used to develop the following second-order ODEs:

$$D\frac{d^2c_a}{dx^2} - U\frac{dc_a}{dx} - k_1c_a = 0$$

$$D\frac{d^2c_b}{dx^2} - U\frac{dc_b}{dx} + k_1c_a - k_2c_b = 0$$

$$D\frac{d^2c_c}{dx^2} - U\frac{dc_c}{dx} + k_2c_b = 0$$

Use the finite-difference approach to solve for the concentration of each reactant as a function of distance given: $D = 0.1$ m²/min, $U = 1$ m/min, $k_1 = 3$/min, $k_2 = 1$/min, $L = 0.5$ m, $c_{a,in} = 10$ mol/L. Employ centered finite-difference approximations with $\Delta x = 0.05$ m to obtain your solutions and assume Danckwerts boundary conditions as described in Prob. 22.12. Also, compute the sum of the reactants as a function of distance. Do your results make sense?

22.14 A biofilm with a thickness L_f (cm), grows on the surface of a solid (Fig. P22.14). After traversing a diffusion layer of thickness L (cm), a chemical compound A diffuses into the biofilm where it is subject to an irreversible first-order reaction that converts it to a product B.

Steady-state mass balances can be used to derive the following ordinary differential equations for compound A:

$$D\frac{d^2c_a}{dx^2} = 0 \qquad 0 \le x < L$$

$$D_f\frac{d^2c_a}{dx^2} - kc_a = 0 \qquad L \le x < L + L_f$$

where D = the diffusion coefficient in the diffusion layer = 0.8 cm²/d, D_f = the diffusion coefficient in the biofilm = 0.64 cm²/d, and k = the first-order rate for the conversion of A to B = 0.1/d. The following boundary conditions hold:

$$c_a = c_{a0} \qquad \text{at } x = 0$$

$$\frac{dc_a}{dx} = 0 \qquad \text{at } x = L + L_f$$

where c_{a0} = the concentration of A in the bulk liquid = 100 mol/L. Use the finite-difference method to compute the steady-state distribution of A from $x = 0$ to $L + L_f$, where $L = 0.008$ cm and $L_f = 0.004$ cm. Employ centered finite differences with $\Delta x = 0.001$ cm.

22.15 A cable is hanging from two supports at A and B (Fig. P22.15). The cable is loaded with a distributed load whose magnitude varies with x as

$$w = w_o\left[1 + \sin\left(\frac{\pi x}{2l_A}\right)\right]$$

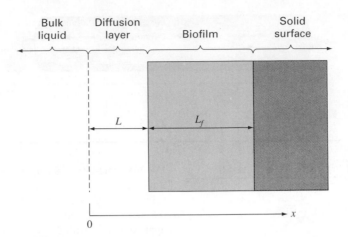

FIGURE P22.14
A biofilm growing on a solid surface.

where $w_o = 450$ N/m. The slope of the cable $(dy/dx) = 0$ at $x = 0$, which is the lowest point for the cable. It is also the point where the tension in the cable is a minimum of T_o. The differential equation which governs the cable is

$$\frac{d^2y}{dx^2} = \frac{w_o}{T_o}\left[1 + \sin\left(\frac{\pi x}{2l_A}\right)\right]$$

Solve this equation using a numerical method and plot the shape of the cable (y versus x). For the numerical solution, the value of T_o is unknown, so the solution must use an iterative technique, similar to the shooting method, to converge on a correct value of h_A for various values of T_o.

22.16 The basic differential equation of the elastic curve for a simply supported, uniformly loaded beam (Fig. P22.16) is given as

$$EI\frac{d^2y}{dx^2} = \frac{wLx}{2} - \frac{wx^2}{2}$$

where E = the modulus of elasticity, and I = the moment of inertia. The boundary conditions are $y(0) = y(L) = 0$. Solve for the deflection of the beam using (**a**) the finite-difference approach ($\Delta x = 0.6$ m) and (**b**) the shooting method. The following parameter values apply: $E = 200$ GPa, $I = 30,000$ cm⁴, $w = 15$ kN/m, and $L = 3$ m. Compare your

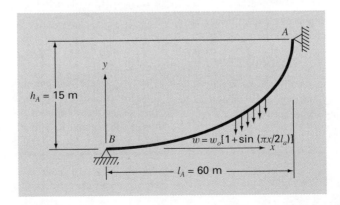

FIGURE P22.15

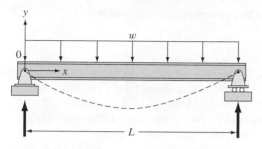

FIGURE P22.16

numerical results to the analytical solution:

$$y = \frac{wLx^3}{12EI} - \frac{wx^4}{24EI} - \frac{wL^3x}{24EI}$$

22.17 In Prob. 22.16, the basic differential equation of the elastic curve for a uniformly loaded beam was formulated as

$$EI\frac{d^2y}{dx^2} = \frac{wLx}{2} - \frac{wx^2}{2}$$

Note that the right-hand side represents the moment as a function of x. An equivalent approach can be formulated in terms of the fourth derivative of deflection as

$$EI\frac{d^4y}{dx^4} = -w$$

For this formulation, four boundary conditions are required. For the supports shown in Fig. P22.16, the conditions are

that the end displacements are zero, $y(0) = y(L) = 0$, and that the end moments are zero, $y''(0) = y''(L) = 0$. Solve for the deflection of the beam using the finite-difference approach ($\Delta x = 0.6$ m). The following parameter values apply: $E = 200$ GPa, $I = 30,000$ cm^4, $w = 15$ kN/m, and $L = 3$ m. Compare your numerical results with the analytical solution given in Prob. 22.16.

22.18 Under a number of simplifying assumptions, the steady-state height of the water table in a one-dimensional, unconfined groundwater aquifer (Fig. P22.18) can be modeled with the following second-order ODE:

$$K\bar{h}\frac{d^2h}{dx^2} + N = 0$$

where x = distance (m), K = hydraulic conductivity (m/d), h = height of the water table (m), $\bar{h}$ = the average height of the water table (m), and N = infiltration rate (m/d).

Solve for the height of the water table for $x = 0$ to 1000 m where $h(0) = 10$ m and $h(1000) = 5$ m. Use the following parameters for the calculation: $K = 1$ m/d and $N = 0.1$ m/d. Set the average height of the water table as the average of the boundary conditions. Obtain your solution with **(a)** the shooting method and **(b)** the finite-difference method ($\Delta x = 100$ m).

22.19 In Prob. 22.18, a linearized groundwater model was used to simulate the height of the water table for an unconfined aquifer. A more realistic result can be obtained by using the following nonlinear ODE:

$$\frac{d}{dx}\left(Kh\frac{dh}{dx}\right) + N = 0$$

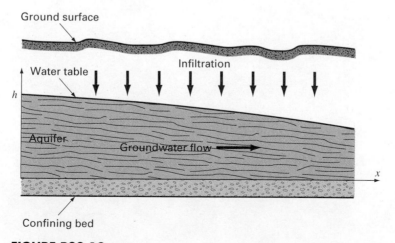

FIGURE P22.18
An unconfined or "phreatic" aquifer.

where x = distance (m), K = hydraulic conductivity (m/d), h = height of the water table (m), and N = infiltration rate (m/d). Solve for the height of the water table for the same case as in Prob. 22.18. That is, solve from $x = 0$ to 1000 m with $h(0) = 10$ m, $h(1000) = 5$ m, $K = 1$ m/d, and $N = 0.1$ m/d. Obtain your solution with (a) the shooting method and (b) the finite-difference method ($\Delta x = 100$ m).

22.20 Just as Fourier's law and the heat balance can be employed to characterize temperature distribution, analogous relationships are available to model field problems in other areas of engineering. For example, electrical engineers use a similar approach when modeling electrostatic fields. Under a number of simplifying assumptions, an analog of Fourier's law can be represented in one-dimensional form as

$$D = -\varepsilon \frac{dV}{dx}$$

where D is called the electric flux density vector, ε = permittivity of the material, and V = electrostatic potential. Similarly, a Poisson equation (see Prob. 22.8) for electrostatic

fields can be represented in one dimension as

$$\frac{d^2V}{dx^2} = -\frac{\rho_v}{\varepsilon}$$

where ρ_v = charge density. Use the finite-difference technique with $\Delta x = 2$ to determine V for a wire where $V(0) = 1000$, $V(20) = 0$, $\varepsilon = 2$, $L = 20$, and $\rho_v = 30$.

22.21 Suppose that the position of a falling object is governed by the following differential equation:

$$\frac{d^2x}{dt^2} + \frac{c}{m}\frac{dx}{dt} - g = 0$$

where c = a first-order drag coefficient = 12.5 kg/s, m = mass = 70 kg, and g = gravitational acceleration = 9.81 m/s^2. Use the shooting method to solve this equation for the boundary conditions:

$$x(0) = 0$$
$$x(12) = 500$$

APPENDIX A

EIGENVALUES

Eigenvalue, or characteristic-value, problems are a special class of problems that are common in engineering and scientific problem contexts involving vibrations and elasticity. In addition, they are used in a wide variety of other areas including the solution of linear differential equations and statistics.

Before describing numerical methods for solving such problems, we will present some general background information. This includes discussion of both the mathematics and the engineering and scientific significance of eigenvalues.

A.1 Mathematical Background

Chapters 8 through 12 dealt with methods for solving sets of linear algebraic equations of the general form

$$[A]\{x\} = \{b\}$$

Such systems are called *nonhomogeneous* because of the presence of the vector $\{b\}$ on the right-hand side of the equality. If the equations comprising such a system are linearly independent (i.e., have a nonzero determinant), they will have a unique solution. In other words, there is one set of x values that will make the equations balance.

In contrast, a *homogeneous* linear algebraic system has the general form

$$[A]\{x\} = 0$$

Although nontrivial solutions (i.e., solutions other than all x's $= 0$) of such systems are possible, they are generally not unique. Rather, the simultaneous equations establish relationships among the x's that can be satisfied by various combinations of values.

Eigenvalue problems associated with engineering are typically of the general form

$$
\begin{aligned}
(a_{11} - \lambda)x_1 + \quad &a_{12}x_2 + \cdots + \quad &a_{1n}x_n = 0 \\
a_{21}x_1 + (a_{22} - \lambda)x_2 + \cdots + \quad &a_{2n}x_n = 0 \\
\vdots \qquad\qquad\qquad\qquad &\vdots \\
a_{n1}x_1 + \quad &a_{n2}x_2 + \cdots + (a_{nn} - \lambda)x_n = 0
\end{aligned}
$$

where λ is an unknown parameter called the *eigenvalue,* or *characteristic value.* A solution $\{x\}$ for such a system is referred to as an *eigenvector.* The above set of equations may also be expressed concisely as

$$\big[[A] - \lambda[I]\big]\{x\} = 0 \tag{A.1}$$

The solution of Eq. (A.1) hinges on determining λ. One way to accomplish this is based on the fact that the determinant of the matrix $\big[[A] - \lambda[I]\big]$ must equal zero for nontrivial solutions to be possible. Expanding the determinant yields a polynomial in λ, which is called the *characteristic polynomial.* The roots of this polynomial are the solutions for the eigenvalues. An example of this approach, called the *polynomial method,* will be provided in Section A.3. Before describing the method, we will first describe how eigenvalues arise in engineering and science.

A.2 Physical Background

The mass-spring system in Fig. A.1a is a simple context to illustrate how eigenvalues occur in physical problem settings. It also will help to illustrate some of the mathematical concepts introduced in Section A.1.

To simplify the analysis, assume that each mass has no external or damping forces acting on it. In addition, assume that each spring has the same natural length l and the same spring constant k. Finally, assume that the displacement of each spring is measured relative to its own local coordinate system with an origin at the spring's equilibrium position (Fig. A.1a). Under these assumptions, Newton's second law can be employed to develop a force balance for each mass:

$$m_1 \frac{d^2 x_1}{dt^2} = -k x_1 + k(x_2 - x_1)$$

FIGURE A.1
Positioning the masses away from equilibrium creates forces in the springs that on release lead to oscillations of the masses. The positions of the masses can be referenced to local coordinates with origins at their respective equilibrium positions.

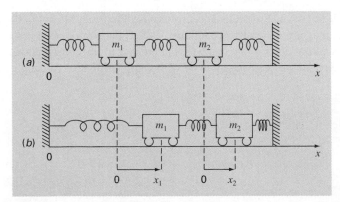

and

$$m_2 \frac{d^2 x_2}{dt^2} = -k(x_2 - x_1) - kx_2$$

where x_i is the displacement of mass i away from its equilibrium position (Fig. A.1*b*). By collecting terms, these equations can be expressed as

$$m_1 \frac{d^2 x_1}{dt^2} - k(-2x_1 + x_2) = 0 \tag{A.2a}$$

$$m_2 \frac{d^2 x_2}{dt^2} - k(x_1 - 2x_2) = 0 \tag{A.2b}$$

From vibration theory, it is known that solutions to Eq. (A.2) can take the form

$$x_i = X_i \sin(\omega t) \tag{A.3}$$

where X_i = the amplitude of the vibration of mass i and ω = the frequency of the vibration, which is equal to

$$\omega = \frac{2\pi}{T_p} \tag{A.4}$$

where T_p is the period. From Eq. (A.3) it follows that

$$x_i'' = -X_i \omega^2 \sin(\omega t) \tag{A.5}$$

Equations (A.3) and (A.5) can be substituted into Eq. (A.2), which, after collection of terms, can be expressed as

$$\left(\frac{2k}{m_1} - \omega^2 \right) X_1 - \frac{k}{m_1} X_2 = 0 \tag{A.6a}$$

$$-\frac{k}{m_1} X_1 + \left(\frac{2k}{m_2} - \omega^2 \right) X_2 = 0 \tag{A.6b}$$

Comparison of Eq. (A.6) with Eq. (A.1) indicates that at this point, the solution has been reduced to an eigenvalue problem. That is, we can determine values of the eigenvalue ω^2 that satisfy the equations. For a two-degree-of-freedom system such as Fig. A.1, there will be two such values. Each of these eigenvalues establishes a unique relationship between the unknowns X called an *eigenvector*. Section A.3 describes a simple approach to determine both the eigenvalues and eigenvectors. It also illustrates the physical significance of these quantities for the mass-spring system.

A.3 The Polynomial Method

As stated at the end of Section A.1, the *polynomial method* consists of expanding the determinant to generate the characteristic polynomial. The roots of this polynomial are the solutions for the eigenvalues. The following example illustrates how it can be used to determine both the eigenvalues and eigenvectors for the mass-spring system (Fig. A.1).

EXAMPLE A.1   The Polynomial Method

Problem Statement.   Evaluate the eigenvalues and the eigenvectors of Eq. (A.6) for the case where $m_1 = m_2 = 40$ kg and $k = 200$ N/m.

Solution.   Substituting the parameter values into Eq. (A.6) yields

$$(10 - \omega^2)X_1 - \qquad\qquad 5X_2 = 0$$
$$-5X_1 + (10 - \omega^2)X_2 = 0$$

The determinant of this system is

$$(\omega^2)^2 - 20\omega^2 + 75 = 0$$

which can be solved by the quadratic formula for $\omega^2 = 15$ and 5 s^{-2}. Therefore, the frequencies for the vibrations of the masses are $\omega = 3.873$ s^{-1} and 2.236 s^{-1}, respectively. These values can be used to determine the periods for the vibrations with Eq. (A.4). For the first mode, $T_p = 1.62$ s, and for the second, $T_p = 2.81$ s.

As stated in Section A.1, a unique set of values cannot be obtained for the unknown amplitudes X. However, their ratios can be specified by substituting the eigenvalues back

FIGURE A.2
The principal modes of vibration of two equal masses connected by three identical springs between fixed walls.

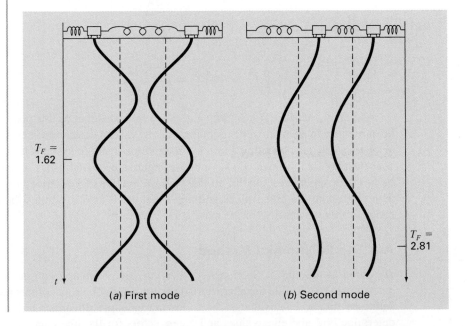

$T_F = 1.62$

$T_F = 2.81$

t

(a) First mode   (b) Second mode

into the equations. For example, for the first mode ($\omega^2 = 15 \text{ s}^{-2}$):

$$(10 - 15)X_1 - \quad\quad 5X_2 = 0$$
$$-5X_1 + (10 - 15)X_2 = 0$$

Thus, we conclude that $X_1 = -X_2$. In a similar fashion for the second mode ($\omega^2 = 5 \text{ s}^{-2}$), $X_1 = X_2$. These relationships are the eigenvectors.

This example provides valuable information regarding the behavior of the system in Fig. A.1. Aside from its period, we know that if the system is vibrating in the first mode, the eigenvector tells us that the amplitude of the second mass will be equal but of opposite sign to the amplitude of the first. As in Fig. A.2a, the masses vibrate apart and then together indefinitely.

In the second mode, the eigenvector specifies that the two masses have equal amplitudes at all times. Thus, as in Fig. A.2b, they vibrate back and forth in unison. We should note that the configuration of the amplitudes provides guidance on how to set their initial values to attain pure motion in either of the two modes. Any other configuration will lead to superposition of the modes.

We should recognize that MATLAB has built-in functions to facilitate the polynomial method. For Example A.1, the `poly` function can be used to generate the characteristic polynomial as in

```
>> A = [10 -5;-5 10];
>> p = poly(A)

p =
    1   -20    75
```

Then, the `roots` function can be employed to compute the eigenvalues:

```
>> roots(p)

ans =
    15
     5
```

A.4 The Power Method

The power method is an iterative approach that can be employed to determine the largest or *dominant eigenvalue*. With slight modification, it can also be employed to determine the smallest value. It has the additional benefit that the corresponding eigenvector is obtained as a by-product of the method.

To implement the power method, the system being analyzed is expressed in the form

$$[A]\{x\} = \lambda\{x\} \tag{A.7}$$

As illustrated by the following example, Eq. (A.7) forms the basis for an iterative solution technique that eventually yields the highest eigenvalue and its associated eigenvector.

EXAMPLE A.2   Power Method for Highest Eigenvalue

Problem Statement.   Using the same approach as in Section A.2, we can derive the following homogeneous set of equations for a three mass–four spring system between two fixed walls:

$$\left(\frac{2k}{m_1} - \omega^2\right)X_1 - \frac{k}{m_1}X_2 = 0$$

$$-\frac{k}{m_2}X_1 + \left(\frac{2k}{m_2} - \omega^2\right)X_2 - \frac{k}{m_2}X_3 = 0$$

$$-\frac{k}{m_3}X_2 + \left(\frac{2k}{m_3} - \omega^2\right)X_3 = 0$$

If all the masses $m = 1$ kg and all the spring constants $k = 20$ N/m, the system can be expressed in the matrix format of Eq. (A.1) as

$$\begin{bmatrix} 40 & -20 & 0 \\ -20 & 40 & -20 \\ 0 & -20 & 40 \end{bmatrix} - \lambda[I] = 0$$

where the eigenvalue λ is the square of the angular frequency ω^2. Employ the power method to determine the highest eigenvalue and its associated eigenvector.

Solution.   The system is first written in the form of Eq. (A.7):

$$40X_1 - 20X_2 = \lambda X_1$$
$$-20X_1 + 40X_2 - 20X_3 = \lambda X_2$$
$$-20X_2 + 40X_3 = \lambda X_3$$

At this point, we can make initial values of the X's and use the left-hand side to compute an eigenvalue and eigenvector. A good first choice is to assume that all the X's on the left-hand side of the equation are equal to one:

$$40(1) - 20(1) = 20$$
$$-20(1) + 40(1) - 20(1) = 0$$
$$-20(1) + 40(1) = 20$$

Next, the right-hand side is normalized by 20 to make the largest element equal to one:

$$\left\{\begin{array}{c} 20 \\ 0 \\ 20 \end{array}\right\} = 20 \left\{\begin{array}{c} 1 \\ 0 \\ 1 \end{array}\right\}$$

Thus, the normalization factor is our first estimate of the eigenvalue (20) and the corresponding eigenvector is $\lfloor 1 \quad 0 \quad 1 \rfloor^T$. This iteration can be expressed concisely in matrix form as

$$\begin{bmatrix} 40 & -20 & 0 \\ -20 & 40 & -20 \\ 0 & -20 & 40 \end{bmatrix}\left\{\begin{array}{c} 1 \\ 1 \\ 1 \end{array}\right\} = \left\{\begin{array}{c} 20 \\ 0 \\ 20 \end{array}\right\} = 20\left\{\begin{array}{c} 1 \\ 0 \\ 1 \end{array}\right\}$$

The next iteration consists of multiplying the matrix by $\lfloor 1 \quad 0 \quad 1 \rfloor^T$ to give

$$\begin{bmatrix} 40 & -20 & 0 \\ -20 & 40 & -20 \\ 0 & -20 & 40 \end{bmatrix} \begin{Bmatrix} 1 \\ 0 \\ 1 \end{Bmatrix} = \begin{Bmatrix} 40 \\ -40 \\ 40 \end{Bmatrix} = 40 \begin{Bmatrix} 1 \\ -1 \\ 1 \end{Bmatrix}$$

Therefore, the eigenvalue estimate for the second iteration is 40, which can be employed to determine the error estimate:

$$|\varepsilon_a| = \left| \frac{40 - 20}{40} \right| \times 100\% = 50\%$$

The process can then be repeated.

Third iteration:

$$\begin{bmatrix} 40 & -20 & 0 \\ -20 & 40 & -20 \\ 0 & -20 & 40 \end{bmatrix} \begin{Bmatrix} 1 \\ -1 \\ 1 \end{Bmatrix} = \begin{Bmatrix} 60 \\ -80 \\ 60 \end{Bmatrix} = -80 \begin{Bmatrix} -0.75 \\ 1 \\ -0.75 \end{Bmatrix}$$

where $|\varepsilon_a| = 150\%$ (which is high because of the sign change).

Fourth iteration:

$$\begin{bmatrix} 40 & -20 & 0 \\ -20 & 40 & -20 \\ 0 & -20 & 40 \end{bmatrix} \begin{Bmatrix} -0.75 \\ 1 \\ -0.75 \end{Bmatrix} = \begin{Bmatrix} -50 \\ 70 \\ -50 \end{Bmatrix} = 70 \begin{Bmatrix} -0.71429 \\ 1 \\ -0.71429 \end{Bmatrix}$$

where $|\varepsilon_a| = 214\%$ (which is high because of the sign change).

Fifth iteration:

$$\begin{bmatrix} 40 & -20 & 0 \\ -20 & 40 & -20 \\ 0 & -20 & 40 \end{bmatrix} \begin{Bmatrix} -0.71429 \\ 1 \\ -0.71429 \end{Bmatrix} = \begin{Bmatrix} -48.5714 \\ 68.5714 \\ -48.5714 \end{Bmatrix} = 68.5714 \begin{Bmatrix} -0.70833 \\ 1 \\ -0.70833 \end{Bmatrix}$$

where $|\varepsilon_a| = 2.08\%$.

Thus, the eigenvalue is converging. After several more iterations, it stabilizes on a value of 68.28427 with a corresponding eigenvector of $\lfloor -0.707107 \quad 1 \quad -0.707107 \rfloor^T$.

Note that there are some instances where the power method will converge to the second-largest eigenvalue instead of to the largest. James, Smith, and Wolford (1985) provide an illustration of such a case. Other special cases are discussed in Fadeev and Fadeeva (1963).

In addition, there are sometimes cases where we are interested in determining the smallest eigenvalue. This can be done by applying the power method to the matrix inverse of $[A]$. For this case, the power method will converge on the largest value of $1/\lambda$—in other words, the smallest value of λ. An application to find the smallest eigenvalue will be left as a problem exercise.

Finally, after finding the largest eigenvalue, it is possible to determine the next highest by replacing the original matrix by one that includes only the remaining eigenvalues. The process of removing the largest known eigenvalue is called *deflation*.

We should mention that although the power method can be used to locate intermediate values, better methods are available for cases where we need to determine all the eigenvalues as described in Section A.5. Thus, the power method is primarily used when we want to locate the largest or the smallest eigenvalue.

A.5 MATLAB Function: `eig`

As might be expected, MATLAB has powerful and robust capabilities for evaluating eigenvalues and eigenvectors. The function `eig`, which is used for this purpose, can be used to generate a vector of the eigenvalues as in

```
>> e = eig(A)
```

where `e` is a vector containing the eigenvalues of a square matrix `A`. Alternatively, it can be invoked as

```
>> [V,D] = eig(A)
```

where `D` is a diagonal matrix of the eigenvalues and `V` is a full matrix whose columns are the corresponding eigenvectors.

EXAMPLE A.3   Use of MATLAB to Determine Eigenvalues and Eigenvectors

Problem Statement.   Use MATLAB to determine all the eigenvalues and eigenvectors for the system described in Example A.2.

Solution.   Recall that the matrix to be analyzed is

$$\begin{bmatrix} 40 & -20 & 0 \\ -20 & 40 & -20 \\ 0 & -20 & 40 \end{bmatrix}$$

The matrix can be entered as

```
>> A = [40 -20 0;-20 40 -20;0 -20 40];
```

If we just desire the eigenvalues we can enter

```
>> e = eig(A)

e =
   11.7157
   40.0000
   68.2843
```

Notice that the highest eigenvalue (68.2843) is consistent with the value previously determined with the power method in Example A.2.

If we want both the eigenvalues and eigenvectors, we can enter

```
>> [v,d] = eig(A)

v =
    0.5000   -0.7071   -0.5000
    0.7071   -0.0000    0.7071
    0.5000    0.7071   -0.5000

d =
   11.7157         0         0
         0   40.0000         0
         0         0   68.2843
```

Again, although the results are scaled differently, the eigenvector corresponding to the highest eigenvalue $\lfloor -0.5 \quad 0.7071 \quad -0.5 \rfloor^T$ is consistent with the value previously determined with the power method in Example A.2: $\lfloor -0.707107 \quad 1 \quad -0.707107 \rfloor^T$.

A.6 CASE STUDY   EIGENVALUES AND EARTHQUAKES

Background. Engineers and scientists use mass-spring models to gain insight into the dynamics of structures under the influence of disturbances such as earthquakes. Figure A.3 shows such a representation for a three-story building. Each floor mass is represented by m_i, and each floor stiffness is represented by k_i for $i = 1$ to 3.

For this case, the analysis is limited to horizontal motion of the structure as it is subjected to horizontal base motion due to earthquakes. Using the same approach as developed in Section A.2, dynamic force balances can be developed for this system as

$$\left(\frac{k_1 + k_2}{m_1} - \omega_n^2 \right) X_1 - \frac{k_2}{m_1} X_2 = 0$$

$$-\frac{k_2}{m_2} X_1 + \left(\frac{k_2 + k_3}{m_2} - \omega_n^2 \right) X_2 - \frac{k_3}{m_2} X_3 = 0$$

$$- \frac{k_3}{m_3} X_2 + \left(\frac{k_3}{m_3} - \omega_n^2 \right) X_3 = 0$$

where X_i represents horizontal floor translations (m), and ω_n is the *natural,* or *resonant, frequency* (radians/s). The resonant frequency can be expressed in Hertz (cycles/s) by dividing it by 2π radians/cycle.

Use MATLAB to determine the eigenvalues and eigenvectors for this system. Graphically represent the modes of vibration for the structure by displaying the amplitudes versus height for each of the eigenvectors. Normalize the amplitudes so that the translation of the third floor is one.

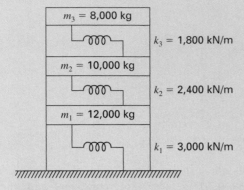

FIGURE A.3

Solution.   The parameters can be substituted into the force balances to give

$$\left(450 - \omega_n^2\right) X_1 - \qquad 200 X_2 \qquad\qquad = 0$$
$$-240 X_1 + \left(420 - \omega_n^2\right) X_2 - \qquad 180 X_3 = 0$$
$$- \qquad 225 X_2 + \left(225 - \omega_n^2\right) X_3 = 0$$

A MATLAB session can be conducted to evaluate the eigenvalues and eigenvectors as

```
>> A=[450 -200 0;-240 420 -180;0 -225 225];
>> [v, d]=eig(A)

v =
  -0.5879  -0.6344   0.2913
   0.7307  -0.3506   0.5725
  -0.3471   0.6890   0.7664

d =
  698.5982        0        0
        0  339.4779        0
        0        0  56.9239
```

Therefore, the eigenvalues are 698.6, 339.5, and 56.92, and the resonant frequencies in Hz are

```
>> wn=sqrt(diag(d))'/2/pi

w =
4.2066  2.9324  1.2008
```

The corresponding eigenvectors are (normalizing so that the amplitude for the third floor is one)

$$\left\{ \begin{array}{c} 1.693748 \\ -2.10516 \\ 1 \end{array} \right\} \quad \left\{ \begin{array}{c} -0.92075 \\ -0.50885 \\ 1 \end{array} \right\} \quad \left\{ \begin{array}{c} 0.380089 \\ 0.746999 \\ 1 \end{array} \right\}$$

A.6 CASE STUDY continued

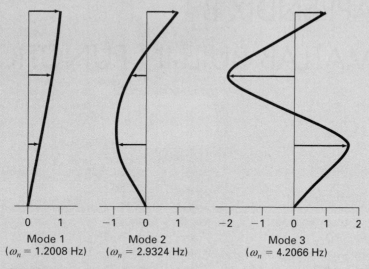

Mode 1
($\omega_n = 1.2008$ Hz)

Mode 2
($\omega_n = 2.9324$ Hz)

Mode 3
($\omega_n = 4.2066$ Hz)

FIGURE A.4

A graph can be made showing the three modes (Fig. A.4). Note that we have ordered them from the lowest to the highest natural frequency as is customary in structural engineering.

Natural frequencies and mode shapes are characteristics of structures in terms of their tendencies to resonate at these frequencies. The frequency content of an earthquake typically has the most energy between 0 to 20 Hz and is influenced by the earthquake magnitude, the epicentral distance, and other factors. Rather than a single frequency, they contain a spectrum of all frequencies with varying amplitudes. Buildings are more receptive to vibration at their lower modes of vibrations due to their simpler deformed shapes and requiring less strain energy to deform in the lower modes. When these amplitudes coincide with the natural frequencies of buildings, large dynamic responses are induced creating large stresses and strains in the structure's beams, columns, and foundations. Based on analyses like the one in this case study, structural engineers can more wisely design buildings to withstand earthquakes with a good factor of safety.

APPENDIX B
MATLAB BUILT-IN FUNCTIONS

abs, 30
acos, 30
ascii, 51
axis square, 35
beep, 62
besselj, 356
ceil, 31, 180
chol, 245
clabel, 180, 466
clear, 50
cond, 257, 338
contour, 180, 466
conv, 157
cumtrapz, 414
dblquad, 418
deconv, 156
det, 217
diag, 574
diff, 413, 460, 462
disp, 47
double, 51
eig, 572
elfun, 30
eps, 90
erf, 445
error, 52
exp, 30
eye, 202
factorial, 40, 59
fix, 365
floor, 31, 180
fminbnd, 178

fminsearch, 181, 326
format bank, 23
format compact, 21
format long, 23
format long e, 23
format long eng, 23
format loose, 21
format short, 23
format short e, 23
format short eng, 23
fplot, 67
fprintf, 48
fzero, 151
gradient, 463
grid, 33
help, 30
help elfun, 30
hist, 291
hold off, 34
hold on, 34
humps, 78, 440
inline, 67
input, 47
interp1, 376
interp2, 381
interp3, 382
inv, 201, 203
isempty, 127
legend, 385, 415
length, 32
linspace, 26
load, 50

log, 30
log10, 30
log2, 30, 126
loglog, 40, 106
logspace, 26
lookfor, 36, 44
lu, 242
max, 31, 226, 290
mean, 31, 290
median, 290
mesh, 62
meshgrid, 180, 466
min, 31, 290
mode, 290
nargin, 57
norm, 257
ode113, 517
ode15s, 529
ode23, 516
ode23s, 529
ode23t, 529
ode23tb, 529
ode45, 517, 546
odeset, 519
ones, 25
optimset, 153, 179, 326
pause, 62
pchip, 374, 377
pi, 23
plot, 33
plot3, 35, 508
poly, 156, 569

polyfit, 307, 325, 339
polyval, 156, 307, 339
prod, 31
quad, 439
quadl, 439
quiver, 465
realmax, 90
realmin, 90
roots, 155, 569
round, 31
save, 50
semilogy, 40
set, 385

sign, 55
sin, 30
size, 202
sort, 31
spline, 374
sqrt, 30
sqrtm, 31
std, 290
subplot, 34
sum, 31, 246
surfc, 180
tan, 30
tanh, 7

tic, 62
title, 33
toc, 62
trapz, 414, 422
triplequad, 418
var, 290
varargin, 70
who, 25
whos, 25
xlabel, 33
ylabel, 33
zeros, 25
zlabel, 180

APPENDIX C
MATLAB M-FILE FUNCTIONS

M-file Name	Description	Page
bisect	Root location with bisection	127
eulode	Integration of a single ordinary differential equation with Euler's method	486
GaussNaive	Solving linear systems with Gauss elimination without pivoting	222
GaussPivot	Solving linear systems with Gauss elimination with partial pivoting	227
GaussSeidel	Solving linear systems with the Gauss-Seidel method	269
goldmin	Minimum of one-dimensional function with golden-section search	176
incsearch	Root location with an incremental search	120
Lagrange	Interpolation with the Lagrange polynomial	349
linregr	Fitting a straight line with linear regression	306
natspline	Cubic spline with natural end conditions	383
Newtint	Interpolation with the Newton polynomial	346
newtmult	Root location for nonlinear systems of equations	276
newtraph	Root location with the Newton-Raphson method	149
rk4sys	Integration of system of ODEs with 4th-order RK method	502
romberg	Integration of a function with Romberg integration	432
TableLook	Table lookup with linear interpolation	364
trap	Integration of a function with the composite trapezoidal rule	404
trapuneq	Integration of unequispaced data with the trapezoidal rule	413
Tridiag	Solving tridiagonal linear systems	229

BIBLIOGRAPHY

Anscombe, F. J., "Graphs in Statistical Analysis," *Am. Stat.,* *27*(1):17–21, 1973.

Bogacki, P. and L. F. Shampine, "A 3(2) Pair of Runge-Kutta Formulas," *Appl. Math. Letters, 2*(1989):1–9, 1989.

Butcher, J. C., "On Runge-Kutta Processes of Higher Order," *J. Austral. Math. Soc., 4*:179, 1964.

Carnahan, B., H. A. Luther, and J. O. Wilkes, *Applied Numerical Methods,* Wiley, New York, 1969.

Chapra, S. C. and R. P. Canale, *Numerical Methods for Engineers,* 5th ed., McGraw-Hill, New York, 2006.

Dormand, J. and P. A. Prince, "A Family of Embedded Runge-Kutta Formulae," *J. Comput. and Appl. Math., 6*:19–26, 1980.

Draper, N. R. and H. Smith, *Applied Regression Analysis,* 2d ed., Wiley, New York, 1981.

Fadeev, D. K. and V. N. Fadeeva, *Computational Methods of Linear Algebra,* Freeman, San Francisco, 1963.

Gander, W. and W. Gautschi, *Adaptive Quadrature– Revisited, BIT Num. Math., 40*:84–101, 2000.

Gerald, C. F. and P. O. Wheatley, *Applied Numerical Analysis,* 3d ed., Addison-Wesley, Reading, MA, 1989.

Hanselman, D. and B. Littlefield, *Mastering MATLAB 7,* Prentice Hall, Upper Saddle River, NJ, 2005.

Hornbeck, R. W., *Numerical Methods,* Quantum, New York, 1975.

James, M. L., G. M. Smith, and J. C. Wolford, *Applied Numerical Methods for Digital Computations with FORTRAN and CSMP,* 3d ed., Harper & Row, New York, 1985.

Moore, H., *MATLAB for Engineers,* Prentice Hall, Upper Saddle River, NJ, 2007.

Ortega, J. M., *Numerical Analysis–A Second Course,* Academic Press, New York, 1972.

Palm, W. J. III, *Introduction to MATLAB 7 for Engineers,* McGraw-Hill, New York, 2005.

Ralston, A., "Runge-Kutta Methods with Minimum Error Bounds," *Match. Comp., 16*:431, 1962.

Ralston, A. and P. Rabinowitz, *A First Course in Numerical Analysis,* 2d ed., McGraw-Hill, New York, 1978.

Recktenwald, G., *Numerical Methods with MATLAB,* Prentice Hall, Englewood Cliffs, NJ, 2000.

Scarborough, I. B., *Numerical Mathematical Analysis,* 6th ed., Johns Hopkins Press, Baltimore, MD, 1966.

Scarborough, J. B., *Numerical Mathematical Analysis,* 6th ed., Johns Hopkins Press, Baltimore, MD, 1966.

Shampine, L. F., *Numerical Solution of Ordinary Differential Equations,* Chapman & Hall, New York, 1994.

White, F. M., *Fluid Mechanics.* McGraw-Hill, New York, 1999.

INDEX

A

Accuracy, 12, 80–81
Adams-Bashforth-Moulton solver, 517
Adaptive integration, 391
Adaptive methods for ODEs, 514–525.
    see also Stiff systems
  error estimates, 524–525
  MATLAB functions for nonstiff
    systems, 516–520
  multistep methods, 477, 481, 521–525
  non-self-starting Heun method, 478,
    521–524
  ode23 function, 516–517
  ode45 function, 517–519, 530, 534–535,
    546–547
  ode113 function, 517
  odeset function, 519–520
  Pliny's intermittent fountain, 532–536
  Runge-Kutta methods, 478, 514–520
  step-size control, 515
  for systems of ODEs, 518–519
Adaptive quadrature, 427, 439–440
Addition, 27, 90
  flops, naive Gauss elimination, 223–224
  of large numbers with small numbers, 92
  of two matrices, 197, 200
Allosteric enzymes, 308, 311
Amplification factor, 485
Analytical (closed-form) solutions, 9
  for heated rod, 544–545
  optimization by root location, 168–169
& (And) logical condition, 53
Anonymous functions, 66–67, 70
Anscombe's data sets, 299–300
Approximations. *see* Truncation errors
Areal integral, 396
Arithmetic manipulations, computer
    numbers, 90–92
Arithmetic mean, 286, 290
Array operations, 29
Arrays, 24–25, 31. *see also* Matrix/Matrices
Arrow keys, up (↑), 30
ASCII files, 50–51
Assignment, 22–27
  arrays, vectors, and matrices, 24–25
  colon operator, 25–26

linspace function, 26, 375
logspace function, 26–27
  scalars, 22–23
  of variable names, automatic, 22
Associative properties, 197, 198
Atmospheric fluid dynamics, 504–509
Augmentation, of a matrix, 199, 202
Automatic assignment of variable names, 22
Avogadro's number, 87
axis square command, 35

B

Backslash operator, 27, 203, 246–247, 325
Back substitution, 219, 220, 224
Backward Euler's method, 527–529
Backward finite-difference approximations,
    100–103, 455
Banded matrix, 197, 227–232
Base-2 (binary) number system, 85, 88–90
Base-8 (octal) number system, 85
Base-10 (decimal) number system, 85
beep command, 62–63
Best fit criteria, 292–294
Bias, 80
Bilinear interpolation, 379–381
Binary (base-2) number system, 85, 88–90
Binary digits (bits), 84, 85
Binary files, 50
Binary search, 364–365
Bins, 289
Bisection, 112, 122–128
  bisect function, 127–128
  error estimates for, 124–127
  false position vs., 130–131
Bits, 84, 85
Blank lines, 21n
Blunders, 108
Boolean variables, 533
Boole's rule, 411, 497
Boundary-value problems, 477, 540–564
  described, 541–542
  in engineering/science, 542–545
  finite-difference methods, 552–559
    with derivative boundary conditions, 554–557
    for nonlinear ODEs, 557–559
  initial-value problems vs., 541–542

  shooting method, 545–552
    with derivative boundary conditions,
      548–550
    for linear ODEs, 546–548
    for nonlinear ODEs, 550–552
Bracketing methods, 112, 114–138
  bisection. *see* Bisection
  defined, 119
  entering arrays in command mode, 24
  false position, 128–131
  graphical methods, 116–117
  greenhouse gases/rainwater example, 132–135
  incremental search, 119–122
  initial guesses and, 117–122
Brent's methods
  optimization, 178
  root finding, 113
BS23 algorithm, 516
Built-in functions, 30–32, 576. *see also*
    MATLAB functions
Bungee jumper problem
  analytical solution, 7–10
  exploratory data analysis, 37–39
  function function, 68–74
  MATLAB solution, 204–205, 531–532
  matrix inverse analysis, 252–253
  Newton-Raphson solution, 148–149
  numerical solution, 10–12
Butcher's fifth-order RK method, 497–498
Butterfly effect, 508

C

Calculator mode, 2, 21
Calculus, 449
Calls, of other functions, 46–47
Carriage return (ENTER key), 24, 30
Case-sensitivity
  of function names, 44
  of variable names, 22
ceil function, 31
Centered finite-difference approximations,
    100–103, 456
Chaotic solutions, 507
Characteristic polynomial, 566
Characteristic value, 566. *see also* Eigenvalues
Charge, conservation of, 13, 14, 205

580

Chemical engineering, conservation laws for, 13, 14
Chemical reactions, 277–279
Cholesky factorization, 244–246
chol function, 245–246
Circuits
  conservation laws for, 13, 14
  currents and voltages in, 205–208
Civil engineering, conservation laws for, 13, 14
Clamped end condition, 373–374
Classical fourth-order RK methods, 495–498, 500–501
clear command, 50
Closed-form solutions, 9, 391
  Newton-Cotes formulas, 398, 411
Closed integration, 391
Closed integration formulas, 398, 411
Coefficient of determination, 298
Coefficient of variation, 287
Colebrook equation, 157–158
Colon operator, 25–26
Colors, specifiers for, 34
Columns, matrix, 195
Column-sum norm, 255
Column vectors, 24, 195, 203
Command window, 21, 43, 44
  anonymous function definition, 66
  nargin function and, 58
Commas, separating commands, 22
Commutative properties, 197, 198
Companion matrix, 155
Complete pivoting, 225
Complex quantities, 28
Composite integration formulas, 401
  Simpson's 1/3 rule, 407–409
  trapezoidal rule, 401–403
Computation effort, accuracy and, 12
Computer-aided calculations, 11
Computer mathematics, 1
Computer number representation, 84–90
  arithmetic manipulations of, 90–92
  floating-point format, 86–89
  IEEE double precision format, 89
  integers, 85–86
  number systems and, 85
  precision of, 89–90
  range of, 89
  signed magnitude method, 85–86
  2s complement technique, 86
cond function, 257–258
Conditional stability, 485
Condition number, matrices, 255–258
Conservation laws, 12–14
Constant of integration, 475–476
Constants, column vector of, 203
Constitutive laws, 451–452
Continuity condition, 366
Contour plots, 180, 466
Control codes, 49
Convergence, 139, 143, 147
  diagonal dominance and, 267
  slow, with Newton-Raphson, 145–146
Corrector equation, 488

Correlation coefficient, 298
Cramer's rule, 214–217
Critical point, phase-plane plots, 506
Ctrl+Break, 63
Ctrl+C, 63
Cubic splines, 359–360, 368–374
  clamped condition, 373–374
  derivation of, 369–373
  end conditions, 373–374
  natural, 372, 373
  not-a-knot condition, 373–374, 378
  piecewise cubic Hermite interpolation (pchip), 377–379
  spline function, 374–376
cumtrapz function, 414–415
Current, electrical
  current balance, 13, 14
  as dependent variable, 115
  Kirchhoff's current (point) rule, 205
  and voltages in circuits, 205–208
Curvature, 450
Curve fitting, 15, 16
  defined, 281
  in engineering/science, 281–283
  general linear least squares, 283, 322–324
  linear regression. *see* Linear regression
  nonlinear regression, 283, 326–327
  polynomial interpolation. *see* Polynomial interpolation
  polynomial regression. *see* Polynomial regression
  sinusoids, 328–331
  splines. *see* Splines
  statistics. *see* Statistics review
Curvilinear interpolation, 282

D

Danckwertz boundary condition, 561
D'Arcy's law, 452
Data analysis, exploratory, 37–39
Data distribution, 289
Data files, 50–51
Data uncertainty, errors and, 109
dblquad function, 418
Debug, Run, 43
Decimal (base-10) number system, 85
Decimal places, 23
Decisions, structured programming. *see* Structured programming
Default value, 57
Definite integrals, 475, 515
Deflation, 572
Degrees of freedom, 287
Dependent variables, 5, 70, 115, 473
Derivative mean-value theorem, 97–98
Derivatives, 10, 97–98, 389, 449–451. *see also* Numerical differentiation
Descent optimization methods, 181
Descriptive statistics, 286–291
Design problems, fundamental principles, 115
Determinants, 214–217
det function, 217
Devices, conservation laws for, 14

Diagonal dominance, 267, 558
Diagonal matrix, 196
Differential equations, 7, 15, 473
  first-order, 474
  higher-order, 474–475
  ordinary. *see* Ordinary differential equations (ODEs)
  partial, 474
  second-order, 474
Differentiation, 15, 16. *see also* Differential equations; Numerical differentiation
diff function, 383, 460–462
Digital computers, size/precision limits, 84
Direct optimization methods, 181
Dirichlet boundary condition, 548, 554
disp function, 47–48
display parameter, 153
Distributed (micro-) variable problems, 190–191
Divergence, 139. *see also* Convergence
Divided difference table, 344
Division, 27
  left, 27, 27n, 246–247, 325
  matrix, 199
  multiplication/division flops, 223, 224
Dominant eigenvalue, 569–572
Dot product of two vectors, 28
Double integral method, 416–418
Double precision, 51, 89
Drag coefficient, 7, 20–21, 37–39, 284–286
Dummy variable, 73

E

Earthquakes, 573
Echo printing, 22
Edit window, 21, 43
Eigenvalues, 155, 255, 565–575
  deflation, 572
  earthquakes and, 573–575
  eig function, 572–573
  largest, determining, 569–572
  mathematical background, 565–566
  physical background, 566–567
  poly function, 569
  polynomial method, 566, 567–569
  power method, 569–572
  roots function, 569
  smallest, determining, 571
  spectral norm, 255
Eigenvectors, 566, 567–568
Electrical engineering, conservation laws for, 13, 14
Element-by-element operations, 29
Element of a matrix, 195
Elimination of unknowns, 217–218
Embedded RK methods, 516
End conditions, cubic splines, 373–374
end statements, 46n
Energy
  conservation of, 14, 206
  energy balance, 13, 115
Engineering practice
  boundary-value problems, 542–545
  curve fitting, 281–283

Engineering practice (*continued*)
  differentiation, 451–452
  eigenvalue problems, 565–567
  integration, 394–396, 412
  linear algebraic equations, 189–191
  roots of equations, 115–116
  table lookup, 364–365
ENTER key (carriage return), 24, 30
Enzyme kinetics, 307–312
Epilimnion, 382
eps function, 90
== (Equal), 53
Equilibrium and minimum potential
    energy, 181–183
Error(s), 79–110
  accuracy and precision, 80–81
  blunders, 108
  data uncertainty and, 109
  defined, 81–84
  error analysis
    for Euler's method, 483–485
    matrix inversion, 253–258
    and system condition, 253–258
  error function, 52
  error messages, 29, 47
  estimates of
    for iterative methods, 83–84, 141
    non-self-starting Heun method, 524–525
  linear least-squares regression and,
      296–300
  local, 483–485, 491, 524–525
  model errors, 108–109
  numerical
    control of, 107–108
    total, 103–108
  numerical differentiation and, 104–107,
      458–459
  quantification of, 81–84
  relative, 82
  round-off. *see* Round-off errors
  standard error of the estimate, 297
  system condition and, 253–258
  true error, 81
  true fractional relative error, 82
  truncation. *see* Truncation errors
Euclidean norm, 254
Euler-Cauchy method, 481
Euler's method, 10, 477, 481–487
  error analysis for, 483–485
  eulode function, 486–487
  explicit, 527–529
  implicit, 527–529
  improvements of, 487–493
    Heun method, 488–492
    midpoint method, 492–493
  stability of, 485–486
  for systems of ODEs, 498–500
Explicit expression, in formulas, 115
  Euler's method, 527–529
Exploratory data analysis, 37–39
Exponential model, 300
Exponentiation, 27
External stimuli, 251
Extrapolation, 351–353
eye function, 202

F
factorial function, 59n
False position, 112, 128–131
  bisection vs., 130–131
  formula, 128–129
fhandle, 66
Fick's law, 452
50th percentile, 286
File, Import Data, 51
File, New, M-file, 43
File management, 50–51
Finish value, loops, 59
Finite-difference methods, 10, 99–103, 478.
      see also Numerical differentiation
  boundary-value problems, 552–559
First-order approximation, 94
First-order differential equations, 474
First-order method, 485
First-order splines, 362–363
Fixed boundary condition, 548, 554.
      see also Dirichlet boundary condition
Fixed-point iteration, 113, 140–144, 270–272
Floating-point operations (flops), 90, 222–225,
      229, 242
Floating-point representation, 86–89
floor function, 31
fminbnd function, 113, 170, 178–179
fminsearch function, 113, 181, 326–327
Force
  as dependent variable, 115
  force balance, 13, 14, 20, 115
Forcing functions, 5, 70, 251
for . . . end structure, 59–60
for loop, 59–60
format bank command, 23
Format codes, 48–49
Format commands, routine, 23
format compact command, 21n
format function, 48
format long command, 23, 91
format loose command, 21n
format short command, 23
Formulas, evaluating, 31–32
Forward elimination of unknowns, 219–220
Forward finite-difference approximations,
      99–101, 102–103, 454
Fourier analysis, 322
Fourier's law, 451, 452, 458, 543, 549
Fourth-order RK methods, 495–498, 500–504
fplot function, 67
fprintf function, 48–49
Friction factor, 157
Frobenius norm, 255
Functions. *see also* MATLAB functions; M-files
  function files, 44–46
  function functions, 66, 67–71
    generic, 69–70, 74
  function handle, 66
  passing parameters, 70
fzero function, 113, 151–154, 160, 178

G
Gauss elimination, 191, 212–235
  GaussPivot function, 226–227
  heated rod example, 229–232

as *LU* factorization, 238–242
  naive. *see* Naive Gauss elimination
  pivoting, 225–227
  Tridiag function, 229, 232
  tridiagonal systems, 227–229, 232
Gauss-Legendre formulas, 433
  three-point, 438
  two-point, 435–438
Gauss quadrature, 391, 427, 432–439
  Gauss-Legendre formulas, 433, 435–438
  higher-point formulas, 438–439
  undetermined coefficients method, 433–435
Gauss-Seidel method, 192, 264–269
  boundary-value problems, 558
  convergence and diagonal dominance, 267
  GaussSeidel function, 268, 269
  relaxation, 268–269
Gear backward differentiation, 529
General linear least squares, 283, 322–324
Generic functions, 69–70, 74
Global optimum, 170
Global truncation error, 484–485
Global variables, 534
Golden ratio, 171
Golden-section search, 113, 171–176
Gradients, 451
  gradient function, 463–467
  optimization methods, 181
Graphical methods, 32–36
  customizing graphs, 33
  graphics window, 21
  hold on/hold off commands, 34
  panes, 34–35
  plot function. *see plot* function
  quiver function, 465–466
  for root location, 116–117
  solving small numbers of equations, 213–214
  subplot function, 34–35
  to visualize two-dimensional functions, 180
> (Greater Than), 53
>= (Greater than or equal to), 53
grid function, 33

H
Half-saturation constant, 307
Heat balance, 115
Heat transfer
  boundary-value problems, 543–545
  Gauss elimination, 229–232
  splines, 382–386
Helix, plot of, 35–36
Help, online, 30
Helpcomments, 44
help elfun function, 30
help function, 30, 36, 44, 45
Heun method, 478, 488–492
  without iteration, 495
  non-self-starting, 478, 521–524
High-accuracy finite-difference formulas, 391,
      452–455
Higher-order differential equations, 474–475
Higher-order polynomials
  ill-conditioned systems, 353–355
  interpolation, dangers of, 353–355
  splines vs., 360–361

Higher-point formulas, Gauss quadrature, 438–439
Hilbert matrix, 256
hist function, 291
Histograms, 289, 291
hold off command, 34
hold on command, 34
Homogeneous systems, 565
H1 line, 44
Hooke's law, 181, 452
humps function, 440
Hypolimnion, 382
Hypothesis testing, 282–283

I
Identity matrix, 196, 199, 202
IEEE double precision format, 89
if structure, 51–52, 56, 73
if . . . else structure, 54, 56
if . . . break statement, 62
if . . . elseif structure, 54–55, 56
Ill-conditioned systems, 213, 214, 253–254
    backslash operator and, 325
    higher-order polynomials, 353–355
    ODEs, 485
    QR factorization and, 325
Implicit expression, in formulas, 115
    Euler's method, 527–529
Import wizard, 51
Imprecision, 80
Inaccuracy, 80, 486
Increment function, 481, 493
incsearch function, 119–122
Indefinite integrals, 475
Indentation, nesting and, 63–66
Independent variables, 5, 70, 115, 473
Index variable, loops, 59
Indoor air pollution, 258–261
Infinite loop, 62, 63
Initial guesses
    bracketing methods and, 117–122
    defined, 119
    incremental search and, 119–122
inline functions, 67. *see also* Anonymous
        functions
Inner product (dot product)
    round-off error and, 92
    of two vectors, 28
input function, 47, 48
Input/output in programming, 47–49
Integer representation, 85–86
Integration, 15, 16. *see also* Numerical
        integration formulas
    closed, 391, 398, 411
    defined, 393–394
    differentiation vs., 389–390
    in engineering/science, 394–396, 412
    open, 391, 398, 416
Interfacing files, with other programs, 50–51
Intermittent fountain, 532–536
Interpolation, 15, 16, 281, 283, 336–339
    interp1 function, 376–379
    interp2 function, 381
    interp3 function, 381
    inverse, 152, 350–351
    inverse quadratic, 113, 152

linear. *see* Linear interpolation
    piecewise, 374–381. *see also* Splines.
    polynomial. *see* Polynomial interpolation
    spline. *see* Splines
Inverse interpolation, 350–351
Inverse quadratic interpolation, 152
Inverse of matrices. *see* Matrix/Matrices
inv function, 201–202
Iterative methods, 82–84, 264–279
    fixed-point iteration, 113, 140–144, 270–272
    for chemical reactions, 277–279
    error estimates for, 83–84, 141
    Gauss-Seidel method, 264–269
    linear systems, 264–269
    Newton-Raphson method, 272–276
    nonlinear systems, 270–279
    power method, for eigenvalues, 569–572
    successive substitution, 270–272

J
Jacobian, of the system, 273, 278
Jacobian matrix, 274–275
Jacobi method, 266–267

K
Kirchhoff's laws, 115, 205–207
Knots, 363

L
Lagrange function, 349–350
Lagrange interpolating polynomial, 283,
        347–350, 457
Large computations, round-off errors and, 91
Least squares criterion, 294. *see also* Linear
        least-squares regression
Left division, 27, 27n, 203, 246–247, 325
length function, 32
< (Less Than), 53
<= (Less than or equal to), 53
Linear algebraic equations, 15, 16, 204–205
    currents/voltages in circuits, 205–208
    defined, 189
    in engineering/science, 189–191
    Gauss elimination solutions for. *see* Gauss
        elimination
    Gauss-Seidel solutions for, 264–269
    LU factorization, 236–248
    MATLAB matrix manipulation, 200–202
    MATLAB solutions for, 203–205
    matrices and, 193–211
    matrix algebra overview, 194–203
    matrix representation, 202–203
    matrix notation, 195–197
    matrix operating rules, 197–202
    polynomial coefficients with, 337–338
    small sets of, solving, 213–218
        Cramer's rule, 214–217
        determinants, 214–217
        elimination of unknowns, 217–218
        graphical method, 213–214
Linear convergence, 141
Linear interpolation, 336, 376
    bracketing method, 128–131
    curve fitting, 282
    Newton's interpolating polynomial, 339–341

Linear Lagrange interpolating polynomial, 347
Linear least-squares regression, 281, 282,
        292–300
    Anscombe's data sets, 299–300
    best fit criteria, 292–294
    enzyme kinetics example, 307–312
    error quantification, 296–300
    general comments on, 304
    general linear least squares, 283, 322–324
    least-squares fit of a straight line, 294–296
    least-squares fit of sinusoids, 328–331
    linregr function, 305–306, 309–310
    polyfit function, 304, 307
    polyval function, 307
Linear regression, 283, 284–315
    computer applications, 304–312
    least-squares. *see* Linear least-squares
        regression
    linregr function, 305–306, 309–310
    multiple linear regression, 283, 320–322
    for nonlinear relationships, 300–304,
        307–312
    statistics review, 286–291
Linear splines, 361–365
    first-order, 362–363
    table lookup, 364–365
Line types, specifiers for, 34
linspace function, 26, 69, 375
load command, 50, 51
Lobatto quadrature, 439
Local error
    Heun method, 491
    truncation error, 524–525
        with Euler's method, 483–485
Local optimum, 170
Local variables, 46
log function, 30
log2 function, 126n
log10 function, 305–306
Logical conditions, 53–54, 61–62
Logical variables, 533
loglog function, 40
logm function, 25
logspace function, 26–27
lookfor function, 36, 44, 45–46
Loops. *see* Structured programming
Lorenz equations, 504–509
Lotka-Volterra equations, 504–509, 518–519
Lower triangular matrix, 197
LU decomposition, 236n. *see also*
        LU factorization
LU factorization, 191–192, 236–248
    Cholesky factorization, 244–246
    chol function, 245–246
    Gauss elimination as, 238–242
    lu function, 242–243
    MATLAB left division, 246–247
    steps in, 238
    substitution steps, 241–242
Lumped (macro-) variable problems, 190–191

M
Machine epsilon, 88–90
Machines, conservation laws for, 13–14
Maclaurin series expansion, 83

Main diagonal, of a matrix, 196
Main functions, 47
Mantissa, 88–89
Mass
  conservation of, 13, 14
  as dependent variable, 115
  mass balance, 13, 14, 115, 258–259
MAT-files, 50
Mathematical models, 5–12
  analytical (closed-form) solutions, 9
  characteristics of, 6
  defined, 5
  numerical methods, 9
Mathematical operations, 27–30
MathWorks, Inc., 36
MATLAB, 20–41. *see also* MATLAB functions;
    M-files
  assignment. *see* Assignment
  built-in functions, 30–32, 576
  defined, 21
  descriptive statistics in, 290–291
  eigenvalues/eigenvectors, 572–573
  environment, 21–22
  graphics with, 32–36
  linear algebraic equations, 203–205
  mathematical operations, 27–30
  matrix manipulation, 200–202
  multidimensional interpolation, 381
  nonlinear regression, 326–327
  norms in, 257–258
  numerical differentiation, 460–464
  other resources, 36
  piecewise interpolation, 374–381
  polynomial regression, 323–324
  polynomial roots, 155–157
  predator-prey model, 518–519
  primary windows, 21
  programming with. *see* Programming with
    MATLAB
MATLAB functions, 576. *see also* M-files
  abs, 25
  acos, 25
  besselj, 356
  built-in, 30–32, 576
  ceil, 31
  chol, 245–246
  cond, 257–258
  conv, 157
  cumtrapz, 414–415
  dblquad, 418
  deconv, 156
  det, 217
  diff, 383, 460–462
  disp, 47–48
  eig, 572–573
  elfun, 30
  eps, 90
  erf, 445
  error, 52
  exp, 25
  eye, 202
  factorial, 59n
  floor, 31
  fminbnd, 113, 178–179

  fminsearch, 113, 181, 326–327
  format, 48
  fplot, 67
  fprintf, 48–49
  fzero, 113, 151–154, 160, 178
  gradient, 463–467
  grid, 33
  help, 30, 36, 44, 45
  help elfun, 30
  hist, 291
  humps, 440
  inline, 67
  input, 47, 48
  interp1, 376–379
  interp2, 381
  interp3, 381
  inv, 201–202
  length, 32
  linspace, 26, 69, 375
  log, 30
  log2, 126n
  log10, 305–306
  loglog, 40
  logspace, 26–27
  lookfor, 36, 44, 45–46
  lu, 242–243
  max, 31, 226
  mean, 31, 290
  median, 290
  min, 31
  mode, 291
  nargin, 57–58
  for nonstiff systems, 516–520
  norm, 257–258
  ode15s, 529
  ode23, 516–517
  ode23s, 529, 530
  ode23t, 529
  ode23tb, 529
  ode45, 517–519, 530, 534–535, 546–547
  ode113, 517
  odeset, 519–520
  ones, 25
  optimset, 153–154, 160, 326
  pchip, 377–379
  pi, 23
  plot, 32–36, 519
  plot3, 35–36, 508–509
  poly, 156, 569
  polyfit, 304, 307, 325, 339, 353–354
  polyval, 307, 339, 353–354
  prod, 31
  quad, 439–440
  quadl, 439–440
  quiver, 465–466
  realmax, 90
  realmin, 90
  roots, 155–157, 569
  round, 31
  semilogy, 40
  sign, 480n
  sin, 25
  size, 202
  sort, 31

  spline, 374–376
  sqrt, 25, 32
  sqrtm, 30–31
  for stiff systems, 529–531
  subplot, 34–35, 38
  sum, 31
  tanh, 7n, 25
  title, 33
  trapz, 414
  triplequad, 418
  who, 25
  whos, 25
  xlabel, 33
  ylabel, 33
  zeros, 25
  zlabel, 180
Matrix/Matrices, 24–25
  algebraic equation representation, 202–203
  built-in functions for, 31
  of coefficients, 203
  condition number, 255–258
  eigenvalues of, 155
  inversion, 192, 199, 201–202, 249–263
    calculating, 249–251
    cond function, 257–258
    error analysis and system condition, 253–258
    ill-conditioned systems, 253–254
    indoor air pollution example, 258–261
    inv function, 201–202
    matrix condition number, 255–258
    norm function, 257–258
    stimulus-response computations, 251–253
    vector and matrix norms, 254–255
  Jacobian, 274–275
  linear algebraic equations and, 193–211
  MATLAB manipulation of, 200–202
  matrix algebra overview, 194–203
  matrix division, 199
  matrix norms, 254–255, 257–258
  matrix notation, 195–197
  multiplication of, 29, 197–198, 201–202
  operating rules, 197–202
  Vandermonde, 338
  vector-matrix calculations, 28–29
max function, 31, 226
Maximum error, minimizing, 293
Maximum likelihood principle, 297
Mean, 286, 297
  for continuous data, 394, 395
  for discrete data, 394, 395
  mean function, 31, 290
Measures of spread, 287–289
Mechanical engineering, conservation
    laws for, 13, 14
Median, 286
median function, 290
Memory locations, 22
Mesh plots, 180
.m file extension, 43
M-files, 43–47, 577. *see also* MATLAB functions
  bisect, 127–128
  bungee jumper velocity example, 71–74
  end statements, 46n
  eulode, 486–487

function files, 44–46
function functions, 68–70
GaussNaive, 222
GaussPivot, 226–227
GaussSeidel, 268, 269
goldmin, 176
incsearch, 119–122
interactive, 48
Lagrange, 349–350
linregr, 305–306, 309–310
.m extension, 43
naming, 44
natspline, 383–386
newtint, 346–347
newtmult, 276, 278
newtraph, 148–149, 159
passing functions to. *see* Passing functions
    to M-files
rk4sys, 502–504, 535–536
romberg, 431–432
script files, 43–44
subfunctions, 46–47, 72
TableLook, 364–365
trap, 403–405
trapuneq, 413
Tridiag, 229, 232
Michaelis-Menten model, 307–312
Midpoint method, 478, 492–493
    Newton-Cotes open integration formulas, 416
    second-order RK methods, 495
Midtest loop, 62
min function, 31
Minimax criterion, 293
Minimum potential energy, equilibrium and,
    181–183
Minors, 215
Mixed operations, with scalars, 29
Modal class interval, 289
Mode, 286–287
mode function, 291
Model errors, 108–109
Modified secant method, 150–151
Moler, Cleve, 352
Momentum, conservation of, 13, 14
Monotonic, 462
Multidimensional interpolation, 379–381
Multimodal optima, 170
Multiple-application integration formulas. *see*
    Composite integration formulas
Multiple integrals, 416–418
Multiple linear regression, 283, 320–322
Multiple roots, 117
Multiplication, 27
    of matrices, 29, 197–198, 201–202
    multiplication/division flops, 223, 224
Multistep methods for ODEs, 477, 481,
    521–525
    error estimates, 524–525
    non-self-starting Heun method, 521–524

N

Naive Gauss elimination, 218–225
    back substitution, 219, 220, 224
    forward elimination of unknowns, 219–220

GaussNaive function, 222
    operation counting, 222–225
nargin function, 57–58, 127
natspline function, 383–386
Natural cubic splines, 372, 373, 383–386
Natural end condition, 369, 373
Natural frequency, 573
Nearest neighbor interpolation, 376, 377–378
Negation, 27
Nelder-Mead method, 181
Nesting and indentation, 63–66
Neumann boundary condition, 548, 555
newtint function, 346–347
newtmult function, 276, 278
Newton-Cotes formulas, 390, 396–422
    Boole's rule, 411, 497
    closed forms, 398, 411
    composite Simpson's 1/3 rule, 407–409
    higher-order, 411–412
    midpoint method, 416
    open forms, 398, 416
    Simpson's 1/3 rule, 390, 405–407, 411
    Simpson's 3/8 rule, 390, 409–411
    trapezoidal rule. *see* Trapezoidal rule
    truncation errors in, 411, 416
Newton linear-interpolation formula, 339–340
Newton-Raphson method, 113, 144–149, 159–160
    for bungee jumper problem, 148–149
    newtmult function, 276, 278
    Newton-Raphson formula, 144
    newtraph function, 148–149, 159
    for nonlinear systems, 272–279
    for slowly converging function, 145–146
Newton's interpolating polynomials, 283, 339–345
    general form, 343–345
    linear interpolation, 339–341
    newtint function, 346–347
    quadratic interpolation, 341–343
Newton's laws of motion, 115
Newton's Second Law, 5–6
Newton's viscosity law, 452
newtraph function, 148–149, 159
Nongradient optimization methods, 181
Nonhomogeneous systems, 565
Nonlinear systems
    iterative methods for. *see* Iterative methods
    linearization of, 300–304
        enzyme kinetics example, 307–312
        exponential model, 300
        power equation, 300, 302–304, 305–306
        saturation-growth-rate equation, 301
    Newton-Raphson method, 272–279
    ordinary differential equations (ODEs), 478,
        550–552, 557–559
    regression, nonlinear, 283, 326–327
Non-self-starting Heun method, 478, 521–524
Nonstiff systems, MATLAB functions for,
    516–520
Normal distribution, 289–290
Normal equations, 294, 323
Normalization, 86, 88–89, 220
Norms
    column-sum, 255
    defined, 254

Euclidean, 254
Frobenius, 255
in MATLAB, 257–258
matrix, 254–255
norm function, 257–258
row-sum, 255
spectral, 255
vector and matrix, 254–255
~ (Not), 53
~= (Not equal), 53
Not-a-knot end condition, 373–374, 378
nth-order approximation, 95
nth-order Taylor series expansion, 95
Number systems, 85
Numerical differentiation, 15, 16, 99–103, 391,
    448–471
    backward finite-difference, 100–103, 455
    centered finite-difference, 100–103, 456
    for data with errors, 458–459
    described, 449–451
    diff function, 383, 460–462
    in engineering/science, 451–452
    error analysis of, 104–107
    forward finite-difference, 99–101, 102–103, 454
    gradient function, 463–467
    high-accuracy formulas, 452–455
    integration vs., 389–390
    with MATLAB, 460–464
    notation for, 450n
    partial derivatives, 450–451, 459–460, 474
    Richardson extrapolation, 455–457
    truncation errors in, 104–107
    for unequally spaced data, 457–458
Numerical integration formulas, 16, 392–425.
        see also Integration; Numerical integration
        of functions
    for data with errors, 458–459
    dblquad function, 418
    multiple integrals, 416–418
    Newton-Cotes formulas. *see* Newton-Cotes
        formulas
    triplequad function, 418
    for unequal segments, 412–415
    work, computing, 419–422
Numerical integration of functions, 426–447.
        see also Numerical integration formulas
    adaptive quadrature, 427, 439–440
    Gauss quadrature. *see* Gauss quadrature
    Lobatto quadrature, 439
    quad function, 439–440
    quadl function, 439–440
    Romberg integration. *see* Romberg integration
    root-mean-square electrical current example,
        440–444
Numerical methods, 1, 9
    covered in this book, 13, 15–16
    MATLAB implementation of, 21. *see also*
        MATLAB.
    reasons for studying, 1–2

O

Octal (base-8) number system, 85
ode15s function, 529
ode23 function, 516–517

ode23s function, 529, 530
ode23t function, 529
ode23tb function, 529
ode45 function, 517–519, 530, 534–535, 546–547
ode113 function, 517
ODEs. *see* Ordinary differential equations (ODEs)
odeset function, 519–520
Ohm's law, 206, 452
One-dimensional optimization, 169, 170–179
  fminbnd function, 170, 178–179
  global optimum, 170
  golden-section search, 113, 171–176
  local optimum, 170
  multimodal optima, 170
  parabolic interpolation, 177–178
  unimodal function, 172
ones function, 25
One-step methods, 477, 481
Online help, 30
Open formulas, Newton-Cotes, 398, 416
Open methods, 112–113, 139–165
  defined, 119
  fzero function, 151–154, 160, 178
  Newton-Raphson, 144–149
  newtraph function, 148–149, 159
  numerical integration, 391, 398, 416
  optimset function, 153–154, 160, 326
  pipe friction example, 157–161
  polynomials, 154–161
  roots function for polynomials, 155–157, 569
  secant methods, 149–151, 152
  simple fixed-point iteration, 113, 140–144,
    270–272
  slow convergence with Newton-Raphson,
    145–146
  two-curve method, 142–143
Operation counting, 90, 222–225, 229, 242
Optimization, 15, 16, 112, 113, 166–187
  described, 167–168
  equilibrium and minimum potential energy,
    181–183
  multidimensional, 179–181
  one-dimensional. *see* One-dimensional
    optimization
  root location solutions, 168–169
  two-dimensional. *see* Two-dimensional
    optimization
  underdetermined systems, 203
optimset function, 153–154, 160, 179, 326
Ordinary differential equations (ODEs), 16,
  473–474
  adaptive methods. *see* Adaptive methods
    for ODEs
  boundary-value problems. *see* Boundary-value
    problems
  initial value problems, 477, 479–513
    atmospheric fluid dynamics, 504–509
    boundary-value problems vs., 541–542
    Euler's method. *see* Euler's method
    eulode function, 486–487
    Heun method, 488–492
    midpoint method, 492–493
    predator-prey models, 504–509
    Runge-Kutta methods. *see* Runge-Kutta
      methods

systems of equations, 478, 498–504
  Euler's method, 498–500
  rk4sys function, 502–504, 535–536
  Runge-Kutta methods, 500–504
  nonlinear, 478, 550–552, 557–559
  shooting method for linear ODEs, 546–548
  stiff systems. *see* Stiff systems
| (Or) logical condition, 53
Oscillations, polynomial interpolation and, 353–355
Outer product of two vectors, 28
Overdetermined systems, 203, 325
Overflow, 87, 90
Overrelaxation, 268

P
Panes, 34–35
Parabolic interpolation, 113, 177–178
Parameters, 5, 252
  of design problems, 115
  odeset function, 519
  passing, 70–71
Parentheses, overriding priority order, 27, 54
Partial derivatives, 450–451, 459–460, 474
Partial differential equations, 474
Partial pivoting, 191, 225–227
Passed function, 67
Passing functions to M-files, 66–71
  anonymous functions, 66–67, 70
  function functions, 66, 67–71
  inline function, 67
  passed function, 67
  passing parameters, 70–71
  varargin argument, 70–71
pause command, 62–63
pause (inf) command, 63
pause (n) command, 62–63
pchip function, 377–379
Periods, prior to operators, 37
Per-step truncation error, 525
Phase-plane plots, 505–509, 519
Φ(Phi), 171
Piecewise interpolation, 374–381. *see also* Splines
  cubic Hermite (pchip), 377–379
pi function, 23
Pi (π), 23, 89
Pipe friction, 157–161
Pivot element, 220
Pivot equation, 220
Pivoting, 225–227
Place value, 85
Planck's constant, 87
Pliny's fountain, 532–536
plot function, 32–36, 519
  exploratory data analysis, 37–38
plot3 function, 35–36, 508–509
Point-slope method, 481
poly function, 156, 569
polyfit function, 339, 353–354
  linear regression, 304, 307
  polynomial regression, 325
Polynomials, 92, 113, 154–161
Polynomial interpolation, 335–358
  basics, 336–339
  defined, 336
  extrapolation, 351–353

higher-order, 353–355
inverse, 350–351
Lagrange function, 349–350
Lagrange interpolating polynomial, 347–350
Newton's. *see* Newton's interpolating
  polynomials
oscillations, 353–355
polyfit function, 304, 307, 339, 353–354
polyval function, 307, 339, 353–354
Runge's function, 353–355
Polynomial method, for eigenvalues, 566,
  567–569
Polynomial regression, 283, 300, 316–320
  with MATLAB, 323–324
  with *polyfit* and left division, 325
polyval function, 307, 339, 353–354
Position, as independent variable, 115
Positional notation, 85
Posttest loop, 62
Potential energy
  as dependent variable, 115
  minimum, equilibrium and, 181–183
Power equation, 300, 302–304, 305–306
Power method, for eigenvalues, 569–572
Preallocation of memory, 60–61
Precision, 23, 51, 80–81
  of computer number representation, 89–90
  double, 51, 89
  IEEE double precision format, 89
  limits of digital computers, 84
  machine, 88
Predator-prey models, 504–509, 518–519
Predefined variables, 23
Predictor-corrector approach, 489
Predictor equation, 488
Pretest loop, 62
Primary functions, 47
Primary windows, MATLAB, 21
Principal diagonal, of a matrix, 196
Priority order, 27, 53–54
prod function, 31
Product, enzyme kinetics, 307
Programming with MATLAB, 3, 42–78
  bungee jumper velocity example, 71–74
  files, creating and accessing, 50–51
  input-output, 47–49
  M-files. *see* M-files
  nesting and indentation, 63–66
  passing functions to M-files. *see* Passing
    functions to M-files
  structured programming. *see* Structured
    programming
Propagated truncation error, 483–484
Proportionality, 252

Q
QR factorization, 247n, 325
quad function, 439–440
quadl function, 439–440
Quadratic convergence, 145
Quadratic interpolation, 341–343
Quadratic splines, 365–368
Quadrature, 393
Quantification, of error, 81–84
quiver function, 465–466

R

Ralston's method, 495
Range
  of computer number representation, 89
  measure of spread, 287
Rate equations, 473
Reactors, conservation laws for, 13, 14
realmax function, 90
realmin function, 90
Regression, 15, 16. *see also* Linear regression;
    Polynomial regression
Regression line, spread and, 297
Relational operators, 53
Relative error, 82
Relaxation, Gauss-Seidel method, 268
Repeating calculations (↑), 30
Residual error, 297–298
Residuals, 292, 297–298
Resolution, of floating-point arithmetic, 90–91
Resonant frequency, 573
Response, of systems, 251
Reynolds number, 158
Richardson extrapolation, 391, 426
  numerical differentiation, 455–457
  Romberg integration, 427–429
RK methods. *see* Runge-Kutta methods
Romberg integration, 391, 426, 427–432
  algorithm for, 429–432
  higher-order corrections, 429
  Richardson extrapolation, 427–429
  romberg function, 431–432
Root-mean-square electrical current, 440–444
Roots of equations, 15, 16, 111
  for analytical optimization, 168–169
  bracketing methods. *see* Bracketing methods
  in engineering/science, 115–116
  graphical approach, 116–117
  open methods. *see* Open methods
  roots function, 155–157, 569
Rosenbrock formula, 529
round function, 31
Round-off errors, 3, 84–92. *see also*
    Ill-conditioned systems
  arithmetic manipulations and, 90–92
  computer number representation. *see* Computer
    number representation
  with Euler's method, 483
  inner products, 92
  large computations, 91
  large numbers added to small numbers, 92
  in numerical differentiation, 105–107
  smearing, 92
  total numerical error and, 103–108
  truncation errors vs., 103–104
Rows, matrix, 195
Row-sum norm, 255
Row vectors, 24, 195
Runge-Kutta methods, 477, 481, 487, 493–498, 529
  adaptive, 478, 514–520
  Butcher's fifth-order method, 497–498
  classical fourth-order method, 495–498,
    500–504
  embedded, 516
  Heun method, without iteration, 495
  midpoint method, 495

Ralston's method, 495
RK-Fehlberg, 516
rk4sys function, 502–504, 535–536
  second-order, 494–495
  for systems of ODEs, 500–504
Runge's function, 353–355, 374–376

S

Saturation-growth-rate equation, 301.
    see also Michaelis-Menten model
save-ascii command, 51
save command, 50
Scalars, 22–23, 27, 29
Scientific applications
  boundary-value problems, 542–545
  curve fitting, 281–283
  differentiation, 451–452
  eigenvalues, 566–567
  integration, 394–396, 412
  linear algebraic equations, 189–191
  roots of equations, 115–116
  table lookup, 364–365
Script files, 43–44
Secant methods, 113, 149–151, 152
Second forward finite difference, 103
Second-order differential equations, 474
Second-order Newton polynomial, 342–343
Second-order polynomials, 316–320, 342–343
Second-order Runge-Kutta methods, 494–495
Semicolons, separating commands, 22
semilogy function, 40
Sensitivity analysis, 70
Sequential performance of instructions, 51
Sequential search, 364
Shooting method, 478, 545–552
  with derivative boundary conditions,
    548–550
  for linear ODEs, 546–548
  for nonlinear ODEs, 550–552
Signed magnitude method, 85–86
sign function, 480n
Significand, 86, 88
Signum function, 480
Simple fixed-point iteration, 140–144,
    270–272
Simple mathematical models, 5–12
Simpson's 1/3 rule, 390, 405–407, 411
Simpson's 3/8 rule, 390, 409–411
Simultaneous equations. *see* Linear algebraic
    equations
Single-line **if** structure, 52, 73
Singular systems, 213, 214
Singular value decomposition, 325
Sinusoids, least-squares fit of, 328–331
size function, 202
Size limits, of digital computers, 84
Smearing, round-off errors and, 92
sort function, 31
Spectral norm, 255
Splines, 283, 359–374
  bilinear interpolation, 379–381
  cubic. *see* Cubic splines
  end conditions, 373
  first-order linear, 362–363
  heat transfer example, 382–386

  higher-order polynomials vs., 360–361
  interp1 function, 376–379
  interp2/interp3 functions, 381
  knots and, 363
  natspline function, 383–386
  piecewise interpolation, 374–381
  quadratic, 365–368
  spline function, 374–376
  table lookup, 364–365
  two-dimensional interpolation, 379–381
Spread
  around the mean, 297
  around the regression line, 297
  measures of, 287–289
Square matrix, 196
Stability, of Euler's method, 485–486
Standard deviation, 287, 290
Standard error of the estimate, 297
States, of systems, 251
Statistics review, 286–291
  descriptive statistics, 286–291
  measures of location, 286–287
  measures of spread, 287–289
  normal distribution, 289–290
Statistics Toolbox, 290n
Steady-state calculation, 12
Step-size reduction
  step halving, 515–516
  total numerical error and, 103–104
Step value, loops, 59
Stiff systems, 478, 525–531
  explicit Euler's method, 527–529
  implicit Euler's method, 527–529
  MATLAB functions for, 529–531
  ode15s function, 529
  ode23s function, 529
  ode23t function, 529
  ode23tb function, 529
  single stiff ODEs, 526–528
  systems of ODEs, 528–529
Stimulus-response computations, 251–253
Stopping criterion, 83
Strange attractors, 508
Structured programming, 51–63
  decisions, 51–58
    complex, 55
    error function, 52
    if structure, 51–52, 56
    if . . . else structure, 54, 56
    if . . . elseif structure, 54–55, 56
    logical conditions, 53–54
    priority order, 53–54
    relational operators, 53
    switch structure, 57–58
    truth tables, 53
    variable argument list, 57–58
  loops, 51, 59–63
    to compute factorials, 59–60, 59n
    index variable, 59
    for loop, 59–60
    midtest loop, 62
    posttest loop, 62
    preallocation of memory, 60–61
    pretest loop, 62
    vectorization, 60

Structured programming (*continued*)
  while . . . break loops, 61–62, 72–73
  while loops, 59, 61–62
  nesting and indentation, 63–66
Structures, conservation laws for, 13, 14
Subfunctions, 46–47, 72
subplot function, 34–35, 38
Subscript notation, 25
Substrate, enzyme kinetics, 307
Subtraction, 27, 90–91, 197, 201
Subtractive cancellation, 91, 103
Successive overrelaxation (SOR), 268
Successive substitution, 270–272
sum function, 31
Sum of absolute values of residuals, 292
Sum of residuals, 292
Sum of the squares of residuals, 293–294
Superposition, 252
Swamee-Jain equation, 158, 160
switch structure, 57–58
Symbols, specifiers for, 34
Symmetric matrix, 196, 244
Systems
  of equations
    initial value problems, 498–504
    iterative methods. *see* Iterative methods
    MATLAB for, 518–519
    overdetermined, 203
    stiffness and, 528–529
    underdetermined, 203
  interactions, 252
  response of, 251
  state of, 251
  stiff. *see* Stiff systems

T
Table lookup, 364–365
tanh function, 7n, 25
Taylor series expansion, 92–103, 452–453
  approximation of function with, 96
  backward finite-difference formulas, 455
  centered finite-difference formulas, 456
  forward finite-difference formulas, 454
  multi-variable, 272
  remainder for, 97–98
  for truncation error estimation, 98–99
Taylor theorem, 92
Terminal velocity, 9, 12, 20
Thermocline, 382
Third-order Newton polynomial, 344–345
Three-dimensional plots, 35–36
Three-point Gauss-Legendre formula, 438
tic command, 63
Time-series plots, 505–509
title function, 33
toc command, 63
tolx parameter, 153

Toolboxes, 30, 290n
Top-down design, 63
Torricelli's law, 532
Total numerical error, 103–108
Transient computations, 12
Transpose, of a matrix, 199, 200
Trapezoidal rule, 390, 398–405, 411
  composite, 401–403
  cumtrapz function, 414–415
  error of, 399–401
  Gauss quadrature and, 432–433
  ode23t function, 529
  single application, 399–401
  trap function, 403–405
  trapuneq function, 413
  trapz function, 414
  with unequal segments, 412
Trend analysis, 282
Trial and error, for root location, 111
Tridiag function, 229, 232
Tridiagonal matrix, 191, 197
Tridiagonal systems, 227–229, 232
triplequad function, 418
True error, 81
True fractional relative error, 82
True value, 81
Truncation errors, 3, 92–103
  estimating, 98–99
  with Euler's method, 483–485
  global, 484–485
  local, 483–485, 524–525
  in Newton-Cotes formulas, 411, 416
  in numerical differentiation, 104–107
  per-step, 525
  propagated, 483–484
  round-off errors vs., 103–104
  Taylor series estimation, 98–99
  total numerical error and, 103–108
Truth tables, 53
Two-curve method, 142–143, 161
Two-dimensional interpolation, 379–381
Two-dimensional optimization, 169, 179–181
  derivative evaluation, required/not
      required, 180
  descent optimization methods, 181
  direct methods, 181
  fminsearch function, 181
  gradient methods, 181
  nongradient methods, 181
  visualizing a two-dimensional function, 180
Two-dimensional plots, 35–36
Two-point Gauss-Legendre formula, 435–438

U
Uncertainty, 80
Unconditionally stable method, 527
Underflow, 87, 90
Underrelaxation, 268

Undetermined coefficients method, 433–435
Unequal segments
  integration with, 412–415
  trapezoidal rule with, 412
Unimodal function, 172
Up (↑) arrow key, 30
Upper triangular matrix, 196

V
Vandermonde matrices, 338
Van der Pol equation, 529–530
varargin argument, 70–71
Variable names
  assigning value to. *see* Assignment
  case-sensitivity of, 22
Variance, 287
Vectors, 24–25
  column, 181
  diff function for evaluating, 462
  norms for, 254–255
  row, 181
  vector fields, visualizing, 465–467
Vectorization, 60
Vector-matrix calculations, 28–29
Velocity
  as dependent variable, 115
  distance computed from, 414–415
Voltage
  and currents in circuits, 205–208
  as dependent variable, 115
  Kirchhoff's voltage (loop) rule, 205–206
  voltage balance, 14
Volume integral, 396

W
while . . . break loops, 72–73, 364
while . . . break structure, 61–62
while loops, 59, 61–62
while structure, 61
who function, 25
whos function, 25
Wild card, colon used as, 26
Words, 84
Work, computing, 419–422
www.mathworks.com, 36

X
xlabel function, 33

Y
ylabel function, 33

Z
Zero-order approximation, 93, 97
zeros function, 25
Zeros of equations, 111
zlabel function, 180